Studies in Systems, Decision and Control

Volume 173

AF832342

Series editor

Janusz Kacprzyk, Polish Academy of Sciences, Warsaw, Poland
e-mail: kacprzyk@ibspan.waw.pl

The series "Studies in Systems, Decision and Control" (SSDC) covers both new developments and advances, as well as the state of the art, in the various areas of broadly perceived systems, decision making and control–quickly, up to date and with a high quality. The intent is to cover the theory, applications, and perspectives on the state of the art and future developments relevant to systems, decision making, control, complex processes and related areas, as embedded in the fields of engineering, computer science, physics, economics, social and life sciences, as well as the paradigms and methodologies behind them. The series contains monographs, textbooks, lecture notes and edited volumes in systems, decision making and control spanning the areas of Cyber-Physical Systems, Autonomous Systems, Sensor Networks, Control Systems, Energy Systems, Automotive Systems, Biological Systems, Vehicular Networking and Connected Vehicles, Aerospace Systems, Automation, Manufacturing, Smart Grids, Nonlinear Systems, Power Systems, Robotics, Social Systems, Economic Systems and other. Of particular value to both the contributors and the readership are the short publication timeframe and the world-wide distribution and exposure which enable both a wide and rapid dissemination of research output.

More information about this series at http://www.springer.com/series/13304

Michael Z. Zgurovsky · Alexander A. Pavlov

Combinatorial Optimization Problems in Planning and Decision Making

Theory and Applications

Springer

Michael Z. Zgurovsky
Kyiv Polytechnic Institute
National Technical
University of Ukraine
Kyiv, Ukraine

Alexander A. Pavlov
Faculty of Informatics
and Computer Science
National Technical
University of Ukraine
Kyiv, Ukraine

ISSN 2198-4182 ISSN 2198-4190 (electronic)
Studies in Systems, Decision and Control
ISBN 978-3-030-07552-1 ISBN 978-3-319-98977-8 (eBook)
https://doi.org/10.1007/978-3-319-98977-8

This Springer imprint is published by the registered company Springer Nature Switzerland AG
The registered company address is: Gewerbestrasse 11, 6330 Cham, Switzerland

Acknowledgements

We express our sincere gratefulness to the students and researchers of the Department of Computer-Aided Management and Data Processing Systems (Faculty of Informatics and Computer Science, National Technical University of Ukraine "Igor Sikorsky Kyiv Polytechnic Institute") who took part in scientific research.

We want to thank the editors of "Springer" Publishing House who prepared this manuscript for publication.

We also want to express our special gratitude to Oleg V. Melnikov for a technical support of this project.

Kyiv, Ukraine
June 30, 2018

Michael Z. Zgurovsky
Alexander A. Pavlov

Contents

Abbreviations

AA	Approximation algorithm
AHP	Analytic hierarchy process method by Thomas L. Saaty
APS	Advanced planning and scheduling
CAC	Computer-aided control
CAM	Computer-aided management
CRP	Capacity resources planning
CSRP	Customer synchronized resource planning
DMU	Decision making unit
DRP	Distribution resources planning
E/T	Earliness/tardiness, sum of total earliness and total tardiness
EBSR	Extraction and backward-shifted reinsertion
EFSR	Extraction and forward-shifted reinsertion
ERP	Enterprise resource planning
ERP II	Enterprise resource and relationship processing
FLM	Four-level model of planning (including operative planning) and decision making
JIT	Just in time
LPP	Linear programming problem
MES	Manufacturing execution system
MPS	Master production schedule
MPSS	Maximum priority subsequence
MRP	Material requirements planning
MRP II	Manufacturing resource planning
MSNSP	Multistage network scheduling problem
MSTAA	Machines start times adjustment algorithm
OPF	Orthogonal polynomial of G.E. Forsythe
PCM	Pairwise comparison matrix

PSC-algorithm	An algorithm which includes: sufficient conditions of a feasible solution optimality for which their checking can be implemented only at the stage of a feasible solution construction, and this construction is carried out by a polynomial algorithm (the first polynomial component of the PSC-algorithm); an approximation algorithm with polynomial complexity (the second polynomial component of the PSC-algorithm); also, for NP-hard combinatorial optimization problems, an exact sub-algorithm if sufficient conditions were found, fulfillment of which during the algorithm execution turns it into a polynomial complexity algorithm
SCM	Supply chain management
SSO(s)	Sufficient sign(s) of optimality of a feasible solution
SSPS	Sufficient sign of polynomial solvability of the current iteration
ST	Scheduling theory
TT	Single machine total tardiness minimization problem
TTP	Parallel machines total tardiness minimization problem with a common due date and common fixed start time of machines
TTPL	Parallel machines total tardiness minimization problem with a common due date and arbitrary fixed start times of machines which are less than the due date
TWCT	Single machine total weighted completion time minimization problem with precedence relations given by an oriented acyclic graph
TWCTZ	The TWCT problem for the case when nonzero weights are specified only for the terminal vertices of the oriented graph
TWT	Single machine total weighted tardiness minimization problem
UML	Unified modeling language

Chapter 1
Introduction

Abstract We formulate the research problems which are a generalization of our earlier results in the field of intractable combinatorial optimization problems. On the basis of these problems, we have created a hierarchical model of planning and decision making for objects with a network representation of technological processes and limited resources (Chap. 9). We say that the problem is intractable if it is NP-hard (NP-hard in the strong sense) or such for which an exact polynomial time solution algorithm has not been yet obtained. To solve such problems efficiently, we developed a methodology of PSC-algorithms construction meaning the algorithms which necessarily include the following: sufficient conditions (signs) of a feasible solution optimality, verification of which can be implemented only at the stage of a feasible solution construction by a polynomial algorithm (the first polynomial component of the PSC-algorithm). The second polynomial component of the PSC-algorithm is an approximation algorithm with polynomial complexity. For NP-hard (NP-hard in the strong sense) combinatorial optimization problems, a PSC-algorithm may include an exact algorithm for its solving in case if sufficient conditions were found, satisfying of which during this algorithm execution turns it into a polynomial complexity algorithm (Chaps. 4 and 5). We also give a brief overview of the monograph's chapters content.

1.1 Scope

The book contains our latest research results, which are the natural extension and development of our researches presented in [1].

In [1], we considered the problem of implementation of efficient planning in systems with a network representation of technological processes and limited resources. These problems, for example, include: planning of all types of small-scale production (70% of the world production), project planning and management, "made to order" production planning, job shop production planning, batch production planning, and planning of complex objects construction. We proposed and substantiated an hierarchical model of scheduling. The hierarchical feature of

M. Z. Zgurovsky and A. A. Pavlov, *Combinatorial Optimization Problems in Planning and Decision Making*, Studies in Systems, Decision and Control 173,
https://doi.org/10.1007/978-3-319-98977-8_1

the planning model (aggregation and disaggregation of the planning process) allowed us to use efficiently in its algorithmic implementation a number of NP-hard (in particular, NP-hard in the strong sense) single-stage scheduling problems. We have proved their efficient solving within the framework of our methodology of PSC-algorithms construction [1–3].

Practical use of the hierarchical model of scheduling proposed in [1] arose certain ambiguity. Also, it did not have an operational planning model in it. The cause of both problems is: the formal model which sets the operational network representation of discrete technological processes was not constructively formalized —it was given qualitatively in general terms. We present the four-level model of planning (including operative planning) and decision making (the FLM) in this monograph which is free from this flaw. Moreover, constructive formalization of the third level—the operational planning level—allowed us to formulate a multi-stage network scheduling problem (MSNSP). Efficient solving of MSNSP is based on new formulations of several single-stage scheduling problems. PSC-algorithms of their solving are given in the first part of the book. The fourth level of the FLM—operative planning level—formalizes procedures of the operational plan adjustment in case of its partial disruption.

The book consists of two parts. In the first part, we formulate the problems of combinatorial optimization which constitute the algorithmic basis of the FLM. Some of the problems are NP-hard; for other problems an exact polynomial solving algorithms were not created so far and their P-solvability was not proved.

Let us introduce the following definition of intractable problems of combinatorial optimization as an area of the PSC-algorithms application.

Definition 1.1 A problem is intractable if it is either NP-hard (NP-hard in the strong sense) or there is currently absent an algorithmically implemented exact polynomial algorithm of its solving (it is not proved that the problem belongs to the class of NP-hard problems, an exact polynomial algorithm of its solving does not exist, and the researcher himself cannot offer such algorithm).

Thus, a task emerges to solve these problems efficiently. Garey and Johnson [4] give their general classification. Today's methods for their exact solving include those commensurable by complexity with an exhaustive search (branch and bound, dynamic programming and other methods based on them). To solve higher dimension problems, various approximate and heuristic methods of local search are used. The main methods to solve the intractable problems of combinatorial optimization are the following [5, 6]:

- branch and bound method (the main enumerative strategy) [7];
- constructive approximation methods: priority dispatch rules [8–10] including fuzzy logic [11], the insertion algorithm of Werner and Winkler [12] with the beam search [13, 14];
- approximation heuristic search procedures [15–17];
- iterative local search methods [18–20] including, in particular, recession vector method [21] and its modifications [22], global equilibrium search [23], shifting bottleneck procedure [24–29];

- iterative methods of artificial intelligence: genetic algorithms [30, 31], genetic local search [32–36]; constraint satisfaction [27, 37–40] including that with neural networks use [41–46]; ant colony optimization [47–49]; large step optimization [50–56]; tabu-search [57–63]; problem space methods and heuristics [64] including greedy algorithms [65]; threshold algorithms: iterative improvements and threshold accepting methods, simulated annealing [18, 29, 32, 66, 67];
- relatively new approaches are logic programming by constraints (constraint programming, Gecode packet) [68–70] where the new results are obtained due to novel algorithms of integral and scheduling constraints propagation; analytical combinatorics [71] which are the analytical methods of the properties' description for large size combinatorial structures, new results can be obtained directly applying these methods for complex problems including "combinatorial monsters" from multi-machine multi-stage (job shop, open shop) planning problems.

In this monograph we solve the task of intractable problems solving with the further development of our methodology of PSC-algorithms construction.

The algorithms presented in Chaps. 2–7 are apparently different and built on the basis of unrelated theoretically substantiated mathematical results. However, they all share a common methodology of constructing and, as a result, have common properties which may be strictly classified. Based on the analysis of existing PSC-algorithms, let us give their most common definition and classification. This will allow, at first, to see what unites so seemingly different algorithms and, secondly, will significantly facilitate the process of new PSC-algorithms creation.

PSC-algorithm [1–3] is an algorithm which contains:

- the first polynomial component which constructs the optimal solution if certain constructively verifiable conditions are satisfied in the process of its execution;
- the second polynomial component—an approximation subalgorithm for the problem solving;
- for NP-hard (NP-hard in the strong sense) problems may contain an exact algorithm of solving for which sufficient conditions were found, satisfying of which during the algorithm execution turns it into a polynomial complexity algorithm.

Construction of the first polynomial component of a PSC-algorithm is based on theoretically proven logical-analytical conditions called the sufficient signs of optimality of a feasible solution (SSOs). In contrast to the known results, verification of the SSOs is possible only during a feasible solution constructing by the first polynomial component of the PSC-algorithm. The aim of the first polynomial component is constructing a feasible solution. If there is a feasible solution satisfying any SSO then either its obtaining by a polynomial algorithm is theoretically proven or it must be obtained statistically significantly. The statistical efficiency of obtaining an exact solution by the first polynomial component of the PSC-algorithm is determined by the following three factors.

The first factor is either determined by a theoretically grounded guaranteed construction of a feasible solution which satisfies the SSOs if such a feasible solution exists (e.g., the PSC-algorithms in Chaps. 4 and 5), or it is implemented as a statistically significant computational experiment in which one generates problem instances, each having a feasible solution which satisfies the SSOs. The first polynomial component must almost always find such feasible solution (see, e.g., Chap. 3), otherwise it is inefficient.

The second factor: when generating an arbitrary statistically significant set of problem instances, the percentage of optimal solutions obtained by the first polynomial component of the PSC-algorithm is quite high.

The third factor is the computational simplicity of the first polynomial component of the PSC-algorithm.

If during the execution of the first polynomial component of PSC-algorithm the sufficient sings of optimality were not satisfied then the problem is solved either by an exact algorithm for which sufficient conditions were found, satisfying of which during the algorithm execution turns it into a polynomial complexity algorithm (Chaps. 4 and 5 of this book); or by an approximation subalgorithm of polynomial complexity (the second polynomial component of PSC-algorithm). Therefore, the integral efficiency of the PSC-algorithm is defined by the efficiency of its first polynomial component verifying the SSOs during a feasible solution construction, as well as the efficiency in the traditional sense of its approximation subalgorithm and in some cases also the efficiency of the exact algorithm.

It follows from the definition of the PSC-algorithm that a necessary and sufficient condition to construct a PSC-algorithm is finding theoretically substantiated sufficient conditions—SSOs, verification of which can be implemented only during a feasible solution constructing which is executed by a polynomial algorithm (the first polynomial component of PSC-algorithm).

Thus, the abbreviation of PSC means that the algorithm must contain a polynomial (P) component which checks the sufficient (S) conditions (C)—sufficient signs of optimality of a feasible solution—and a polynomial approximation subalgorithm of the problem solving. For NP-hard (NP-hard in the strong sense) combinatorial optimization problems a PSC-algorithm may contain an exact exponential algorithm of its solving if sufficient conditions were found, satisfying of which during the algorithm execution turns it into a polynomial complexity algorithm (Chaps. 4 and 5 of this book).

PSC-algorithm may appear a necessary computational procedure for P-solvable combinatorial optimization problems when:

- an exact polynomial algorithm for the problem solving is unknown (P-solvability of the problem is proven qualitatively);
- an exact polynomial algorithm is known but its computational complexity is much higher than the computational complexity of the efficient first polynomial component of the PSC-algorithm.

Despite the unambiguous definition of PSC-algorithm and a common methodology of its construction for various combinatorial optimization problems, their qualitatively different theoretical properties lead to a variety of modifications of PSC-algorithms inside a single definition. Let us introduce the classification of the PSC-algorithms based on the analysis of the results from Chaps. 1–4 of [1] as well as from Chaps. 2–7 of this book.

The first polynomial component of PSC-algorithm which constructs strictly optimal solution may be:

- a part of the exact subalgorithm or a part of the approximation subalgorithm with polynomial complexity;
- to coincide fully with a polynomial approximation subalgorithm—in this case, if during the algorithm execution an intermediate feasible solution satisfies any SSO then the calculations stop, an optimal solution was obtained by the first polynomial component, otherwise, an approximate solution will be obtained by the second polynomial component;
- a separate polynomial subalgorithm which does not coincide with the exact subalgorithm or the polynomial approximation subalgorithm.

PSC-algorithms may currently be divided into two classes. PSC-algorithms belonging to the first class include two polynomial components and an exponential subalgorithm for which sufficient conditions were found, satisfying of which during the algorithm execution turns it into a polynomial complexity algorithm (Chaps. 4 and 5 of this book). We obtain a problem solution either by the first polynomial component of the algorithm, or by an exact subalgorithm if it is a part of the PSC-algorithm and the constraints on the time of its execution are satisfied, or we get an approximate solution by the second polynomial component of the PSC-algorithm.

PSC-algorithms of the second class contain two polynomial components. As a result of a problem solving, we obtain an optimal solution by the first polynomial component of the PSC-algorithm or an approximate solution with estimation of the deviation from optimum by the second polynomial component of the PSC-algorithm.

To improve an overall practical efficiency of PSC-algorithms, the concept of a heuristic sign of optimality is introduced.

Definition 1.2 A heuristic sign of optimality of a feasible solution are logical-analytical conditions, satisfying of which for the feasible solution statistically significantly leads to an optimal solution.

These properties of a feasible solution are qualitatively proven and confirmed by computational experiments. Since exact proof of these properties is missing, verification of such heuristic sign of optimality is done by a polynomial approximation subalgorithm which is a part of the second polynomial component of the PSC-algorithm (Chap. 5 of this book).

Introducing the heuristic sign of optimality of a feasible solution clearly defines paths for further theoretical studies: an exact proof of this sign (or, more likely, of its particular case) will allow to include the sign together with corresponding polynomial subalgorithm to the first polynomial component of the PSC-algorithm, significantly increasing its theoretical efficiency.

1.2 Content

In the second chapter, "Optimal Scheduling for Two Criteria for a Single Machine With Arbitrary Due Dates of Tasks," we consider the problem of constructing a feasible (in which all tasks complete before their due dates) schedule for a single machine with arbitrary due dates and maximum start time of the machine or minimum total earliness of the tasks completion times in relation to their due dates. It is shown that for the criterion of maximum start time of the machine the problem is polynomially solvable, we give a polynomial algorithm of its solving. With a fixed start time of the machine the problem is polynomially solvable for the criterion of minimizing the total earliness of the completion times of the tasks if qualitatively proven and statistically significant properties of an optimal solution (heuristics 1 and 2) are met. The problem with an arbitrary start time of the machine to construct an optimal schedule minimizing the total earliness of the tasks completion times is intractable: an exact polynomial algorithm for its solving is not known. For the case if the heuristics 1 and 2 are true for an arbitrary start time of the machine, we develop an efficient PSC-algorithm. For the opposite case, this PSC-algorithm is an efficient approximation algorithm for the problem solving.

In the third chapter, "Optimal scheduling for vector or scalar criterion on parallel machines with arbitrary due dates of tasks," we examine the problem of constructing a feasible schedule for parallel machines of equal or various productivities with various due dates of tasks and arbitrary start times of machines. We consider two optimality criteria: maximizing the earliest start time of machines (scalar criterion) or obtaining optimal start times of machines subject to a direct lexicographical order of their assignment (vector criterion). In fact, in this chapter we examine four intractable combinatorial optimization problems (for each one there are no efficient polynomial algorithms for its solving). Each of four presented PSC-algorithms contains separate SSOs, the first polynomial component which verifies the SSOs, and an approximation algorithm. Since the formulated problems are quite complex, each component of the PSC-algorithms contains many subalgorithms, each of which implements a separate original heuristic. We give examples of the problems solving.

In the fourth chapter, "The Total Weighted Tardiness of Tasks Minimization on a Single Machine," we solve the problem of constructing a schedule for a single machine with various due dates and a fixed start time of the machine which minimizes the sum of weighted tardiness of tasks in relation to their due dates. The problem is NP-hard in the strong sense and is one of the most known intractable combinatorial optimization problems [4]. Unlike other PSC-algorithms in this monograph, in this chapter we present an efficient PSC-algorithm which, in addition to the first and second polynomial components (the first one contains twelve SSOs) includes exact subalgorithm for its solving. We obtained the sufficient conditions which are constructively verified in the process of its execution. If the conditions are true, the exact subalgorithm becomes polynomial. We give statistical studies of the developed algorithm and show the solving of well-known examples of the problem. We present an approximation algorithm (the second polynomial component) based on the exact algorithm. Average statistical estimate of deviation of an approximate solution from the optimum does not exceed 5% of it.

In the fifth chapter, "The Total Earliness/Tardiness Minimization on a Single Machine With Arbitrary Due Dates," we solve a single machine problem of constructing a schedule of tasks with arbitrary due dates minimizing the total earliness/tardiness in relation to their due dates. This problem is solved in three different formulations:

1. The start time of the machine is fixed. In this case the problem is NP-hard.
2. The start time of the machine belongs to a specified time segment. The problem is intractable because there is no exact polynomial algorithm for its solving.
3. The start time of the machine is arbitrary. The problem is intractable because there is no exact polynomial algorithm for its solving.

For the first two problems we give PSC-algorithms, each of them contains SSOs and is based on the optimal solution for the single machine problem to minimize the total tardiness of tasks in relation to their various due dates (with equal weights). We show that the PSC-algorithm of its solving is a simplified modification of the PSC-algorithm presented in Chap. 4. For the third problem we build an efficient approximation algorithm.

In the sixth chapter, "The Total Tardiness of Tasks Minimization on Identical Parallel Machines With Arbitrary Fixed Times of Their Start and a Common Due Date," we solve a NP-hard problem of constructing a schedule for identical parallel machines which minimizes the total tardiness of tasks in relation to a common due date in case when the start times of machines are fixed at arbitrary time points less than the due date. We present an efficient PSC-algorithm of its solving which is a generalization of our previously developed results: for the problem with equal start times of machines we have derived two SSOs and constructed two PSC-algorithms. Each of the algorithms verifies one of these SSOs. In this chapter we propose a generalized PSC-algorithm for equal start times of machines which combines the best properties of both PSC-algorithms. We have obtained a modification of the generalized PSC-algorithm for the case of arbitrary start times of machines, its complexity is determined by $O(n^2m)$ function. The first polynomial component of the PSC-algorithm coincides with its second polynomial component. We obtain an efficient estimate of the deviation from an optimal solution for an approximation algorithm of the problem solving. We also present the statistical studies of the PSC-algorithm which showed its high efficiency (efficient exact solutions have been obtained for problems with tens of thousands of variables which is unique for NP-hard combinatorial optimization problems).

In the seventh chapter, "The total weighted completion time of tasks minimization with precedence relations on a single machine," we consider the problem of constructing a schedule for a single machine minimizing the total weighted completion time of tasks when the restrictions on their processing order are given by an arbitrary oriented acyclic graph. The problem is NP-hard in the strong sense. Efficient polynomial algorithms for its solving are known only for cases when the oriented acyclic graph is a tree or a series-parallel graph [4]. We give a new efficient PSC-algorithm of its solving. It is based on our earlier theoretical and practical results [1, 2, 72] and solves the problem with precedence relations specified by an oriented acyclic graph of the general form. The first polynomial component of the PSC-algorithm contains sixteen SSOs. One of them will be statistically significantly satisfied at each iteration of the algorithm when solving randomly generated problem instances. In case when the SSOs fail, the PSC-algorithm is an efficient approximation algorithm. If any SSO is satisfied at each iteration then the algorithm becomes exact. We present the empirical properties of the PSC-algorithm on the basis of statistical studies.

Chapters 8 and 9 constitute the second part of the book, entitled "Hierarchical planning and decision making in network systems with limited resources." The results of the second part generalize and develop the ideas underlying our three-level hierarchical model of planning and decision making described in [1].

The eighth chapter, "The four-level model of planning and decision making," provides an overview of known models, methods and software for scheduling and operational planning for objects with a network representation of technological processes and limited resources. The problem stated in [1] was an operational plan construction to produce a potential portfolio of orders which is the best in terms of criteria defined by the customer. We make the conclusion that an immediate solution of this problem (which is MSNSP) is inefficient. To solve it for one of five basic criteria or their arbitrary linear combination, we proposed in [1] an hierarchical system of interconnected aggregated models of scheduling and decision making (combined into a single framework of formal models and methods with expert decisions), which has significantly simplified the efficient solutions obtaining. These aggregated models may be implemented for arbitrary network representations of technological processes. This was both an advantage and a disadvantage of the three-level hierarchical scheduling model proposed in [1]: there was no formal procedure for the operational plan constructing. It implied that this problem is obviously solved heuristically using the results of coordinated planning at the aggregate level [1]. However, in this case, we also could not present the formal methods of operative planning—the operational plan adjustment in case of its disruption during execution. The result of analysis is the four-level model of planning (including operative planning) and decision making (FLM), in which we formalize formal procedures both for obtaining an operational schedule and for its operative adjustment. The FLM includes the combinatorial optimization problems presented in Chaps. 2–7 as well as the Decision Making Unit—a subsystem which performs decision making functions in case if various events appear during planning. In the Decision Making Unit we use our modified Analytic Hierarchy Process

(AHP) which is based on the research of empirical matrices of pairwise comparisons by combinatorial models of optimization with weighted components of the additive functional. To solve the problem of an overall goal reconstruction from the experimental results (an implementation of modified AHP), we suggest to use our solution to one of two problems:

- the true function reconstruction from redundant representation in the form of $\sum_{i=1}^{L} a_i f_i(\bar{x})$ where a_i are unknown coefficients and $f_i(\bar{x})$ is an arbitrary known function (the gist of the problem is to determine which of the coefficients $a_i \equiv 0$ and to find the values of the remaining coefficients);
- a stochastic analogue: multivariate polynomial regression reconstruction from its redundant representation. These results also have an independent significance for the next problem: to reconstruct from the experimental results not an interpolation model of the researched phenomenon but a true function which determines its properties.

In the ninth chapter, "Algorithms and software of the four-level model of planning and decision making," we give an interrelated description for all algorithms which implement a procedure of scheduling and operative planning in the FLM on the basis of formal procedures and expert solutions. We have achieved this result sequentially solving the following problems:

- we formalize the network representation of a technological process, develop a language which implements the dynamics of the process functioning, define a set of practical optimality criteria for the operational plan (five basic criteria and all possible linear combinations of them);
- we obtain a formal representation of the technological process, adapt the general procedures of its two-level aggregation [1]: combining jobs to the aggregated units (the first level of aggregation) and combining critical paths of products with intersection on their common vertices of the precedence graph with combining the set of resources into a single unit (the second level of aggregation);
- for the second level of aggregation, we substantiate the scheduling problem reduction to the approximating problem of the total weighted completion times minimization on a single machine with precedence relations between the tasks; we propose an algorithm for its solving based on the ideas of PSC-algorithm for this problem from the seventh chapter of the book;
- we formalize the procedure of the coordinated planning which is executed on the aggregated network of the first level of aggregation; in this procedure, the solution obtained at the second level of aggregation defines the processing order of products (or product groups) corresponding to a single basic criterion;
- based on results of the coordinated planning (the second level of the planning model), we formalize the process of an operational plan construction (the third level of the FLM):

  - for a basic optimality criterion, we formulate and solve the operational planning problem as a MSNSP (the completion times of products (product groups) become the due dates in the scheduling problem). An efficient method of its solution uses PSC-algorithms for single-stage scheduling problems (Chaps. 2 and 3 of the monograph);
  - for a synthetic optimality criterion (a linear combination of basic criteria), we propose the method implemented by four alternative algorithms for operational plan construction;

- we develop two algorithms for an operational plan adjustment (the fourth level of the model—operative planning level):

  - the first algorithm is a modification of the algorithmic procedure of planning on the levels 1–3 of the model. We solve the problem for one of four basic criteria or their linear combinations. It should be noted that customer sets again the weight coefficients of basic optimality criteria for the operative planning problem. We need to distinguish the problem of an operational plan construction to execute the potential order portfolio and setting weight coefficients as the tardiness costs with respect to due dates of products (product groups) on the already approved operational plan. In the first case, due dates of products or product groups may be adjusted during coordination planning, with the profit correction corresponding to the optimality criterion;
  - the second algorithm of operative planning contains within single framework a formal algorithm and expert planning methods. If its implementation is possible, it efficiently (in terms of the optimization criterion) solves the problem of operative adjustment of the operational schedule obtained at the third level of planning. The formal algorithm includes PSC-algorithms for intractable combinatorial optimization problems (Chaps. 4–6 of this book);

- we substantiate a procedure for a decision maker for the expert evaluation and adjustment both of a potential order portfolio and the due dates of the products (or product groups) included in it, for the final operational plan approval on the basis of analysis of plans obtained by formal methods.

The last section describes an informational software system implementing the hierarchical model in a specific practical area: for solving the problems of scheduling and planning for small-series productions.

1.3 Audience

The book will be interesting for scientists, graduate students, master students, system engineers who work in the following fields of computer science: combinatorial optimization, scheduling theory, decision theory, regression analysis, computer-aided production management systems. It will be helpful for those who are interested in intractable problems of combinatorial optimization, decision making with non-formalized overall goals, constructing a multiple regression from its redundant representation. The book can also serve as a quick introduction into our methodology of PSC-algorithms construction, which are a new class of efficient methods for intractable problems of combinatorial optimization. It may also be useful for practitioners and software developers who implement advanced methods of the production organization in the fields of planning (including operative planning) and decision making for various types of systems with a network representation of technological processes. The content of the book can also be used in lecture courses for bachelors and masters in discrete mathematics, scheduling theory, decision theory, information systems and technologies.

References

1. Zgurovsky, M.Z., Pavlov, A.A.: Prinyatie Resheniy v Setevyh Sistemah s Ogranichennymi Resursami (Принятие решений в сетевых системах с ограниченными ресурсами; Decision Making in Network Systems with Limited Resources). Naukova dumka, Kyiv (2010) (in Russian)
2. Pavlov, A.A. (ed.): Konstruktivnye Polinomialnye Algoritmy Resheniya Individualnyh Zadach iz Klassa NP (Конструктивные полиномиальные алгоритмы решения индивидуальных задач из класса NP; Constructive Polynomial Algorithms for Solving Individual Problems from the Class NP). Tehnika, Kyiv (1993) (in Russian)
3. Pavlov, A.A., Pavlova, L.A.: PDC-algorithms for intractable combinatorial problems. Theory and methodology of design. "Karpatskij region" shelf #15, Uzhhorod (1998)
4. Garey, M.R., Johnson, D.S.: Computers and Intractability: A Guide to the Theory of NP-Completeness. W.H. Freeman and Co, San Francisco (1979). https://doi.org/10.1137/1024022
5. Jain, A.S., Meeran, S.: Deterministic job-shop scheduling: past, present and future. Eur. J. Oper. Res. **113**, 390–434 (1999). https://doi.org/10.1016/S0377-2217(98)00113-1
6. Pinedo, M.L.: Scheduling: Theory, Algorithms, and Systems. Springer, Cham (2016). https://doi.org/10.1007/978-3-319-26580-3
7. Lawler, E.L., Lenstra, J.K., Rinnooy Kan, A.H.G., Shmoys, D.B.: Sequencing and scheduling: Algorithms and complexity. In: Graves, S.C., Rinnooy Kan, A.H.G., Zipkin, P.H. (eds.) Logistics of Production and Inventory. Handbook in Operations Research and Management Science, vol. 4, pp. 445–522. North-Holland, Amsterdam (1993). https://doi.org/10.1016/s0927-0507(05)80189-6
8. Fisher, H., Thompson, G.L.: Probabilistic learning combinations of local job-shop scheduling rules. In: Muth, J.F., Thompson, G.L. (eds.) Industrial Scheduling, pp. 225–251. Prentice-Hall, Englewood Cliffs (1963)
9. Lawrence, S.: Supplement to resource constrained project scheduling: an experimental investigation of heuristic scheduling techniques. GSIA, Carnegie-Mellon University, Pittsburgh (1984)
10. Panwalkar, S.S., Iskander, W.: A survey of scheduling rules. Oper. Res. **25**(1), 45–61 (1977). https://doi.org/10.1287/opre.25.1.45
11. Grabot, B., Geneste, L.: Dispatching rules in scheduling: a fuzzy approach. Int. J. Prod. Res. **32**(4), 903–915 (1994). https://doi.org/10.1080/00207549408956978
12. Werner, F., Winkler, A.: Insertion techniques for the heuristic solution of the job-shop problem. Discrete Appl. Math. **58**(2), 191–211 (1995). https://doi.org/10.1016/0166-218X(93)E0127-K

13. Morton, T.E., Pentico, D.W.: Heuristic Scheduling Systems: With Applications to Production Systems and Project Management. Wiley, New York (1993)
14. Sabuncuoglu, I., Bayiz, M.: A beam search based algorithm for the job shop scheduling problem. Research Report IEOR-9705. Bilkent University, Bilkent (1997)
15. Fisher, M.L., Rinnooy Kan, A.H.G.: The design, analysis and implementation of heuristics. Manag. Sci. **34**(3), 263–265 (1988). https://doi.org/10.1287/mnsc.34.3.263
16. Glover, F., Greenberg, H.J.: New approaches for heuristic search: a bilateral linkage with artificial intelligence. Eur. J. Oper. Res. **39**(2), 119–130 (1989). https://doi.org/10.1016/0377-2217(89)90185-9
17. Rodammer, F.A., White, K.P., Jr.: A recent survey of production scheduling. IEEE Trans. Syst. Man Cybern. **18**(6), 841–851 (1988). https://doi.org/10.1109/21.23085
18. Sergienko, I.V., Kapitonova, Y.V., Lebedeva, T.T.: Informatika v Ukraine: Stanovlenie, Razvitie, Problemy (Информатика в Украине: становление, развитие, проблемы; Informatics in Ukraine: Formation, Development, Problems). Naukova dumka, Kyiv (1999) (in Russian)
19. Johnson, D.S., Papadimitriou, C.H., Yannakakis, M.: How easy is local search? J. Comput. Syst. Sci. **37**(1), 79–100 (1988). https://doi.org/10.1016/0022-0000(88)90046-3
20. Yannakakis, M.: The analysis of local search problems and their heuristics. In: Choffrut, C., Lengauer, T. (eds.) STACS 90. Lecture Notes in Computer Science, vol. 415, pp. 298–311. Springer, Berlin (1990). https://doi.org/10.1007/3-540-52282-4_52
21. Sergienko, I.V.: O primenenii metoda vektora spada dlya resheniya zadach optimizacii kombinatornogo tipa (О применении метода вектора спада для решения задач оптимизации комбинаторного типа; On the application of the recession vector method to solve optimization problems of combinatorial type). Upravlyayuschie sistemy i mashiny **2**, 86–94 (1975) (in Russian)
22. Bykov, A.Y., Artamonova, A.Y.: Modifikaciya metoda vektora spada dlya optimizacionno-imitacionnogo podhoda k zadacham proektirovaniya sistem zaschity informacii (Модификация метода вектора спада для оптимизационно-имитационного подхода к задачам проектирования систем защиты информации; A modified recession vector method based on the optimization-simulation approach to design problems of information security systems). Nauka i Obrazovanie of Bauman MSTU, vol. 1, pp. 158–175 (2015). https://doi.org/10.7463/0115.0754845 (in Russian)
23. Shilo, V.P.: Metod globalnogo ravnovesnogo poiska (Метод глобального равновесного поиска; Global equilibrium search method). Cybern. Syst. Anal. **35**(1), 74–81 (1999) (in Russian)
24. Adams, J., Balas, E., Zawack, D.: The shifting bottleneck procedure for job-shop scheduling. Manag. Sci. **34**(3), 391–401 (1988). https://doi.org/10.1287/mnsc.34.3.391
25. Balas, E., Lancia, G., Serafini, P., et al.: Job-shop scheduling with deadlines. J. Comb. Optim. **1**(4), 329–353 (1998). https://doi.org/10.1023/A:1009750409895
26. Balas, E., Vazacopoulos, A.: Guided local search with shifting bottleneck for job-shop scheduling. Manag. Sci. **44**(2), 262–275 (1998). https://doi.org/10.1287/mnsc.44.2.262
27. Caseau, Y., Laburthe, F.: Disjunctive scheduling with task intervals. LIENS Technical Report 95-25. Laboratoire d'Informatique de l' Ecole Normale Superieure, Paris (1995)
28. Demirkol, E., Mehta, S., Uzsoy, R.: A computational study of shifting bottleneck procedures for shop scheduling problems. J. Heuristics **3**(2), 111–137 (1997). https://doi.org/10.1023/A:1009627429878
29. Yamada, T., Nakano, R.: Job-shop scheduling by simulated annealing combined with deterministic local search. In: Osman, I.H., Kelly, J.P. (eds.) Meta-Heuristics: Theory and Applications, pp. 237–248. Springer, Boston (1996). https://doi.org/10.1007/978-1-4613-1361-8_15
30. Falkenauer, E., Bouffouix, S.: A genetic algorithm for job-shop. In: Proceedings of the IEEE International Conference on Robotics and Automation, Sacramento, 9–11 April 1991. https://doi.org/10.1109/robot.1991.131689

31. Nakano, R., Yamada, T.: Conventional genetic algorithm for job-shop problems. In: Kenneth, M.K., Booker, L.B. (eds.) Proceedings of the 4th International Conference on Genetic Algorithms and their Applications, San Diego (1991)
32. Aarts, E.H.L., Van Laarhooven, P.J.M., Ulder, N.L.J.: Local search based algorithms for job-shop scheduling. Working Paper. University of Technology, Eindhoven (1991)
33. Dorndorf, U., Pesch, E.: Evolution based learning in a job-shop scheduling environment. Comp. Oper. Res. **22**(1), 25–40 (1995). https://doi.org/10.1016/0305-0548(93)E0016-M
34. Grefenstette, J.J.: Incorporating problem specific knowledge into genetic algorithms. In: Davis, L. (ed.) Genetic Algorithms and Simulated Annealing, pp. 42–60. Pitman, London (1987)
35. Moscato, P.: On evolution, search, optimization, genetic algorithms and martial arts: towards memetic algorithms. C3P Report 826: Caltech Concurrent Computation Program, Caltech (1989)
36. Ulder, N.L.J., Aarts, E.H.L., Bandelt, H.-J., et al.: Genetic local search algorithm for the travelling salesman problem. Lect. Notes Comput. Sci. **496**, 109–116 (1991)
37. Fox, M.S.: Constraint-directed search: a case study of job shop scheduling. Dissertation, Carnegy Mellon University, Pittsburgh (1983)
38. Nuijten, W.P.M., Le Pape, C.: Constraint-based job-shop scheduling with ILOG SCHEDULER. J. Heuristics **3**(4), 271–286 (1998). https://doi.org/10.1023/A:1009687210594
39. Pesch, E., Tetzlaff, U.A.W.: Constraint propagation based scheduling of job shops. INFORMS J. Comput. **8**(2), 144–157 (1996). https://doi.org/10.1287/ijoc.8.2.144
40. Sadeh, N.: Look-ahead techniques for micro-opportunistic job shop scheduling. Dissertation, Carnegie Mellon University, Pittsburgh (1991)
41. Foo, S.Y., Takefuji, Y.: Integer linear programming neural networks for job-shop scheduling. In: IEEE International Conference on Neural Networks, San Diego, 24–27 July 1988
42. Foo, S.Y., Takefuji, Y.: Stochastic neural networks for solving job-shop scheduling: Part 1. problem representation. In: IEEE International Conference on Neural Networks, San Diego, 24–27 July 1988
43. Foo, S.Y., Takefuji, Y.: Stochastic neural networks for solving job-shop scheduling: part 2. Architecture and simulations. IEEE International Conference on Neural Networks, San Diego, 24–27 July 1988
44. Sabuncuoglu, I., Gurgun, B.: A neural network model for scheduling problems. Eur. J. Oper. Res. **93**(2), 288–299 (1996). https://doi.org/10.1016/0377-2217(96)00041-0
45. Zhou, D.N., Cherkassky, V., Baldwin, T.R., et al.: Scaling neural networks for job-shop scheduling. In: Proceedings of International Joint Conference on Neural Networks (IJCNN'90), San Diego, 17–21 June 1990, pp. 889–894. https://doi.org/10.1109/ijcnn.1990.137947
46. Zhou, D.N., Cherkassky, V., Baldwin, T.R., et al.: A neural network approach to job-shop scheduling. IEEE Trans. Neural. Netw. **2**(1), 175–179 (1991). https://doi.org/10.1109/72.80311
47. Dorigo, M.: Optimization, learning and natural algorithms. Dissertation, Politecnico di Milano (1992)
48. Donati, A.V., Darley, V., Ramachandran, B.: An ant-bidding algorithm for multistage flowshop scheduling problem: optimization and phase transitions. In: Siarry, P., Michalewicz, Z. (eds.) Advances in Metaheuristics for Hard Optimization, pp. 111–136. Springer, Berlin (2008). https://doi.org/10.1007/978-3-540-72960-0_6
49. Prabhakar, B., Dektar, K.N., Gordon, D.M.: The regulation of ant colony foraging activity without spatial information. PLoS Comput. Biol. **8**(8), e1002670 (2012). https://doi.org/10.1371/journal.pcbi.1002670
50. Brucker, P., Hurink, J., Werner, F.: Improving local search heuristics for some scheduling problems—I. Discrete Appl. Math. **65**(1–3), 97–122 (1996). https://doi.org/10.1016/0166-218X(95)00030-U

51. Brucker, P., Hurink, J., Werner, F.: Improving local search heuristics for some scheduling problems. Part II. Discrete Appl. Math. **72**(1–2), 47–69 (1997). https://doi.org/10.1016/S0166-218X(96)00036-4
52. Lourenco, H.R.D.: A computational study of the job-shop and the flow-shop scheduling problems. Dissertation, Cornell University (1993)
53. Lourenco, H.R.D.: Job-shop scheduling: computational study of local search and large-step optimization methods. Eur. J. Oper. Res. **83**(2), 347–364 (1995). https://doi.org/10.1016/0377-2217(95)00012-F
54. Lourenco, H.R.D., Zwijnenburg, M.: Combining the large-step optimization with tabu-search: Application to the job-shop scheduling problem. In: Osman, I.H., Kelly, J.P. (eds.) Meta-Heuristics: Theory and Applications, pp. 219–236. Springer, Boston (1996). https://doi.org/10.1007/978-1-4613-1361-8_14
55. Martin, O., Otto, S.W., Felten, E.W.: Large-step Markov chains for traveling salesman problem. Complex Syst. **5**(3), 299–326 (1989)
56. Martin, O., Otto, S.W., Felten, E.W.: Large-step Markov chains for TSP incorporating local search heuristics. Oper. Res. Lett. **11**(4), 219–224 (1992). https://doi.org/10.1016/0167-6377(92)90028-2
57. Glover, F.: Future paths for integer programming and links to artificial intelligence. Comput. Oper. Res. **13**(5), 533–549 (1986). https://doi.org/10.1016/0305-0548(86)90048-1
58. Glover, F.: Heuristics for integer programming using surrogate constraints. Decis. Sci. **8**(1), 156–166 (1977). https://doi.org/10.1111/j.1540-5915.1977.tb01074.x
59. Glover, F.: Tabu search—part I. ORSA J. Comput. **1**(3), 190–206 (1989). https://doi.org/10.1287/ijoc.1.3.190
60. Glover, F.: Tabu search—part II. ORSA J. Comput. **2**(1), 4–32 (1990). https://doi.org/10.1287/ijoc.2.1.4
61. Glover, F., Laguna, M.: Tabu Search. Springer, Boston (1997). https://doi.org/10.1007/978-1-4615-6089-0
62. Nowicki, E., Smutnicki, C.: A fast taboo search algorithm for the job-shop problem. Manag. Sci. **42**(6), 797–813 (1996). https://doi.org/10.1287/mnsc.42.6.797
63. Taillard, E.: Parallel taboo search technique for the job-shop scheduling problem. Internal Research Report ORWP89/11. Ecole Polytechnique Federale de Lausanne, Lausanne (1989)
64. Edelkamp, S., Schrödl, S.: Heuristic Search: Theory and Applications. Morgan Kaufmann Publishers, Waltham, MA (2012). https://doi.org/10.1016/c2009-0-16511-x
65. Resende, M.G.C.: A GRASP for job shop scheduling. In: INFORMS National Meeting, pp. 23–31, San Diego, CA, 4–7 May 1997
66. Matsuo, H., Suh, C.J., Sullivan, R.S.: A controlled search simulated annealing method for the general job-shop scheduling problem. Working Paper. University of Texas, Austin (1988)
67. Van Laarhoven, P.J.M., Aarts, E.H.L., Lenstra, J.K.: Job-shop scheduling by simulated annealing. Report OS-R8809. Centrum voor Wiskunde en Informatica, Amsterdam (1988)
68. Dechter, R.: Constraint Processing. Morgan Kaufmann, San Francisco (2003)
69. Michel, L., Hentenryck, P.V.: Activity-based search for black-box constraint programming solvers. In: Beldiceanu, N., Jussien, N., Pinson, É. (eds.) Integration of AI and OR Techniques in Constraint Programming for Combinatorial Optimization Problems. CPAIOR 2012. Lecture Notes in Computer Science, vol. 7298, pp. 228–243. Springer, Berlin (2012). https://doi.org/10.1007/978-3-642-29828-8_15
70. Schulte, C., Tack, G., Lagerkvist, M.Z.: Modeling and programming with Gecode. http://www.gecode.org/doc-latest/MPG.pdf (2018). Accessed 02 Apr 2018.
71. Flajolet, P., Sedgewick, R.: Analytic Combinatorics. Cambridge University Press, Cambridge (2009). https://doi.org/10.1017/cbo9780511801655
72. Pavlov, A.A. (ed.): Osnovy Sistemnogo Analiza i Proektirovaniya ASU (Основы системного анализа и проектирования АСУ; Fundamentals of System Analysis and Design of Automated Control Systems). Vyshcha Shkola, Kyiv (1991) (in Russian)

Part I
Intractable Combinatorial Optimization Problems. PSC-algorithms

Chapter 2
Optimal Scheduling for Two Criteria for a Single Machine with Arbitrary Due Dates of Tasks

Abstract We consider the problem of constructing a feasible (in which all tasks complete before their due dates) schedule for a single machine with arbitrary due dates and maximum start time of the machine or minimum total earliness of the tasks completion times in relation to their due dates. It is shown that for the criterion of maximum start time of the machine the problem is polynomially solvable, we give a polynomial algorithm for its solving. With a fixed start time of the machine the problem is polynomially solvable for the criterion of minimizing the total earliness of the completion times of the tasks if qualitatively proven and statistically significant properties of an optimal solution (Heuristics 1 and 2) are met. The problem with an arbitrary start time of the machine to construct an optimal schedule minimizing the total earliness of the tasks completion times is intractable: an exact polynomial algorithm for its solving is not known. For the case if the Heuristics 1 and 2 are true for an arbitrary start time of the machine, we develop an efficient PSC-algorithm. For the opposite case, this PSC-algorithm is an efficient approximation algorithm for the problem solving.

2.1 The Problem Formulation

This chapter contains our new original results based on our earlier works [1–4].

The Problem Statement Given a set of tasks $J = \{1, 2, \ldots, n\}$ each consisting of a single operation. We know a processing time l_j and a due date d_j for each task $j \in J$. All tasks become available at the same time point for processing on a single machine. Interruptions during a task processing are not allowed. The process is continuous: after the first task completes, processing of the second task starts immediately, and so on, until all the tasks are completed. We need to find a feasible schedule in which either the start time of tasks s is as late as possible or the total earliness of the tasks' completion times in relation to their due dates is minimal.

Definition 2.1 A schedule is called *feasible* if all due dates are met in it (that is, if all tasks are not tardy).

M. Z. Zgurovsky and A. A. Pavlov, *Combinatorial Optimization Problems in Planning and Decision Making*, Studies in Systems, Decision and Control 173,
https://doi.org/10.1007/978-3-319-98977-8_2

Definition 2.2 For a given schedule, we call the *start time of tasks* (or the *start time of the machine*) a time point where the processing of the first task in this schedule begins.

2.2 Construction of a Feasible Schedule with Maximum Start Time of Tasks

Let us use the following notation: let C_j be the completion time of a task j; $E_j = \max(0, d_j - C_j)$ the earliness of a task j; $E(\sigma)$ the total earliness in a schedule σ; $T_j = \max(0, C_j - d_j)$ the tardiness of a task j. Let s denote the start time of tasks. In a feasible schedule we have $d_j \geq C_j \; \forall j \in J$, that is, $E_j = d_j - C_j$. Let j_i denote the number of a task in position i in a feasible schedule. And let $\overline{a,b}$ denote the interval of integer numbers from a to b, that is, $\overline{a,b} = \mathbb{Z} \cap [a,b] = a, a+1, \ldots, b$.

Theorem 2.1 (Tanaev and Shkurba [5], Chap. 3) *Suppose that there exists a feasible sequence for some start time of tasks on a single machine. Then the sequence of the tasks in non-decreasing order of their due dates is also feasible.*

Corollary 2.1 (Tanaev and Shkurba [5]) If a schedule with zero start time and the tasks in non-decreasing order of their due dates is not feasible, then a feasible schedule does not exist.

Corollary 2.2 We have got new logical-analytical conditions for the fulfillment of the first polynomial component of the PSC-algorithm for the single machine total weighted tardiness minimization criterion: if the schedule with zero start time of tasks in non-decreasing order of due dates is feasible, then it is optimal; otherwise, the total weighted tardiness in an optimal schedule will be strictly greater than zero.

Proof obviously follows from Corollary 2.1. ∎

Let us give an algorithm to determine the latest start time of tasks ($s \to \max$) at which the schedule remains feasible. Here and below we will denote by $s_{\max}$ the value of the latest start time of tasks of the set J.

Algorithm A has the following three steps.

1. Build a schedule $\sigma = \{1, 2, \ldots, n\}$ in which tasks are in non-decreasing order of due dates, according to Theorem 2.1: $d_1 \leq d_2 \leq \cdots \leq d_n$.
2. Determine the start time of tasks $s = d_{\max} - \sum_{i=1}^{n} l_i$ where $d_{\max} = \max_{j \in J} d_j$.

   Determine the tardiness values of tasks in the schedule σ.
3. If there are no tardy tasks, then the algorithm terminates. Otherwise, determine a task l that has the maximum tardiness and calculate the new start time of tasks $s_{\max} = s - T_l$. Tardiness of the task l will become zero in the schedule with the start time $s_{\max}$, and other tasks will become non-tardy.

We would construct a feasible schedule σ^{feas} in result of this procedure.

Remark 2.1 We should always select the start point of time axis so that $s_{\max} > 0$ is always satisfied and that acceptable $s < s_{\max}$ are always positive.

Examples to Algorithm A Consider two examples of how the algorithm works.

Example 2.1 Given $n = 4$, the initial sequence is in Table 2.1.

Determine the start time of tasks $s = 37 - 29 = 8$ and the tardiness values T_j of each task in the schedule σ. At the start time 8, the sequence will take the following form (Table 2.2).

Task 1 has the maximum tardiness. So, the new start time of tasks is $s_{\max} = 8 - 7 = 1$. The tardiness of task 1 became zero, and other tasks became non-tardy. We have obtained a feasible schedule with the maximum start time (Table 2.3).

We illustrate this example at Fig. 2.1.

Example 2.2 Given $n = 7$, the initial sequence is in Table 2.4.

Determine the start time of tasks $s = 400 - 270 = 130$ and the tardiness values T_j of each task in the schedule σ. At the start time 130, the sequence will take the following form (Table 2.5).

Task 3 has the maximum tardiness 85. The new start time of tasks is $s_{\max} = 130 - 85 = 45$. The tardiness of task 3 became zero, and other tasks became non-tardy. We have obtained a feasible schedule with the maximum start time (Table 2.6).

Theorem 2.2 ([2]) Algorithm A *constructs a feasible schedule in which the start time of tasks (also called the start time of the machine) is the latest possible.*

Proof is given in [2]. ■

Theorem 2.3 ([2]) *A feasible schedule with the start time* $s_{\max}$ *built by arranging tasks in non-decreasing order of due dates is optimal by the criterion of the total earliness minimization if the following conditions are fulfilled:*

$$d_{j_1} \le d_{j_2} \le \cdots \le d_{j_n}, \quad l_{j_1} \ge l_{j_2} \ge \cdots \ge l_{j_n} \tag{2.1}$$

where j_i *is the number of task occupying the position i in the feasible schedule. The system of inequalities* (2.1) *is the sufficient condition of optimality for the two criteria. If a schedule* σ *does not satisfy the conditions* (2.1), *then its optimality for the criteria of the total earliness minimization is not guaranteed.*

Table 2.1 Initial sequence σ for Example 2.1

j	l_j	d_j
1	9	10
2	5	16
3	8	27
4	7	37

j the task number

Table 2.2 The sequence at $s = 8$

j	l_j	d_j	C_j	T_j
1	9	10	17	7
2	5	16	22	6
3	8	27	30	3
4	7	37	37	0

Table 2.3 The sequence at start time $s_{\max} = 1$

j	l_j	d_j	C_j	T_j
1	9	10	10	0
2	5	16	15	0
3	8	27	23	0
4	7	37	30	0

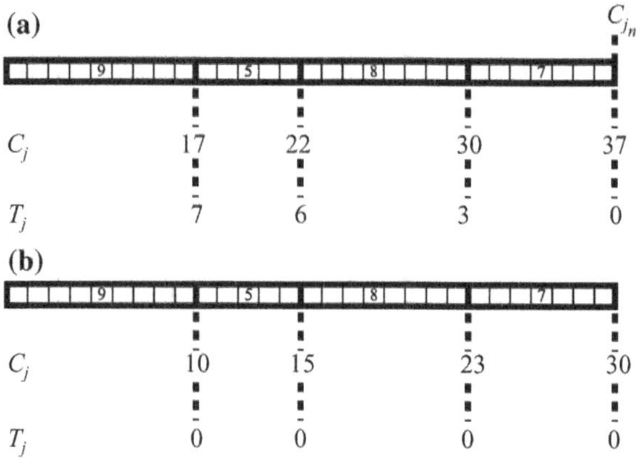

Fig. 2.1 **a** Initial schedule with $s = 8$, **b** Final schedule with $s = s_{\max} = 1$

Table 2.4 Initial sequence σ for Example 2.2

j	l_j	d_j
1	30	115
2	40	125
3	45	160
4	25	210
5	25	250
6	45	270
7	60	400

Proof is given in [2]. ∎

Remark 2.2 Suppose that there are tasks with $C_k \le d_i$ for an arbitrary C_k in a feasible schedule $\sigma = \{1, 2, \ldots, n\}$ with the maximum start time of tasks. Then, arranging these tasks in non-increasing order of their processing times yields a feasible schedule with the maximum start time of tasks and a smaller total earliness

Table 2.5 The sequence at $s = 130$

j	l_j	d_j	C_j	T_j
1	30	115	160	45
2	40	125	200	75
3	45	160	245	85
4	25	210	270	60
5	25	250	295	45
6	45	270	340	70
7	60	400	400	0

Table 2.6 The sequence at $s_{\max} = 45$

j	l_j	d_j	C_j	T_j
1	30	115	75	0
2	40	125	115	0
3	45	160	160	0
4	25	210	185	0
5	25	250	205	0
6	45	270	250	0
7	60	400	310	0

(under the assumption that there are at least two such tasks in σ with different processing times which are not in non-increasing order of their values).

Example 2.3 to Theorem 2.3 Consider some schedule σ (Table 2.7) that satisfies the system of conditions (2.1).

The total earliness $E(\sigma) = 70$. Consider another feasible schedule σ_1 (Table 2.8) that differs from the schedule σ in the order of tasks in positions 4–7.

The total earliness $E(\sigma_1) = 90$. Tasks 1, 2, 3 are in positions corresponding to the non-decreasing order of processing times in this schedule. Task 4 is in position $k_4 = 5 \neq 4$, so it is not in its proper position. We move it into position 4 and obtain a schedule σ_2 (Table 2.9). The change in the total earliness is $\Delta_4 = \sum_{i=4}^{5-1} l_{j_i} - (5-4)l_4 = 30 - 40 = -10 < 0$.

The total earliness $E(\sigma_2) = 80$. Tasks 1, … , 5 are in positions corresponding to the non-decreasing order of processing times in the schedule σ_2. Task 6 is in

Table 2.7 The schedule σ with start time of tasks $s = s_{\max} = 80$

j	l_j	d_j	C_j	E_j
1	70	160	150	10
2	60	220	210	10
3	50	260	260	0
4	40	330	300	30
5	30	340	330	10
6	20	360	350	10
7	10	360	360	0

Table 2.8 The schedule σ_1

j	l_j	d_j	C_j	E_j
1	70	160	150	10
2	60	220	210	10
3	50	260	260	0
5	30	340	290	50
4	40	330	330	0
7	10	360	340	20
6	20	360	360	0

Table 2.9 The schedule σ_2

j	l_j	d_j	C_j	E_j
1	70	160	150	10
2	60	220	210	10
3	50	260	260	0
4	40	330	300	30
5	30	340	330	10
7	10	360	340	20
6	20	360	360	0

position $k_6 = 7 \neq 6$, so it is not in its proper position. We move it into position 6 and obtain a schedule that coincides with the schedule σ (Table 2.7). The change in the total earliness is $\Delta_6 = \sum_{i=6}^{7-1} l_{j_i} - (7-6)l_6 = 10 - 20 = -10 < 0$. The total earliness is minimal in the schedule σ.

Suppose we have obtained a schedule σ by Algorithm A. This schedule is feasible (all tasks are not tardy in it) and has the latest possible start time of tasks. If the schedule σ also satisfies the condition

$$l_{j_1} \geq l_{j_2} \geq \cdots \geq l_{j_n}, \tag{2.2}$$

then it is also optimal by the total earliness minimization criterion, according to Theorem 2.3. If the schedule σ does not satisfy the condition (2.2), then its optimality for the criteria of the total earliness minimization is not guaranteed.

2.3 Construction of a Feasible Schedule That Minimizes the Total Earliness of Tasks with Maximum Start Time of the Machine $s_{\max}$

Suppose that a schedule σ does not satisfy the conditions (2.1). In this case for this problem, there may exist another feasible schedule in which the start time of tasks is also the latest but the total earliness is smaller. Let us state the problem: to construct a feasible schedule with a minimal total earliness of tasks for a given start time of the machine $s_{\max}$.

Heuristic 1 Suppose that there exists a feasible schedule σ with the latest start time. Then there is such a feasible schedule that: has the latest start time; is optimal by the criterion of minimizing the total earliness of tasks in regard to their due dates; contains at the last position a task K that satisfies both the conditions:

1. The task K is processed last if it does not violate the feasibility of the schedule (it is not tardy): $d_K \geq \sum_{j=1}^{n} l_j + s_{\max}$.
2. The processing time of the task K is minimal among processing times of all tasks which are not tardy when processed last:

$$l_K = \min_{r:d_r \geq \sum_{j=1}^{n} l_j + s_{\max}} \{l_r\} .$$

Justification of Heuristic 1 is based on Theorem 2.3 and the following fact. Consider a schedule σ with the following properties:

- the schedule is feasible (all tasks are not tardy);
- the start time of tasks is the latest possible;
- the task K that is processed last satisfies the conditions 1 and 2 of Heuristic 1.

Let us show that a move of the task K from the last position will cause some task to be tardy or increase the total earliness (if the task K is the only one).

Remark 2.3 If there is more than one task that satisfies both the conditions of Heuristic 1 (for example, there exists a task J with the processing time $l_J = l_K$), then interchanging their positions does not change the total earliness of tasks.

Suppose that a task J satisfies the condition 1 of Heuristic 1 and does not satisfy the condition 2, i.e., $l_J > l_K$. Let us show that interchanging of tasks J and K will lead to an increase in the total earliness of tasks (Fig. 2.2).

Suppose that a task J is in position ν ($1 \leq \nu \leq n$) in the schedule σ. Interchanging the tasks J and K, we obtain the schedule σ′. The completion times of the tasks in positions 1, 2, …, $\nu - 1$ coincide in these schedules, and the completion times of the tasks in positions ν, ν + 1, …, $n - 1$ differ in these schedules.

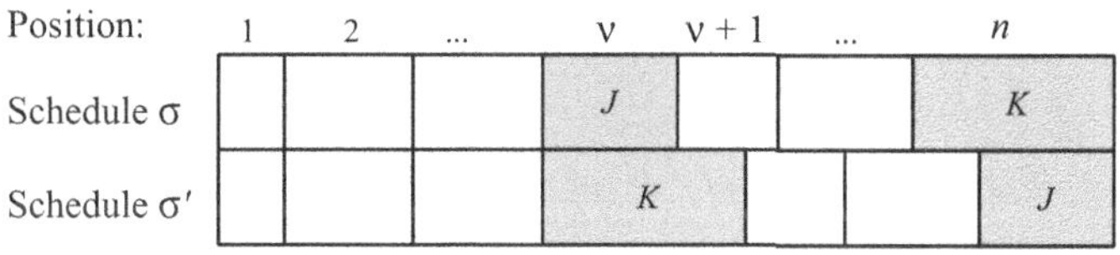

Fig. 2.2 Schedules σ and σ′

Let us introduce some notation. Common values for the schedules σ and σ': $b = \sum_{i=\nu+1}^{n-1} l_{j_i}$ is the total processing time of tasks in positions between the tasks J and K in the schedules σ and σ' (i.e., in positions $\nu + 1$, $\nu + 2$, …, $n - 1$) where l_{j_i} is the processing time of the task in position i in the schedule; $g = \sum_{i=1}^{\nu-1} l_{j_i}$ is the total processing time of tasks in positions 1, 2, …, $\nu - 1$ in the schedules σ and σ'; $f = C_{j_n}(\sigma) = C_{j_n}(\sigma')$ is the completion time of the task in the last position n (these are the tasks K and J in the schedules σ and σ', correspondingly).

For the schedule σ: let $E_i(\sigma)$ be the earliness of a task i in the schedule σ; $E(\sigma)$ be the total earliness of tasks in the schedule σ; $E_{j_n}(\sigma) = E_K(\sigma) = d_K - f$; $E_{j_\nu}(\sigma) = E_J(\sigma) = d_J - (g + l_J)$.

For the schedule σ': let $E_i(\sigma')$ be the earliness of a task i in the schedule σ; $E(\sigma')$ be the total earliness of tasks in the schedule σ'; $E_{j_i}(\sigma') = E_{j_i}(\sigma)$, i = 1, 2, …, $\nu - 1$; $E_{j_i}(\sigma') = E_{j_i}(\sigma) + (l_J - l_K)$, $i = \nu + 1$, …, $n - 1$; $E_{j_n}(\sigma') = E_J(\sigma') = d_K - f$; $E_{j_\nu}(\sigma') = E_K(\sigma') = d_K - (g + l_K)$.

Let us determine the difference between the total earliness of tasks in the schedules σ' and σ:

$$\begin{aligned} E(\sigma') - E(\sigma) &= \sum_{i=1}^{n} E_{j_i}(\sigma') - \sum_{i=1}^{n} E_{j_i}(\sigma) = \sum_{i=\nu}^{n} E_{j_i}(\sigma') - \sum_{i=\nu}^{n} E_{j_i}(\sigma) \\ &= E_K(\sigma') - E_J(\sigma) + \sum_{i=\nu+1}^{n-1} E_{j_i}(\sigma') - \sum_{i=\nu+1}^{n-1} E_{j_i}(\sigma) + E_J(\sigma') - E_K(\sigma) \\ &= d_K - g - l_K - d_J + g + l_J + (n - \nu - 1)(l_J - l_K) + d_J - f - d_K + f \\ &= (n - \nu)(l_J - l_K). \end{aligned}$$

We see that the interchanging of the tasks J and K leads to increase in the total earliness by $(n - \nu)(l_J - l_K)$. ∎

Hence, the last position in a feasible optimal by the total earliness minimization criterion schedule with the start time $s_{\max}$ should occupy the task with the minimum processing time, provided that this does not violate the feasibility of the schedule.

Heuristic 2 Let σ be a feasible schedule for n tasks with the latest start time $s_{\max}$. Suppose that a feasible schedule $\sigma_1 = \{j_1, j_2, \ldots, j_n\}$ (where j_i is the number of task in position i in σ_1) was constructed by the following rule:

$$l_{j_n} = \min_{r: d_r \geq \sum_{j=1}^{n} l_j + s_{\max}} \{l_r\} \quad ; \tag{2.3}$$

$$l_{j_i} = \min_{r: d_r \geq \sum_{j=1}^{n} l_j - \sum_{l=0}^{n-i-1} l_{j_{n-l}} + s_{\max}} \{l_r\} \quad , \quad i = \overline{n-1, 1} \tag{2.4}$$

Then, the schedule σ_1 is optimal by the total earliness minimization criterion for the latest start time of tasks s_{max}.

Justification The verity of Heuristic 2 follows from the verity of Heuristic 1. Indeed, from the verity of Heuristic 1 and the additivity of the functional of the total earliness minimization follows the fulfillment of Bellman's principle of optimality: if a feasible schedule is optimal by the total earliness minimization criterion, then any beginning part of the schedule is also optimal by this criterion.

According to the recurrent Exs. (2.3) and (2.4), the final form of the feasible schedule σ_1 in which the total earliness reaches a minimum while keeping the latest start time of tasks, is found recurrently by selecting the last, the one before last, etc. task. In other words, we sequentially build the feasible schedule σ_1 from the schedule σ, starting with the last task determination and ending with the first task determination. This procedure can always be executed because σ is a feasible schedule. In this case, tasks in schedule σ_1 either remain in their positions they occupied in the schedule σ or are shifted (shorter tasks to the left, longer tasks to the right) without violating their due dates. ■

Remark 2.4 A schedule built according to the recurrent Exs. (2.3) and (2.4), for any given start time $\forall s < s_{max}$, has a minimum total earliness for the given s.

2.4 The Polynomial Algorithm for Finding an Optimal Schedule for the Total Earliness Criterion for a Given Start Time of the Machine for the Case if the Heuristics 1 and 2 Are True

Algorithm A1 The algorithm to construct an optimal schedule with a given start time of tasks s for the total earliness criterion has the following scheme.

1. Select the task with the smallest processing time from the set of tasks for which the due dates are greater than or equal to the completion time of all tasks. Move this task into the last position (if there are several such tasks, then select one of them arbitrarily):

$$l_{j_n} = \min_{r: d_r \geq \sum_{j=1}^{n} l_j + s} \{l_r\}\quad, \tag{2.5}$$

2. Continue with the position $i = n - 1$.
3. Determine the task to be processed in the position i:

   3.1 Find the completion time of the task $j_{[i]}$ (that *will occupy* the position i in the schedule):

$$C_{j_{[i]}} = \sum_{j=1}^{n} l_j - \sum_{l=0}^{n-i-1} l_{j_{n-l}} + s \tag{2.6}$$

3.2 Select the task with the smallest processing time from the set of tasks for which the due dates are not less than $C_{j_{[i]}}$:

$$l_{j_{[i]}} = \min_{r:d_r \geq C_{j_{[i]}}} \{l_r\}. \tag{2.7}$$

3.3 Move the task $j_{[i]}$ (found in step 3.2) into the position i.

4. Continue with the next position: $i = i - 1$. If $i = 0$, then the algorithm terminates. Otherwise, go to step 3.

If we need to construct a schedule optimal by the total earliness minimization criterion for the latest start time $s_{\max}$, then we can reach it replacing s with $s_{\max}$ in Exs. (2.5)–(2.7).

The following example illustrates the process of construction of a schedule optimal by the two criteria.

Example 2.4 Given: $n = 6$, initial data for tasks are in Table 2.10.

We have found a schedule σ with the latest possible start time $s_{\max} = 65$ in accordance with Algorithm A. We present it in Table 2.11.

The total earliness at the latest possible start time is $E(\sigma) = 310$. We select the task with the smallest processing time from the set of tasks for which the due dates are greater than or equal to the completion time of all tasks. Then we move it into the last position:

$$l_{j_{[6]}} = \min_{r:d_r \geq \sum_{i=1}^{6} l_i + s_{\max}} \{l_r\} = \min_{r:d_r \geq 505} \{l_r\} = \min\{l_4; l_5\} = 50.$$

Tasks 4 and 5 have a greater due date value than 505. But task 5 has a smaller processing time ($l_5 = 50 < l_4 = 100$), so we select the task 5 for the last position. As a result, we obtain the subsequence of the last tasks of an optimal schedule shown in Table 2.12. We determine the completion time of a task before last task in Table 2.12 this way:

$$C_{j_{[5]}} = \sum_{j=1}^{6} l_j - l_{j_6} + s_{\max} = 440 - 50 + 65 = 455.$$

Table 2.10 Initial data for Example 2.4

j	1	2	3	4	5	6
l_j	110	30	70	100	50	80
d_j	250	280	255	520	510	430

Table 2.11 The schedule σ with the latest possible start time $s_{\max} = 65$

i	j	l_j	d_j	C_j	E_j
1	1	110	250	175	75
2	3	70	255	245	10
3	2	30	280	275	5
4	6	80	455	355	100
5	5	50	510	405	105
6	4	100	520	505	15

Table 2.12 Subsequence of the last tasks of an optimal schedule (task 5 is in the last position)

i	j	l_j	d_j	C_j	E_j
				455	
6	5	50	510	505	5

We select the task with the smallest processing time from the set of tasks for which the due dates are not less than $C_{j_{[5]}}$.

$$l_{j_{[5]}} = \min_{r:d_r \geq C_{j_{[5]}}}\{l_r\} = \min_{r:d_r \geq 455}\{l_r\} = \min\{l_4, l_6\} = 80.$$

We select for the position before last the task 6 and obtain the subsequence of the last tasks of an optimal schedule shown in Table 2.13. We determine the completion time of a task before the task 6 in Table 2.13 this way:

$$C_{j_{[4]}} = \sum_{j=1}^{6} l_j - l_{j_6} - l_{j_5} + s_{\max} = 440 - 50 - 80 + 65 = 375.$$

According to Algorithm A1, there will not be any further change in positions of tasks 1, 2, 3, and 4. Therefore, we have obtained an optimal schedule σ^{opt} with the latest possible start time $s_{\max} = 65$ and the minimum total earliness $E(\sigma^{opt}) = 240$ (Table 2.14). We have got the solution for the problem.

Table 2.13 Subsequence of the last tasks of an optimal schedule (task 6 is in the position before last)

i	j	l_j	d_j	C_j	E_j
				375	
5	6	80	455	455	0
6	5	50	510	505	5

Table 2.14 A schedule optimal by the total earliness minimization criterion at $s_{max} = 65$

i	j	l_j	d_j	C_j	E_j
1	1	110	250	175	75
2	3	70	255	245	10
3	2	30	280	275	5
4	4	100	520	375	145
5	6	80	455	455	0
6	5	50	510	505	5

2.5 Construction of a Feasible Schedule Which Is Optimal for the Criterion of the Total Earliness of Tasks Minimization

Let us consider a problem similar to the bi-criteria problem considered above. We remind that there we needed to find a feasible schedule in which the start time of tasks is the latest possible (*Criterion 1*) and the total earliness of tasks in regard to their due dates is minimal (*Criterion* 2). Within the framework of the bi-criteria problem solving, we assumed that Criterion 1 is the main criterion. In some practical problems, the main criterion for a schedule evaluation is Criterion 2 (the total earliness minimization). We will now consider such bi-criteria problem. Because of this, the formulation of the problem will change as follows.

The Problem Statement Given a set of independent tasks $J = \{1, 2, \ldots, n\}$ each consisting of a single operation. We know a processing time l_j and a due date d_j for each task $j \in J$. All tasks become available at the same time point for processing on a single machine. Interruptions during a task processing are not allowed. The process is continuous: after the first task completes, processing of the second task starts immediately, and so on, until all the tasks are completed. We need to find a feasible schedule in which the total earliness of the completion times of the tasks in regard to their due dates is minimal.

Note that if a schedule gets worse by Criterion 1 (if we shift it to the left), we can get a schedule that is better than the previous one by Criterion 2. The following example illustrates this statement.

Example 2.5 Given $n = 6$ tasks, processing times and due dates are in Table 2.15.

We have found that $s_{max} = 19$, in accordance with Algorithm A. By means of Algorithm A1 we have built an optimal by Criterion 2 schedule with the total earliness equal to 500 (Fig. 2.3).

After that, using Algorithm A1, we have built an optimal by Criterion 2 schedule for the start time $s = 18$ with the total earliness equal to 6 (Fig. 2.4).

Table 2.15 Initial data for Example 2.5

j	1	2	3	4	5	6
l_j	1	1	1	1	1	101
d_j	120	121	122	123	124	125

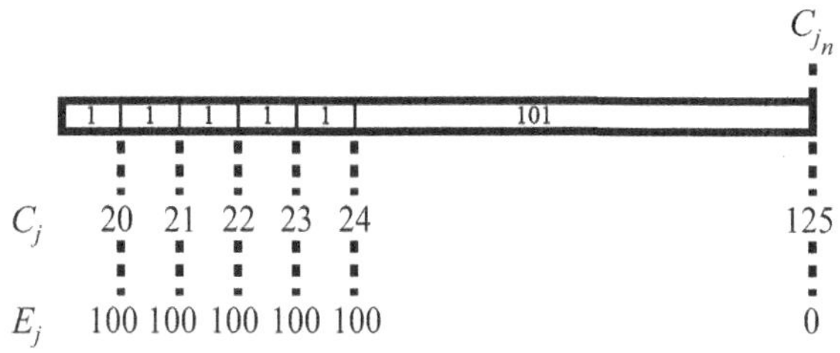

Fig. 2.3 The schedule at the latest start time $s_{\max} = 19$

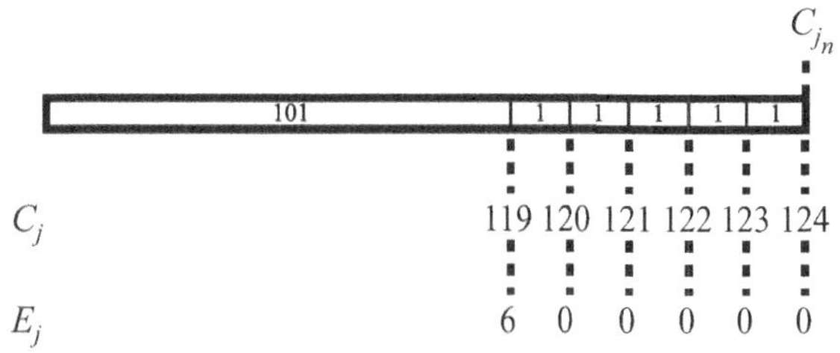

Fig. 2.4 The schedule at the start time $s = 18$

As we can see, worsening the value of Criterion 1 by one has improved the value of Criterion 2 by 500 − 6 = 494.

PSC-algorithms development [6] is one of the approaches to the efficient solving of intractable (NP-hard) combinatorial optimization problems as well as polynomially solvable combinatorial optimization problems in some cases. We solve this problem by the PSC-algorithm of the second class: in addition to the first polynomial component, it contains an approximation algorithm that yields an approximate solution if the sufficient signs of optimality are not fulfilled.

Sufficient signs of optimality of a feasible solution (SSOs) for the case when the structure of an optimal schedule at arbitrary s satisfies Heuristics 1 and 2.

Let $E(s)$ denote the value of Criterion 2 (the minimum total earliness at the given start time s) in a schedule constructed according to Heuristics 1 and 2. Let s_{opt} be the start time in the schedule with the minimum total earliness among the schedules with all possible start times.

Suppose we have a schedule constructed in accordance with Heuristics 1 and 2 for the start time $s_{\max}$. Obviously, if we decrease the start time by an arbitrary value Δ (shift the obtained sequence of tasks by Δ to the left), then the total earliness will increase by $n\Delta$. Further increasing the value of Δ, we may come to some start time $s = s_{\max} - \Delta \geq 0$ when the schedule structure (the tasks order) changes. In this case, $E(s) < E(s^+)$ where $E(s^+)$ are points of discontinuity in the graph of the dependence of $E(s)$ on s, shown in Fig. 2.5.

SSO #1 Suppose we have a schedule with the start time $s_{\max}$ that satisfies

$$E(s_{\max}) + ns_{\max} = E(0), \tag{2.8}$$

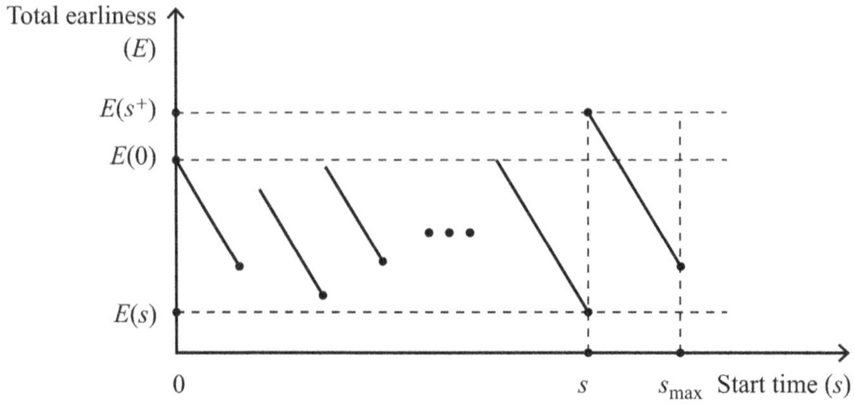

Fig. 2.5 Change of the total earliness in dependence of the start time

where $E(s_{\max})$ and $E(0)$ are the values of Criterion 2 for the start times $s_{\max}$ and 0, respectively. Then, this schedule is optimal for both criteria, i.e., $s_{opt} = s_{\max}$.

We can also note that fulfilling of inequality $E(s) + ns < E(0)$ is impossible for any s (including $s_{\max}$).

SSO #2 Suppose that (2.8) is not satisfied but there is a fixed start time $s < s_{\max}$ for which the following is satisfied:

$$E(s) + ns = E(0) \tag{2.9}$$

Then, the optimal start time s_{opt} corresponding to an optimal total earliness satisfies the inequality [3]:

$$s \le s_{opt} \le s_{\max}.$$

Proof of both SSOs will be carried out by contradiction.

Suppose that SSO #1 is fulfilled but $s_{opt} \ne s_{\max}$. This means that there exists a start time $s_{opt} < s_{\max}$ for which $E(s_{opt}) < E(s_{\max})$. Since $E(s_{\max}) + ns_{\max} = E(0)$, we have: $ns_{\max} > ns_{opt}$. Therefore,

$$E(s_{opt}) < E(s_{\max}),$$

$$E(s_{opt}) + ns_{opt} < E(s_{\max}) + ns_{\max} = E(0).$$

We have arrived at a contradiction, so we have proven the SSO #1.

Suppose that SSO #2 is fulfilled but $s_{opt} < s$, then $E(s_{opt}) < E(s)$. Therefore, $E(s_{opt}) + ns_{opt} < E(s) + ns = E(0)$ which is impossible. This proves the SSO #2. ∎

Outline of PSC-algorithm for the case when the structure of an optimal schedule at arbitrary s satisfies Heuristics 1 and 2.

First we build an optimal by Criterion 2 schedule at the start time s_{max}. If it satisfies SSO #1, i.e., if $E(s_{max}) + ns_{max} = E(0)$, then the obtained schedule is optimal. If SSO #1 is not satisfied, then we apply polynomial algorithm with $kO(n^2)$ complexity where k is a given limit for the total number of iterations of the PSC-algorithm. By means of the polynomial algorithm, we vary the start times from s_{max} to zero and apply the recurrent procedure from Heuristic 2. This yields a series of optimal schedules for start times

$$s_{max}, \quad s_1, s_2, \ldots, s_k$$

where $s_1 = s_{max} - \Delta_1$, $s_2 = s_1 - \Delta_2$, …, $s_k = s_{k-1} - \Delta_k$; $s_1, s_2, \ldots, s_k$ are the discrete time points of a change in the structure of an optimal by Criterion 2 schedule where the position of at least one task is changed.

Consider a l-th iteration, $l = \overline{1,k}$. Determine for it the length of the interval Δ_l where the structure of the schedule is unchanged (if $\Delta < \Delta_l$, then all tasks remain in their positions; if $\Delta \geq \Delta_l$, then at least one task moves into a later position). According to the recurrent procedure from Heuristic 2, we can determine the length of Δ_l this way:

$$\Delta_l = \min_{2 \leq i \leq n} \left\{ C_{j_i}^{s_{l-1}} - \max_{r: l_r < l_{j_i}^{s_{l-1}}, d_r < C_{j_i}^{s_{l-1}}} \{d_r\} \right\}. \tag{2.10}$$

Here, $C_{j_i}^{s_{l-1}}$ is the completion time of a task j_i that occupies an i-th position in an optimal schedule for a start time s_{l-1}; $l_{j_i}^{s_{l-1}}$ is the processing time of this task j_i in this optimal schedule; $\max\limits_{r: l_r < l_{j_i}^{s_{l-1}}, d_r < C_{j_i}^{s_{l-1}}} \{d_r\}$ is the maximum due date of the tasks which are shorter than the task j_i and have less due date than the completion time of j_i.

The procedure terminates if we cannot determine the maximum in (2.10) for any i, i.e., if there are no tasks r satisfying the conditions: $l_r < l_{j_i}^{s_{l-1}}$, $d_r < C_{j_i}^{s_{l-1}}$.

The structure of an optimal schedule is unchanged $\forall s \in [s_l, s_{l-1})$, $l = \overline{1,k}$ ($s_0 = s_{max}$). And $E(s_l - \Delta) = E(s_l) + n\Delta$ is true for all $0 < \Delta < s_l - s_{l+1}$.

We have built an optimal schedule if we have found the first s_l, $l \leq k$, such that

$$E(s_l) + ns_l = E(0) \tag{2.11}$$

This means that SSO #2 is fulfilled, and s_{opt} is one of the values from $s_l, s_{l-1}, \ldots, s_{max}$ if $l \leq k$. s_{opt} is equal to s_p corresponding to $\min\limits_{p=\overline{0,l}, s_0 = s_{max}} E(s_p)$.

The following theorem obviously follows from the above results.

Theorem 2.4 *Suppose that* s_{max}, s_1, …, s_p *are all the time points where the structure of an optimal schedule for the total earliness minimization criterion changes. Then, subject to* Heuristics 1 *and* 2 *fulfillment,* $E(s_p) + ns_p = E(0)$ *and, consequently,* $s_{opt} \in \{s_{max}, s_1, \ldots, s_p\}$.

Corollary 2.3 We have $E(s_j) + ns_j > E(0)$ for $\forall s_j \in \{s_{\max}, s_1, \ldots, s_{p-1}\}$ (if $E(s_l) + ns_l = E(0)$, then $l = p$).

Algorithm A2 The first polynomial component of the PSC-algorithm to find an optimal schedule for the total earliness minimization criterion.

1. Construct optimal by Criterion 2 schedules for the start times $s_{\max}$ and zero. Determine their corresponding total earliness values $E(s_{\max})$ and $E(0)$. If $E(s_{\max}) + ns_{\max} = E(0)$, then the algorithm terminates: the schedule obtained for the start time $s_{\max}$ is optimal. Otherwise, set $l = 0$ (the serial number of a change in the structure of the schedule); $s_0 = s_{\max}$.

2. Do the following:

   2.1 Set $l = l + 1$.
   2.2 Find the length Δ_l of the l-th interval using Eq. (2.10).
   2.3 Shift the current sequence of tasks to the left by Δ_l: $s_l = s_{l-1} - \Delta_l$.
   2.4 Construct an optimal by Criterion 2 schedule for the obtained start time s_l.
   2.5 If SSO #2 is fulfilled, i.e., $E(s_l) + ns_l = E(0)$, then determine $E(s_{opt}) = \min_{1 \le l \le k} E(s_l)$, the algorithm terminates. Otherwise, go to step 2.1.

The complexity of the first polynomial component is $kO(n^2)$ where k is a given number of changes in the schedule's structure. Let us show it. The most complex part of the algorithm is sub-step 2.2 of step 2. It consists in finding the length Δ_l of the l-th interval (in which the schedule structure is unchanged) and, accordingly, the start time $s_l = s_{l-1} - \Delta_l$. After applying Algorithm A1, at least one task in an optimal by criterion 2 schedule for the start time s_l will change its position in comparison to the schedule for the start time s_{l-1}. For a fixed pair of tasks r and j_i, the inequalities $l_r < l_{j_i}$, $d_r < C_{j_i}$ are checked in time $\Theta(1)$. The total number of existing pairs is $n(n-1)/2$. Thus, the upper bound for the number of compared pairs is $O(n^2)$. Let the total number of structure changes be k. Then, the complexity of the first polynomial component of the algorithm is $kO(n^2)$.

If SSO #2 is fulfilled at some stage for a start time s_l, then the exact solution is within the time segment $[s_l, s_{\max}]$, it is one of the values from s_l, s_{l-1}, …, $s_{\max}$ if $l \le k$. If the SSO #2 is not fulfilled for k changed structures of the schedule, then we find the minimum value of the total earliness for all discrete points of the structure changes. Thus, we will obtain an approximate solution which is exact for $\forall s \in [s_l, s_{\max}]$. So, this is an approximation subalgorithm. We have constructed the PSC-algorithm of the second class.

Remark 2.5 If Heuristics 1 and 2 are not fulfilled, then the obtained schedule is approximate and is statistically guaranteed to be optimal for random generation of the problem instances.

Let us give some examples for finding an optimal by Criterion 2 schedule.

Example 2.6 Given $n = 6$ tasks, processing times and due dates are in Table 2.16.

We have found a schedule with the latest possible start time $s_{\max} = 65$, in accordance with Algorithm A. The minimal total earliness after applying Algorithm A1 at the latest start time is $E(s_{\max}) = 310$. The final form of the schedule with the minimum total earliness at the latest start time is in Table 2.17.

$E(s_{\max}) = E(65) = 310$. The schedule with the minimum total earliness at zero start time is in Table 2.18.

$E(0) = 700$. In this problem instance, the first step of Algorithm A2 yields an optimal by Criterion 2 schedule. Let us check it: $E(s_{\max}) + ns_{\max} = E(0)$: $310 + 6 \times 65 = 700$. We have obtained an optimal by Criterion 2 schedule because SSO #1 is fulfilled, $E(s_{\max}) = E(s_{opt})$. We have solved the problem.

Example 2.7 Let us find an optimal by Criterion 2 schedule for the following problem. Given $n = 6$, processing times and due dates are in Table 2.19.

We determine the total earliness with zero start time by means of Algorithm A2. Table 2.20 shows an optimal by Criterion 2 schedule with zero start time.

$E(0) = 196$. We apply Algorithm A1 to find the maximum start time. Table 2.21 presents an optimal by Criterion 2 schedule for $s_{\max} = 34$.

$E(s_{\max}) = 156$. SSO #1 is not fulfilled: $E(s_{\max}) + ns_{\max} \neq E(0)$, $156 + 6 \times 34 = 360 \neq 196$. Let us find the minimum value among all possible time points where the schedule structure changes:

$$\Delta_1 = \min_{2 \le i \le n} \left\{ \underbrace{110 - \max\{108, 104, 96, 60, 86\}}_{\text{correspondsto } i=6}; \underbrace{68 - 60}_{\text{corr.to } i=4}; \underbrace{62 - 60}_{\text{corr.to } i=3} \right\} \tag{2.12}$$

The minimum in Eq. (2.12) corresponds to position 6: $\Delta_1 = 110 - 108 = 2$. Components of Eq. (2.12) corresponding to positions 2 and 5 were not determined because the tasks in these positions do not satisfy the conditions (2.10). We shift the sequence of tasks to the left by 2: $s_1 = s_{\max} - \Delta_1 = 34 - 2 = 32$. Next, we build an optimal by Criterion 2 schedule for the start time $s_1 = 32$ (Table 2.22).

$E(s_1) = 124$. SSO #2 is not fulfilled: $E(s_1) + ns_1 \neq E(0)$, $124 + 6 \times 32 = 316 \neq 196$. Let us find the minimum value among all possible time points where the schedule structure changes:

$$\Delta_2 = \min_{2 \le i \le n} \left\{ \underbrace{106 - \max\{104, 60, 96, 86\}}_{\text{correspondsto } i=5}; \underbrace{66 - 60}_{\text{corr.to } i=4} \right\} \tag{2.13}$$

Table 2.16 Initial data for Example 2.6

j	1	2	3	4	5	6
l_j	110	30	70	100	50	80
d_j	250	280	255	520	510	430

Table 2.17 The schedule optimal by the total earliness minimization criterion at $s_{max} = 65$

i	j	l_j	d_j	C_j	E_j
1	1	110	250	175	75
2	3	70	255	245	10
3	2	30	280	275	5
4	4	80	455	355	100
5	6	50	510	405	105
6	5	100	520	505	15

Table 2.18 The schedule optimal by the total earliness minimization criterion at zero start time

i	j	l_j	d_j	C_j	E_j
1	1	110	250	110	140
2	3	70	255	180	75
3	2	30	280	210	70
4	4	80	455	290	165
5	6	50	510	340	170
6	5	100	520	440	80

Table 2.19 Initial data for Example 2.7

j	1	2	3	4	5	6
l_j	4	14	10	6	2	40
d_j	60	86	96	104	108	112

Table 2.20 Optimal schedule with zero start time

i	j	l_j	d_j	C_j	E_j
1	6	40	112	40	72
2	2	14	86	54	32
3	1	4	60	58	2
4	3	10	96	68	28
5	4	6	104	74	30
6	5	2	108	76	32

Table 2.21 Optimal schedule with the start time $s_{max} = 34$

i	j	l_j	d_j	C_j	E_j
1	2	14	86	48	38
2	1	4	60	52	8
3	3	10	96	62	34
4	4	6	104	68	36
5	5	2	108	70	38
6	6	40	112	110	2

Table 2.22 Optimal schedule with the start time $s_1 = 32$

i	j	l_j	d_j	C_j	E_j
1	2	14	86	46	40
2	3	10	96	56	40
3	1	4	60	60	0
4	4	6	104	66	38
5	6	40	112	106	6
6	5	2	108	108	0

The minimum in Eq. (2.13) corresponds to position 5: $\Delta_2 = 106 - 104 = 2$. Components of Eq. (2.13) corresponding to positions 2, 3 and 5 were not determined because the tasks in these positions do not satisfy the conditions (2.10). We shift the sequence of tasks to the left by 2: $s_2 = s_1 - \Delta_2 = 32 - 2 = 30$. Next, we build an optimal by Criterion 2 schedule for the start time $s_2 = 30$ (Table 2.23).

$E(s_2) = 102$. SSO #2 is not fulfilled: $E(s_2) + ns_2 \neq E(0)$, $102 + 6 \times 30 = 282 \neq 196$. Let us find the minimum value among all possible time points where the schedule structure changes:

$$\Delta_3 = \min_{2 \le i \le n} \left\{ \underbrace{104 - 60}_{\text{corr.to } i=5}; \underbrace{98 - \max\{60, 96, 84\}}_{\text{correspondsto } i=4} \right\} \tag{2.14}$$

The minimum in Eq. (2.14) corresponds to position 4: $\Delta_3 = 98 - 96 = 2$. Components of Eq. (2.14) corresponding to positions 2, 3 and 6 were not determined because the tasks in these positions do not satisfy the conditions (2.10). We shift the sequence of tasks to the left by 2: $s_3 = s_2 - \Delta_3 = 30 - 2 = 28$. Next, we build an optimal by Criterion 2 schedule for the start time $s_3 = 28$ (Table 2.24).

$E(s_3) = 74$. SSO #2 is not fulfilled: $E(s_3) + ns_3 \neq E(0)$, $74 + 6 \times 28 = 242 \neq 196$. Let us find the minimum value among all possible time points where the schedule structure changes:

$$\Delta_4 = \min_{2 \le i \le n} \left\{ \underbrace{102 - 60}_{\text{corr.to } i=5}; \underbrace{96 - 60}_{\text{corr.to } i=4}; \underbrace{86 - 60}_{\text{corr.to } i=3}; \underbrace{72 - 60}_{\text{corr.to } i=2} \right\} \tag{2.15}$$

The minimum in Eq. (2.15) corresponds to position 2: $\Delta_4 = 72 - 60 = 12$. The component of Eq. (2.15) corresponding to position 6 was not determined because

Table 2.23 Optimal schedule with the start time $s_2 = 30$

i	j	l_j	d_j	C_j	E_j
1	2	14	86	44	42
2	3	10	96	54	42
3	1	4	60	58	2
4	6	40	112	98	14
5	4	6	104	104	0
6	5	2	108	106	2

Table 2.24 Optimal schedule with the start time $s_3 = 28$

i	j	l_j	d_j	C_j	E_j
1	1	4	60	32	28
2	6	40	112	72	40
3	2	14	86	86	0
4	3	10	96	96	0
5	4	6	104	102	2
6	5	2	108	104	4

Table 2.25 Optimal schedule with the start time $s_4 = 16$

i	j	l_j	d_j	C_j	E_j
1	6	40	112	56	56
2	1	4	60	60	0
3	2	14	86	74	12
4	3	10	96	84	12
5	4	6	104	90	14
6	5	2	108	92	16

the task in this positions does not satisfy the conditions (2.10). We shift the sequence of tasks to the left by 12: $s_4 = s_3 - \Delta_4 = 28 - 12 = 16$. Next, we build an optimal by Criterion 2 schedule for the start time $s_4 = 16$ (Table 2.25).

$E(s_4) = 110$. SSO #2 is not fulfilled: $E(s_4) + ns_4 \neq E(0)$, $110 + 6 \times 16 = 206 \neq 196$. Let us find the minimum value among all possible time points where the schedule structure changes:

$$\Delta_5 = \min_{2 \le i \le n} \left\{ \underbrace{90 - 60}_{\text{corr.to } i=5}; \underbrace{84 - 60}_{\text{corr.to } i=4}; \underbrace{74 - 60}_{\text{corr.to } i=3} \right\} \tag{2.16}$$

The minimum in Eq. (2.16) corresponds to position 3: $\Delta_5 = 74 - 60 = 14$. The components of Eq. (2.16) corresponding to positions 6 and 2 were not determined because the tasks in these positions do not satisfy the conditions (2.10). We shift the sequence of tasks to the left by 14: $s_5 = s_4 - \Delta_5 = 16 - 14 = 2$. Next, we build an optimal by Criterion 2 schedule for the start time $s_5 = 2$ (Table 2.26).

Table 2.26 Optimal schedule with the start time $s_5 = 2$

i	j	l_j	d_j	C_j	E_j
1	6	40	112	42	70
2	2	14	86	56	30
3	1	4	60	60	0
4	3	10	96	70	26
5	4	6	104	76	28
6	5	2	108	78	30

$E(s_5) = 184$. SSO #2 is fulfilled: $E(s_5) + ns_5 = E(0)$, $184 + 6 \times 2 = 196$. We have built an optimal schedule. s_{opt} is within the time segment [2, 34]. We can find $E(s_{opt}) = \min_{1 \le l \le 5} E(s_l) = 74$. Hence, the optimal by Criterion 2 schedule coincides with the schedule for the start time s_3 (Table 2.27).

We will illustrate the process of Example 2.7 solving graphically at Fig. 2.6.

Table 2.27 The schedule optimal by the total earliness minimization criterion

i	j	l_j	d_j	C_j	E_j
1	1	4	60	32	28
2	6	40	112	72	40
3	2	14	86	86	0
4	3	10	96	96	0
5	4	6	104	102	2
6	5	2	108	104	4

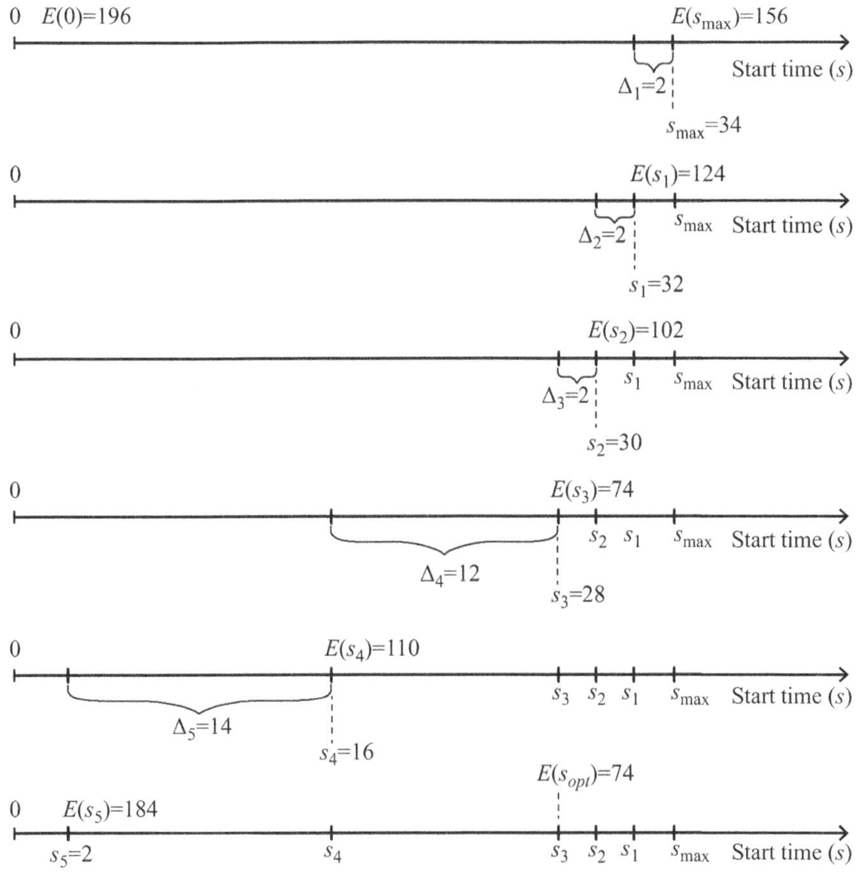

Fig. 2.6 Step by step process of finding an optimal schedule for the total earliness minimization criterion

References

1. Pavlov, A.A., Misura, E.B., Khalus, E.A.: Polinomial'nyi algoritm polucheniya dopustimogo raspisaniya s maksimal'no pozdnim momentom nachala vypolneniya odnim priborom nezavisimyh zadaniy proizvol'noi dlitel'nosti s raznymi direktivnymi srokami, pri kotorom vse zadaniya ostayutsya nezapazdyvayuschimi (Полиномиальный алгоритм получения допустимого расписания с максимально поздним моментом начала выполнения одним прибором независимых заданий произвольной длительности с разными директивными сроками, при котором все задания остаются незапаздывающими; A polynomial algorithm for obtaining a feasible schedule with the latest processing start time on a single machine for independent tasks with arbitrary processing times and different due dates when all tasks remain non-tardy). Paper presented at the 1st international conference Informaciini tehnologiї yak innovaciynyi shlyah rozvitku Ukrainy u XXI stolitti, Transcarpathian State University, Uzhhorod, 6–8 Dec 2012 (in Russian)
2. Pavlov, A.A., Misura, E.B., Khalus, E.A.: Issledovanie svoistv zadachi kalendarnogo planirovaniya dlya odnogo pribora po kriteriyu minimizacii summarnogo operejeniya zadaniy pri uslovii dopustimosti raspisaniya (Исследование свойств задачи календарного планирования для одного прибора по критерию минимизации суммарного опережения заданий при условии допустимости расписания; Properties' research of the scheduling problem for a single machine by minimizing the total earliness of tasks with the condition of the schedule feasibility). Visnyk NTUU KPI Inform. Oper. Comput. Sci. **56**, 98–102 (2012) (in Russian)
3. Pavlov, A.A., Khalus, E.A.: Sostavlenie dopustimogo raspisaniya vypolneniya rabot na odnom pribore, optimal'nogo po kriteriyu minimizacii summarnogo operejeniya rabot (Составление допустимого расписания выполнения работ на одном приборе, оптимального по критерию минимизации суммарного опережения работ; Drawing up a feasible schedule of jobs on one machine in order to minimize the total earliness of jobs). Visnyk NTUU KPI Inform. Oper. Comput. Sci. **61**, 27–34 (2014) (in Russian)
4. Pavlov, A.A., Misura, E.B., Khalus, E.A.: Skladannya rozkladu vikonannya zavdan' na odnomu priladi z metoyu minimizacii sumarnogo vyperedjennya ta znahodjennya maksimal'nogo pizn'ogo momentu pochatku vikonannya zavdan' v dopustimomu rozkladi (Складання розкладу виконання завдань на одному приладі з метою мінімізації сумарного випередження та знаходження максимального пізнього моменту початку виконання завдань в допустимому розкладі; Single machine scheduling to minimize the total earliness and find the latest start time of tasks in a feasible schedule). Paper presented at the 21st international conference on automatic control Automatics-2014, National Technical University of Ukraine, Kyiv, 23–27 Sept 2014 (in Ukrainian)
5. Tanaev, V.S., Shkurba, V.V.: Vvedenie v Teoriju Raspisaniy (Введение в теорию расписаний; Introduction to Scheduling Theory). Nauka, Moscow (1975) (in Russian)
6. Zgurovsky, M.Z., Pavlov, A.A.: Prinyatie Resheniy v Setevyh Sistemah s Ogranichennymi Resursami (Принятие решений в сетевых системах с ограниченными ресурсами; Decision Making in Network Systems with Limited Resources), Naukova dumka, Kyiv (2010) (in Russian)

Chapter 3
Optimal Scheduling for Vector or Scalar Criterion on Parallel Machines with Arbitrary Due Dates of Tasks

Abstract We examine the problem of constructing a feasible schedule for parallel machines of equal or various productivities with various due dates of tasks and arbitrary start times of machines. We consider two optimality criteria: maximizing the earliest start time of machines (scalar criterion) or obtaining optimal start times of machines subject to a direct lexicographical order of their assignment (vector criterion). In fact, in this chapter we examine four intractable combinatorial optimization problems (for each one there are no efficient polynomial algorithms for its solving). Each of four presented PSC-algorithms contains separate sufficient signs of optimality of a feasible solution, the first polynomial component that checks the sufficient signs of optimality, and an approximation algorithm. Since the formulated problems are quite complex, each component of the PSC-algorithms contain many subalgorithms, each of which implements a separate original heuristic. We give examples of the problems solving.

3.1 Introduction

In this chapter, we formulate new one-stage scheduling problems for independent tasks processing on parallel machines with equal (Problem 1) or various (Problem 2) productivities for a vector (lexicographical) and a scalar criteria. For each problem, we have found and proved sufficient signs of optimality of feasible solutions (SSO) and on their basis developed efficient PSC-algorithms classified as belonging to the second class. We have created a total of 39 algorithms that implement the first and the second polynomial components of the PSC-algorithms including approximation algorithms with an estimate of deviation from an optimal solution. In this case, the first polynomial component is implemented by separate polynomial algorithms that do not coincide with the second polynomial component. The individual problem is solved by each algorithm, the best solution is chosen. Depending on the initial data, one algorithm may give a better solution than others. It is not known in advance which algorithm will appear the best. We give examples of individual problems solving.

M. Z. Zgurovsky and A. A. Pavlov, *Combinatorial Optimization Problems in Planning and Decision Making*, Studies in Systems, Decision and Control 173,
https://doi.org/10.1007/978-3-319-98977-8_3

The problems presented in this chapter were not investigated on what class of problems (P or not easier than NP-complete) they belong to. However, they are intractable problems of combinatorial optimization: it is not proved that they are NP-hard, but exact polynomial algorithms of their solving are not known.

This chapter contains our new original results based on our earlier paper [1].

3.2 Problem 1. Machines with Equal Productivities

Definition 3.1 A schedule is called feasible if it does not contain tardy tasks. A task is called feasible if it is not tardy.

Let $\overline{a,b}$ denote the interval of integer numbers from a to b, that is, $\overline{a,b} = \mathbb{Z} \cap [a, b] = a,\ a + 1,\ \ldots,\ b$.

The Problem Statement Given m independent parallel machines with equal productivities to process n tasks without interruptions. l_i is the processing time of task i, $i = \overline{1,n}$, its due date is d_i. The start times of machines are arbitrary. We need to build a feasible schedule that maximizes one of the two criteria.

Criterion 1 A feasible schedule is optimal by Criterion 1 if it satisfies

$$\begin{gathered} s_{j_1} = \max\{\min_{j=\overline{1,m}} s_j\} \\ s_{j_l} = \max\{\min_j s_j, j = \overline{1,m}, j \neq j_k, k = \overline{1,l-1}\}, \quad l = \overline{2,m} \end{gathered} \tag{3.1}$$

where s_j is the start time of machine j; j_1 is the number of machine with the earliest start time in an optimal schedule (it is the latest for all feasible schedules); j_l, $l = \overline{2,m}$ is the number of machine with the next earliest start time after machines j_k, $k = \overline{1,l-1}$ (it is the latest for all feasible schedules with fixed s_{j_k}, $k = \overline{1,l-1}$).

Definition 3.2 A feasible schedule for which start times of machines are s_{j_l}, $l = \overline{1,m}$, is called optimal in the direct lexicographical order.

Remark 3.1 A feasible schedule is called optimal in the *inverse* lexicographical order if it satisfies

$$\begin{gathered} s_{j_1} = \max\{\max_{j=\overline{1,m}} r_j\} \\ s_{j_l} = \max\{\max_j s_j, j = \overline{1,m}, j \neq j_k, k = \overline{1,l-1}\}, \quad l = \overline{2,m}. \end{gathered}$$

Criterion 2 A feasible schedule is optimal by Criterion 2 if it satisfies

$$s_{j_1} = \max\{\min_{j=\overline{1,m}} s_j\}. \tag{3.2}$$

3.2.1 *Signs of Optimality of a Feasible Solution*

3.2.1.1 SSOs for Vector Criterion 1

Denote by I the set of all tasks $\{i{:}\ i = \overline{1,n}\}$. It is obvious that $d_i - l_i \geq 0,\ i = \overline{1,n}$. Let us renumber tasks in non-decreasing order of $d_i - l_i$, and let the following inequalities are satisfied: $d_1 - l_1 < d_2 - l_2 < \cdots < d_n - l_n$.

Let us formulate and prove SSOs #1 and #2 which will allow us to construct the first polynomial component of PSC-algorithm for the considered problem.

SSO #1 for Criterion 1 §1. Assign to the first machine the task with index 1 as the first (earliest) task. The start time of machine 1 is $s_1 = d_1 - l_1$. The number s_1 corresponds to the maximum possible value of s_{j_1} in Eq. (3.1). Suppose that the following inequalities hold: $d_i + l_k > d_k,\ i = \overline{1,m},\ k = \overline{i+1,m}$. Then assign to machine k, $k = \overline{2,m}$, as the first task the one with index k and the start time of the machine $s_k = d_k - l_k$.

Statement 3.1 An arbitrary feasible schedule for which §1 is fulfilled is optimal by criterion (3.1).

Proof Indeed, the violation of the algorithmic procedure of §1 in an arbitrary feasible schedule immediately leads to the satisfaction of

$$\begin{pmatrix} S_1 \\ \vdots \\ S_m \end{pmatrix} \not\succcurlyeq \forall \begin{pmatrix} S_{l_1} \\ \vdots \\ S_{l_m} \end{pmatrix}$$

in accordance with a preorder—a lexicographical order where the vector $(s_1 \ \ldots \ s_m)^{\mathrm{T}}$ corresponds to an arbitrary feasible schedule for which §1 is fulfilled, and $(s_{l1} \ \ldots \ s_{lm})^{\mathrm{T}}$ is the vector of start times of machines for an arbitrary feasible schedule. ∎

SSO #2 for Criterion 1 §2. Assign to the first machine as the first task the one with index 1. $s_1 = d_1 - l_1$ [$d_1 - l_1$ is the maximum possible value of s_{j_1} in Eq. (3.1)]. Suppose that $k_2 - 1$ is the maximum natural number for which the following inequalities hold:

$$d_1 + \sum_{i=2}^{l} l_i \leq d_l, \quad l = \overline{1, k_2 - 1}. \tag{3.3}$$

Then to the first machine successively assign tasks with indices 1, 2, …, $k_2 - 1$. The task with index k_2 is assigned as the first task to the second machine, $s_2 = d_{k_2} - l_{k_2}$. If the inequality

$$\min\{d_1 + \sum_{i=2}^{k_2-1} l_i + l_{k_2+1},\ d_{k_2} + l_{k_2+1}\} \le d_{k_2+1}$$

does not hold, then assign to the third machine as the first task the one with index $k_2 + 1$ at the time moment $s_3 = d_{k_2+1} - l_{k_2+1}$. In this case, $k_3 = k_2 + 1$. Otherwise, these inequalities must hold:

$$\min\{d_1 + \sum_{i=2}^{k_2-1} l_i,\ d_{k_2} + l_{k_2+1}\} \le d_{k_2+1}, \tag{3.4}$$

$$\max\{d_1 + \sum_{i=2}^{k_2-1} l_i,\ d_{k_2} + l_{k_2+1}\} > d_{k_2+1}. \tag{3.5}$$

Then the task with index $k_2 + 1$ is assigned to a machine corresponding to a minimum in (3.4). If the inequality (3.4) is true but the inequality (3.5) is false, then SSO #2 for this particular problem is not satisfied. Similarly, to the first or the second machine assign sequentially the tasks with indices $\overline{k_2 + 2, k_3 - 1}$ (here k_3 is the maximum possible natural number for which the task with index $k_3 - 1$ is not tardy). In this case, if current task can be assigned to a machine with lower start time, then assigning it to a machine with higher start time should lead to violation of its due date [an analogue to the fulfillment of inequalities (3.4), (3.5)]. The task with index k_3 is assigned as the first task to the third machine at the time moment $s_3 = d_{k_3} - l_{k_3}$.

In a similar way, we sequentially assign other tasks with indices $k_3 + i$ to machines. A necessary condition for the fulfillment of SSO #2 is the requirement that a task can be assigned only to one machine (with the minimum release time) from the current set of machines to which the tasks are assigned. If the task cannot be assigned to any machine from the current set of machines, then it is assigned as the first task to the next machine at the time that equals to its due date minus its processing time. The distribution of tasks terminates either when all tasks are assigned to l machines ($l < m$) or with assignment to machine m as the first task the one with index k_m, $s_m = d_{k_m} - l_{k_m}$. Assignment in accordance with §2 is complete.

Let I_1 be a set of all tasks with indices from set $\{\overline{1, k_m}\}$ or $\{\overline{1, k_l}\}$ if the tasks are distributed to $l < m$ machines. We require the following condition. Renumber all tasks from I_1 in the order of their assignment to machines. Then for each task with index i, $i = \overline{2, k_m \vee k_l}$, the following must hold:

$$d_i - C_i < l_p, \quad p = \overline{i + 1, k_m \vee k_l}, \quad d_i - C_i < \min_{i=\overline{2,k_m \vee k_l}} l_i \tag{3.6}$$

where C_i is the completion time of task i. Then the following statement is true.

Statement 3.2 Any arbitrary feasible schedule for which §1 or §2 is fulfilled and the condition (3.6) holds is optimal by criterion (3.1), i.e.

$$\begin{pmatrix} s_1 \\ s_2 \\ \vdots \\ s_m \end{pmatrix} \succcurlyeq \forall \begin{pmatrix} s_{q_1} \\ s_{q_2} \\ \vdots \\ s_{q_m} \end{pmatrix} \quad (3.7)$$

in accordance with a preorder—a lexicographical order where $(s_{q_1} \;\; \ldots \;\; s_{q_m})^{\mathrm{T}}$ is the vector of start times of machines for an arbitrary feasible schedule; components of the vector are in direct lexicographical order, i.e. the first component is the lowest, the second is the next lowest, etc.; $(s_1 \;\; \ldots \;\; s_m)^{\mathrm{T}}$ corresponds to a feasible schedule for which §1 or §2 is fulfilled and (3.6) holds (components of $(s_1 \;\; \ldots \;\; s_m)^{\mathrm{T}}$ are in direct lexicographical order by construction).

Proof Indeed, from the logic of tasks distribution to machines (§2) in combination with fulfillment of conditions (3.6), (3.4), (3.5), and their analogs at the subsequent stages of the distribution, it follows that any change in the order of assignments (except for cases when the minimum in the inequalities (3.4), (3.5), and their analogs is not unique) leads to worsening (decreasing) of machines start times in the lexicographical order (3.7). ∎

The condition (3.7) holds if either $s_1 > s_{q_1}$, or $s_1 = s_{q_1}$ and $s_2 > s_{q_2}$, or $s_1 = s_{q_1}$ and $s_2 = s_{q_2}$ but $s_3 > s_{q_3}$, etc., or $s_j = s_{q_j}, j = \overline{1, m}$.

Similarly to (3.7), we can compare two arbitrary vectors by a direct lexicographical order.

Remark 3.2 If, in accordance with §2, only l machines, $l < m$, became loaded, then in (3.7) we formally use $+\infty$ for the values of $s_{l+j}, j = \overline{1, m-l}$.

Remark 3.3 If *all* tasks are distributed to l machines, $l < m$, then we have constructed an optimal schedule according to criterion (3.1).

3.2.1.2 SSO for Scalar Criterion 2

Assign to the first machine as the first task the one with index 1. $s_1 = d_1 - l_1$. The number s_1 corresponds to the maximum possible value of s_{j_1} in (3.2).

Statement 3.3 An arbitrary feasible schedule for which $s_j \geq s_1, j = \overline{2, m}$, is optimal by criterion (3.2).

3.2.2 PSC-Algorithm for Problem 1 Solving by Criterion 1 (Algorithm A)

3.2.2.1 The First Polynomial Component of PSC-Algorithm A

The first polynomial component of the PSC-algorithm consists of four algorithms (A1.1, A1.2, A1.2a, A1.3). They are to be executed sequentially in an arbitrary order. The first algorithm that constructed a feasible schedule for which SSO #1 or #2 is satisfied, obtains an optimal solution of the problem by criterion (3.1), and the first polynomial component of the PSC-algorithm is considered to be fulfilled. If a feasible solution is not obtained by any algorithm of the first polynomial component, we search it using algorithms of the second polynomial component.

Algorithm A1.1 The algorithm implements the first polynomial component for a feasible schedule construction if an initial schedule was constructed according to SSO #1 or #2. As a result of this, the initial set of tasks I is divided into two subsets: I_1 is the set of previously assigned tasks and I'_1 is the set of not yet assigned tasks. We sequence the tasks from I'_1 in non-decreasing order of $d_i - l_i$ (let σ^{ord} denotes this ordered sequence). For each machine k, we know the release time C_k (the completion time of all tasks assigned to it). The further assignment of tasks from the sequence σ^{ord} is carried out as follows. Assign each next task to the release time of the machine with the minimum release time. If the task is not tardy, recalculate the release time, the iteration terminates: continue with the next task from σ^{ord}. If the task is tardy, the first polynomial component is not fulfilled, the algorithm terminates. If after such distribution all tasks are not tardy, then the first polynomial component is fulfilled.

Algorithms A1.2, A1.2a are used for the case when an initial schedule satisfies SSO #1.

Algorithms A1.2, A1.2a are, respectively, Algorithms A2.1, A2.1a (see below) with the following Procedure A.

Procedure A An algorithm distributes all tasks except those belonging to the initial schedule. After the algorithm execution, we verify the following condition: the start time of each machine is not less than the completion time of tasks in the initial schedule on this machine. If this is true, then, to obtain an optimal schedule, we need to shift the approximate schedule (obtained by an approximation algorithm) on each machine to the left so that the start time of the first task of the approximate schedule on each machine coincides with the completion time of the tasks in the initial schedule (to avoid interruptions on all machines). If the above condition is not satisfied, we should consider a possibility of moving some tasks of the approximate schedule (one or more if required) with the earlier start time than the completion time of the tasks in the initial schedule, to other machines, so that the start time of tasks in the approximate schedule on each machine becomes not less than the completion time of the initial schedule. If this possibility was used, then, to obtain an optimal schedule, we need to shift the tasks of the approximate schedule to the left to avoid interruptions on all machines. Otherwise, the first polynomial component is not fulfilled in this algorithm.

Algorithm A1.3 is used for the case when an initial schedule satisfies SSO #1.

Algorithm A1.3 is Algorithm A2.2 (see below) with Procedure A.

3.2.2.2 The Second Polynomial Component of PSC-Algorithm A

The second polynomial component consists of four approximation algorithms (A2.1, A2.1a, A2.2, A2.3), we solve the problem by each of them and choose the best solution according to the direct lexicographical order (3.7).

Algorithm A2.1 is based on the principle of sequential group assignment using a locally optimal inverse lexicographical order.

The algorithm is a series of similar iterations. Each iteration, except the first one, includes the preliminary stage and the tasks assignment on a group basis: we assign groups of m tasks, one task to each machine. At the first iteration of the algorithm, we assign one task to each machine. The completion time of the task should coincide with its due date, and the start times of machines should correspond to the global optimum in the reverse lexicographical order: for each task we find the machine on which the start time for this task is the latest; the first task is assigned to the machine with the latest start time, then the next task to the machine with the next (earlier) latest start time, etc., until we assign m tasks to m machines.

Remark 3.4 A set of machines, from which we find the machine with the next latest start time, does not include machines to which we have assigned a task at the first iteration.

At current iteration (any iteration except the first one), we distribute the tasks that are not yet assigned to machines at previous iterations. We assign a task so that its completion time coincides with the start time of the task assigned to this machine at the previous iteration. At the preliminary stage we find tasks for each of which there is at least one machine on which it is feasible (the task would be not tardy if assigned to this machine). Then to machines for which there are feasible tasks (the *first group of machines*) we assign the feasible tasks in accordance with the locally optimal inverse lexicographical order: first we assign a feasible task to the machine with the latest start time, then the next task to the machine with the next (earlier) latest start time, etc., until one feasible task is assigned to each machine of the first group. After the assignment, we eliminate the task from the sets of feasible tasks on another machines.

Remark 3.5 A set of machines, from which we find the machine with the next latest start time, does not include machines to which we have assigned a task at the current iteration.

If we have assigned one task to each of m machines, the current iteration terminates.

After such distribution, tasks may remain for each of which there are no machines on which they are feasible. They are combined to the second group of tasks for the current iteration. Similarly to feasible tasks assignment, we assign tasks to the *second group of machines* (machines for which there are no feasible tasks) locally optimally in accordance with the inverse lexicographical order.

Remark 3.6 Unlike the first group of tasks, we assign a task to a machine of the second group with the shift of the task (and all tasks already assigned to this machine on previous iterations) to the left by the amount of the task's tardiness.

Remark 3.7 A set of machines, from which we find the machine with the next latest start time, does not include machines to which we have assigned a task at the current iteration.

Iterations continue until we get start times of the first tasks on each machine.

To improve the efficiency of Algorithm A2.1, we use as its last step the following adjustment algorithm.

Machines Start Times Adjustment Algorithm (MSTAA) has the following four steps.

1. Choose the machine j with the minimum start time $s_j = s_{\min}$.
2. If processing time of the first task in the schedule of machine j is greater than that of the second task, check the possibility to move the first task to the right as close as possible to its due date (to do this, the first task's due date must be greater than or equal to the completion time of the task, after which it is put). If this possibility is available, move the first task to the right and again check processing times of the first tasks on machine j and the possibility of shifting them to the right as long as it exists.
3. Try to move successively the first task from machine j to each other machine (if the task is tardy, then shift it and all tasks already assigned to this machine on previous iterations to the left by the amount of the task's tardiness). Determine the machine on which this task will have the latest start time. If it is greater than $s_{\min}$, then move the task to this machine, calculate new start times of machines, go to step 1.
4. If no permutations have been made, choose machine j with the next minimum start time, go to step 2. If all machines are considered, the algorithm terminates.

Remark 3.8 MSTAA is the last step of all algorithms of the *second* polynomial component of PSC-algorithms for Problems 1 and 2.

Rationale for the Group Assignment Principle of Algorithm A2.1 It is logical to assign tasks to machines at each iteration locally optimally in the direct lexicographical order. However, an implementation of such strategy will lead to exhaustive search of all assignment positions. Therefore, we do the heuristic assignment locally optimally in the inverse lexicographical order under assumption that it is a little different from the locally optimal assignment in the direct lexicographical order. In addition, the efficiency of such assignment is confirmed by the following three statements.

Statement 3.4 Consider the current iteration of the algorithm and the set of machines J. Suppose that $\forall j \in J$ a non-empty set of feasible tasks I_j exists. Feasible tasks from I_j at this iteration are assigned to machine j without tardiness, $\forall j \in J$. For each machine $j \in J$ we find a task $i \in I_j$ with the maximum start time, let it be task $i_j^{\max}$. If all tasks $i_j^{\max}$ are different, then the start times of the tasks $i_j^{\max}$ that we assigned at the current iteration to machine j are optimal both in inverse and in direct lexicographical order in comparison with any other set of feasible tasks assigned to machines $j \in J$ in any feasible order.

Proof The proof evidently follows from the rules for finding the task $i_j^{\max}$. ∎

Statement 3.5 Suppose that for each machine of the second group we find a task with the latest start time and that these tasks are different on all these machines. Then the assignment for these machines is the same in direct and in reverse lexicographical orders.

Proof The proof follows from the rules of schedule construction. ∎

Statement 3.6 If start times of feasible tasks assigned by Algorithm A2.1 at some iteration are optimal in the reverse lexicographical order, then they are at the same time better in the direct lexicographical order comparing with start times of any other set of feasible tasks assigned to the same machines that are chosen by Algorithm A2.1 at this iteration.

Proof Suppose $C = B \backslash A \neq \varnothing$ where A is the set of feasible tasks assigned by Algorithm A2.1, B is an arbitrary set of feasible tasks assigned to the same machines that are chosen by Algorithm A2.1 for the tasks from the set A. Suppose that the latest start time of each feasible task from the set C does not belong to the set of start times of the tasks from the set A. This proves the statement. ∎

Supposedly, more tasks complete just in time after solving the problem with Algorithm A2.1 than after its solving with other approximation algorithms.

Algorithm A2.1a The difference of Algorithm A2.1a from Algorithm A2.1 is that after the last step of any current iteration we check the inequality

$$\max_{j=\overline{1,m}} s_j^{\mathrm{curr}} - \min_{j=\overline{1,m}} s_j^{\mathrm{curr}} > K_{\exp}$$

where s_j^{curr} is the start time of machine j after the current iteration. An acceptable value of $K_{\exp}$ is determined experimentally. If the inequality is satisfied, then we execute MSTAA for the current schedule constructed by iterations from the first to the current inclusively, after this we go to the next iteration.

Algorithm A2.2 is based on explicit construction of a locally optimal inverse lexicographical order.

The algorithm consists of the preliminary stage and the optimization stage. The preliminary stage is an initial schedule construction. This stage completely coincides with the first iteration of Algorithm A2.1.

At the optimization stage we distribute the tasks not yet assigned to machines at the preliminary stage. For the current machine with the latest start time we choose from the set of feasible tasks for this machine and assign the task with the latest start time on this machine. If there are no feasible tasks, then we determine the task with the latest start time considering the shift of the entire schedule assigned to this machine to the left by the task's tardiness. The last step is MSTAA.

Statement 3.7 Suppose that for each machine we have found tasks with the latest start time, and let these tasks be different for all machines. Then the distribution is the same both in direct and in reverse lexicographical order.

Proof The proof follows from the rules of schedule construction. ■

Algorithm A2.3 is a polynomial complexity approximation algorithm to construct a feasible schedule based on an initial schedule constructed according to SSO #1 or #2. If we could not build such initial schedule, then we build it according to SSO #2 without taking into account conditions (3.4), (3.5) or their analogs, as well as the condition (3.6), and, in this case, strict inequalities are not necessary: $d_1 - l_1 \leq d_2 - l_2 \leq \cdots \leq d_n - l_n$.

As a result of the initial schedule construction, the initial set of tasks I is divided into two subsets: I_1 is the set of previously assigned tasks and I_1' is the set of not yet assigned tasks. We order the tasks from I_1' in non-decreasing order of $d_i - l_i$ (sequence σ^{ord}, strict inequalities are not necessary). For each machine k, we know the release time C_k (the completion time of all assigned tasks to it). The further assignment of tasks is carried out as follows. Choose the next task from the sequence σ^{ord} and assign to the release time of the machine with the minimum release time. If the task is not tardy, then the iteration terminates, recalculate the release time, continue with the next task from σ^{ord}. If the task is tardy, then assign it to each machine, shifting this and all assigned to the machine tasks to the left by the tardiness of the assigned task. We get m schedules corresponding to m vectors of current start times of all machines. Choose the best vector in the direct lexicographical order (3.7) and assign the tardy task to the machine corresponding to the best by (3.7) vector of start times, shifting the schedule by the task's tardiness. The last step is MSTAA.

Remark 3.9 (for PSC-Algorithm A) After execution of Algorithms A2.1, A2.2, A2.3, machines in the obtained schedule are reordered in non-decreasing order of start times to correspond to the lexicographical order (3.7).

3.2.3 PSC-Algorithm for Problem 1 Solving by Criterion 2 (Algorithm B)

We considered Criterion 2 (3.2) and a sufficient sign of optimality of a feasible solution earlier.

3.2.3.1 The First Polynomial Component of PSC-Algorithm B

The first polynomial component of the PSC-algorithm consists of four algorithms (B1.1, B1.2, B1.2a, B1.3). They are to be executed sequentially in an arbitrary order. The first algorithm that constructed a feasible schedule for which the SSO is satisfied, obtains an optimal solution of the problem by criterion (3.2), and the first polynomial component of the PSC-algorithm is considered to be fulfilled. If a feasible solution is not obtained by any algorithm of the first polynomial component, we search it using algorithms of the second polynomial component.

Algorithm B1.1 implements the first polynomial component for a feasible schedule construction if an initial schedule was constructed according to SSO #2 for Criterion 1 (3.1) without taking into account conditions (3.4), (3.5) or their analogs, as well as the condition (3.6) wherein the strict inequalities are not necessary: $d_1 - l_1 \leq d_2 - l_2 \leq \cdots \leq d_n - l_n$. As a result of this, the initial set of tasks I is divided into two subsets: I_1 is the set of previously assigned tasks and I_1' is the set of not yet assigned tasks. We order the tasks from I_1' in non-decreasing order of $d_i - l_i$ (sequence σ^{ord}). For each machine k, we know the release time C_k (the completion time of all assigned tasks to it). The further assignment of tasks is carried out as follows. Choose the next task from the sequence σ^{ord}. Assign it to the release time of the machine with the minimum release time if the task is feasible on this machine. If the machine with the minimum release time has index 1, then we assign only feasible task to it; if the task is tardy on the first machine, then we assign it to another machine with the next minimum release time. If the task is tardy on the machine with the minimum release time, then we shift the schedule of all tasks assigned to this machine to the left by the task's tardiness and check the new start time of the machine s^{new}. If $s^{new} \geq s_1$, then we assign the task to the release time of the machine and continue to a similar assignment of the next task from σ^{ord}. If $s^{new} < s_1$, then the first polynomial component is not fulfilled by this algorithm, the algorithm terminates. Hence, we execute the algorithm either until an optimal schedule for Criterion 2 (3.2) is constructed or we find that it cannot be constructed by this algorithm.

Algorithms B1.2, B1.2a are Algorithms A2.1, A2.1a, respectively, with the following Procedure B.

Procedure B The initial schedule consists of a first task assigned to the first machine with start time s_1 (see SSO for Criterion 2). After executing an algorithm that distributes all tasks except the first one, we check the following conditions in the obtained schedule:

1. The start time of the first task of the approximate schedule on the first machine is not less than d_1 (the completion time of the first task of the initial schedule on the first machine assigned in accordance with the SSO).
2. The start times of the first tasks of the approximate schedule on other machines are not less than s_1.

If both conditions are satisfied, then, to obtain an optimal schedule, we need to shift the schedule obtained by an approximation algorithm on the first machine to the left so that the start time of the first task of the approximate schedule coincides with the completion time of the first task of the initial schedule d_1. If the first condition is not satisfied but the second one is true, then we put the first task from the first machine of the approximate schedule successively to another machines and assign it to the machine where the start time of this task will be the greatest and not less than s_1. If such machine is not found, then the first polynomial component is not fulfilled. If the first task is moved to another machine, then we check the condition: if the start time of the second task on the first machine is not less than d_1, then, shifting the schedule on the first machine so that the start time equals to d_1, we get an optimal schedule. If the condition is not satisfied, we try to move the second task from the first machine of the approximate schedule to another machine. The described procedure is repeated until either an optimal schedule is obtained or the first polynomial component is not fulfilled by this algorithm.

Algorithm B1.3 is Algorithm A2.2 with Procedure B.

3.2.3.2 The Second Polynomial Component of PSC-Algorithm B

The second polynomial component contains four algorithms: approximation algorithms A2.1, A2.1a, A2.2 (see above) because we adjust the start times of machines using MSTAA, and also Algorithm B2.1. We solve the problem by each of these algorithms and choose a solution corresponding to the best of $\max\{\min_j s_j\}$, $j = \overline{1,m}$.

Algorithm B2.1 is a modification of Algorithm B1.1 (see above) with the following difference. If the new start time of the machine $s^{new} < s_1$, then we assign the tardy task to the release time of the first machine and shift the schedule of the first machine to the left by the task's tardiness. After that, we assign this task either to the first machine or to the machine with the current start time s^{new}, depending on where the current start time is greater. For further iterations, this machine becomes the analogue of the first machine. The assignment procedure continues similarly until all tasks are distributed. After the algorithm we apply MSTAA.

Estimation of Deviation of a Solution from Optimum for Problem 1 All algorithms of the second polynomial component are the approximation algorithms for Criterion 1 or 2. If the first part of a SSO is fulfilled (an initial schedule was constructed), then there is an estimate of deviation from optimum for the solution obtained by the approximation algorithm.

For Criterion 1: for this purpose we compare vectors using the condition (3.7).

Remark 3.10 Suppose $s_l > s_{q_l}$ where l is the number of the component for which the first positive difference is reached, i.e., $s_j = s_{q_j}, j = \overline{1, l-1}$. Then the signs of the differences of the remaining components $(j = \overline{l+1, m})$ may be arbitrary. The decision maker analyzes the solution and chooses either the optimal or the approximate solution based on practical considerations.

For Criterion 2: we determine the estimate as the difference $s_1 - s_{q_1} \geq 0$ (all Algorithms B2.1, A2.1, A2.1a, A2.2 have an estimate of deviation from optimum for Criterion 2).

Remark 3.11 For Criterion 1, Algorithms A2.1, A2.1a, A2.2 have an estimate of deviation from optimum if an initial schedule was built according to SSO #1. A2.3 has an estimate of deviation from optimum if an initial schedule was built in accordance with SSO #1 or #2. If an initial schedule was not constructed, then we compare the solution with zero vector.

3.3 Illustrative Examples for Problem 1

Algorithm A1.1 Initial data: m = 3, n = 16. Sequence σ^{ord} is in Table 3.1. Initial schedule: assign the first tasks of the sequence to each machine according to SSO #2 (Table 3.2).

Then successively assign tasks 6–16, each to the machine with the minimum completion time of the tasks. Since all tasks are assigned and are not tardy, we get the optimal schedule (Table 3.3).

Table 3.1 Sequence σ^{ord} for Algorithm A1.1

i	l_i	d_i	$d_i - l_i$	i	l_i	d_i	$d_i - l_i$	i	l_i	d_i	$d_i - l_i$
1	2	6	4	7	1	17	16	12	9	34	25
2	1	7	6	8	1	18	17	13	39	65	26
3	9	18	9	9	1	19	18	14	20	50	30
4	6	18	12	10	2	22	20	15	16	52	36
5	3	16	13	11	17	37	20	16	4	54	50
6	4	20	16								

Table 3.2 Initial schedule

i	l_i	d_i	s_i	C_i	i	l_i	d_i	s_i	C_i	i	l_i	d_i	s_i	C_i
Machine 1, $s_1 = 4$					Machine 2, $s_2 = 12$					Machine 3, $s_3 = 13$				
1	2	6	4	6	4	6	18	12	18	5	3	16	13	16
2	1	7	6	7										
3	9	18	7	16										

s_i start time of task i, C_i its completion time, $s_j, j = \overline{1, 3}$, start times of machines 1–3

Table 3.3 Optimal schedule

i	l_i	d_i	s_i	C_i	i	l_i	d_i	s_i	C_i	i	l_i	d_i	s_i	C_i
Machine 1, $s_1 = 4$					Machine 2, $s_2 = 12$					Machine 3, $s_3 = 13$				
1	2	6	4	6	4	6	18	12	18	5	3	16	13	16
2	1	7	6	7	10	2	22	18	20	7	1	17	16	17
3	9	18	7	16	13	39	65	20	59	8	1	18	17	18
6	4	20	16	20						9	1	19	18	19
12	9	34	20	29						11	17	37	19	36
14	20	50	29	49						15	16	52	36	52
16	4	54	49	53										

Table 3.4 Sequence σ^{ord} for Algorithm A1.2

i	l_i	d_i	$d_i - l_i$	i	l_i	d_i	$d_i - l_i$	i	l_i	d_i	$d_i - l_i$
1	3	8	5	7	1	17	16	12	9	34	25
2	2	8	6	8	1	18	17	13	39	65	26
3	2	9	7	9	1	19	18	14	20	50	30
4	6	19	13	10	2	22	20	15	16	52	36
5	3	16	13	11	17	37	20	16	4	54	50
6	4	20	16								

Algorithm A1.2 Initial data: $m = 3$, $n = 16$. Sequence σ^{ord} is in Table 3.4.

Initial schedule: assign the first three tasks, one to each machine, according to SSO #1 (Table 3.5).

Execute Algorithm A2.1. The first iteration. We assign the first tasks from the set of unassigned tasks to the last position in the schedule of each machine. Determine the set of feasible tasks I_j for the next iteration by the condition: the task's due date is not less than the start time of the machine (Table 3.6).

Table 3.5 Initial schedule

i	l_i	d_i	s_i	C_i	i	l_i	d_i	s_i	C_i	i	l_i	d_i	s_i	C_i
Machine 1, $s_1 = 5$					Machine 2, $s_2 = 6$					Machine 3, $s_3 = 7$				
1	3	8	5	8	2	2	8	6	8	3	2	9	7	9

Table 3.6 The first iteration

i	l_i	d_i	s_i	C_i	i	l_i	d_i	s_i	C_i	i	l_i	d_i	s_i	C_i
Machine 1, $s_1 = 50$, $I_1 = \{13\}$					Machine 2, $s_2 = 36$, $I_2 = \{11, 13\}$					Machine 3, $s_3 = 30$, $I_3 = \{11, 12, 13\}$				
1	3	8	5	8	2	2	8	6	8	3	2	9	7	9
16	4	54	50	54	15	16	52	36	52	14	20	50	30	50

Table 3.7 The second iteration

i	l_i	d_i	s_i	C_i	i	l_i	d_i	s_i	C_i	i	l_i	d_i	s_i	C_i
Machine 1, $s_1 = 11$, $I_1 = \{4, 5, 6, 7, 8, 9, 10\}$					Machine 2, $s_2 = 19$, $I_2 = \{4, 6, 9, 10\}$					Machine 3, $s_3 = 21$, $I_3 = \{10\}$				
1	3	8	5	8	2	2	8	6	8	3	2	9	7	9
13	39	65	11	50	11	17	37	19	36	12	9	34	21	30
16	4	54	50	54	15	16	52	36	52	14	20	50	30	50

Iteration 2. Assign tasks 13, 11, and 12 (if there are several feasible tasks, we choose the shortest of them, in the case of the same processing time—the task with the greatest due date). Delete the assigned tasks from the sets of feasible tasks and find new sets of feasible tasks (Table 3.7).

Iteration 3. Assign tasks 10, 9, 8 (Table 3.8).

Iteration 4. Assign tasks 6, 4, 7 (Table 3.9).

Iteration 5. Assign feasible task 5 to machine 3. All tasks are assigned (Table 3.10).

Execute MSTAA for the tasks assigned by Algorithm A2.1: put task 9 before task 4 on machine 2, move task 7 from machine 1 to machine 3 (Table 3.11).

Execute Procedure A: sort the initial schedule in non-decreasing order of completion times C_i (in this case, the order remains the same). Sort machines with the tasks assigned from the end in non-decreasing order of the start times s_j (machines 2 and 3 are interchanged). Remove gaps between tasks on each machine. We get the optimal schedule (Table 3.12).

Algorithm A1.2a Initial data: $m = 3$, $n = 16$, $K_{\exp} = 8$. Sequence σ^{ord} is in Table 3.13.

Initial schedule: assign the first three tasks, one to each machine, according to SSO #1 (Table 3.14).

Table 3.8 The third iteration

i	l_i	d_i	s_i	C_i	i	l_i	d_i	s_i	C_i	i	l_i	d_i	s_i	C_i
Machine 1, $s_1 = 10$, $I_1 = \{4, 5, 6, 7\}$					Machine 2, $s_2 = 18$, $I_2 = \{4, 6\}$					Machine 3, $s_3 = 19$, $I_3 = \{4, 6\}$				
1	3	8	5	8	2	2	8	6	8	3	2	9	7	9
8	1	18	10	11	9	1	19	18	19	10	2	22	19	21
13	39	65	11	50	11	17	37	19	36	12	9	34	21	30
16	4	54	50	54	15	16	52	36	52	14	20	50	30	50

Table 3.9 Iteration 4

i	l_i	d_i	s_i	C_i	i	l_i	d_i	s_i	C_i	i	l_i	d_i	s_i	C_i
Machine 1, $s_1 = 9$, $I_1 = \varnothing$					Machine 2, $s_2 = 12$, $I_2 = \varnothing$					Machine 3, $s_3 = 15$, $I_3 = \{5\}$				
1	3	8	5	8	2	2	8	6	8	3	2	9	7	9
7	1	17	9	10	4	6	19	12	18	6	4	20	15	19
8	1	18	10	11	9	1	19	18	19	10	2	22	19	21
13	39	65	11	50	11	17	37	19	36	12	9	34	21	30
16	4	54	50	54	15	16	52	36	52	14	20	50	30	50

Table 3.10 Iteration 5

i	l_i	d_i	s_i	C_i	i	l_i	d_i	s_i	C_i	i	l_i	d_i	s_i	C_i
Machine 1, $s_1 = 9$					Machine 2, $s_2 = 12$					Machine 3, $s_3 = 12$				
1	3	8	5	8	2	2	8	6	8	3	2	9	7	9
7	1	17	9	10	4	6	19	12	18	5	3	16	12	15
8	1	18	10	11	9	1	19	18	19	6	4	20	15	19
13	39	65	11	50	11	17	37	19	36	10	2	22	19	21
16	4	54	50	54	15	16	52	36	52	12	9	34	21	30
										14	20	50	30	50

Table 3.11 The schedule after MSTAA executing

i	l_i	d_i	s_i	C_i	i	l_i	d_i	s_i	C_i	i	l_i	d_i	s_i	C_i
Machine 1, $s_1 = 10$					Machine 2, $s_2 = 12$					Machine 3, $s_3 = 11$				
1	3	8	5	8	2	2	8	6	8	3	2	9	7	9
8	1	18	10	11	9	1	19	12	13	7	1	17	11	12
13	39	65	11	50	4	6	19	13	19	5	3	16	12	15
16	4	54	50	54	11	17	37	19	36	6	4	20	15	19
					15	16	52	36	52	10	2	22	19	21
										12	9	34	21	30
										14	20	50	30	50

Table 3.12 Optimal schedule

i	l_i	d_i	s_i	C_i	i	l_i	d_i	s_i	C_i	i	l_i	d_i	s_i	C_i
Machine 1, $s_1 = 5$					Machine 2, $s_2 = 6$					Machine 3, $s_3 = 7$				
1	3	8	5	8	2	2	8	6	8	3	2	9	7	9
8	1	18	8	9	7	1	17	8	9	9	1	19	9	10
13	39	65	9	48	5	3	16	9	12	4	6	19	10	16
16	4	54	48	52	6	4	20	12	16	11	17	37	16	33
					10	2	22	16	18	15	16	52	33	49
					12	9	34	18	27					
					14	20	50	27	47					

Table 3.13 Sequence σ^{ord} for Algorithm A1.2a

i	l_i	d_i	$d_i - l_i$	i	l_i	d_i	$d_i - l_i$	i	l_i	d_i	$d_i - l_i$
1	3	8	5	7	1	17	16	12	9	34	25
2	2	8	6	8	1	18	17	13	39	65	26
3	2	9	7	9	1	19	18	14	20	50	30
4	6	19	13	10	2	22	20	15	16	52	36
5	3	16	13	11	17	37	20	16	4	54	50
6	4	20	16								

Table 3.14 Initial schedule

i	l_i	d_i	s_i	C_i	i	l_i	d_i	s_i	C_i	i	l_i	d_i	s_i	C_i
Machine 1, $s_1 = 5$					Machine 2, $s_2 = 6$					Machine 3, $s_3 = 7$				
1	3	8	5	8	2	2	8	6	8	3	2	9	7	9

Execute Algorithm A2.1a. The first iteration. We assign the first tasks from the set of unassigned tasks to the last position in the schedule of each machine. Determine the set of feasible tasks I_j for the next iteration by the condition: the task's due date is not less than the start time of the machine (Table 3.15).

Iteration 2. Assign tasks 13, 11, and 12. $s_{\max} - s_{\min} = 10 > 8$, we execute MSTAA, but it does not change the schedule. Determine new sets of feasible tasks (Table 3.16).

Iteration 3. Assign tasks 10, 9, 8 (Table 3.17).

$s_{\max} - s_{\min} = 9 > 8$, execute MSTAA. Move task 8 from machine 1 to machine 3 with a shift of all tasks on machine 3 to the left by the task's tardiness (1 unit). We get the schedule shown in Table 3.18.

Table 3.15 The first iteration

i	l_i	d_i	s_i	C_i	i	l_i	d_i	s_i	C_i	i	l_i	d_i	s_i	C_i
Machine 1, $s_1 = 50$, $I_1 = \{13\}$					Machine 2, $s_2 = 36$, $I_2 = \{11, 13\}$					Machine 3, $s_3 = 30$, $I_3 = \{11, 12, 13\}$				
1	3	8	5	8	2	2	8	6	8	3	2	9	7	9
16	4	54	50	54	15	16	52	36	52	14	20	50	30	50

Table 3.16 The second iteration

i	l_i	d_i	s_i	C_i	i	l_i	d_i	s_i	C_i	i	l_i	d_i	s_i	C_i
Machine 1, $s_1 = 11$, $I_1 = \{4, 5, 6, 7, 8, 9, 10\}$					Machine 2, $s_2 = 19$, $I_2 = \{4, 6, 9, 10\}$					Machine 3, $s_3 = 21$, $I_3 = \{10\}$				
1	3	8	5	8	2	2	8	6	8	3	2	9	7	9
13	39	65	11	50	11	17	37	19	36	12	9	34	21	30
16	4	54	50	54	15	16	52	36	52	14	20	50	30	50

Table 3.17 The third iteration

i	l_i	d_i	s_i	C_i	i	l_i	d_i	s_i	C_i	i	l_i	d_i	s_i	C_i
Machine 1, $s_1 = 10$, $I_1 = \{4, 5, 6, 7\}$					Machine 2, $s_2 = 18$, $I_2 = \{4, 6\}$					Machine 3, $s_3 = 19$, $I_3 = \{4, 6\}$				
1	3	8	5	8	2	2	8	6	8	3	2	9	7	9
8	1	18	10	11	9	1	19	18	19	10	2	22	19	21
13	39	65	11	50	11	17	37	19	36	12	9	34	21	30
16	4	54	50	54	15	16	52	36	52	14	20	50	30	50

Table 3.18 The third iteration, after MSTAA executing

i	l_i	d_i	s_i	C_i	i	l_i	d_i	s_i	C_i	i	l_i	d_i	s_i	C_i
Machine 1, $s_1 = 11$, $I_1 = \{4, 5, 6, 7\}$					Machine 2, $s_2 = 18$, $I_2 = \{4, 6\}$					Machine 3, $s_3 = 17$, $I_3 = \{4, 6, 7\}$				
1	3	8	5	8	2	2	8	6	8	3	2	9	7	9
13	39	65	11	50	9	1	19	18	19	8	1	18	17	18
16	4	54	50	54	11	17	37	19	36	10	2	22	18	20
					15	16	52	36	52	12	9	34	20	29
										14	20	50	29	49

Iteration 4. Assign tasks 6, 7, 5. $s_{\max} - s_{\min} = 8$, MSTAA is not executed (Table 3.19).

Iteration 5. Assign feasible task 4 to machine 3. All tasks are assigned (Table 3.20).

Execute MSTAA for the tasks assigned by Algorithm A2.1a: put task 4 after task 8 on machine 3, move task 5 from machine 1 to machine 2 (Table 3.21).

Execute Procedure A: sort the initial schedule in non-decreasing order of completion times C_i (in this case, the order remains the same). Sort machines with the tasks assigned from the end in non-decreasing order of the start times s_j (machines 1 and 3 are interchanged). Remove gaps between tasks on each machine. We get the optimal schedule (Table 3.22).

Algorithm A1.3 Initial data: $m = 3$, $n = 16$. Sequence σ^{ord} is in Table 3.23.

Initial schedule: assign the first three tasks, one to each machine, according to SSO #1 (Table 3.24).

Execute Algorithm A2.2. Preliminary stage. We assign the first tasks from the set of unassigned tasks to the last position in the schedule of each machine (Table 3.25).

Optimization stage. Successively assign feasible tasks: 13 to machine 1; 11 to machine 2; 12 and 10 to machine 3; 9 to machine 2; 4 to machine 3; 8, 7, 5, and 6 to machine 2. All tasks are assigned. We get the schedule shown in Table 3.26.

Execute MSTAA for the tasks assigned by Algorithm A2.2: put task 6 and then task 5 after task 9 on machine 2, move task 7 from machine 2 to machine 3 (Table 3.27).

Execute Procedure A: sort the initial schedule in non-decreasing order of completion times C_i (in this case, the order remains the same). Sort machines with

Table 3.19 Iteration 4

i	l_i	d_i	s_i	C_i	i	l_i	d_i	s_i	C_i	i	l_i	d_i	s_i	C_i
Machine 1, $s_1 = 8$, $I_1 = \{4\}$					Machine 2, $s_2 = 14$, $I_2 = \{4\}$					Machine 3, $s_3 = 16$, $I_3 = \{4\}$				
1	3	8	5	8	2	2	8	6	8	3	2	9	7	9
5	3	16	8	11	6	4	20	14	18	7	1	17	16	17
13	39	65	11	50	9	1	19	18	19	8	1	18	17	18
16	4	54	50	54	11	17	37	19	36	10	2	22	18	20
					15	16	52	36	52	12	9	34	20	29
										14	20	50	29	49

Table 3.20 Iteration 5

i	l_i	d_i	s_i	C_i	i	l_i	d_i	s_i	C_i	i	l_i	d_i	s_i	C_i
Machine 1, $s_1 = 8$					Machine 2, $s_2 = 14$					Machine 3, $s_3 = 10$				
1	3	8	5	8	2	2	8	6	8	3	2	9	7	9
5	3	16	8	11	6	4	20	14	18	4	6	19	10	16
13	39	65	11	50	9	1	19	18	19	7	1	17	16	17
16	4	54	50	54	11	17	37	19	36	8	1	18	17	18
					15	16	52	36	52	10	2	22	18	20
										12	9	34	20	29
										14	20	50	29	49

Table 3.21 The schedule after MSTAA executing

i	l_i	d_i	s_i	C_i	i	l_i	d_i	s_i	C_i	i	l_i	d_i	s_i	C_i
Machine 1, $s_1 = 11$					Machine 2, $s_2 = 11$					Machine 3, $s_3 = 10$				
1	3	8	5	8	2	2	8	6	8	3	2	9	7	9
13	39	65	11	50	5	3	16	11	14	7	1	17	10	11
16	4	54	50	54	6	4	20	14	18	8	1	18	11	12
					9	1	19	18	19	4	6	19	12	18
					11	17	37	19	36	10	2	22	18	20
					15	16	52	36	52	12	9	34	20	29
										14	20	50	29	49

Table 3.22 Optimal schedule

i	l_i	d_i	s_i	C_i	i	l_i	d_i	s_i	C_i	i	l_i	d_i	s_i	C_i
Machine 1, $s_1 = 5$					Machine 2, $s_2 = 6$					Machine 3, $s_3 = 7$				
1	3	8	5	8	2	2	8	6	8	3	2	9	7	9
7	1	17	8	9	5	3	16	8	11	13	39	65	9	48
8	1	18	9	10	6	4	20	11	15	16	4	54	48	52
4	6	19	10	16	9	1	19	15	16					
10	2	22	16	18	11	17	37	16	33					
12	9	34	18	27	15	16	52	33	49					
14	20	50	27	47										

Table 3.23 Sequence σ^{ord} for Algorithm A1.3

i	l_i	d_i	$d_i - l_i$	i	l_i	d_i	$d_i - l_i$	i	l_i	d_i	$d_i - l_i$
1	3	8	5	7	1	17	16	12	9	34	25
2	2	8	6	8	1	18	17	13	39	65	26
3	2	9	7	9	1	19	18	14	20	50	30
4	6	19	13	10	2	22	20	15	16	52	36
5	3	16	13	11	17	37	20	16	4	54	50
6	4	20	16								

Table 3.24 Initial schedule

i	l_i	d_i	s_i	C_i	i	l_i	d_i	s_i	C_i	i	l_i	d_i	s_i	C_i
Machine 1, $s_1 = 5$					Machine 2, $s_2 = 6$					Machine 3, $s_3 = 7$				
1	3	8	5	8	2	2	8	6	8	3	2	9	7	9

Table 3.25 Preliminary stage

i	l_i	d_i	s_i	C_i	i	l_i	d_i	s_i	C_i	i	l_i	d_i	s_i	C_i
Machine 1, $s_1 = 50$					Machine 2, $s_2 = 36$					Machine 3, $s_3 = 30$				
1	3	8	5	8	2	2	8	6	8	3	2	9	7	9
16	4	54	50	54	15	16	52	36	52	14	20	50	30	50

Table 3.26 Optimization stage

i	l_i	d_i	s_i	C_i	i	l_i	d_i	s_i	C_i	i	l_i	d_i	s_i	C_i
Machine 1, $s_1 = 11$					Machine 2, $s_2 = 9$					Machine 3, $s_3 = 13$				
1	3	8	5	8	2	2	8	6	8	3	2	9	7	9
13	39	65	11	50	6	4	20	9	13	4	6	19	13	19
16	4	54	50	54	5	3	16	13	16	10	2	22	19	21
					7	1	17	16	17	12	9	34	21	30
					8	1	18	17	18	14	20	50	30	50
					9	1	19	18	19					
					11	17	37	19	36					
					15	16	52	36	52					

Table 3.27 The schedule after MSTAA executing

i	l_i	d_i	s_i	C_i	i	l_i	d_i	s_i	C_i	i	l_i	d_i	s_i	C_i
Machine 1, $s_1 = 11$					Machine 2, $s_2 = 10$					Machine 3, $s_3 = 12$				
1	3	8	5	8	2	2	8	6	8	3	2	9	7	9
13	39	65	11	50	8	1	18	10	11	7	1	17	12	13
16	4	54	50	54	9	1	19	11	12	4	6	19	13	19
					5	3	16	12	15	10	2	22	19	21
					6	4	20	15	19	12	9	34	21	30
					11	17	37	19	36	14	20	50	30	50
					15	16	52	36	52					

Table 3.28 Optimal schedule

i	l_i	d_i	s_i	C_i	i	l_i	d_i	s_i	C_i	i	l_i	d_i	s_i	C_i
Machine 1, $s_1 = 5$					Machine 2, $s_2 = 6$					Machine 3, $s_3 = 7$				
1	3	8	5	8	2	2	8	6	8	3	2	9	7	9
8	1	18	8	9	13	39	65	8	47	7	1	17	9	10
9	1	19	9	10	16	4	54	47	51	4	6	19	10	16
5	3	16	10	13						10	2	22	16	18
6	4	20	13	17						12	9	34	18	27
11	17	37	17	34						14	20	50	27	47
15	16	52	34	50										

Table 3.29 Tasks data for Algorithm A2.1

i	l_i	d_i	i	l_i	d_i	i	l_i	d_i
1	2	6	7	1	17	12	9	34
2	1	7	8	1	18	13	39	65
3	9	18	9	1	19	14	20	50
4	5	19	10	2	22	15	16	52
5	2	16	11	17	37	16	4	54
6	5	20						

the tasks assigned from the end in non-decreasing order of the start times s_j (machines 1 and 2 are interchanged). Remove gaps between tasks on each machine. We get the optimal schedule (Table 3.28).

Algorithm A2.1 Initial data: $m = 3$, $n = 16$ (Table 3.29).

The first iteration. To built an initial schedule, we assign the first tasks to the last position in the schedule of each machine. Determine the set of feasible tasks I_j for the next iteration by the condition: the task's due date is not less than the start time of the machine (Table 3.30).

Iteration 2. Assign tasks 13, 11, and 12 (if there are several feasible tasks, we choose the shortest of them, in the case of the same processing time—the task with the greatest due date). Delete the assigned tasks from the sets of feasible tasks and find new sets of feasible tasks (Table 3.31).

Iteration 3. Assign tasks 10, 9, 8 (Table 3.32).

Iteration 4. Assign tasks 6, 4, 7 (Table 3.33).

Iteration 5. Assign tasks 5 and 3. No tasks are assigned to the first machine. The second group of tasks are tasks 1 and 2. We choose from them the task with the maximum due date (in the case of equal due dates, choose a task with a shorter processing time). This is task 2, assign it to machine 1, shifting all tasks on machine 1 to the left by the task's tardiness (2 units). The result is in Table 3.34.

Iteration 6. Assign feasible task 1 to machine 1. All tasks are assigned (Table 3.35).

Execute MSTAA: put task 4 before task 3 on machine 2 and move it from machine 2 to machine 3. The final schedule is in Table 3.36.

Table 3.30 The first iteration

i	l_i	d_i	s_i	C_i	i	l_i	d_i	s_i	C_i	i	l_i	d_i	s_i	C_i
Machine 1, $s_1 = 50$, $I_1 = \{13\}$					Machine 2, $s_2 = 36$, $I_2 = \{11, 13\}$					Machine 3, $s_3 = 30$, $I_3 = \{11, 12, 13\}$				
16	4	54	50	54	15	16	52	36	52	14	20	50	30	50

Table 3.31 The second iteration

i	l_i	d_i	s_i	C_i	i	l_i	d_i	s_i	C_i	i	l_i	d_i	s_i	C_i
Machine 1, $s_1 = 11$, $I_1 = \{3, 4, 5, 6, 7, 8, 9, 10\}$					Machine 2, $s_2 = 19$, $I_2 = \{4, 6, 9, 10\}$					Machine 3, $s_3 = 21$, $I_3 = \{10\}$				
13	39	65	11	50	11	17	37	19	36	12	9	34	21	30
16	4	54	50	54	15	16	52	36	52	14	20	50	30	50

Table 3.32 The third iteration

i	l_i	d_i	s_i	C_i	i	l_i	d_i	s_i	C_i	i	l_i	d_i	s_i	C_i
Machine 1, $s_1 = 10$, $I_1 = \{3, 4, 5, 6, 7\}$					Machine 2, $s_2 = 18$, $I_2 = \{3, 4, 6\}$					Machine 3, $s_3 = 19$, $I_3 = \{4, 6\}$				
8	1	18	10	11	9	1	19	18	19	10	2	22	19	21
13	39	65	11	50	11	17	37	19	36	12	9	34	21	30
16	4	54	50	54	15	16	52	36	52	14	20	50	30	50

Table 3.33 Iteration 4

i	l_i	d_i	s_i	C_i	i	l_i	d_i	s_i	C_i	i	l_i	d_i	s_i	C_i
Machine 1, $s_1 = 9$, $I_1 = \{3, 5\}$					Machine 2, $s_2 = 13$, $I_2 = \{3, 5\}$					Machine 3, $s_3 = 14$, $I_3 = \{3, 5\}$				
7	1	17	9	10	4	5	19	13	18	6	5	20	14	19
8	1	18	10	11	9	1	19	18	19	10	2	22	19	21
13	39	65	11	50	11	17	37	19	36	12	9	34	21	30
16	4	54	50	54	15	16	52	36	52	14	20	50	30	50

Algorithm A2.1a Initial data: $m = 3$, $n = 16$, $K_{\exp} = 8$ (Table 3.37).

The first iteration. To built an initial schedule, we assign the first tasks to the last position in the schedule of each machine. Determine the set of feasible tasks I_j for the next iteration by the condition: the task's due date is not less than the start time of the machine (Table 3.38).

Iteration 2. Assign tasks 13, 11, and 12. $s_{\max} - s_{\min} = 10 > 8$, we execute MSTAA, but it does not change the schedule. Determine new sets of feasible tasks (Table 3.39).

Iteration 3. Assign tasks 10, 9, 8 (Table 3.40).

Table 3.34 Iteration 5

i	l_i	d_i	s_i	C_i	i	l_i	d_i	s_i	C_i	i	l_i	d_i	s_i	C_i
Machine 1, $s_1 = 6$, $I_1 = \{1\}$					Machine 2, $s_2 = 4$, $I_2 = \varnothing$					Machine 3, $s_3 = 12$, $I_3 = \varnothing$				
2	1	7	6	7	3	9	18	4	13	5	2	16	12	14
7	1	17	7	8	4	5	19	13	18	6	5	20	14	19
8	1	18	8	9	9	1	19	18	19	10	2	22	19	21
13	39	65	9	48	11	17	37	19	36	12	9	34	21	30
16	4	54	48	52	15	16	52	36	52	14	20	50	30	50

Table 3.35 Iteration 6

i	l_i	d_i	s_i	C_i	i	l_i	d_i	s_i	C_i	i	l_i	d_i	s_i	C_i
Machine 1, $s_1 = 4$					Machine 2, $s_2 = 4$					Machine 3, $s_3 = 12$				
1	2	6	4	6	3	9	18	4	13	5	2	16	12	14
2	1	7	6	7	4	5	19	13	18	6	5	20	14	19
7	1	17	7	8	9	1	19	18	19	10	2	22	19	21
8	1	18	8	9	11	17	37	19	36	12	9	34	21	30
13	39	65	9	48	15	16	52	36	52	14	20	50	30	50
16	4	54	48	52										

Table 3.36 Final schedule

i	l_i	d_i	s_i	C_i	i	l_i	d_i	s_i	C_i	i	l_i	d_i	s_i	C_i
Machine 1, $s_1 = 4$					Machine 2, $s_2 = 9$					Machine 3, $s_3 = 7$				
1	2	6	4	6	3	9	18	9	18	4	5	19	7	12
2	1	7	6	7	9	1	19	18	19	5	2	16	12	14
7	1	17	7	8	11	17	37	19	36	6	5	20	14	19
8	1	18	8	9	15	16	52	36	52	10	2	22	19	21
13	39	65	9	48						12	9	34	21	30
16	4	54	48	52						14	20	50	30	50

Table 3.37 Tasks data for Algorithm A2.1a

i	l_i	d_i	i	l_i	d_i	i	l_i	d_i
1	2	6	7	1	17	12	9	34
2	1	7	8	1	18	13	39	65
3	9	18	9	1	19	14	20	50
4	5	19	10	2	22	15	16	52
5	2	16	11	17	37	16	4	54
6	5	20						

Table 3.38 The first iteration

i	l_i	d_i	s_i	C_i	i	l_i	d_i	s_i	C_i	i	l_i	d_i	s_i	C_i
Machine 1, s_1 = 50, I_1 = {13}					Machine 2, s_2 = 36, I_2 = {11, 13}					Machine 3, s_3 = 30, I_3 = {11, 12, 13}				
16	4	54	50	54	15	16	52	36	52	14	20	50	30	50

Table 3.39 The second iteration

i	l_i	d_i	s_i	C_i	i	l_i	d_i	s_i	C_i	i	l_i	d_i	s_i	C_i
Machine 1, s_1 = 11, I_1 = {3, 4, 5, 6, 7, 8, 9, 10}					Machine 2, s_2 = 19, I_2 = {4, 6, 9, 10}					Machine 3, s_3 = 21, I_3 = {10}				
13	39	65	11	50	11	17	37	19	36	12	9	34	21	30
16	4	54	50	54	15	16	52	36	52	14	20	50	30	50

Table 3.40 The third iteration

i	l_i	d_i	s_i	C_i	i	l_i	d_i	s_i	C_i	i	l_i	d_i	s_i	C_i
Machine 1, s_1 = 10					Machine 2, s_2 = 18					Machine 3, s_3 = 19				
8	1	18	10	11	9	1	19	18	19	10	2	22	19	21
13	39	65	11	50	11	17	37	19	36	12	9	34	21	30
16	4	54	50	54	15	16	52	36	52	14	20	50	30	50

$s_{max} - s_{min} = 9 > 8$, we execute MSTAA: move task 8 from machine 1 to machine 3 (Table 3.41).

Iteration 4. Assign tasks 6, 4, 7. $s_{max} - s_{min} = 3 < 8$, MSTAA is not executed (Table 3.42).

Iteration 5. Assign tasks 5 and 3. No tasks are assigned to the first machine. The second group of tasks are tasks 1 and 2. We choose from them the task with the maximum due date. This is task 2, assign it to machine 1, shifting all tasks on machine 1 to the left by the task's tardiness (3 units). $s_{max} - s_{min} = 8$, MSTAA is not executed (Table 3.43).

Iteration 6. Assign feasible task 1 to machine 1. All tasks are assigned (Table 3.44).

Table 3.41 Iteration 3, after MSTAA executing

i	l_i	d_i	s_i	C_i	i	l_i	d_i	s_i	C_i	i	l_i	d_i	s_i	C_i
Machine 1, s_1 = 11, I_1 = {3, 4, 5, 6, 7}					Machine 2, s_2 = 18, I_2 = {3, 4, 6}					Machine 3, s_3 = 17, I_3 = {3, 4, 6, 7}				
13	39	65	11	50	9	1	19	18	19	8	1	18	17	18
16	4	54	50	54	11	17	37	19	36	10	2	22	18	20
					15	16	52	36	52	12	9	34	20	29
										14	20	50	29	49

Table 3.42 Iteration 4

i	l_i	d_i	s_i	C_i	i	l_i	d_i	s_i	C_i	i	l_i	d_i	s_i	C_i
Machine 1, $s_1 = 9$, $I_1 = \{3, 5\}$					Machine 2, $s_2 = 13$, $I_2 = \{3, 5\}$					Machine 3, $s_3 = 12$, $I_3 = \{3, 5\}$				
7	1	17	9	10	6	5	20	13	18	4	5	19	12	17
13	39	65	11	50	9	1	19	18	19	8	1	18	17	18
16	4	54	50	54	11	17	37	19	36	10	2	22	18	20
					15	16	52	36	52	12	9	34	20	29

Table 3.43 Iteration 5

i	l_i	d_i	s_i	C_i	i	l_i	d_i	s_i	C_i	i	l_i	d_i	s_i	C_i
Machine 1, $s_1 = 6$, $I_1 = \{1\}$					Machine 2, $s_2 = 11$, $I_2 = \varnothing$					Machine 3, $s_3 = 3$, $I_3 = \varnothing$				
2	1	7	6	7	5	2	16	11	13	3	9	18	3	12
7	1	17	7	8	6	5	20	13	18	4	5	19	12	17
13	39	65	8	47	9	1	19	18	19	8	1	18	17	18
16	4	54	47	51	11	17	37	19	36	10	2	22	18	20
					15	16	52	36	52	12	9	34	20	29
										14	20	50	29	49

Table 3.44 Iteration 6

i	l_i	d_i	s_i	C_i	i	l_i	d_i	s_i	C_i	i	l_i	d_i	s_i	C_i
Machine 1, $s_1 = 4$					Machine 2, $s_2 = 11$					Machine 3, $s_3 = 3$				
1	2	6	4	6	5	2	16	11	13	3	9	18	3	12
2	1	7	6	7	6	5	20	13	18	4	5	19	12	17
7	1	17	7	8	9	1	19	18	19	8	1	18	17	18
13	39	65	8	47	11	17	37	19	36	10	2	22	18	20
16	4	54	47	51	15	16	52	36	52	12	9	34	20	29
										14	20	50	29	49

Execute MSTAA: put task 3 after task 8 on machine 3 and task 8 after task 4. Move task 8 and then task 4 from machine 3 to machine 2. The final schedule is in Table 3.45.

Algorithm A2.2 Initial data: $m = 3$, $n = 16$ (Table 3.46).

Preliminary stage. We assign the first tasks from the set of unassigned tasks to the last position in the schedule of each machine (Table 3.47).

Optimization stage. Successively assign feasible tasks: 13 to machine 1; 11 to machine 2; 12 and 10 to machine 3; 9 to machine 2; 4 to machine 3; 8, 7, 5, and 6 to machine 2; 3 to machine 3. We get the schedule shown in Table 3.48.

The first machine has the maximum start time 11 but there are no feasible tasks. Remaining unassigned tasks are 1 and 2. Assign task 2 to machine 1, shifting the

Table 3.45 Final schedule

i	l_i	d_i	s_i	C_i	i	l_i	d_i	s_i	C_i	i	l_i	d_i	s_i	C_i
Machine 1, $s_1 = 4$					Machine 2, $s_2 = 5$					Machine 3, $s_3 = 9$				
1	2	6	4	6	4	5	19	5	10	3	9	18	9	18
2	1	7	6	7	8	1	18	10	11	10	2	22	18	20
7	1	17	7	8	5	2	16	11	13	12	9	34	20	29
13	39	65	8	47	6	5	20	13	18	14	20	50	29	49
16	4	54	47	51	9	1	19	18	19					
					11	17	37	19	36					
					15	16	52	36	52					

Table 3.46 Tasks data for Algorithm A2.2

i	l_i	d_i	i	l_i	d_i	i	l_i	d_i
1	2	6	7	1	17	12	9	34
2	1	7	8	1	18	13	39	65
3	9	18	9	1	19	14	20	50
4	5	19	10	2	22	15	16	52
5	2	16	11	17	37	16	4	54
6	5	20						

Table 3.47 Preliminary stage

i	l_i	d_i	s_i	C_i	i	l_i	d_i	s_i	C_i	i	l_i	d_i	s_i	C_i
Machine 1, $s_1 = 50$					Machine 2, $s_2 = 36$					Machine 3, $s_3 = 30$				
16	4	54	50	54	15	16	52	36	52	14	20	50	30	50

Table 3.48 Optimization stage

i	l_i	d_i	s_i	C_i	i	l_i	d_i	s_i	C_i	i	l_i	d_i	s_i	C_i
Machine 1, $s_1 = 11$					Machine 2, $s_2 = 9$					Machine 3, $s_3 = 5$				
13	39	65	11	50	6	5	20	9	14	3	9	18	5	14
16	4	54	50	54	5	2	16	14	16	4	5	19	14	19
					7	1	17	16	17	10	2	22	19	21
					8	1	18	17	18	12	9	34	21	30
					9	1	19	18	19	14	20	50	30	50
					11	17	37	19	36					

schedule by the amount of its tardiness (4 units). Now the second machine has the maximum start time 9. Assign the remaining task 1 to it with a shift by the amount of its tardiness (3 units). All tasks are assigned, MSTAA does not improve the functional value. The final schedule is in Table 3.49.

Table 3.49 Final schedule

i	l_i	d_i	s_i	C_i	i	l_i	d_i	s_i	C_i	i	l_i	d_i	s_i	C_i
Machine 1, $s_1 = 6$					Machine 2, $s_2 = 4$					Machine 3, $s_3 = 5$				
2	1	7	6	7	1	2	6	4	6	3	9	18	5	14
13	39	65	7	46	6	5	20	6	11	4	5	19	14	19
16	4	54	46	50	5	2	16	11	13	10	2	22	19	21
					7	1	17	13	14	12	9	34	21	30
					8	1	18	14	15	14	20	50	30	50
					9	1	19	15	16					
					11	17	37	16	33					
					15	16	52	33	49					

Algorithm A2.3, example 1 For the initial data given for Algorithm A1.1 (Table 3.1), Algorithm A2.3 builds the same schedule as Algorithm A1.1 because all tasks are assigned without tardiness.

Algorithm A2.3, example 2 Initial data: $m = 3$, $n = 16$. Sequence σ^{ord} is in Table 3.50.

Initial schedule: assign the first three tasks of the sequence to each machine according to SSO #1 (Table 3.51).

Successively assign non-tardy tasks 4–14, each to the machine with the minimum completion time of the tasks (Table 3.52).

Task 15 is tardy on all machines. Assign it to the machine 1 that corresponds to the best vector of start times (4, 6, 7) according to the lexicographical order. Shift all tasks on machine 1 to the left by the tardiness of task 15 (1 unit). Then assign task 16 to the machine 2 that corresponds to the best vector of start times (4, 4, 7) in lexicographical order, shifting all tasks on machine 2 to the left by the tardiness of task 16 (2 units). All tasks are assigned, the final schedule is in Table 3.53.

Algorithm B1.1 Initial data: $m = 3$, $n = 16$. Sequence σ^{ord} is in Table 3.54.

Table 3.50 Sequence σ^{ord} for Algorithm A2.3

i	l_i	d_i	$d_i - l_i$	i	l_i	d_i	$d_i - l_i$	i	l_i	d_i	$d_i - l_i$
1	3	8	5	7	1	17	16	12	9	34	25
2	2	8	6	8	1	18	17	13	39	65	26
3	2	9	7	9	1	19	18	14	20	48	28
4	6	19	13	10	2	22	20	15	16	46	30
5	3	16	13	11	17	37	20	16	4	45	41
6	4	20	16								

Table 3.51 Initial schedule

i	l_i	d_i	s_i	C_i	i	l_i	d_i	s_i	C_i	i	l_i	d_i	s_i	C_i
Machine 1, $s_1 = 5$					Machine 2, $s_2 = 6$					Machine 3, $s_3 = 7$				
1	3	8	5	8	2	2	8	6	8	3	2	9	7	9

Table 3.52 The schedule after tasks 4–14 assignment

i	l_i	d_i	s_i	C_i	i	l_i	d_i	s_i	C_i	i	l_i	d_i	s_i	C_i
Machine 1, $s_1 = 5$					Machine 2, $s_2 = 6$					Machine 3, $s_3 = 7$				
1	3	8	5	8	2	2	8	6	8	3	2	9	7	9
4	6	19	8	14	5	3	16	8	11	6	4	20	9	13
11	17	37	14	31	7	1	17	11	12	10	2	22	13	15
					8	1	18	12	13	13	39	65	15	54
					9	1	19	13	14					
					12	9	34	14	23					
					14	20	48	23	43					

Table 3.53 Final schedule

i	l_i	d_i	s_i	C_i	i	l_i	d_i	s_i	C_i	i	l_i	d_i	s_i	C_i
Machine 1, $s_1 = 4$					Machine 2, $s_2 = 4$					Machine 3, $s_3 = 7$				
1	3	8	4	7	2	2	8	4	6	3	2	9	7	9
4	6	19	7	13	5	3	16	6	9	6	4	20	9	13
11	17	37	13	30	7	1	17	9	10	10	2	22	13	15
15	16	46	30	46	8	1	18	10	11	13	39	65	15	54
					9	1	19	11	12					
					12	9	34	12	21					
					14	20	48	21	41					
					16	4	45	41	45					

Table 3.54 Sequence σ^{ord} for Algorithm B1.1

i	l_i	d_i	$d_i - l_i$	i	l_i	d_i	$d_i - l_i$	i	l_i	d_i	$d_i - l_i$
1	2	6	4	7	1	17	16	12	9	34	25
2	1	7	6	8	1	18	17	13	39	65	26
3	9	18	9	9	1	19	18	14	20	48	28
4	6	18	12	10	2	22	20	15	16	46	30
5	3	16	13	11	17	37	20	16	5	55	50
6	4	20	16								

Initial schedule: assign the first tasks of the sequence to each machine according to SSO #2 for Criterion 1 (Table 3.55).

Successively assign non-tardy tasks 6–13, each to the machine with the minimum completion time of the tasks (Table 3.56).

Assign task 14 to machine 3, shifting all tasks assigned to this machine to the left by the amount of its tardiness (8 units). Since the start time of machine 3 becomes equal to 5 > 4 then this assignment is feasible (Table 3.57).

Table 3.55 Initial schedule

i	l_i	d_i	s_i	C_i	i	l_i	d_i	s_i	C_i	i	l_i	d_i	s_i	C_i
Machine 1, $s_1 = 4$					Machine 2, $s_2 = 12$					Machine 3, $s_3 = 13$				
1	2	6	4	6	4	6	18	12	18	5	3	16	13	16
2	1	7	6	7										
3	9	18	7	16										

Table 3.56 The schedule after tasks 6–13 assignment

i	l_i	d_i	s_i	C_i	i	l_i	d_i	s_i	C_i	i	l_i	d_i	s_i	C_i
Machine 1, $s_1 = 4$					Machine 2, $s_2 = 12$					Machine 3, $s_3 = 13$				
1	2	6	4	6	4	6	18	12	18	5	3	16	13	16
2	1	7	6	7	10	2	22	18	20	7	1	17	16	17
3	9	18	7	16	13	39	65	20	59	8	1	18	17	18
6	4	20	16	20						9	1	19	18	19
12	9	34	20	29						11	17	37	19	36

Table 3.57 The schedule after task 14 assignment

i	l_i	d_i	s_i	C_i	i	l_i	d_i	s_i	C_i	i	l_i	d_i	s_i	C_i
Machine 1, $s_1 = 4$					Machine 2, $s_2 = 12$					Machine 3, $s_3 = 5$				
1	2	6	4	6	4	6	18	12	18	5	3	16	5	8
2	1	7	6	7	10	2	22	18	20	7	1	17	8	9
3	9	18	7	16	13	39	65	20	59	8	1	18	9	10
6	4	20	16	20						9	1	19	10	11
12	9	34	20	29						11	17	37	11	28
										14	20	48	28	48

Tasks 15 and 16 are feasible on machine 1, so we assign them to it. All tasks are assigned, we obtain the optimal schedule (Table 3.58).

Algorithm B1.3 Initial data: $m = 3$, $n = 16$. Sequence σ^{ord} is in Table 3.59.

Initial schedule: assign the first task to the first machine according to SSO for Criterion 2 (Table 3.60).

Execute Algorithm A2.2. Preliminary stage. We assign the first tasks from the set of unassigned tasks to the last position in the schedule of each machine (Table 3.61).

Optimization stage. Successively assign feasible tasks: 13 to machine 1; 11 to machine 2; 12 and 10 to machine 3; 9 to machine 2; 4 to machine 3; 8, 7, 5, and 6 to machine 2. We get the schedule shown in Table 3.62.

Assign tardy task 3 to machine 3 with a shift of all tasks assigned to machine 3 to the left by the amount of its tardiness (4 units). Assign remaining task 2 to machine 1, shifting all tasks assigned to machine 1 to the left by 3 units. MSTAA does not change the start times of machines (Table 3.63).

Table 3.58 Optimal schedule

i	l_i	d_i	s_i	C_i	i	l_i	d_i	s_i	C_i	i	l_i	d_i	s_i	C_i
Machine 1, $s_1 = 4$					Machine 2, $s_2 = 12$					Machine 3, $s_3 = 5$				
1	2	6	4	6	4	6	18	12	18	5	3	16	5	8
2	1	7	6	7	10	2	22	18	20	7	1	17	8	9
3	9	18	7	16	13	39	65	20	59	8	1	18	9	10
6	4	20	16	20						9	1	19	10	11
12	9	34	20	29						11	17	37	11	28
15	16	46	29	45						14	20	48	28	48
16	5	55	45	50										

Table 3.59 Sequence σ^{ord} for Algorithm B1.3

i	l_i	d_i	$d_i - l_i$	i	l_i	d_i	$d_i - l_i$	i	l_i	d_i	$d_i - l_i$
1	3	8	5	7	1	17	16	12	9	34	25
2	2	8	6	8	1	18	17	13	39	65	26
3	2	9	7	9	1	19	18	14	20	50	30
4	6	19	13	10	2	22	20	15	16	52	36
5	3	16	13	11	17	37	20	16	4	54	50
6	4	20	16								

Table 3.60 Initial schedule

i	l_i	d_i	s_i	C_i
Machine 1, $s_1 = 5$				
1	3	8	5	8

Table 3.61 Preliminary stage

i	l_i	d_i	s_i	C_i	i	l_i	d_i	s_i	C_i	i	l_i	d_i	s_i	C_i
Machine 1, $s_1 = 50$					Machine 2, $s_2 = 36$					Machine 3, $s_3 = 30$				
1	3	8	5	8	15	16	52	36	52	14	20	50	30	50
16	4	54	50	54										

Table 3.62 Optimization stage

i	l_i	d_i	s_i	C_i	i	l_i	d_i	s_i	C_i	i	l_i	d_i	s_i	C_i
Machine 1, $s_1 = 11$					Machine 2, $s_2 = 9$					Machine 3, $s_3 = 13$				
1	3	8	5	8	6	4	20	9	13	4	6	19	13	19
13	39	65	11	50	5	3	16	13	16	10	2	22	19	21
16	4	54	50	54	7	1	17	16	17	12	9	34	21	30
					8	1	18	17	18	14	20	50	30	50
					9	1	19	18	19					
					11	17	37	19	36					
					15	16	52	36	52					

Table 3.63 The schedule after all tasks assignment

i	l_i	d_i	s_i	C_i	i	l_i	d_i	s_i	C_i	i	l_i	d_i	s_i	C_i
Machine 1, $s_1 = 6$					Machine 2, $s_2 = 9$					Machine 3, $s_3 = 7$				
1	3	8	5	8	6	4	20	9	13	3	2	9	7	9
2	2	8	6	8	5	3	16	13	16	4	6	19	9	15
13	39	65	8	47	7	1	17	16	17	10	2	22	15	17
16	4	54	47	51	8	1	18	17	18	12	9	34	17	26
					9	1	19	18	19	14	20	50	26	46
					11	17	37	19	36					
					15	16	52	36	52					

Execute Procedure B: since the start time of task 2 on machine 1 is less than the completion time of task 1 of the initial schedule, we move task 2 to machine 2, shifting all tasks assigned to machine 2 to the left by the amount of its tardiness (1 unit). As a result, the start time of machine 2 becomes equal to 6 > 5. We get an optimal schedule (Table 3.64).

Algorithm B2.1 Initial data: $m = 3$, $n = 16$. Sequence σ^{ord} is in Table 3.65.

Initial schedule: assign the first tasks of the sequence to each machine according to SSO #2 for Criterion 1 (Table 3.66).

Successively assign non-tardy tasks 6–13, each to the machine with the minimum completion time of the tasks (Table 3.67).

Assign task 14 to machine 3, shifting all tasks assigned to this machine to the left by the amount of its tardiness (10 units). The start time of machine 1 becomes equal to 3 < 4. But alternative assignment of task 14 to machine 1 with a shift by 3 units makes the start time equal to 1, so we assign task 14 to machine 3 and consider machine 3 as analogue of the first machine (Table 3.68).

Tasks 15 and 16 are feasible on machine 1, so we assign them to it. All tasks are assigned, we obtain a final schedule (Table 3.69).

Table 3.64 Optimal schedule

i	l_i	d_i	s_i	C_i	i	l_i	d_i	s_i	C_i	i	l_i	d_i	s_i	C_i
Machine 1, $s_1 = 5$					Machine 2, $s_2 = 6$					Machine 3, $s_3 = 7$				
1	3	8	5	8	2	2	8	6	8	3	2	9	7	9
13	39	65	8	47	6	4	20	8	12	4	6	19	9	15
16	4	54	47	51	5	3	16	12	15	10	2	22	15	17
					7	1	17	15	16	12	9	34	17	26
					8	1	18	16	17	14	20	50	26	46
					9	1	19	17	18					
					11	17	37	18	35					
					15	16	52	35	51					

Table 3.65 Sequence σ^{ord} for Algorithm B2.1

i	l_i	d_i	$d_i - l_i$	i	l_i	d_i	$d_i - l_i$	i	l_i	d_i	$d_i - l_i$
1	2	6	4	7	1	17	16	12	9	34	25
2	1	7	6	8	1	18	17	13	39	65	26
3	9	18	9	9	1	19	18	14	20	46	26
4	6	18	12	10	2	22	20	15	16	46	30
5	3	16	13	11	17	37	20	16	5	55	50
6	4	20	16								

Table 3.66 Initial schedule

i	l_i	d_i	s_i	C_i	i	l_i	d_i	s_i	C_i	i	l_i	d_i	s_i	C_i
Machine 1, $s_1 = 4$					Machine 2, $s_2 = 12$					Machine 3, $s_3 = 13$				
1	2	6	4	6	4	6	18	12	18	5	3	16	13	16
2	1	7	6	7										
3	9	18	7	16										

Table 3.67 The schedule after tasks 6–13 assignment

i	l_i	d_i	s_i	C_i	i	l_i	d_i	s_i	C_i	i	l_i	d_i	s_i	C_i
Machine 1, $s_1 = 4$					Machine 2, $s_2 = 12$					Machine 3, $s_3 = 13$				
1	2	6	4	6	4	6	18	12	18	5	3	16	13	16
2	1	7	6	7	10	2	22	18	20	7	1	17	16	17
3	9	18	7	16	13	39	65	20	59	8	1	18	17	18
6	4	20	16	20						9	1	19	18	19
12	9	34	20	29						11	17	37	19	36

Table 3.68 The schedule after task 14 assignment

i	l_i	d_i	s_i	C_i	i	l_i	d_i	s_i	C_i	i	l_i	d_i	s_i	C_i
Machine 1, $s_1 = 4$					Machine 2, $s_2 = 12$					Machine 3, $s_3 = 3$				
1	2	6	4	6	4	6	18	12	18	5	3	16	3	6
2	1	7	6	7	10	2	22	18	20	7	1	17	6	7
3	9	18	7	16	13	39	65	20	59	8	1	18	7	8
6	4	20	16	20						9	1	19	8	9
12	9	34	20	29						11	17	37	9	26
										14	20	46	26	46

3.4 Problem 2. Machines with Various Productivities

The Problem Statement Given m independent parallel machines with various productivities to process n tasks without interruptions. l_i^j is the processing time of a task i on a machine j. The due date of a task i is d_i. Processing times of tasks are

Table 3.69 Final schedule

i	l_i	d_i	s_i	C_i	i	l_i	d_i	s_i	C_i	i	l_i	d_i	s_i	C_i
Machine 1, $s_1 = 4$					Machine 2, $s_2 = 12$					Machine 3, $s_3 = 3$				
1	2	6	4	6	4	6	18	12	18	5	3	16	3	6
2	1	7	6	7	10	2	22	18	20	7	1	17	6	7
3	9	18	7	16	13	39	65	20	59	8	1	18	7	8
6	4	20	16	20						9	1	19	8	9
12	9	34	20	29						11	17	37	9	26
15	16	46	29	45						14	20	46	26	46
16	5	55	45	50										

independent (it is possible that one task is processed faster than another on some machine but slower than another on a different machine). The start times of machines are arbitrary. We need to build a feasible schedule that maximizes Criterion 1 or Criterion 2.

3.4.1 *Signs of Optimality of a Feasible Solution*

Denote by I the set of all tasks $\{i\colon i = \overline{1,n}\}$. For Problem 2, the SSO #1 for Problem 1 is generalized in an obvious way.

3.4.1.1 Construction of SSO #1 for Criterion 1

Consider the following monotonically non-decreasing sequence of numbers:

$$d_{i_1} - l_{i_1}^{j_1},\ d_{i_2} - l_{i_2}^{j_2}, \ldots,\ d_{i_m} - l_{i_m}^{j_m}$$

where

$$d_{i_1} - l_{i_1}^{j_1} = \min_i\{(d_i - \min_j l_i^j),\ i = \overline{1,n},\ j = \overline{1,m}\}, \tag{3.8}$$

$$d_{i_p} - l_{i_p}^{j_p} = \min_i\{(d_i - \min_j l_i^j),\ i = \overline{1,n},\ j = \overline{1,m}, \tag{3.9}$$

$$i \notin \{\overline{i_1, i_{p-1}}\}, j \notin \{\overline{j_1, j_{p-1}}\},\ p = \overline{2,m}. \tag{3.10}$$

Suppose that only one minimum is reached on each machine. Let all the inequalities hold:

$$d_{i_l} + l_{i_p}^{j_l} > d_{i_p} \quad \forall p = \overline{l+1, m}, \quad l = \overline{1, m-1}. \tag{3.11}$$

Then the following statement is true.

Statement 3.8 If in arbitrary feasible schedule on the machine j_l, $l = \overline{1,m}$, the first assigned task is i_l at the time moment $s_{j_l} = d_{i_l} - l_{i_l}^{j_l}$, then this is the optimal schedule for criterion (3.1).

3.4.1.2 Heuristic SSO #2 for Criterion 1: An Initial Schedule Construction

Let σ^{ord} is the sequence of tasks i_1, i_2, …, i_n defined by formulas (3.8)–(3.10) without taking into account the condition $j \notin \{\overline{j_1, j_{p-1}}\}, p = \overline{2,m}$, and if there are several pairs (i_p, j_p), then any one is chosen. Ascribe to each task i_l the machine j_l. Assign the first task i_1 to ascribed machine j_1 with start time s_{j_1}. Proceed to the next task from σ^{ord} and check if it can be assigned to machine j_1: if the task is not tardy, assign it to machine j_1. A tardy task is assigned to its ascribed machine so that its completion time equals to its due date. Continue to assign the next task from σ^{ord} in the same order as the first tasks were assigned, starting from machine j_1. If we could not assign the task to any of these machines without tardiness, we assign it to its ascribed machine so that its completion time equals to its due date. If the first task was already assigned to this machine, then all unassigned tasks from σ^{ord} including the current task and all the machines that do not contain assigned tasks are reordered according to the rules of σ^{ord} construction. We obtain new sequence σ^{ord}. Choose the next task from new σ^{ord} and assign it to its ascribed machine. Further, this procedure is repeated until the first task is assigned to the last machine.

Additional Condition 1 Start times of all machines are not less than $s_{i_1} = d_{i_1} - l_{i_1}^{j_1}$.

Additional Condition 2 The initial schedule satisfies the condition: according to the order of the first tasks assignment, the start times of machines ascribed to them are a numerical non-decreasing sequence.

Remark 3.12 Additional Condition 2 includes Additional Condition 1.

3.4.1.3 Construction of SSO #1 for Criterion 2

The SSO is: task i_1 is assigned to machine j_1 with start time $s_{j_1} = d_{i_1} - l_{i_1}^{j_1}$ and there is a feasible schedule for which $s_k \geq s_{j_1}$, $k = \overline{1,m}$.

An initial schedule always exists for Criterion 2, since it is always possible to implement (3.8)–(3.10) to find only the task i_1 for the machine j_1.

3.4.1.4 Heuristic SSO #2 for Criterion 2

It coincides with the heuristic SSO #2 for Criterion 1 without taking into account the fulfillment of Additional Condition 2.

3.4.2 PSC-Algorithm for Problem 2 Solving by Criterion 1 (Algorithm C)

3.4.2.1 The First Polynomial Component of PSC-Algorithm C

The first polynomial component of the PSC-algorithm consists of six algorithms (C1.1, C1.2, C1.2a, C1.3, C1.4, C1.4a). They are to be executed sequentially in an arbitrary order. The first algorithm that constructed a feasible schedule for which a SSO is satisfied, obtains an optimal solution of the problem by criterion (3.1), and the first polynomial component of the PSC-algorithm is considered to be fulfilled. If a feasible solution is not obtained by any algorithm of the first polynomial component, we search it using algorithms of the second polynomial component.

Algorithm C1.1 The algorithm implements the first polynomial component for a feasible schedule construction if an initial schedule was constructed according to SSO #1. As a result of this, the initial set of tasks I is divided into two subsets: I_1 is the set of previously assigned tasks and I_1' is the set of not yet assigned tasks. We sequence the tasks from I_1' in non-decreasing order of *due dates* (sequence σ^{ord}; in further considerations, σ^{ord} is the sequence of tasks ordered in non-decreasing order of due dates). The further assignment of tasks is carried out as follows. Assign each next task from the sequence σ^{ord} to the release time of the machine where it will have the maximum earliness in regard to its due date. A similar procedure is repeated for each task. If at some iteration the task is tardy on all machines, then the algorithm terminates, the first polynomial component is not fulfilled. Otherwise, the assignment continues from the task following the assigned one in σ^{ord}, until all tasks of the sequence are distributed. If all tasks are assigned without tardiness, then the first polynomial component is fulfilled.

Algorithm C1.2, C1.2a are Algorithms A2.1, A2.1a, respectively, with Procedure A for the case when an initial schedule was constructed in accordance with SSO #1.

Algorithm C1.3 is Algorithm A2.2 with Procedure A for the case when an initial schedule was constructed in accordance with SSO #1.

Algorithm C1.4, C1.4a are, respectively, Algorithms C2.5, C2.5a (see below) with Procedure A for the case when an initial schedule was constructed in accordance with SSO #1.

3.4.2.2 The Second Polynomial Component of PSC-Algorithm C

The second polynomial component consists of seven approximation Algorithms (C2.1, C2.2, C2.3, C2.3a, C2.4, C2.5, C2.5a), we solve the problem by each of them and choose the best solution according to the direct lexicographical order (3.7).

Algorithm C2.1 is a polynomial complexity approximation algorithm to construct a feasible schedule based on an initial schedule constructed according to SSO #1 for Criterion 1.

Algorithm C2.1 is Algorithm C1.1 with the following change. If a current task from the sequence σ^{ord} is tardy on each machine, then assign it to each machine, shifting this and all assigned to the machine tasks to the left by the tardiness of the current task. We get m schedules corresponding to m vectors of start times of all machines. Choose the best vector in the direct lexicographical order (3.7) and assign the current task from σ^{ord} to the machine corresponding to the best by (3.7) vector of start times. The last step is MSTAA.

Algorithm C2.2 is a polynomial complexity algorithm to construct a feasible schedule based on an initial schedule constructed according to heuristic SSO #2 for Criterion 1. Current iteration of Algorithm C2.2 coincides with current iteration of Algorithm C2.1.

Algorithm C2.3 is Algorithm A2.1 for machines with various productivities.

Algorithm A2.1, its theoretical substantiation (Statements 3.4–3.6), as well as MSTAA, given for parallel machines with equal productivities, are multipurpose and valid also for machines with various productivities.

Algorithm C2.3a is Algorithm A2.1a for machines with various productivities.

The difference of Algorithm C2.3a from Algorithm C2.3 is that after the last step of any current iteration we check the inequality

$$\max_{j=\overline{1,m}} s_j^{\text{curr}} - \min_{j=\overline{1,m}} s_j^{\text{curr}} > K_{\text{exp}}.$$

If the inequality is satisfied, then we execute MSTAA for the current schedule constructed by iterations from the first to the current inclusively, after this we go to the next iteration.

Algorithm C2.4 is Algorithm A2.2 for machines with various productivities.

Algorithm A2.2 and its theoretical substantiation (Statement 3.7), given for parallel machines with equal productivities, are multipurpose and valid also for machines with various productivities.

Algorithm C2.5 is an adaptation of Algorithm A2.1 for machines with various productivities with this modification: at current iteration we assign feasible tasks in non-increasing order of their processing times to the most advantageous for them feasible machines (where they have the minimum processing times).

The Algorithm C2.5 is a series of similar iterations. Each iteration, except the first one, includes the preliminary stage and the tasks assignment on a group basis: we assign groups of m tasks, one task to each machine. At the first iteration of the algorithm, we assign one task to each machine. The completion time of the task should coincide with its due date, and the start times of machines should correspond to the global optimum in the reverse lexicographical order: for each task we find the machine where the start time for this task is the latest; the first task is assigned to the machine with the latest start time, then the next task to the machine with the next (earlier) latest start time, etc., until we assign m tasks to m machines.

Remark 3.13 A set of machines, from which we find the machine with the next latest start time, does not include machines to which we have assigned a task at the first iteration.

At the current iteration, we distribute the tasks that are not yet assigned to machines at previous iterations. We assign a task so that its completion time coincides with the start time of the task assigned to this machine at the previous iteration. At the preliminary stage we find tasks for each of which there is at least one machine on which it is feasible (the task would be not tardy if assigned to this machine). Then to machines for which there are feasible tasks (the first group of machines) we assign the feasible tasks this way: choose a feasible task for which the shortest of its processing times is the greatest; assign it to the machine where it has the shortest processing time. After the assignment, eliminate the task from the sets of feasible tasks on another machines. Repeat the described procedure until one feasible task is assigned to each machine of the first group.

Remark 3.14 A set of machines for the tasks assignment does not include the machines to which we have assigned a task at the current iteration.

If we have assigned one task to each of m machines, the current iteration terminates.

After such distribution, there may remain the tasks for each of which there are no machines on which they are feasible. They are combined to the second group of tasks for the current iteration. We assign tasks to the second group of machines (machines for which there are no feasible tasks) locally optimally in accordance with the inverse lexicographical order, shifting the task (and all tasks already assigned to this machine on previous iterations) to the left by the amount of the task's tardiness, until one task is assigned to each machine of the second group.

Remark 3.15 A set of machines, from which we find the machine with the next latest start time, does not include machines to which we have assigned a task at the current iteration.

Iterations continue until we get start times of the first tasks on each machine. The last step is MSTAA.

Rationale for the Heuristic of Feasible Tasks Assignment Taking into account that at the first iteration we assign the feasible task for which the shortest of its processing times is the greatest, we assume that the number of tardy tasks will be minimal on subsequent iterations of the assignment.

Algorithm C2.5a is an adaptation of Algorithm A2.1a for machines with various productivities with this modification: at current iteration we assign feasible tasks in non-increasing order of their processing times to the most advantageous for them feasible machines (where they have the minimum processing times).

The difference of Algorithm C2.5a from Algorithm C2.5 is that after the last step of any current iteration we check the inequality

$$\max_{j=\overline{1,m}} s_j^{\text{curr}} - \min_{j=\overline{1,m}} s_j^{\text{curr}} > K_{\exp}.$$

If the inequality is satisfied, then we execute MSTAA for the current schedule constructed by iterations from the first to the current inclusively, after this we go to the next iteration.

3.4.3 PSC-Algorithm for Problem 2 Solving by Criterion 2 (Algorithm D)

We considered the Criterion 2 (3.2) and sufficient signs of optimality of a feasible solution earlier.

3.4.3.1 The First Polynomial Component of PSC-Algorithm D

The first polynomial component of the PSC-algorithm consists of six algorithms (D1.1, D1.2, D1.2a, D1.3, D1.4, D1.4a). They are to be executed sequentially in an arbitrary order. The first algorithm that constructed a feasible schedule for which a SSO is satisfied, obtains an optimal solution of the problem by criterion (3.2), and the first polynomial component of the PSC-algorithm is considered to be fulfilled. If a feasible solution is not obtained by any algorithm of the first polynomial component, we search it using algorithms of the second polynomial component.

Algorithm D1.1 implements the first polynomial component for a feasible schedule construction if an initial schedule was constructed according to heuristic SSO #2 for Criterion 2. As a result of this, the initial set of tasks I is divided into two subsets: I_1 is the set of previously assigned tasks and I_1' is the set of not yet assigned tasks. We sequence the tasks from I_1' in non-decreasing order of due dates (the sequence σ^{ord}). The further assignment of tasks is carried out as follows. Assign each next task from σ^{ord} to the release time of the machine where it will have either the maximum earliness in regard to its due date or the minimum tardiness if it is not feasible on all machines. We assign only feasible task to machine j_1; if the task has the minimum tardiness on machine j_1, then we assign it to another machine where the task has the next (greater) minimum tardiness. If the task is assigned to the machine where it has the minimum tardiness, then we shift this task and all tasks assigned to the machine to the left by the task's tardiness and check the new start time of the machine s^{new}. If $s^{new} \geq s_{j_1}$, then the task's assignment is accepted, continue to a similar assignment of the next task from σ^{ord}. If $s^{new} < s_{j_1}$, then the first polynomial component is not fulfilled by this algorithm, terminate. If all tasks are assigned without tardiness, then we obtain an optimal schedule.

Algorithms D1.2, D1.2a are algorithms C2.3, C2.3a, respectively, with the following Procedure C.

Procedure C An algorithm distributes all tasks except task i_1. The initial schedule consists of one task i_1 assigned to machine j_1 with start time s_{j_1}. For the schedule obtained by the approximation algorithm, the following conditions are verified:

1. The start time of the first task of the approximate schedule on machine j_1 is not less than d_{i_1} (the completion time of the first task of the initial schedule on machine j_1 assigned in accordance with a SSO).
2. The start times of the first tasks of the approximate schedule on other machines are not less than s_{j_1}.

If both conditions are satisfied, then, to obtain an optimal schedule, we need to shift the schedule obtained by an approximation algorithm on machine j_1 to the left so that the start time of the first task of the approximate schedule coincides with the completion time of the first task of the initial schedule d_{i_1}. If the first condition is not satisfied but the second one is true, then we put the first task from machine j_1 of the approximate schedule successively to another machines and assign it to the machine where the start time of this task will be the greatest and not less than s_{j_1}. If such machine is not found, then the first polynomial component is not fulfilled. If the first task is moved to another machine, then we check the condition: if the start time of the second task on machine j_1 is not less than d_{i_1}, then, by shifting the schedule on machine j_1 so that the start time equals to d_{i_1}, we get an optimal schedule. If the condition is not satisfied, we try to move the second task of the approximate schedule from machine j_1 to another machine. The described procedure is repeated until either an optimal schedule is obtained or the first polynomial component is not fulfilled by this algorithm.

Algorithm D1.3 is Algorithm C2.4 with Procedure C.

Algorithms D1.4, D1.4a are algorithms C2.5, C2.5a, respectively, with Procedure C.

3.4.3.2 The Second Polynomial Component of PSC-Algorithm D

The second polynomial component contains six algorithms. As approximation algorithms, we use Algorithms C2.3, C2.3a, C2.4, C2.5, C2.5a (see above) because we adjust the start times of tasks using MSTAA. Also we use Algorithm D2.1 (see below). We solve the problem by each of these algorithms and choose a solution corresponding to the minimum deviation from s_{j_1}.

Algorithm D2.1 is the modification of Algorithm D1.1 (see above) with the following difference of the current iteration. If new start time of the machine $s^{new} < s_{j_1}$, then we assign the tardy task to the release time of machine j_1 and shift the schedule of this machine to the left by the task's tardiness. After that, we assign this task either to machine j_1 or to the machine with the current start time s^{new}, depending on where the current start time is greater. For further iterations, this machine becomes the analogue of machine j_1. The assignment procedure continues similarly until all tasks are distributed. After the algorithm execution, we apply MSTAA.

Estimation of Deviation of a Solution from Optimum for Problem 2 All algorithms of the second polynomial component are the approximation algorithms for Criterion 1 or 2. If the first part of a SSO is fulfilled (an initial schedule was constructed), then there is an estimate of deviation from optimum for the solution obtained by the approximation algorithm.

For Criterion 1: for this purpose we compare vectors using the condition (3.7).

Remark 3.16 Suppose $s_l > s_{q_l}$ where l is the number of the component for which the first positive difference is reached, i.e., $s_j = s_{q_j}, j = \overline{1, l-1}$. Then the signs of the differences of the remaining components $(j = \overline{l+1, m})$ may be arbitrary. The decision maker analyzes the solution and chooses either the optimal or approximate solution based on practical considerations.

For Criterion 2: we determine the estimate as the difference $s_1 - s_{q_1} \geq 0$ (Algorithms C2.3, C2.3a, C2.4, C2.5, C2.5a, D2.1 have an estimate of deviation from optimum for Criterion 2; in all other algorithms of the second polynomial component for Criterion 2 we compare the solution with zero).

Remark 3.17 For Criterion 1, Algorithms C2.1, C2.3, C2.3a, C2.4, C2.5, C2.5a have an estimation of deviation from optimum if an initial schedule was built according to SSO #1. In all other algorithms of the second polynomial component of PSC-algorithms for Criterion 1 we compare the solution with zero vector.

Remark 3.18 If in the final schedule obtained by any algorithm of the second polynomial component, at least one of machines has the negative start time, then the solution of the problem is not obtained by this algorithm.

3.5 Illustrative Examples for Problem 2

Algorithm C1.1 Initial data: m = 3, n = 13. Sequence σ^{ord} is in Table 3.70. Initial schedule: assign one task to each machine, according to SSO #1 (Table 3.71).

Table 3.70 Sequence σ^{ord} for Algorithm C1.1

i	l_i^1	l_i^2	l_i^3	d_i	i	l_i^1	l_i^2	l_i^3	d_i	i	l_i^1	l_i^2	l_i^3	d_i
1	2	3	4	5	6	6	9	15	15	10	4	8	16	29
2	4	5	6	7	7	1	12	18	20	11	3	6	12	32
3	3	6	8	13	8	2	4	8	21	12	1	2	4	32
4	4	8	12	14	9	3	6	12	27	13	4	8	20	33
5	5	7	9	15										

Table 3.71 Initial schedule

i	l_i^j	d_i	s_i	C_i	i	l_i^j	d_i	s_i	C_i	i	l_i^j	d_i	s_i	C_i
Machine 1, $s_1 = 3$					Machine 2, $s_2 = 2$					Machine 3, $s_3 = 0$				
1	2	5	3	5	2	5	7	2	7	6	15	15	0	15

Successively assign tasks 3–5, 7–13, each to the machine where it will have the maximum earliness regarding its due date. If there are several such machines, then assign the task to the least loaded machine. Since all tasks are assigned and are not tardy, we get the optimal schedule (Table 3.72).

Algorithm C1.2 Initial data: $m = 3$, $n = 13$. Sequence σ^{ord} is in Table 3.73.

Initial schedule: assign the first three tasks, one to each machine, according to SSO #1 (Table 3.74).

Execute Algorithm A2.1. The first iteration. We assign tasks 12, 11, and 9 from the set of unassigned tasks to the last position in the schedule of each machine. Determine the set of feasible tasks I_j for the next iteration by the condition: the task's due date is not less than the start time of the machine (Table 3.75).

Iteration 2. Assign task 13 to machine 2 and task 8 to machine 3. There are no feasible tasks for machine 1. Assign task 10 to it, shifting all assigned by Algorithm A2.1 tasks on machine 1 to the left by the task's tardiness (2 units). We get the schedule shown in Table 3.76.

Table 3.72 Optimal schedule

i	l_i^j	d_i	s_i	C_i	i	l_i^j	d_i	s_i	C_i	i	l_i^j	d_i	s_i	C_i
Machine 1, $s_1 = 3$					Machine 2, $s_2 = 2$					Machine 3, $s_3 = 0$				
1	2	5	3	5	2	5	7	2	7	6	15	15	0	15
3	3	13	5	8	5	7	15	7	14					
4	4	14	8	12	10	8	29	14	22					
7	1	20	12	13										
8	2	21	13	15										
9	3	27	15	18										
11	3	32	18	21										
12	1	32	21	22										
13	4	33	22	26										

Table 3.73 Sequence σ^{ord} for Algorithm C1.2

i	l_i^1	l_i^2	l_i^3	d_i	i	l_i^1	l_i^2	l_i^3	d_i	i	l_i^1	l_i^2	l_i^3	d_i
1	2	3	4	5	6	6	9	15	15	10	4	8	16	29
2	4	5	6	7	7	1	10	18	20	11	3	6	12	32
3	1	6	8	13	8	2	4	4	21	12	1	2	4	32
4	4	8	12	14	9	3	6	8	27	13	4	8	16	28
5	5	7	9	15										

Table 3.74 Initial schedule

i	l_i^j	d_i	s_i	C_i	i	l_i^j	d_i	s_i	C_i	i	l_i^j	d_i	s_i	C_i
Machine 1, $s_1 = 3$					Machine 2, $s_2 = 2$					Machine 3, $s_3 = 0$				
1	2	5	3	5	2	5	7	2	7	6	15	15	0	15

Table 3.75 The first iteration

i	l_i^j	d_i	s_i	C_i	i	l_i^j	d_i	s_i	C_i	i	l_i^j	d_i	s_i	C_i
Machine 1, $s_1 = 31$, $I_1 = \varnothing$					Machine 2, $s_2 = 26$, $I_2 = \{10, 13\}$					Machine 3, $s_3 = 19$, $I_3 = \{5, 8\}$				
1	2	5	3	5	2	5	7	2	7	6	15	15	0	15
12	1	32	31	32	11	6	32	26	32	9	8	27	19	27

Iteration 3. Assign feasible task 7 to machine 2 and task 5 to machine 1, shifting all assigned by previous iterations tasks on machine 1 to the left by the tardiness of task 5 (10 units). There are no feasible tasks on machine 3 (Table 3.77).

Iteration 4. Assign feasible task 3 to machine 1. There are no feasible tasks on machines 2 and 3 (Table 3.78).

Iteration 5. Assign feasible task 4 to machine 1. All tasks are assigned (Table 3.79).

Execute MSTAA for the tasks assigned by Algorithm A2.1: put task 4 after task 3 on machine 1. But the start times of machines are not improved (Table 3.80).

Execute Procedure A: remove the gap between tasks on machine 2. We get the optimal schedule (Table 3.81).

Algorithm C1.2a Initial data: $m = 3$, $n = 13$, $K_{\exp} = 6$ (Table 3.82).

Table 3.76 The second iteration

i	l_i^j	d_i	s_i	C_i	i	l_i^j	d_i	s_i	C_i	i	l_i^j	d_i	s_i	C_i
Machine 1, $s_1 = 25$, $I_1 = \varnothing$					Machine 2, $s_2 = 18$, $I_2 = \{7\}$					Machine 3, $s_3 = 15$, $I_3 = \varnothing$				
1	2	5	3	5	2	5	7	2	7	6	15	15	0	15
10	4	29	25	29	13	8	28	18	26	8	4	21	15	19
12	1	32	29	30	11	6	32	26	32	9	8	27	19	27

Table 3.77 Iteration 3

i	l_i^j	d_i	s_i	C_i	i	l_i^j	d_i	s_i	C_i	i	l_i^j	d_i	s_i	C_i
Machine 1, $s_1 = 10$, $I_1 = \{3, 4\}$					Machine 2, $s_2 = 8$, $I_2 = \{3\}$					Machine 3, $s_3 = 15$, $I_3 = \varnothing$				
1	2	5	3	5	2	5	7	2	7	6	15	15	0	15
5	5	15	10	15	7	10	20	8	18	8	4	21	15	19
10	4	29	15	19	13	8	28	18	26	9	8	27	19	27
12	1	32	19	20	11	6	32	26	32					

Table 3.78 Iteration 4

i	l_i^j	d_i	s_i	C_i	i	l_i^j	d_i	s_i	C_i	i	l_i^j	d_i	s_i	C_i
Machine 1, $s_1 = 9$, $I_1 = \{4\}$					Machine 2, $s_2 = 8$, $I_2 = \varnothing$					Machine 3, $s_3 = 15$, $I_3 = \varnothing$				
1	2	5	3	5	2	5	7	2	7	6	15	15	0	15
3	1	13	9	10	7	10	20	8	18	8	4	21	15	19
5	5	15	10	15	13	8	28	18	26	9	8	27	19	27
10	4	29	15	19	11	6	32	26	32					
12	1	32	19	20										

Table 3.79 Iteration 5

i	l_i^j	d_i	s_i	C_i	i	l_i^j	d_i	s_i	C_i	i	l_i^j	d_i	s_i	C_i
Machine 1, $s_1 = 5$					Machine 2, $s_2 = 8$					Machine 3, $s_3 = 15$				
1	2	5	3	5	2	5	7	2	7	6	15	15	0	15
4	4	14	5	9	7	10	20	8	18	8	4	21	15	19
3	1	13	9	10	13	8	28	18	26	9	8	27	19	27
5	5	15	10	15	11	6	32	26	32					
10	4	29	15	19										
12	1	32	19	20										

Table 3.80 The schedule after MSTAA executing

i	l_i^j	d_i	s_i	C_i	i	l_i^j	d_i	s_i	C_i	i	l_i^j	d_i	s_i	C_i
Machine 1, $s_1 = 5$					Machine 2, $s_2 = 8$					Machine 3, $s_3 = 15$				
1	2	5	3	5	2	5	7	2	7	6	15	15	0	15
3	1	13	5	6	7	10	20	8	18	8	4	21	15	19
4	4	14	6	10	13	8	28	18	26	9	8	27	19	27
5	5	15	10	15	11	6	32	26	32					
10	4	29	15	19										
12	1	32	19	20										

Table 3.81 Optimal schedule

i	l_i^j	d_i	s_i	C_i	i	l_i^j	d_i	s_i	C_i	i	l_i^j	d_i	s_i	C_i
Machine 1, $s_1 = 3$					Machine 2, $s_2 = 2$					Machine 3, $s_3 = 0$				
1	2	5	3	5	2	5	7	2	7	6	15	15	0	15
3	1	13	5	6	7	10	20	7	17	8	4	21	15	19
4	4	14	6	10	13	8	28	17	25	9	8	27	19	27
5	5	15	10	15	11	6	32	25	31					
10	4	29	15	19										
12	1	32	19	20										

Table 3.82 Sequence σ^{ord} for Algorithm C1.2a

i	l_i^1	l_i^2	l_i^3	d_i	i	l_i^1	l_i^2	l_i^3	d_i	i	l_i^1	l_i^2	l_i^3	d_i
1	2	3	4	5	6	6	9	15	15	10	4	8	16	29
2	4	5	6	7	7	1	10	18	20	11	3	6	12	32
3	1	6	8	13	8	2	4	4	21	12	1	2	4	32
4	4	7	12	14	9	3	6	8	27	13	4	8	16	28
5	5	7	9	15										

Initial schedule: assign the first three tasks, one to each machine, according to SSO #1 (Table 3.83).

Execute Algorithm A2.1a. The first iteration. We assign tasks 12, 11, and 9 from the set of unassigned tasks to the last position in the schedule of each machine. Determine the set of feasible tasks I_j for the next iteration by the condition: the task's due date is not less than the start time of the machine (Table 3.84).

Iteration 2. Assign task 13 to machine 2 and task 8 to machine 3. There are no feasible tasks for machine 1. Assign task 10 to it, shifting all assigned by Algorithm A2.1a tasks on machine 1 to the left by the task's tardiness (2 units). We get the schedule shown in Table 3.85.

Table 3.83 Initial schedule

i	l_i^j	d_i	s_i	C_i	i	l_i^j	d_i	s_i	C_i	i	l_i^j	d_i	s_i	C_i
Machine 1, $s_1 = 3$					Machine 2, $s_2 = 2$					Machine 3, $s_3 = 0$				
1	2	5	3	5	2	5	7	2	7	6	15	15	0	15

Table 3.84 The first iteration

i	l_i^j	d_i	s_i	C_i	i	l_i^j	d_i	s_i	C_i	i	l_i^j	d_i	s_i	C_i
Machine 1, $s_1 = 31$, $I_1 = \varnothing$					Machine 2, $s_2 = 26$, $I_2 = \{10, 13\}$					Machine 3, $s_3 = 19$, $I_3 = \{5, 8\}$				
1	2	5	3	5	2	5	7	2	7	6	15	15	0	15
12	1	32	31	32	11	6	32	26	32	9	8	27	19	27

Table 3.85 The second iteration

i	l_i^j	d_i	s_i	C_i	i	l_i^j	d_i	s_i	C_i	i	l_i^j	d_i	s_i	C_i
Machine 1, $s_1 = 25$, $I_1 = \varnothing$					Machine 2, $s_2 = 18$, $I_2 = \{7\}$					Machine 3, $s_3 = 15$, $I_3 = \varnothing$				
1	2	5	3	5	2	5	7	2	7	6	15	15	0	15
10	4	29	25	29	13	8	28	18	26	8	4	21	15	19
12	1	32	29	30	11	6	32	26	32	9	8	27	19	27

$s_{max} - s_{min} = 10 > 6$, execute MSTAA. Move task 8 from machine 3 to machine 1 with a shift of all tasks on machine 1 to the left by 4 units (Table 3.86).

Iteration 3. Assign feasible task 7 to machine 3 and task 5 to machine 1, shifting all assigned by previous iterations tasks on machine 1 to the left by the tardiness of task 5 (4 units). Assign task 4 to machine 2 with a shift by 4 units (Table 3.87).

$s_{max} - s_{min} = 9 > 6$, execute MSTAA: move task 7 from machine 3 to machine 1 (Table 3.88).

Iteration 4. Assign feasible task 3 to machine 1. All tasks are assigned. MSTAA does not improve the start times of machines (Table 3.89).

Execute Procedure A: remove gaps between tasks on machine 1 and 3. We get the optimal schedule (Table 3.90).

Algorithm C1.3 Initial data: $m = 3$, $n = 13$ (Table 3.91).

Initial schedule: assign one task to each machine, according to SSO #1 (Table 3.92).

Table 3.86 Iteration 2, after executing MSTAA

i	l_i^j	d_i	s_i	C_i	i	l_i^j	d_i	s_i	C_i	i	l_i^j	d_i	s_i	C_i
Machine 1, $s_1 = 19$, $I_1 = \{7\}$					Machine 2, $s_2 = 18$, $I_2 = \{7\}$					Machine 3, $s_3 = 19$, $I_3 = \{7\}$				
1	2	5	3	5	2	5	7	2	7	6	15	15	0	15
8	2	21	19	21	13	8	28	18	26	9	8	27	19	27
10	4	29	21	25	11	6	32	26	32					
12	1	32	25	26										

Table 3.87 Iteration 3

i	l_i^j	d_i	s_i	C_i	i	l_i^j	d_i	s_i	C_i	i	l_i^j	d_i	s_i	C_i
Machine 1, $s_1 = 10$, $I_1 = \{3\}$					Machine 2, $s_2 = 7$, $I_2 = \{3\}$					Machine 3, $s_3 = 1$, $I_3 = \varnothing$				
1	2	5	3	5	2	5	7	2	7	6	15	15	0	15
5	5	15	10	15	4	7	14	7	14	7	18	20	1	19
8	2	21	15	17	13	8	28	14	22	9	8	27	19	27
10	4	29	17	21	11	6	32	22	28					
12	1	32	21	22										

Table 3.88 The third iteration, after MSTAA executing

i	l_i^j	d_i	s_i	C_i	i	l_i^j	d_i	s_i	C_i	i	l_i^j	d_i	s_i	C_i
Machine 1, $s_1 = 9$, $I_1 = \{3\}$					Machine 2, $s_2 = 7$, $I_2 = \{3\}$					Machine 3, $s_3 = 19$, $I_3 = \varnothing$				
1	2	5	3	5	2	5	7	2	7	6	15	15	0	15
7	1	20	9	10	4	7	14	7	14	9	8	27	19	27
5	5	15	10	15	13	8	28	14	22					
8	2	21	15	17	11	6	32	22	28					
10	4	29	17	21										
12	1	32	21	22										

Table 3.89 Iteration 4

i	l_i^j	d_i	s_i	C_i	i	l_i^j	d_i	s_i	C_i	i	l_i^j	d_i	s_i	C_i
Machine 1, $s_1 = 8$					Machine 2, $s_2 = 7$					Machine 3, $s_3 = 19$				
1	2	5	3	5	2	5	7	2	7	6	15	15	0	15
3	1	13	8	9	4	7	14	7	14	9	8	27	19	27
7	1	20	9	10	13	8	28	14	22					
5	5	15	10	15	11	6	32	22	28					
8	2	21	15	17										
10	4	29	17	21										
12	1	32	21	22										

Table 3.90 Optimal schedule

i	l_i^j	d_i	s_i	C_i	i	l_i^j	d_i	s_i	C_i	i	l_i^j	d_i	s_i	C_i
Machine 1, $s_1 = 3$					Machine 2, $s_2 = 2$					Machine 3, $s_3 = 0$				
1	2	5	3	5	2	5	7	2	7	6	15	15	0	15
3	1	13	5	6	4	7	14	7	14	9	8	27	15	23
7	1	20	6	7	13	8	28	14	22					
5	5	15	7	12	11	6	32	22	28					
8	2	21	12	14										
10	4	29	14	18										
12	1	32	18	19										

Table 3.91 Sequence σ^{ord} for Algorithm C1.3

i	l_i^1	l_i^2	l_i^3	d_i	i	l_i^1	l_i^2	l_i^3	d_i	i	l_i^1	l_i^2	l_i^3	d_i
1	2	3	4	5	6	6	9	15	15	10	4	8	16	29
2	4	5	6	7	7	1	10	18	20	11	3	6	12	32
3	1	6	8	13	8	2	4	4	21	12	1	2	4	32
4	4	8	12	14	9	3	6	8	27	13	4	8	16	28
5	4	7	9	15										

Table 3.92 Initial schedule

i	l_i^j	d_i	s_i	C_i	i	l_i^j	d_i	s_i	C_i	i	l_i^j	d_i	s_i	C_i
Machine 1, $s_1 = 3$					Machine 2, $s_2 = 2$					Machine 3, $s_3 = 0$				
1	2	5	3	5	2	5	7	2	7	6	15	15	0	15

Execute Algorithm A2.2. Preliminary stage. We assign the first tasks from the set of unassigned tasks to the last position in the schedule of each machine (Table 3.93).

Optimization stage. Successively assign: task 10 to machine 1 (with a shift by its tardiness, 2 units); task 13 to machine 2 (without a shift); task 8 to machine 1 (with a shift by 4 units); task 7 to machine 1 (without a shift); task 5 to machine 3 (with a shift by 4 units); task 4 to machine 1 (with a shift by 4 units); task 3 to machine 2 (with a shift by 5 units). We get the schedule shown in Table 3.94.

Execute MSTAA for the tasks assigned by Algorithm A2.2: move task 5 from machine 3 to machine 1 (Table 3.95).

Execute Procedure A: remove gaps between the tasks assigned by Algorithm A2.2 and the initial schedule on machine 1. We get the optimal schedule (Table 3.96).

Algorithm C1.4 Initial data: $m = 3$, $n = 12$ (Table 3.97).

Initial schedule: assign one task to each machine, according to SSO #1 (Table 3.98).

Execute Algorithm C2.5. The first iteration. We assign tasks 11, 10, and 9 from the set of unassigned tasks to the last position in the schedule of each machine. Determine the set of feasible tasks I_j for the next iteration by the condition: the task's due date is not less than the start time of the machine (Table 3.99).

Table 3.93 Preliminary stage

i	l_i^j	d_i	s_i	C_i	i	l_i^j	d_i	s_i	C_i	i	l_i^j	d_i	s_i	C_i
Machine 1, $s_1 = 31$					Machine 2, $s_2 = 26$					Machine 3, $s_3 = 19$				
1	2	5	3	5	2	5	7	2	7	6	15	15	0	15
12	1	32	31	32	11	6	32	26	32	9	8	27	19	27

Table 3.94 Optimization stage

i	l_i^j	d_i	s_i	C_i	i	l_i^j	d_i	s_i	C_i	i	l_i^j	d_i	s_i	C_i
Machine 1, $s_1 = 10$					Machine 2, $s_2 = 7$					Machine 3, $s_3 = 6$				
1	2	5	3	5	2	5	7	2	7	6	15	15	0	15
4	4	14	10	14	3	6	13	7	13	5	9	15	6	15
7	1	20	14	15	13	8	28	13	21	9	8	27	15	23
8	2	21	15	17	11	6	32	21	27					
10	4	29	17	21										
12	1	32	21	22										

Table 3.95 The schedule after MSTAA executing

i	l_i^j	d_i	s_i	C_i	i	l_i^j	d_i	s_i	C_i	i	l_i^j	d_i	s_i	C_i
Machine 1, $s_1 = 6$					Machine 2, $s_2 = 7$					Machine 3, $s_3 = 15$				
1	2	5	3	5	2	5	7	2	7	6	15	15	0	15
5	4	15	6	10	3	6	13	7	13	9	8	27	15	23
4	4	14	10	14	13	8	28	13	21					
7	1	20	14	15	11	6	32	21	27					
8	2	21	15	17										
10	4	29	17	21										
12	1	32	21	22										

Table 3.96 Optimal schedule

i	l_i^j	d_i	s_i	C_i	i	l_i^j	d_i	s_i	C_i	i	l_i^j	d_i	s_i	C_i
Machine 1, $s_1 = 3$					Machine 2, $s_2 = 2$					Machine 3, $s_3 = 0$				
1	2	5	3	5	2	5	7	2	7	6	15	15	0	15
5	4	15	5	9	3	6	13	7	13	9	8	27	15	23
4	4	14	9	13	13	8	28	13	21					
7	1	20	13	14	11	6	32	21	27					
8	2	21	14	16										
10	4	29	16	20										
12	1	32	20	21										

Table 3.97 Sequence σ^{ord} for Algorithm C1.4

i	l_i^1	l_i^2	l_i^3	d_i	i	l_i^1	l_i^2	l_i^3	d_i	i	l_i^1	l_i^2	l_i^3	d_i
1	2	3	4	5	5	5	7	9	15	9	3	6	8	27
2	4	5	6	7	6	6	9	15	15	10	3	6	12	32
3	1	6	8	13	7	1	10	18	20	11	1	2	4	32
4	4	7	12	14	8	2	4	4	21	12	4	8	16	28

Table 3.98 Initial schedule

i	l_i^j	d_i	s_i	C_i	i	l_i^j	d_i	s_i	C_i	i	l_i^j	d_i	s_i	C_i
Machine 1, $s_1 = 3$					Machine 2, $s_2 = 2$					Machine 3, $s_3 = 0$				
1	2	5	3	5	2	5	7	2	7	6	15	15	0	15

Table 3.99 The first iteration

i	l_i^j	d_i	s_i	C_i	i	l_i^j	d_i	s_i	C_i	i	l_i^j	d_i	s_i	C_i
Machine 1, $s_1 = 31$, $I_1 = \varnothing$					Machine 2, $s_2 = 26$, $I_2 = \{12\}$					Machine 3, $s_3 = 19$, $I_3 = \{7, 8, 12\}$				
1	2	5	3	5	2	5	7	2	7	6	15	15	0	15
11	1	32	31	32	10	6	32	26	32	9	8	27	19	27

Iteration 2. Assign task 12 to machine 2 and task 8 to machine 3 (we assign tasks only to machines of the first group and only one task to each machine). There are no feasible tasks for machine 1. Assign task 7 to it, shifting all assigned by Algorithm C2.5 tasks on machine 1 to the left by the task's tardiness (11 units). We get the schedule shown in Table 3.100.

Iteration 3. Assign feasible task 5 to machine 3; task 3 to machine 1, shifting all assigned by previous iterations tasks on machine 1 to the left by the tardiness of task 3 (6 units); task 4 to machine 2 with a shift by 4 units. All tasks are assigned (Table 3.101).

Execute MSTAA for the tasks assigned by Algorithm C2.5: move task 5 from machine 3 to machine 1 and put it to machine 1 after task 1 (Table 3.102).

Execute Procedure A: remove the gap between tasks on machine 1. We get the optimal schedule (Table 3.103).

Algorithm C1.4a Initial data: $m = 3$, $n = 12$, $K_{\text{exp}} = 3$ (Table 3.104).

Initial schedule: assign one task to each machine, according to SSO #1 (Table 3.105).

Table 3.100 The second iteration

i	l_i^j	d_i	s_i	C_i	i	l_i^j	d_i	s_i	C_i	i	l_i^j	d_i	s_i	C_i
Machine 1, $s_1 = 19$, $I_1 = \varnothing$					Machine 2, $s_2 = 18$, $I_2 = \varnothing$					Machine 3, $s_3 = 15$, $I_3 = \{5\}$				
1	2	5	3	5	2	5	7	2	7	6	15	15	0	15
7	1	20	19	20	12	8	28	18	26	8	4	21	15	19
11	1	32	20	21	10	6	32	26	32	9	8	27	19	27

Table 3.101 Iteration 3

i	l_i^j	d_i	s_i	C_i	i	l_i^j	d_i	s_i	C_i	i	l_i^j	d_i	s_i	C_i
Machine 1, $s_1 = 12$					Machine 2, $s_2 = 7$					Machine 3, $s_3 = 6$				
1	2	5	3	5	2	5	7	2	7	6	15	15	0	15
3	1	13	12	13	4	7	14	7	14	5	9	15	6	15
7	1	20	13	14	12	8	28	14	22	8	4	21	15	19
11	1	32	14	15	10	6	32	22	28	9	8	27	19	27

Table 3.102 The schedule after MSTAA executing

i	l_i^j	d_i	s_i	C_i	i	l_i^j	d_i	s_i	C_i	i	l_i^j	d_i	s_i	C_i
Machine 1, $s_1 = 7$					Machine 2, $s_2 = 7$					Machine 3, $s_3 = 15$				
1	2	5	3	5	2	5	7	2	7	6	15	15	0	15
3	1	13	7	8	4	7	14	7	14	8	4	21	15	19
7	1	20	8	9	12	8	28	14	22	9	8	27	19	27
11	1	32	9	10	10	6	32	22	28					
5	5	15	10	15										

Table 3.103 Optimal schedule

i	l_i^j	d_i	s_i	C_i	i	l_i^j	d_i	s_i	C_i	i	l_i^j	d_i	s_i	C_i
Machine 1, $s_1 = 3$					Machine 2, $s_2 = 2$					Machine 3, $s_3 = 0$				
1	2	5	3	5	2	5	7	2	7	6	15	15	0	15
3	1	13	5	6	4	7	14	7	14	8	4	21	15	19
7	1	20	6	7	12	8	28	14	22	9	8	27	19	27
11	1	32	7	8	10	6	32	22	28					
5	5	15	8	13										

Table 3.104 Sequence σ^{ord} for Algorithm C1.4a

i	l_i^1	l_i^2	l_i^3	d_i	i	l_i^1	l_i^2	l_i^3	d_i	i	l_i^1	l_i^2	l_i^3	d_i
1	2	3	4	5	5	5	7	9	15	9	3	6	8	27
2	4	5	6	7	6	6	9	15	15	10	3	6	12	32
3	1	6	8	13	7	1	10	18	20	11	1	2	4	32
4	3	7	12	14	8	2	4	4	21	12	4	8	16	28

Table 3.105 Initial schedule

i	l_i^j	d_i	s_i	C_i	i	l_i^j	d_i	s_i	C_i	i	l_i^j	d_i	s_i	C_i
Machine 1, $s_1 = 3$					Machine 2, $s_2 = 2$					Machine 3, $s_3 = 0$				
1	2	5	3	5	2	5	7	2	7	6	15	15	0	15

Execute Algorithm C2.5a. The first iteration. We assign tasks 11, 10, 9 from the set of unassigned tasks to the last position in the schedule of each machine. Determine the set of feasible tasks I_j for the next iteration by the condition: the task's due date is not less than the start time of the machine (Table 3.106).

Iteration 2. Assign task 12 to machine 2 and task 8 to machine 3 (we assign tasks only to machines of the first group and only one task to each machine). There are no feasible tasks for machine 1. Assign task 7 to it, shifting all assigned by Algorithm C2.5a tasks on machine 1 to the left by the task's tardiness (11 units). We get the schedule shown in Table 3.107.

$s_{\max} - s_{\min} = 4 > 3$, execute MSTAA: move task 8 to machine 1 and put it after task 11 (Table 3.108).

Table 3.106 The first iteration

i	l_i^j	d_i	s_i	C_i	i	l_i^j	d_i	s_i	C_i	i	l_i^j	d_i	s_i	C_i
Machine 1, $s_1 = 31$, $I_1 = \varnothing$					Machine 2, $s_2 = 26$, $I_2 = \{12\}$					Machine 3, $s_3 = 19$, $I_3 = \{7, 8, 12\}$				
1	2	5	3	5	2	5	7	2	7	6	15	15	0	15
11	1	32	31	32	10	6	32	26	32	9	8	27	19	27

Table 3.107 The second iteration

i	l_i^j	d_i	s_i	C_i	i	l_i^j	d_i	s_i	C_i	i	l_i^j	d_i	s_i	C_i
Machine 1, $s_1 = 19$, $I_1 = \varnothing$					Machine 2, $s_2 = 18$, $I_2 = \varnothing$					Machine 3, $s_3 = 15$, $I_3 = \varnothing$				
1	2	5	3	5	2	5	7	2	7	6	15	15	0	15
7	1	20	19	20	12	8	28	18	26	8	4	21	15	19
11	1	32	20	21	10	6	32	26	32	9	8	27	19	27

Table 3.108 Iteration 2, after MSTAA executing

i	l_i^j	d_i	s_i	C_i	i	l_i^j	d_i	s_i	C_i	i	l_i^j	d_i	s_i	C_i
Machine 1, $s_1 = 17$, $I_1 = \varnothing$					Machine 2, $s_2 = 18$, $I_2 = \varnothing$					Machine 3, $s_3 = 19$, $I_3 = \varnothing$				
1	2	5	3	5	2	5	7	2	7	6	15	15	0	15
7	1	20	17	18	12	8	28	18	26	9	8	27	19	27
11	1	32	18	19	10	6	32	26	32					
8	2	21	19	21										

Iteration 3. There are no feasible tasks. Assign task 5 to machine 3, shifting all assigned by previous iterations tasks on machine 3 to the left by the tardiness of task 5 (6 units). Assign task 3 to machine 2 with a shift by 5 units and task 4 to machine 1 with a shift by 3 units. All tasks are assigned (Table 3.109).

Execute MSTAA for the tasks assigned by Algorithm C2.5a: move task 5 from machine 3 to machine 1 and put it to machine 1 after task 7 (Table 3.110).

Execute Procedure A: remove the gap between tasks on machine 1. We get the optimal schedule (Table 3.111).

Algorithm C2.1 Initial data: $m = 3$, $n = 13$. Sequence σ^{ord} is in Table 3.112.

Initial schedule: assign one task to each machine, according to SSO #1 (Table 3.113).

Successively assign tasks 3–5, 7–11, each to the machine where it will have the maximum earliness regarding its due date. If there are several such machines, then assign the task to the least loaded machine (Table 3.114).

Task 12 is tardy on all machines. Assign it to the machine 1 that corresponds to the best vector of start times (0, 2, 2) according to the lexicographical order.

Table 3.109 Iteration 3

i	l_i^j	d_i	s_i	C_i	i	l_i^j	d_i	s_i	C_i	i	l_i^j	d_i	s_i	C_i
Machine 1, $s_1 = 11$					Machine 2, $s_2 = 7$					Machine 3, $s_3 = 6$				
1	2	5	3	5	2	5	7	2	7	6	15	15	0	15
4	3	14	11	14	3	6	13	7	13	5	9	15	6	15
7	1	20	14	15	12	8	28	13	21	9	8	27	15	23
11	1	32	15	16	10	6	32	21	27					
8	2	21	16	18										

Table 3.110 The schedule after MSTAA executing

i	l_i^j	d_i	s_i	C_i	i	l_i^j	d_i	s_i	C_i	i	l_i^j	d_i	s_i	C_i
Machine 1, $s_1 = 6$					Machine 2, $s_2 = 7$					Machine 3, $s_3 = 15$				
1	2	5	3	5	2	5	7	2	7	6	15	15	0	15
4	3	14	6	9	3	6	13	7	13	9	8	27	15	23
7	1	20	9	10	12	8	28	13	21					
5	5	15	10	15	10	6	32	21	27					
11	1	32	15	16										
8	2	21	16	18										

Table 3.111 Optimal schedule

i	l_i^j	d_i	s_i	C_i	i	l_i^j	d_i	s_i	C_i	i	l_i^j	d_i	s_i	C_i
Machine 1, $s_1 = 3$					Machine 2, $s_2 = 2$					Machine 3, $s_3 = 0$				
1	2	5	3	5	2	5	7	2	7	6	15	15	0	15
4	3	14	5	8	3	6	13	7	13	9	8	27	15	23
7	1	20	8	9	12	8	28	13	21					
5	5	15	9	14	10	6	32	21	27					
11	1	32	14	15										
8	2	21	15	17										

Table 3.112 Sequence σ^{ord} for Algorithm C2.1

i	l_i^1	l_i^2	l_i^3	d_i	i	l_i^1	l_i^2	l_i^3	d_i	i	l_i^1	l_i^2	l_i^3	d_i
1	2	3	4	5	6	6	9	15	15	10	4	8	16	29
2	4	5	6	7	7	1	12	18	20	11	3	6	12	32
3	3	6	8	13	8	2	4	8	21	12	12	12	20	32
4	4	8	12	14	9	3	6	12	27	13	4	12	20	33
5	5	7	9	15										

Table 3.113 Initial schedule

i	l_i^j	d_i	s_i	C_i	i	l_i^j	d_i	s_i	C_i	i	l_i^j	d_i	s_i	C_i
Machine 1, $s_1 = 3$					Machine 2, $s_2 = 2$					Machine 3, $s_3 = 0$				
1	2	5	3	5	2	5	7	2	7	6	15	15	0	15

Similarly, task 13 is tardy on all machines, we assign it to machine 2. The final schedule is in Table 3.115.

Algorithm C2.3 Initial data: $m = 3$, $n = 12$ (Table 3.116).

The first iteration. Initial schedule: assign the first tasks to the last position in the schedule of each machine. Determine the set of feasible tasks I_j for the next iteration

Table 3.114 The schedule after 11 tasks assignment

i	l_i^j	d_i	s_i	C_i	i	l_i^j	d_i	s_i	C_i	i	l_i^j	d_i	s_i	C_i
Machine 1, $s_1 = 3$					Machine 2, $s_2 = 2$					Machine 3, $s_3 = 0$				
1	2	5	3	5	2	5	7	2	7	6	15	15	0	15
3	3	13	5	8	5	7	15	7	14					
4	4	14	8	12	10	8	29	14	22					
7	1	20	12	13										
8	2	21	13	15										
9	3	27	15	18										
11	3	32	18	21										

Table 3.115 Final schedule

i	l_i^j	d_i	s_i	C_i	i	l_i^j	d_i	s_i	C_i	i	l_i^j	d_i	s_i	C_i
Machine 1, $s_1 = 2$					Machine 2, $s_2 = 1$					Machine 3, $s_3 = 0$				
1	2	5	2	4	2	5	7	1	6	6	15	15	0	15
3	3	13	4	7	5	7	15	6	13					
4	4	14	7	11	10	8	29	13	21					
7	1	20	11	12	13	12	33	21	33					
8	2	21	12	14										
9	3	27	14	17										
11	3	32	17	20										
12	12	32	20	32										

Table 3.116 Sequence σ^{ord} for Algorithm C2.3

i	l_i^1	l_i^2	l_i^3	d_i	i	l_i^1	l_i^2	l_i^3	d_i	i	l_i^1	l_i^2	l_i^3	d_i
1	1	2	4	2	5	5	10	20	15	9	4	8	16	29
2	2	4	8	6	6	7	14	28	20	10	3	6	12	32
3	3	6	12	15	7	2	4	8	21	11	1	2	4	32
4	4	8	16	16	8	3	6	12	27	12	4	8	16	28

by the condition: the task's due date is not less than the start time of the machine (Table 3.117).

Iteration 2. Assign task 12 to machine 2 and task 7 to machine 3. There are no feasible tasks for machine 1. Assign to it the task 9 that has the maximum start time accounting a shift by the task's tardiness (2 units). Shift all assigned tasks on machine 1 to the left by 2 units. We get the schedule shown in Table 3.118.

Iteration 3. Assign task 6 to machine 2. There are no feasible tasks on machines 1 and 3. Assign task 4 to machine 1 with a shift by its tardiness (9 units). We cannot assign any task to machine 3 because processing times of the remaining unassigned tasks on machine 3 exceed the start time of the machine (Table 3.119).

Table 3.117 Initial schedule

i	l_i^j	d_i	s_i	C_i	i	l_i^j	d_i	s_i	C_i	i	l_i^j	d_i	s_i	C_i
Machine 1, $s_1 = 31$, $I_1 = \varnothing$					Machine 2, $s_2 = 26$, $I_2 = \{9, 12\}$					Machine 3, $s_3 = 15$, $I_3 = \{3, 7\}$				
11	1	32	31	32	10	6	32	26	32	8	12	27	15	27

Table 3.118 The second iteration

i	l_i^j	d_i	s_i	C_i	i	l_i^j	d_i	s_i	C_i	i	l_i^j	d_i	s_i	C_i
Machine 1, $s_1 = 25$, $I_1 = \varnothing$					Machine 2, $s_2 = 18$, $I_2 = \{6\}$					Machine 3, $s_3 = 7$, $I_3 = \varnothing$				
9	4	29	25	29	12	8	28	18	26	7	8	21	7	15
11	1	32	29	30	10	6	32	26	32	8	12	27	15	27

Table 3.119 Iteration 3

i	l_i^j	d_i	s_i	C_i	i	l_i^j	d_i	s_i	C_i	i	l_i^j	d_i	s_i	C_i
Machine 1, $s_1 = 12$, $I_1 = \{3, 5\}$					Machine 2, $s_2 = 4$, $I_2 = \{2\}$					Machine 3, $s_3 = 7$, $I_3 = \varnothing$				
4	4	16	12	16	6	14	20	4	18	7	8	21	7	15
9	4	29	16	20	12	8	28	18	26	8	12	27	15	27
11	1	32	20	21	10	6	32	26	32					

Iteration 4. Assign task 3 to machine 1 and task 2 to machine 2. There are no feasible tasks on machine 3 (Table 3.120).

Iteration 5. Assign task 5 to machine 1. There are no feasible tasks on machines 2 and 3 (Table 3.121).

Iteration 6. There are no feasible tasks. Assign the last unassigned task 1 to machine 1, shifting all tasks on machine 1 to the left by its tardiness (2 units). MSTAA does not improve the final schedule which is shown in Table 3.122.

Algorithm C2.3a Initial data: $m = 3$, $n = 12$, $K_{\text{exp}} = 6$ (Table 3.123).

The first iteration. Initial schedule: assign the first tasks to the last position in the schedule of each machine. Determine the set of feasible tasks I_j for the next iteration by the condition: the task's due date is not less than the start time of the machine (Table 3.124).

Table 3.120 Iteration 4

i	l_i^j	d_i	s_i	C_i	i	l_i^j	d_i	s_i	C_i	i	l_i^j	d_i	s_i	C_i
Machine 1, $s_1 = 9$, $I_1 = \{5\}$					Machine 2, $s_2 = 0$, $I_2 = \varnothing$					Machine 3, $s_3 = 7$, $I_3 = \varnothing$				
3	3	15	9	12	2	4	6	0	4	7	8	21	7	15
4	4	16	12	16	6	14	20	4	18	8	12	27	15	27
9	4	29	16	20	12	8	28	18	26					
11	1	32	20	21	10	6	32	26	32					

Table 3.121 Iteration 5

i	l_i^j	d_i	s_i	C_i	i	l_i^j	d_i	s_i	C_i	i	l_i^j	d_i	s_i	C_i
Machine 1, $s_1 = 4$					Machine 2, $s_2 = 0$					Machine 3, $s_3 = 7$				
5	5	15	4	9	2	4	6	0	4	7	8	21	7	15
3	3	15	9	12	6	14	20	4	18	8	12	27	15	27
4	4	16	12	16	12	8	28	18	26					
9	4	29	16	20	10	6	32	26	32					
11	1	32	20	21										

Table 3.122 Final schedule

i	l_i^j	d_i	s_i	C_i	i	l_i^j	d_i	s_i	C_i	i	l_i^j	d_i	s_i	C_i
Machine 1, $s_1 = 1$					Machine 2, $s_2 = 0$					Machine 3, $s_3 = 7$				
1	1	2	1	2	2	4	6	0	4	7	8	21	7	15
5	5	15	2	7	6	14	20	4	18	8	12	27	15	27
3	3	15	7	10	12	8	28	18	26					
4	4	16	10	14	10	6	32	26	32					
9	4	29	14	18										
11	1	32	18	19										

Table 3.123 Sequence σ^{ord} for Algorithm C2.3a

i	l_i^1	l_i^2	l_i^3	d_i	i	l_i^1	l_i^2	l_i^3	d_i	i	l_i^1	l_i^2	l_i^3	d_i
1	1	2	4	2	5	5	10	20	15	9	4	8	16	29
2	2	4	8	6	6	7	14	28	20	10	3	6	12	32
3	3	6	12	15	7	2	4	8	21	11	1	2	4	32
4	4	8	16	16	8	3	6	12	27	12	4	8	16	28

Table 3.124 Initial schedule

i	l_i^j	d_i	s_i	C_i	i	l_i^j	d_i	s_i	C_i	i	l_i^j	d_i	s_i	C_i
Machine 1, $s_1 = 31$, $I_1 = \varnothing$					Machine 2, $s_2 = 26$, $I_2 = \{9, 12\}$					Machine 3, $s_3 = 15$, $I_3 = \{3, 7\}$				
11	1	32	31	32	10	6	32	26	32	8	12	27	15	27

Iteration 2. Assign task 12 to machine 2 and task 7 to machine 3. There are no feasible tasks for machine 1. Assign to it the task 9 that has the maximum start time accounting a shift by the task's tardiness (2 units). Shift all assigned tasks on machine 1 to the left by 2 units. We get the schedule shown in Table 3.125.

$s_{max} - s_{min} = 18 > 6$, execute MSTAA: move task 7 from machine 3 to machine 1, shifting all tasks assigned to machine 1 to the left by 4 units (Table 3.126).

Table 3.125 The second iteration

i	l_i^j	d_i	s_i	C_i	i	l_i^j	d_i	s_i	C_i	i	l_i^j	d_i	s_i	C_i
Machine 1, $s_1 = 25, I_1 = \varnothing$					Machine 2, $s_2 = 18, I_2 = \{6\}$					Machine 3, $s_3 = 7, I_3 = \varnothing$				
9	4	29	25	29	12	8	28	18	26	7	8	21	7	15
11	1	32	29	30	10	6	32	26	32	8	12	27	15	27

Table 3.126 Iteration 2, after MSTAA executing

i	l_i^j	d_i	s_i	C_i	i	l_i^j	d_i	s_i	C_i	i	l_i^j	d_i	s_i	C_i
Machine 1, $s_1 = 19, I_1 = \{6\}$					Machine 2, $s_2 = 18, I_2 = \{6\}$					Machine 3, $s_3 = 15$, $I_3 = \{3, 6\}$				
7	2	21	19	21	12	8	28	18	26	8	12	27	15	27
9	4	29	21	25	10	6	32	26	32					
11	1	32	25	26										

Iteration 3. Assign task 6 to machine 2 and task 3 to machine 3. There are no feasible tasks on machine 2. Assign task 4 to machine 2 with a shift by its tardiness (2 units). We get the schedule shown in Table 3.127.

$s_{\max} - s_{\min} = 9 > 6$, execute MSTAA: move task 3 from machine 3 to machine 1 (Table 3.128).

Iteration 4. Assign task 5 to machine 1. There are no feasible tasks on machines 2 and 3. Assign task 2 to machine 2 with a shift by its tardiness (2 units). We cannot assign any task to machine 3 because after such assignment its start time will be negative (Table 3.129).

$s_{\max} - s_{\min} = 13 > 6$, execute MSTAA: put task 5 after task 3 on machine 1. The start times of machines are not changed (Table 3.130).

Iteration 5. Assign the last unassigned task 1 to machine 1 with a shift to the left by 2 units. Execute MSTAA, it does not change anything. We get the final schedule (Table 3.131).

Algorithm C2.4 Initial data: $m = 3$, $n = 13$. Sequence σ^{ord} is in Table 3.132.

Table 3.127 Iteration 3

i	l_i^j	d_i	s_i	C_i	i	l_i^j	d_i	s_i	C_i	i	l_i^j	d_i	s_i	C_i
Machine 1, $s_1 = 12, I_1 = \{5\}$					Machine 2, $s_2 = 8, I_2 = \varnothing$					Machine 3, $s_3 = 3, I_3 = \varnothing$				
6	7	20	12	19	4	8	16	8	16	3	12	15	3	15
7	2	21	19	21	12	8	28	16	24	8	12	27	15	27
9	4	29	21	25	10	6	32	24	30					
11	1	32	25	26										

Table 3.128 Iteration 3, after MSTAA executing

i	l_i^j	d_i	s_i	C_i	i	l_i^j	d_i	s_i	C_i	i	l_i^j	d_i	s_i	C_i
Machine 1, $s_1 = 9$, $I_1 = \{5\}$					Machine 2, $s_2 = 8$, $I_2 = \varnothing$					Machine 3, $s_3 = 15$, $I_3 = \varnothing$				
3	3	15	9	12	4	8	16	8	16	8	12	27	15	27
6	7	20	12	19	12	8	28	16	24					
7	2	21	19	21	10	6	32	24	30					
9	4	29	21	25										
11	1	32	25	26										

Table 3.129 Iteration 4

i	l_i^j	d_i	s_i	C_i	i	l_i^j	d_i	s_i	C_i	i	l_i^j	d_i	s_i	C_i
Machine 1, $s_1 = 4$, $I_1 = \varnothing$					Machine 2, $s_2 = 2$, $I_2 = \varnothing$					Machine 3, $s_3 = 15$, $I_3 = \varnothing$				
5	5	15	4	9	2	4	6	2	6	8	12	27	15	27
3	3	15	9	12	4	8	16	6	14					
6	7	20	12	19	12	8	28	14	22					
7	2	21	19	21	10	6	32	22	28					
9	4	29	21	25										
11	1	32	25	26										

Table 3.130 Iteration 4, after MSTAA executing

i	l_i^j	d_i	s_i	C_i	i	l_i^j	d_i	s_i	C_i	i	l_i^j	d_i	s_i	C_i
Machine 1, $s_1 = 4$, $I_1 = \varnothing$					Machine 2, $s_2 = 2$, $I_2 = \varnothing$					Machine 3, $s_3 = 15$, $I_3 = \varnothing$				
3	3	15	4	7	2	4	6	2	6	8	12	27	15	27
5	5	15	7	12	4	8	16	6	14					
6	7	20	12	19	12	8	28	14	22					
7	2	21	19	21	10	6	32	22	28					
9	4	29	21	25										
11	1	32	25	26										

Table 3.131 Final schedule

i	l_i^j	d_i	s_i	C_i	i	l_i^j	d_i	s_i	C_i	i	l_i^j	d_i	s_i	C_i
Machine 1, $s_1 = 1$					Machine 2, $s_2 = 2$					Machine 3, $s_3 = 15$				
1	1	2	1	2	2	4	6	2	6	8	12	27	15	27
3	3	15	2	5	4	8	16	6	14					
5	5	15	5	10	12	8	28	14	22					
6	7	20	10	17	10	6	32	22	28					
7	2	21	17	19										
9	4	29	19	23										
11	1	32	23	24										

Table 3.132 Sequence σ^{ord} for Algorithm C2.4

i	l_i^1	l_i^2	l_i^3	d_i	i	l_i^1	l_i^2	l_i^3	d_i	i	l_i^1	l_i^2	l_i^3	d_i
1	1	2	4	2	6	6	12	24	15	10	4	8	16	29
2	2	4	8	6	7	7	14	28	20	11	3	6	12	32
3	3	6	12	15	8	2	4	8	21	12	1	2	4	32
4	4	8	16	16	9	3	6	12	27	13	4	8	16	28
5	5	10	20	15										

Preliminary stage. Initial schedule: we assign the first tasks to the last position in the schedule of each machine (Table 3.133).

Optimization stage. Successively assign feasible tasks: 13 and 8 to machine 2; 3 to machine 3; 10, 2, and 1 to machine 2. We get the schedule shown in Table 3.134.

There are no feasible tasks. We sequentially assign to machine 1, with a shift by their tardiness, task 8 (tardiness 10) and task 4 (tardiness 3). Tasks 5 and 6 become feasible on machine 1. Assign at first task 5, then task 6. All tasks are assigned. The adjustment algorithm does not change the start times of machines. The final schedule is shown in Table 3.135.

Algorithm C2.5 Initial data: $m = 3$, $n = 12$ (Table 3.136).

The first iteration. Initial schedule: assign the first tasks to the last position in the schedule of each machine. Determine the set of feasible tasks I_j for the next iteration by the conditions: the task's due date is not less than the start time of the machine; the task's processing time on the machine is not greater than its start time (Table 3.137).

Iteration 2. Tasks 9 and 12 are the two feasible tasks on machine 2. Both are feasible only on machine 2 and have the same processing time. Assign task 12 to machine 2. Task 3 is feasible on machine 3, assign it to machine 3. There are no

Table 3.133 Initial schedule

i	l_i^j	d_i	s_i	C_i	i	l_i^j	d_i	s_i	C_i	i	l_i^j	d_i	s_i	C_i
Machine 1, $s_1 = 31$					Machine 2, $s_2 = 26$					Machine 3, $s_3 = 15$				
12	1	32	31	32	11	6	32	26	32	9	12	27	15	27

Table 3.134 Optimization stage

i	l_i^j	d_i	s_i	C_i	i	l_i^j	d_i	s_i	C_i	i	l_i^j	d_i	s_i	C_i
Machine 1, $s_1 = 31$					Machine 2, $s_2 = 0$					Machine 3, $s_3 = 3$				
12	1	32	31	32	1	2	2	0	2	3	12	15	3	15
					2	4	6	2	6	9	12	27	15	27
					10	8	29	6	14					
					8	4	21	14	18					
					13	8	28	18	26					
					11	6	32	26	32					

Table 3.135 Final schedule

i	l_i^j	d_i	s_i	C_i	i	l_i^j	d_i	s_i	C_i	i	l_i^j	d_i	s_i	C_i
Machine 1, $s_1 = 1$					Machine 2, $s_2 = 0$					Machine 3, $s_3 = 3$				
6	6	15	1	7	1	2	2	0	2	3	12	15	3	15
5	5	15	7	12	2	4	6	2	6	9	12	27	15	27
4	4	16	12	16	10	8	29	6	14					
8	2	21	16	18	8	4	21	14	18					
12	1	32	18	19	13	8	28	18	26					
					11	6	32	26	32					

Table 3.136 Sequence σ^{ord} for Algorithm C2.5

i	l_i^1	l_i^2	l_i^3	d_i	i	l_i^1	l_i^2	l_i^3	d_i	i	l_i^1	l_i^2	l_i^3	d_i
1	1	2	4	2	5	5	10	20	15	9	4	8	16	29
2	2	4	8	6	6	6	12	24	15	10	3	6	12	32
3	3	6	12	15	7	7	14	28	20	11	1	2	4	32
4	4	8	16	16	8	3	6	12	27	12	4	8	16	28

Table 3.137 Initial schedule

i	l_i^j	d_i	s_i	C_i	i	l_i^j	d_i	s_i	C_i	i	l_i^j	d_i	s_i	C_i
Machine 1, $s_1 = 31$, $I_1 = \varnothing$					Machine 2, $s_2 = 26$, $I_2 = \{9, 12\}$					Machine 3, $s_3 = 15$, $I_3 = \{3\}$				
11	1	32	31	32	10	6	32	26	32	8	12	27	15	27

Table 3.138 Iteration 2

i	l_i^j	d_i	s_i	C_i	i	l_i^j	d_i	s_i	C_i	i	l_i^j	d_i	s_i	C_i
Machine 1, $s_1 = 25$, $I_1 = \varnothing$					Machine 2, $s_2 = 18$, $I_2 = \{7\}$					Machine 3, $s_3 = 3$, $I_3 = \varnothing$				
9	4	29	25	29	12	8	28	18	26	3	12	15	3	15
11	1	32	29	30	10	6	32	26	32	8	12	27	15	27

feasible tasks on machine 1. Assign to it the task 9 that has the maximum start time accounting the shift by its tardiness (2 units). Shift all tasks on machine 1 to the left by 2 units. We get the schedule shown in Table 3.138.

Iteration 3. Assign task 7 to machine 2. There are no feasible tasks on machines 1 and 3. Assign task 4 to machine 1 with a shift by its tardiness (9 units). We cannot assign any task to machine 3 because processing times of the remaining unassigned tasks on machine 3 exceed the start time of the machine (Table 3.139).

Iteration 4. Assign task 6 to machine 1 and task 2 to machine 2. There are no feasible tasks on machine 3 (Table 3.140).

Table 3.139 Iteration 3

i	l_i^j	d_i	s_i	C_i	i	l_i^j	d_i	s_i	C_i	i	l_i^j	d_i	s_i	C_i
Machine 1, $s_1 = 12$, $I_1 = \{5, 6\}$					Machine 2, $s_2 = 4$, $I_2 = \{2\}$					Machine 3, $s_3 = 3$, $I_3 = \varnothing$				
4	4	16	12	16	7	14	20	4	18	3	12	15	3	15
9	4	29	16	20	12	8	28	18	26	8	12	27	15	27
11	1	32	20	21	10	6	32	26	32					

Table 3.140 Iteration 4

i	l_i^j	d_i	s_i	C_i	i	l_i^j	d_i	s_i	C_i	i	l_i^j	d_i	s_i	C_i
Machine 1, $s_1 = 6$, $I_1 = \{5\}$					Machine 2, $s_2 = 0$, $I_2 = \varnothing$					Machine 3, $s_3 = 3$, $I_3 = \varnothing$				
6	6	15	6	12	2	4	6	0	4	3	12	15	3	15
4	4	16	12	16	7	14	20	4	18	8	12	27	15	27
9	4	29	16	20	12	8	28	18	26					
11	1	32	20	21	10	6	32	26	32					

Table 3.141 Iteration 5

i	l_i^j	d_i	s_i	C_i	i	l_i^j	d_i	s_i	C_i	i	l_i^j	d_i	s_i	C_i
Machine 1, $s_1 = 1$, $I_1 = \{1\}$					Machine 2, $s_2 = 0$, $I_2 = \varnothing$					Machine 3, $s_3 = 3$, $I_3 = \varnothing$				
5	5	15	1	6	2	4	6	0	4	3	12	15	3	15
6	6	15	6	12	7	14	20	4	18	8	12	27	15	27
4	4	16	12	16	12	8	28	18	26					
9	4	29	16	20	10	6	32	26	32					
11	1	32	20	21										

Iteration 5. Assign task 5 to machine 1. There are no feasible tasks on machines 2 and 3 (Table 3.141).

Iteration 6. Assign the last unassigned task 1 to machine 1. Execute MSTAA, it does not change anything. We get the final schedule (Table 3.142).

Table 3.142 Final schedule

i	l_i^j	d_i	s_i	C_i	i	l_i^j	d_i	s_i	C_i	i	l_i^j	d_i	s_i	C_i
Machine 1, $s_1 = 0$					Machine 2, $s_2 = 0$					Machine 3, $s_3 = 3$				
1	1	2	0	1	2	4	6	0	4	3	12	15	3	15
5	5	15	1	6	7	14	20	4	18	8	12	27	15	27
6	6	15	6	12	12	8	28	18	26					
4	4	16	12	16	10	6	32	26	32					
9	4	29	16	20										
11	1	32	20	21										

Algorithm C2.5a Initial data: $m = 3$, $n = 12$, $K_{exp} = 6$ (Table 3.143).

The first iteration. Initial schedule: assign the first tasks to the last position in the schedule of each machine. Determine the set of feasible tasks I_j for the next iteration by the conditions: the task's due date is not less than the start time of the machine; the task's processing time on the machine is not greater than its start time (Table 3.144).

Iteration 2. Tasks 9 and 12 are the two feasible tasks on machine 2. Both are feasible only on machine 2 and have the same processing time. Assign task 12 to machine 2. Both tasks 3 and 7 are feasible only on machine 3. The shortest processing time of task 3 is greater than that of task 7, so we assign task 3 to machine 3. There are no feasible tasks on machine 1. Assign to it the task 9 that has the maximum start time accounting the shift by its tardiness (2 units). Shift tasks on machine 1 to the left by 2 units. We get the schedule shown in Table 3.145.

$s_{max} - s_{min} = 22 > 6$, execute MSTAA: move task 3 from machine 3 to machine 1, shifting all tasks assigned to machine 1 to the left by 10 units (Table 3.146).

Iteration 3. Assign task 6 to machine 1 and task 7 to machine 2. There are no feasible tasks on machine 3 (Table 3.147).

$s_{max} - s_{min} = 10 > 6$, execute MSTAA: put task 6 after task 12 on machine 1, move task 3 from machine 1 to machine 2; put task 9 after task 11 on machine 1, move task 11 from machine 1 to machine 3 (Table 3.148).

Table 3.143 Sequence σ^{ord} for Algorithm C2.5a

i	l_i^1	l_i^2	l_i^3	d_i	i	l_i^1	l_i^2	l_i^3	d_i	i	l_i^1	l_i^2	l_i^3	d_i
1	1	2	4	2	5	6	12	24	15	9	4	8	16	29
2	2	4	8	6	6	7	14	28	20	10	3	6	12	32
3	3	6	12	15	7	2	4	8	21	11	1	2	4	32
4	4	8	16	16	8	3	6	12	27	12	4	8	16	28

Table 3.144 Initial schedule

i	l_i^j	d_i	s_i	C_i	i	l_i^j	d_i	s_i	C_i	i	l_i^j	d_i	s_i	C_i
Machine 1, $s_1 = 31$, $I_1 = \varnothing$					Machine 2, $s_2 = 26$, $I_2 = \{9, 12\}$					Machine 3, $s_3 = 15$, $I_3 = \{3, 7\}$				
11	1	32	31	32	10	6	32	26	32	8	12	27	15	27

Table 3.145 The second iteration

i	l_i^j	d_i	s_i	C_i	i	l_i^j	d_i	s_i	C_i	i	l_i^j	d_i	s_i	C_i
Machine 1, $s_1 = 25$					Machine 2, $s_2 = 18$					Machine 3, $s_3 = 3$				
9	4	29	25	29	12	8	28	18	26	3	12	15	3	15
11	1	32	29	30	10	6	32	26	32	8	12	27	15	27

Table 3.146 Iteration 2, after MSTAA executing

i	l_i^j	d_i	s_i	C_i	i	l_i^j	d_i	s_i	C_i	i	l_i^j	d_i	s_i	C_i
Machine 1, $s_1 = 12$, $I_1 = \{4, 5, 6, 7\}$					Machine 2, $s_2 = 18$, $I_2 = \{6, 7\}$					Machine 3, $s_3 = 15$, $I_3 = \{7\}$				
3	3	15	12	15	12	8	28	18	26	8	12	27	15	27
9	4	29	15	19	10	6	32	26	32					
11	1	32	19	20										

Table 3.147 Iteration 3

i	l_i^j	d_i	s_i	C_i	i	l_i^j	d_i	s_i	C_i	i	l_i^j	d_i	s_i	C_i
Machine 1, $s_1 = 5$					Machine 2, $s_2 = 14$					Machine 3, $s_3 = 15$				
6	7	20	5	12	7	4	21	14	18	8	12	27	15	27
3	3	15	12	15	12	8	28	18	26					
9	4	29	15	19	10	6	32	26	32					
11	1	32	19	20										

Table 3.148 Iteration 3, after MSTAA executing

i	l_i^j	d_i	s_i	C_i	i	l_i^j	d_i	s_i	C_i	i	l_i^j	d_i	s_i	C_i
Machine 1, $s_1 = 9$, $I_1 = \{4, 5\}$					Machine 2, $s_2 = 8$, $I_2 = \{4\}$					Machine 3, $s_3 = 11$, $I_3 = \varnothing$				
9	4	29	9	13	3	6	15	8	14	11	4	32	11	15
6	7	20	13	20	7	4	21	14	18	8	12	27	15	27
					12	8	28	18	26					
					10	6	32	26	32					

Iteration 4. Assign task 5 to machine 1 and task 4 to machine 2. There are no feasible tasks on machines 3 (Table 3.149).

$s_{max} - s_{min} = 11 > 6$, execute MSTAA: put task 5 after task 9 on machine 1 and task 4 after task 3 on machine 2. The start times of machines are not changed (Table 3.150).

Iteration 5. Assign task 2 to machine 1. There are no feasible tasks on machines 2 and 3. $s_{max} - s_{min} = 11 > 6$, MSTAA does not change anything (Table 3.151).

Iteration 6. Assign task 1 to machine 1. All tasks are assigned. Execute MSTAA, it does not change anything. We get the final schedule (Table 3.152).

Algorithm D1.1 Initial data: $m = 3$, $n = 13$. Sequence σ^{ord} is in Table 3.153.

Initial schedule in accordance with SSO #2 is shown in Table 3.154.

Successively assign tasks 7–13, each to the machine where it will have the maximum earliness regarding its due date. Since all tasks are assigned and are not tardy, we get the optimal schedule (Table 3.155).

Table 3.149 Iteration 4

i	l_i^j	d_i	s_i	C_i	i	l_i^j	d_i	s_i	C_i	i	l_i^j	d_i	s_i	C_i
Machine 1, $s_1 = 3$					Machine 2, $s_2 = 0$					Machine 3, $s_3 = 11$				
5	6	15	3	9	4	8	16	0	8	11	4	32	11	15
9	4	29	9	13	3	6	15	8	14	8	12	27	15	27
6	7	20	13	20	7	4	21	14	18					
					12	8	28	18	26					
					10	6	32	26	32					

Table 3.150 Iteration 4, after MSTAA executing

i	l_i^j	d_i	s_i	C_i	i	l_i^j	d_i	s_i	C_i	i	l_i^j	d_i	s_i	C_i
Machine 1, $s_1 = 3$, $I_1 = \{2\}$					Machine 2, $s_2 = 0$, $I_2 = \varnothing$					Machine 3, $s_3 = 11$, $I_3 = \varnothing$				
9	4	29	3	7	3	6	15	0	6	11	4	32	11	15
5	6	15	7	13	4	8	16	6	14	8	12	27	15	27
6	7	20	13	20	7	4	21	14	18					
					12	8	28	18	26					
					10	6	32	26	32					

Table 3.151 Iteration 5

i	l_i^j	d_i	s_i	C_i	i	l_i^j	d_i	s_i	C_i	i	l_i^j	d_i	s_i	C_i
Machine 1, $s_1 = 1$, $I_1 = \{1\}$					Machine 2, $s_2 = 0$, $I_2 = \varnothing$					Machine 3, $s_3 = 11$, $I_3 = \varnothing$				
2	2	6	1	3	3	6	15	0	6	11	4	32	11	15
9	4	29	3	7	4	8	16	6	14	8	12	27	15	27
5	6	15	7	13	7	4	21	14	18					
6	7	20	13	20	12	8	28	18	26					
					10	6	32	26	32					

Table 3.152 Final schedule

i	l_i^j	d_i	s_i	C_i	i	l_i^j	d_i	s_i	C_i	i	l_i^j	d_i	s_i	C_i
Machine 1, $s_1 = 0$					Machine 2, $s_2 = 0$					Machine 3, $s_3 = 11$				
1	1	2	0	1	3	6	15	0	6	11	4	32	11	15
2	2	6	1	3	4	8	16	6	14	8	12	27	15	27
9	4	29	3	7	7	4	21	14	18					
5	6	15	7	13	12	8	28	18	26					
6	7	20	13	20	10	6	32	26	32					

Table 3.153 Sequence σ^{ord} for Algorithm D1.1

i	l_i^1	l_i^2	l_i^3	d_i	i	l_i^1	l_i^2	l_i^3	d_i	i	l_i^1	l_i^2	l_i^3	d_i
1	3	4	5	5	6	5	3	1	9	10	16	8	4	29
2	5	4	6	7	7	12	8	18	20	11	6	3	12	32
3	3	5	7	8	8	2	4	8	21	12	1	2	4	32
4	6	1	8	8	9	12	6	3	27	13	3	8	15	33
5	1	4	6	9										

Table 3.154 Initial schedule

i	l_i^j	d_i	s_i	C_i	i	l_i^j	d_i	s_i	C_i	i	l_i^j	d_i	s_i	C_i
Machine 1, $s_1 = 2$					Machine 2, $s_2 = 3$					Machine 3, $s_3 = 8$				
1	3	5	2	5	2	4	7	3	7	6	1	9	8	9
3	3	8	5	8	4	1	8	7	8					
5	1	9	8	9										

Table 3.155 The optimal schedule

i	l_i^j	d_i	s_i	C_i	i	l_i^j	d_i	s_i	C_i	i	l_i^j	d_i	s_i	C_i
Machine 1, $s_1 = 2$					Machine 2, $s_2 = 3$					Machine 3, $s_3 = 8$				
1	3	5	2	5	2	4	7	3	7	6	1	9	8	9
3	3	8	5	8	4	1	8	7	8	9	3	27	9	12
5	1	9	8	9	7	8	20	8	16	10	4	29	12	16
8	2	21	9	11	12	2	32	16	18					
11	6	32	11	17										
13	3	33	17	20										

Algorithm D1.3 Initial data: $m = 3$, $n = 13$ (Table 3.156).

Initial schedule consists of one task assigned to the first machine (Table 3.157).

Execute Algorithm C2.4. Preliminary stage. We assign the first tasks to the last position in the schedule of each machine (Table 3.158).

Optimization stage. Successively assign feasible tasks: 13 and 8 to machine 2; 3 to machine 3; 10 and 2 to machine 2. We get the schedule shown in Table 3.159.

There are no feasible tasks. We sequentially assign to machine 1, with a shift by their tardiness, task 8 (tardiness 10) and task 4 (tardiness 3). Tasks 5 and 6 become feasible on machine 1. Assign at first task 6, then task 5. All tasks are assigned. MSTAA does not change the start times of machines (Table 3.160).

Execute Procedure C: remove the gap between the tasks assigned by Algorithm C2.4 and the initial schedule on machine 1. We get the optimal schedule (Table 3.161).

Algorithm D2.1 Initial data: $m = 3$, $n = 13$. Sequence σ^{ord} is in Table 3.162.

Table 3.156 Sequence σ^{ord} for Algorithm D1.3

i	l_i^1	l_i^2	l_i^3	d_i	i	l_i^1	l_i^2	l_i^3	d_i	i	l_i^1	l_i^2	l_i^3	d_i
1	1	2	4	2	6	4	12	24	15	10	4	8	16	29
2	2	4	8	6	7	7	14	28	20	11	3	6	12	32
3	3	6	12	15	8	2	4	8	21	12	1	2	4	32
4	4	8	16	16	9	3	6	12	27	13	4	8	16	28
5	5	10	20	15										

Table 3.157 Initial schedule

i	l_i^j	d_i	s_i	C_i
Machine 1, $s_1 = 1$				
1	1	2	1	2

Table 3.158 Preliminary stage

i	l_i^j	d_i	s_i	C_i	i	l_i^j	d_i	s_i	C_i	i	l_i^j	d_i	s_i	C_i
Machine 1, $s_1 = 31$					Machine 2, $s_2 = 26$					Machine 3, $s_3 = 15$				
1	1	2	1	2	11	6	32	26	32	9	12	27	15	27
12	1	32	31	32										

Table 3.159 Optimization stage

i	l_i^j	d_i	s_i	C_i	i	l_i^j	d_i	s_i	C_i	i	l_i^j	d_i	s_i	C_i
Machine 1, $s_1 = 31$					Machine 2, $s_2 = 2$					Machine 3, $s_3 = 3$				
1	1	2	1	2	2	4	6	2	6	3	12	15	3	15
12	1	32	31	32	10	8	29	6	14	9	12	27	15	27
					8	4	21	14	18					
					13	8	28	18	26					
					11	6	32	26	32					

Table 3.160 The schedule after all tasks assignment

i	l_i^j	d_i	s_i	C_i	i	l_i^j	d_i	s_i	C_i	i	l_i^j	d_i	s_i	C_i
Machine 1, $s_1 = 3$					Machine 2, $s_2 = 2$					Machine 3, $s_3 = 3$				
1	1	2	1	2	2	4	6	2	6	3	12	15	3	15
6	4	15	3	7	10	8	29	6	14	9	12	27	15	27
5	5	15	7	12	8	4	21	14	18					
4	4	16	12	16	13	8	28	18	26					
8	2	21	16	18	11	6	32	26	32					
12	1	32	18	19										

Table 3.161 Optimal schedule

i	l_i^j	d_i	s_i	C_i	i	l_i^j	d_i	s_i	C_i	i	l_i^j	d_i	s_i	C_i
Machine 1, $s_1 = 1$					Machine 2, $s_2 = 2$					Machine 3, $s_3 = 3$				
1	1	2	1	2	2	4	6	2	6	3	12	15	3	15
6	4	15	2	6	10	8	29	6	14	9	12	27	15	27
5	5	15	6	11	8	4	21	14	18					
4	4	16	11	15	13	8	28	18	26					
8	2	21	15	17	11	6	32	26	32					
12	1	32	17	18										

Table 3.162 Sequence σ^{ord} for Algorithm D2.1

i	l_i^1	l_i^2	l_i^3	d_i	i	l_i^1	l_i^2	l_i^3	d_i	i	l_i^1	l_i^2	l_i^3	d_i
1	3	4	5	5	6	5	3	1	9	10	16	8	4	29
2	5	4	6	7	7	12	8	18	20	11	6	3	12	32
3	3	5	7	8	8	9	4	8	21	12	1	2	4	32
4	6	1	8	8	9	21	16	12	22	13	3	8	15	33
5	1	4	6	9										

Initial schedule in accordance with SSO #2 is shown in Table 3.163.

Assign tasks 7 and 8, each to the machine where it will have the maximum earliness regarding its due date (Table 3.164).

Task 9 is tardy on all machines. Assign it to the machine 3 where it has the minimum tardiness, shifting all tasks on machine 3 to the left by the amount of the task's tardiness (7 units). The start time of machine 3 becomes equal to $1 < 2$. But alternative assignment of task 9 to the machine 1 that has the minimum start time makes even lesser start time. Therefore, we assign task 9 to machine 3 and consider machine 3 as analogue of the first machine (Table 3.165).

Successively assign tasks 10–13, each to the machine where it will have the maximum earliness regarding its due date. All these tasks are not tardy. MSTAA does not change the start times. We get the optimal schedule (Table 3.166).

Table 3.163 Initial schedule

i	l_i^j	d_i	s_i	C_i	i	l_i^j	d_i	s_i	C_i	i	l_i^j	d_i	s_i	C_i
Machine 1, $s_1 = 2$					Machine 2, $s_2 = 3$					Machine 3, $s_3 = 8$				
1	3	5	2	5	2	4	7	3	7	6	1	9	8	9
3	3	8	5	8	4	1	8	7	8					
5	1	9	8	9										

Table 3.164 The schedule after 8 tasks assignment

i	l_i^j	d_i	s_i	C_i	i	l_i^j	d_i	s_i	C_i	i	l_i^j	d_i	s_i	C_i
Machine 1, $s_1 = 2$					Machine 2, $s_2 = 3$					Machine 3, $s_3 = 8$				
1	3	5	2	5	2	4	7	3	7	6	1	9	8	9
3	3	8	5	8	4	1	8	7	8	8	8	21	9	17
5	1	9	8	9	7	8	20	8	16					

Table 3.165 The schedule after task 9 assignment

i	l_i^j	d_i	s_i	C_i	i	l_i^j	d_i	s_i	C_i	i	l_i^j	d_i	s_i	C_i
Machine 1, $s_1 = 2$					Machine 2, $s_2 = 3$					Machine 3, $s_3 = 1$				
1	3	5	2	5	2	4	7	3	7	6	1	9	1	2
3	3	8	5	8	4	1	8	7	8	8	8	21	2	10
5	1	9	8	9	7	8	20	8	16	9	12	22	10	22

Table 3.166 Optimal schedule

i	l_i^j	d_i	s_i	C_i	i	l_i^j	d_i	s_i	C_i	i	l_i^j	d_i	s_i	C_i
Machine 1, $s_1 = 2$					Machine 2, $s_2 = 3$					Machine 3, $s_3 = 1$				
1	3	5	2	5	2	4	7	3	7	6	1	9	1	2
3	3	8	5	8	4	1	8	7	8	8	8	21	2	10
5	1	9	8	9	7	8	20	8	16	9	12	22	10	22
11	6	32	9	15						10	4	29	22	26
12	1	32	15	16										
13	3	33	16	19										

Reference

1. Zgurovsky, M.Z., Pavlov, A.A., Misura, E.B.: Minimizaciya leksikograficheskogo kriteriya dlya dopustimogo raspisaniya na nezavisimyh parallelnyh priborah s proizvolnymi direktivnymi srokami (Минимизация лексикографического критерия для допустимого расписания на независимых параллельных приборах с произвольными директивными сроками; The lexicographical criterion minimization for a feasible schedule on parallel machines with arbitrary due dates). Visnyk NTUU KPI Inform. Oper. Comput. Sci. **61**, 4–17 (2014) (in Russian)

Chapter 4
The Total Weighted Tardiness of Tasks Minimization on a Single Machine

Abstract We solve the problem of constructing a schedule for a single machine with various due dates and a fixed start time of the machine that minimizes the sum of weighted tardiness of tasks in relation to their due dates. The problem is NP-hard in the strong sense and is one of the most known intractable combinatorial optimization problems. Unlike other PSC-algorithms in this monograph, in this chapter we present an efficient PSC-algorithm which, in addition to the first and second polynomial components (the first one contains twelve sufficient signs of optimality of a feasible schedule) includes exact subalgorithm for its solving. We have obtained the sufficient conditions that are constructively verified in the process of its execution. If the conditions are true, the exact subalgorithm becomes polynomial. We give statistical studies of the developed algorithm and show the solving of well-known examples of the problem. We present an approximation algorithm (the second polynomial component) based on the exact algorithm. Average statistical estimate of deviation of an approximate solution from the optimum does not exceed 5% of it.

4.1 The Problem Statement

This chapter contains our new original results based on our earlier works [1–6].

The Total Weighted Tardiness (TWT) Problem Statement Given a set of independent tasks $J = \{j_1, j_2, \ldots, j_n\}$, each task consists of a single operation. For each task j, we know its processing time $l_j > 0$, weight coefficient $\omega_j > 0$ and due date $d_j \geq 0$. (We do not consider the cases with $\omega_j = 0$, since the tasks with zero weights will occupy the last positions in an optimal sequence and, therefore, we eliminate them from optimization.) All tasks become available at the same time $r_j = 0$, $j \in \{1, \ldots, n\}$. Interruptions are not allowed. We need to build a schedule for one machine that minimizes the total weighted tardiness of the tasks:

$$f = \sum_{j=1}^{n} \omega_j \max\left(0, C_j - d_j\right) \tag{4.1}$$

where C_j is the completion time of task j.

M. Z. Zgurovsky and A. A. Pavlov, *Combinatorial Optimization Problems in Planning and Decision Making*, Studies in Systems, Decision and Control 173,
https://doi.org/10.1007/978-3-319-98977-8_4

This problem is NP-hard in the strong sense [7]. It attracted the attention of scientists for many years. But because of severe computational complexity, obtaining an exact solution for a 50 tasks problem instance had been almost impossible challenge until relatively recent times.

Lawler [7] proposed a pseudo-polynomial algorithm for the problem with equal weights. Various enumerative solution methods have been proposed for both the weighted and unweighted cases [8]. Emmons [9] derives several dominance rules that restrict the search for an optimal solution to the unweighted problem. Emmons' rules are used in both branch and bound and dynamic programming algorithms [10–12]). Rinnooy Kan et al. [13] extended these results to the weighted tardiness problem. Rachamadugu [14] identified a condition characterizing the location of two adjacent tasks in an optimal schedule of the TWT problem [15].

Chambers et al. [16] have developed new heuristic dominance rules and a flexible decomposition heuristic [17]. Abdul-Razaq et al. [18] show that the most promising lower bound both in quality and time consumption is the linear lower bound method by Potts and Van Wassenhove [11], which is obtained from Lagrangian relaxation of machine capacity constraints. Hoogeveen and Van de Velde [19] reformulate the problem by using slack variables and show that better Lagrangian lower bounds can be obtained.

Szwarc and Liu [20] present a two-stage decomposition mechanism for the TWT problem when tardiness penalties are proportional to the processing times of tasks. Akturk and Yildirim [17] propose a new dominance rule for the most general case of the TWT problem. The proposed rule covers and extends the Emmons' results and generalizations of Rinnooy Kan et al. by considering the time dependent orderings between each pair of jobs.

The most efficient of known exact algorithms for the TWT problem is the dynamic programming algorithm of Tanaka et al. [21], it solves instances of up to 300 tasks in average 350 s on a PC with 2.4 GHz Pentium 4 CPU. They consider dominance rules involving up to 4 tasks. Karakostas et al. [22] investigate particular cases of the problem with restrictions on input data, in many cases they obtain exact polynomial time solving algorithms or polynomial time approximation schemes.

Our PSC-algorithm for the problem solving is based on the results published in [3] and on our PSC-algorithm to solve the problem of minimizing the total tardiness of tasks on a single machine (TT) presented in [1].

4.2 Main Theoretical Propositions

Let us introduce the following notation and definitions [3]. Let i or j_i be the number of task corresponding to the indexation specified by the functional; $j_{[i]}$ be the number of a task in position i in a feasible schedule. Let $\overline{a, b}$ denote the interval of integer numbers from a to b, that is, $\overline{a, b} = \mathbb{Z} \cap [a, b] = a,\ a + 1,\ \ldots,\ b$.

Definition 4.1 $R_{j_{[i]}} = d_{j_{[i]}} - C_{j_{[i]}} > 0$ is called the time *reserve* of task $j_{[i]}$.

Definition 4.2 $p_{j_{[i]}} = \omega_{j_{[i]}} / l_{j_{[i]}}$ is called the *priority* of task $j_{[i]}$.

Definition 4.3 *Permutation* (also known as EFSR, extraction and forward-shifted reinsertion [3]) is a procedure of moving a task $j_{[g]}$ into a position $k > g$ and, at the same time, tasks in positions $g + 1,\ g + 2,\ \ldots,\ k - 1,\ k$ to positions $g,\ g + 1,\ \ldots,\ k - 2,\ k - 1$, respectively.

Definition 4.4 *Interval of permutation* of a task $j_{[g]}$ into a position k in a sequence σ is the time interval defined by the sum of the processing times of tasks in positions $g + 1,\ g + 2,\ \ldots,\ k - 1,\ k$ in this sequence.

Definition 4.5 *Insertion* (also known as EBSR, extraction and backward-shifted reinsertion [3]) is a procedure of moving a task $j_{[g]}$ into a position $p < g$ and, at the same time, tasks in positions $p,\ p + 1,\ \ldots,\ g - 2,\ g - 1$ to positions $p + 1,\ p + 2,\ \ldots,\ g - 1,\ g$, respectively.

Definition 4.6 *Interval of insertion* $I_{j_{[g]}}$ of a task $j_{[g]}$ into a position $p < g$ in a sequence σ is the time interval defined by the sum of the processing times of tasks in positions $p,\ p + 1,\ \ldots,\ g - 1$ in σ. The position p is defined by condition

$$\sum_{i=p-1}^{g-1} l_{j_{[i]}} < C_{j_{[g]}} - d_{j_{[g]}} \le \sum_{i=p}^{g-1} l_{j_{[i]}}, \tag{4.2}$$

and if (4.2) is false for every position, then $p = 1$. Thus, the tardiness of task j in the position p must be zero or minimal.

Definition 4.7 A task $j_{[g]}$ is called *tardy* in a sequence σ if $d_{j_{[g]}} < C_{j_{[g]}}$.

Definition 4.8 *Sequence* σ^{ord} (sigma ordered) is a sequence of tasks j, $j = \overline{1, n}$, that are in non-increasing order of priorities p_j, i.e., $\forall j, i,\ j < i$: $p_j \ge p_i$, and if $p_j = p_i$, then $d_j \le d_i$.

Definition 4.9 A *free permutation* is the permutation of a task $j_{[k]}$ into such position $q > k$ that $d_{j_{[k]}} \ge C_{j_{[q]}}$ and $d_{j_{[k]}} < C_{j_{[q+1]}}$ if for at least one task within the interval $\overline{k+1, q}$ we have: $d_{j_{[i]}} < C_{j_{[i]}}$, $i \in \overline{k+1, q}$.

So, we perform a free permutation of a task k into the maximum number position where the task k will not be tardy. It is obvious that free permutations in the sequence σ^{ord} will decrease the value of the objective function.

One free permutation decreases the objective function value by

$$\sum_{i \in Z} \omega_{j_{[i]}} \min\left(l_{j_{[k]}}, C_{j_{[i]}} - d_{j_{[i]}}\right)$$

where

$$Z = \left\{ j_{[i]} : C_{j_{[i]}} > d_{j_{[i]}}, \ i = \overline{k+1, q} \right\}.$$

Remark 4.1 (to Definition 4.9) In order to avoid looping when performing several free permutations, it is necessary to start with the task that has the maximum due date, i.e., to consider the non-tardy tasks in descending order of their due dates.

Definition 4.10 The tasks sequence obtained as a result of performing all free permutations in the sequence σ^{ord} is called σ^{fp} (sigma free permutations).

Theorem 4.1 ([23, Chap. 3]) *The total tardiness is zero in an optimal schedule (for the total tardiness minimization criterion) if and only if it is zero in the schedule in which the tasks are in non-decreasing order of their due dates.*

Corollary 4.1, Sufficient Sign of Optimality of a Feasible[1] Solution (SSO) #1 If all tasks in a schedule are in non-decreasing order of due dates and there are no tardy tasks, then the total weighted tardiness is also zero and such schedule is optimal (see Corollary 2.2).

Corollary 4.2 If SSO #1 is not satisfied, then the total weighted tardiness in an optimal schedule is greater than zero.

During the problem solving, using directional permutations and insertions, we sequentially develop from the sequence σ^{ord} to the sequence σ^{fp} and then to the current subsequences. As a result of the optimization iterations, we construct an optimal sequence on the entire set of tasks.

In the following statements, we formulate the properties of each considered feasible sequence, prove the sufficient signs of their optimality, and justify the rules of permutations.

4.2.1 Sequence σ^{ord}

Statement 4.1, SSO #2 If tasks with a time reserve do not precede the tardy tasks in the sequence σ^{ord}, then there are no moves (permutations or insertions) of tasks that lead to a decrease in the functional value, and the sequence σ^{ord} is optimal on the entire set of tasks.

[1]In this chapter, all schedules obtained during the problem solving are feasible including those with tardy tasks.

Proof Suppose $C_{j_{[i]}} - d_{j_{[i]}} \geq 0$ in σ^{ord} for tasks within positions $i = \overline{1,s}$ and $C_{j_{[i]}} - d_{j_{[i]}} < 0$ for tasks within positions $i = \overline{s+1,n}$. Obviously, the moves that decrease the functional value are possible only for tasks within positions $i \in \overline{1,s}$.

Let us prove that there are no *permutations* that decrease the functional value. Suppose that in a sequence σ some positions $l, g \in \overline{1,s}$, $l < g$. Perform a permutation of the task $j_{[l]}$ into the position g. The functional value for positions $\overline{l,g}$ before the permutation is

$$f = \sum_{i=l}^{g} \omega_{j_{[i]}} \left(C_{j_{[i]}} - d_{j_{[i]}} \right).$$

After the permutation, taking into account the assumption that the tasks within positions $\overline{l+1,g}$ became non-tardy (i.e., assuming that inequalities $C_{j_{[i]}} - d_{j_{[i]}} \leq l_{j_{[l]}}$, $i = \overline{l+1,g}$, were true before the permutation), we have:

$$f' = \omega_{j_{[l]}} \left(C_{j_{[l]}} + \sum_{i=l+1}^{g} l_{j_{[i]}} - d_{j_{[l]}} \right)$$

where $C_{j_{[l]}}$ is the completion time of the task $j_{[l]}$ before the permutation; $C_{j_{[l]}} + \sum_{i=l+1}^{g} l_{j_{[i]}}$ is its completion time after the permutation.

We need to prove that $f - f' \leq 0$.

$$\begin{aligned} f - f' &= \omega_{j_{[l]}} \left(C_{j_{[l]}} - d_{j_{[l]}} \right) + \sum_{i=l+1}^{g} \omega_{j_{[i]}} \left(C_{j_{[i]}} - d_{j_{[i]}} \right) - \omega_{j_{[l]}} \left(C_{j_{[l]}} + \sum_{i=l+1}^{g} l_{j_{[i]}} - d_{j_{[l]}} \right) \\ &= \sum_{i=l+1}^{g} \omega_{j_{[i]}} \left(C_{j_{[i]}} - d_{j_{[i]}} \right) - \sum_{i=l+1}^{g} \omega_{j_{[l]}} l_{j_{[i]}} \leq \sum_{i=l+1}^{g} \omega_{j_{[i]}} l_{j_{[l]}} - \sum_{i=l+1}^{g} \omega_{j_{[l]}} l_{j_{[i]}} \\ &= \sum_{i=l+1}^{g} \left(\omega_{j_{[i]}} l_{j_{[l]}} - \omega_{j_{[l]}} l_{j_{[i]}} \right) \leq 0 \end{aligned}$$

because $p_{j_{[i]}} \leq p_{j_{[l]}}, i = \overline{l+1,g}$.

Let us prove that there are no *insertions* that increase the functional value. The value of the objective function before the insertion is

$$f = \sum_{i=p}^{g} \omega_{j_{[i]}} \left(C_{j_{[i]}} - d_{j_{[i]}} \right), \quad g, p \in \overline{1,s}.$$

After the insertion of the task $j_{[g]}$ into the position p, taking into account the fact that the task $j_{[g]}$ will be non-tardy after the insertion, i.e., $C_{j_{[g]}} - d_{j_{[g]}} \le \sum_{i=p}^{g-1} l_{j_{[i]}}$, the functional value will be equal to

$$f' = \sum_{i=p}^{g-1} \omega_{j_{[i]}} \left(C_{j_{[i]}} + l_{j_{[g]}} - d_{j_{[i]}} \right)$$

where $C_{j_{[i]}}$ is the completion time of a task $j_{[i]}$, $i \in \overline{p, g-1}$, before the insertion of the task $j_{[g]}$; $C_{j_{[i]}} + l_{j_{[g]}}$ is its completion time after the insertion.

It is necessary to prove that $f - f' \le 0$.

$$\begin{aligned} f - f' &= \omega_{j_{[g]}} \left(C_{j_{[g]}} - d_{j_{[g]}} \right) + \sum_{i=p}^{g-1} \left(C_{j_{[i]}} - d_{j_{[i]}} \right) - \sum_{i=p}^{g-1} \omega_{j_{[i]}} \left(C_{j_{[i]}} + l_{j_{[g]}} - d_{j_{[i]}} \right) \\ &= \omega_{j_{[g]}} \left(C_{j_{[g]}} - d_{j_{[g]}} \right) - \sum_{i=p}^{g-1} \omega_{j_{[i]}} l_{j_{[g]}} \le \omega_{j_{[g]}} \sum_{i=p}^{g-1} l_{j_{[i]}} - \sum_{i=p}^{g-1} \omega_{j_{[i]}} l_{j_{[g]}} \\ &= \sum_{i=p}^{g-1} \left(\omega_{j_{[g]}} l_{j_{[i]}} - \omega_{j_{[i]}} l_{j_{[g]}} \right) \le 0 \end{aligned}$$

because $p_{j_{[g]}} \le p_{j_{[i]}} \; \forall i = \overline{p, g-1}$. ∎

Statement 4.2 Moving a tardy task $j_{[g]}$ into a position $f < p$, where p is the position of its insertion, cannot decrease the functional value.

Proof According to Definition 4.6, the position p is defined as the position at which the weighted tardiness of the task $j_{[g]}$ becomes minimal or equal to zero. Insertion of the task $j_{[g]}$ to any position $f < p$ increases its reserve and moves the tasks in positions $f, f+1, \ldots, p-1$ into positions $f+1, f+2, \ldots, p$, respectively. Such move increases the functional value if for any task $j_{[i]}$, $i \in \overline{f, p-1}$, at least one of the following conditions is true:

1. $C_{j_{[i]}} \ge d_{j_{[i]}}$, i.e., the task $j_{[i]}$ is tardy. The insertion increases the total weighted tardiness by $\omega_{j_{[i]}} l_{j_{[g]}}$.
2. $C_{j_{[i]}} < d_{j_{[i]}}$ and $d_{j_{[i]}} - C_{j_{[i]}} < l_{j_{[g]}}$, i.e., the task $j_{[i]}$ is not tardy before the insertion of the task $j_{[g]}$ but it will be tardy in consequence of the insertion having the weighted tardiness $\omega_{j_{[i]}} \left(C_{j_{[i]}} - d_{j_{[i]}} + l_{j_{[g]}} \right)$.

If both these conditions are not satisfied for any task $j_{[i]}$ within positions $i \in \overline{f, p-1}$, then the functional value does not increase. But it also cannot be reduced in consequence of the insertion of the task $j_{[g]}$ into a position $f < p$ since the weighted tardiness of the task $j_{[g]}$ in position p is zero. ∎

Statement 4.3 Suppose that we have $R_{j_{[i]}} \le 0 \; \forall j_{[i]},\; i = \overline{1, p-1}$, in the sequence σ^{ord}. Denote such sequence by σ^{ord1}. A tardy task $j_{[g]}$ can take an earlier position in the sequence σ^{ord1} in consequence of its insertion, which will decrease the functional value, only if at least one of the tasks within the interval of insertion $I_{j_{[g]}} = \overline{p, g-1}$ in the sequence σ^{ord1} has a positive time reserve.

Proof To prove this statement, we derive the conditions under which the functional value decreases in consequence of the insertion, i.e., $f - f' > 0$.

Suppose that $d_{j_{[z]}} > C_{j_{[z]}}$ for a task $j_{[z]}$ $(p \le z < g)$ and $d_{j_{[i]}} \le C_{j_{[i]}}$ for other tasks $i \in I_{j_{[g]}}$. According to Statements 4.1 and 4.2, we should insert the task $j_{[g]}$ into the position z in order to decrease the functional value. In this case, we have:

$$f = \sum_{i=p}^{z-1} \omega_{j_{[i]}} \left(C_{j_{[i]}} - d_{j_{[i]}} \right) + \sum_{i=z+1}^{g} \omega_{j_{[i]}} \left(C_{j_{[i]}} - d_{j_{[i]}} \right);$$

$$f' = \omega_{j_{[g]}} \max\left(0, C_{j_{[g]}} - \sum_{i=z}^{g-1} l_{j[i]} - d_{j_{[g]}} \right) + \sum_{i=p}^{z-1} \omega_{j_{[i]}} \left(C_{j_{[i]}} - d_{j_{[i]}} \right)$$
$$+ \sum_{i=z+1}^{g-1} \omega_{j_{[i]}} \left(C_{j_{[i]}} + l_{j_{[g]}} - d_{j_{[i]}} \right) + \omega_{j_{[z]}} \max\left(0, C_{j_{[z]}} + l_{j_{[g]}} - d_{j_{[z]}} \right);$$

$$\begin{aligned}
f - f' &= \sum_{i=p}^{z-1} \omega_{j_{[i]}} \left(C_{j_{[i]}} - d_{j_{[i]}} \right) + \sum_{i=z+1}^{g} \omega_{j_{[i]}} \left(C_{j_{[i]}} - d_{j_{[i]}} \right) \\
&\quad - \sum_{i=p}^{z-1} \omega_{j_{[i]}} \left(C_{j_{[i]}} - d_{j_{[i]}} \right) - \sum_{i=z+1}^{g-1} \omega_{j_{[i]}} \left(C_{j_{[i]}} - d_{j_{[i]}} \right) - \sum_{i=z+1}^{g-1} \omega_{j_{[i]}} l_{j_{[g]}} \\
&\quad - \omega_{j_{[z]}} \max\left(0, C_{j_{[z]}} + l_{j_{[g]}} - d_{j_{[z]}} \right) - \omega_{j_{[g]}} \max\left(0, C_{j_{[g]}} - \sum_{i=z}^{g-1} l_{j_{[i]}} - d_{j_{[g]}} \right) \\
&= - \sum_{i=z+1}^{g-1} \omega_{j_{[i]}} l_{j_{[g]}} + \omega_{j_{[g]}} \left(C_{j_{[g]}} - d_{j_{[g]}} \right) \\
&\quad - \omega_{j_{[g]}} \max\left(0, C_{j_{[g]}} - \sum_{i=z}^{g-1} l_{j_{[i]}} - d_{j_{[g]}} \right) - \omega_{j_{[z]}} \max\left(0, C_{j_{[z]}} + l_{j_{[g]}} - d_{j_{[z]}} \right) \\
&= \omega_{j_{[g]}} \min\left(C_{j_{[g]}} - d_{j_{[g]}}, \sum_{i=z}^{g-1} l_{j_{[i]}} \right) - \sum_{i=z+1}^{g-1} \omega_{j_{[i]}} l_{j_{[g]}} - \omega_{j_{[z]}} \max\left(0, C_{j_{[z]}} + l_{j_{[g]}} - d_{j_{[z]}} \right) \\
&= \omega_{j_{[g]}} \min\left(C_{j_{[g]}} - d_{j_{[g]}}, \sum_{i=z}^{g-1} l_{j_{[i]}} \right) - \sum_{i=z}^{g-1} \omega_{j_{[i]}} \max\left[0, \min\left(l_{j_{[g]}} - R_{j_{[i]}}, l_{j_{[g]}} \right) \right].
\end{aligned}$$

So, the insertion of the task $j_{[g]}$ into the position z decreases the functional value by

$$f - f' = \omega_{j_{[g]}} \min\left(C_{j_{[g]}} - d_{j_{[g]}}, \sum_{i=z}^{g-1} l_{j_{[i]}}\right) - \sum_{i=z}^{g-1} \omega_{j_{[i]}} \max\left[0, \min\left(l_{j_{[g]}} - R_{j_{[i]}}, l_{j_{[g]}}\right)\right]. \tag{4.3}$$

Consequently, a decrease in the functional value is possible only if a task with a positive reserve exists within the interval of insertion $I_{j_{[g]}}$. If reserves are absent, Eq. (4.3) has the negative value because the priority of the task $j_{[g]}$ is less than priorities of tasks within the interval of insertion. ∎

Note that if there are several tasks $j_{[z_k]}$ with time reserves within $\overline{p, g-1}$, then Eq. (4.3) for insertion of the task $j_{[g]}$ into any of positions z_k will take the form

$$f - f' = \omega_{j_{[g]}} \min\left(C_{j_{[g]}} - d_{j_{[g]}}, \sum_{i=z_k}^{g-1} l_{j_{[i]}}\right) - \sum_{i=z_k}^{g-1} \omega_{j_{[i]}} \max\left[0, \min\left(l_{j_{[g]}} - R_{j_{[i]}}, l_{j_{[g]}}\right)\right]. \tag{4.4}$$

Statement 4.4, SSO #3 If for any tardy task $j_{[g]}$ in the sequence σ^{ord} there are no preceding tasks $j_{[s]}$, $s = \overline{1, g-1}$, for which $d_{j_{[s]}} - C_{j_{[s]}} > 0$ and $d_{j_{[s]}} > d_{j_{[g]}} - l_{j_{[g]}}$, then there are no permutations or insertions of tasks that decrease the functional value, and the sequence σ^{ord} is optimal on the entire set of tasks.

Proof

1. *Insertions*. Insert the task $j_{[g]}$ into the position of insertion p (Definition 4.6). Considering the condition of the statement that there are no preceding tasks $j_{[s]}$, $s = \overline{1, g-1}$, for which $d_{j_{[s]}} - C_{j_{[s]}} > 0$ and $d_{j_{[s]}} > d_{j_{[g]}} - l_{j_{[g]}}$, we find that the task $j_{[g]}$ should be inserted into the interval $\overline{p, g-1}$ where we have $C_{j_{[i]}} \geq d_{j_{[i]}}$ for any task $j_{[i]}$, $i = \overline{p, g-1}$.

   Suppose $C_{j_{[i]}} - d_{j_{[i]}} = 0$, $i = \overline{p, g-1}$. Then

$$f = \omega_{j_{[g]}}\left(C_{j_{[g]}} - d_{j_{[g]}}\right), f' = \sum_{i=p}^{g-1} \omega_{j_{[i]}}\left(C_{j_{[i]}} + l_{j_{[g]}} - d_{j_{[i]}}\right) = \sum_{i=p}^{g-1} \omega_{j_{[i]}} l_{j_{[g]}},$$

$$f - f' = \omega_{j_{[g]}}\left(C_{j_{[g]}} - d_{j_{[g]}}\right) - \sum_{i=p}^{g-1} \omega_{j_{[i]}} l_{j_{[g]}}$$
$$\leq \omega_{j_{[g]}} \sum_{i=p}^{g-1} l_{j_{[i]}} - \sum_{i=p}^{g-1} \omega_{j_{[i]}} l_{j_{[g]}} = \sum_{i=p}^{g-1}\left(\omega_{j_{[g]}} l_{j_{[i]}} - \omega_{j_{[i]}} l_{j_{[g]}}\right) \leq 0$$

because $p_{j_{[g]}} \leq p_{j_{[i]}}$ $\forall i < g$.

Proof for the case $C_{j_{[i]}} - d_{j_{[i]}} > 0$ is similar.

2. *Permutations.* Permutations of tasks with zero time reserve and with a nonzero reserve are possible. Let us consider all possible cases.

(a) Suppose $d_{j_{[s]}} - C_{j_{[s]}} > 0$, $d_{j_{[s]}} \leq d_{j_{[g]}} - l_{j_{[g]}}$, and the task $j_{[g]}$ is tardy after the permutation, i.e., $C_{j_{[g]}} - l_{j_{[s]}} > d_{j_{[g]}}$. In this case, $f = \omega_{j_{[g]}}\left(C_{j_{[g]}} - d_{j_{[g]}}\right)$,

$$f' = \omega_{j_{[g]}}\left(C_{j_{[g]}} - l_{j_{[s]}} - d_{j_{[g]}}\right) + \omega_{j_{[s]}}\left(C_{j_{[g]}} - d_{j_{[s]}}\right).$$

$$\begin{aligned} f - f' &= \omega_{j_{[g]}}\left(C_{j_{[g]}} - d_{j_{[g]}}\right) - \omega_{j_{[g]}}\left(C_{j_{[g]}} - l_{j_{[s]}} - d_{j_{[g]}}\right) - \omega_{j_{[s]}}(C_{j_{[g]}} - d_{j_{[s]}}) \\ &= \omega_{j_{[g]}} l_{j_{[s]}} - \omega_{j_{[s]}}\left(C_{j_{[g]}} - d_{j_{[s]}}\right) \leq \omega_{j_{[g]}} l_{j_{[s]}} - \omega_{j_{[s]}}\left(C_{j_{[g]}} - d_{j[g]} + l_{j_{[g]}}\right) \\ &\leq \omega_{j_{[g]}} l_{j_{[s]}} - \omega_{j_{[s]}} l_{j_{[g]}} \leq 0 \end{aligned}$$

because $p_{j_{[g]}} \leq p_{j_{[s]}} \; \forall s < g$.

(b) Suppose $d_{j_{[s]}} - C_{j_{[s]}} > 0$, $d_{j_{[s]}} \leq d_{j_{[g]}} - l_{j_{[g]}}$, and the task $j_{[g]}$ is not tardy after the permutation, i.e., $C_{j_{[g]}} - d_{j_{[g]}} \leq l_{j_{[s]}}$. Then

$$f' = \omega_{j_{[s]}}\left(C_{j_{[g]}} - d_{j_{[s]}}\right),$$

$$\begin{aligned} f - f' &= \omega_{j_{[g]}}\left(C_{j_{[g]}} - d_{j_{[g]}}\right) - \omega_{j_{[s]}}\left(C_{j_{[g]}} - d_{j_{[s]}}\right) \leq \omega_{j_{[g]}}\left(C_{j_{[g]}} - \left(C_{j_{[g]}} - l_{j_{[s]}}\right)\right) \\ &- \omega_{j_{[s]}}\left(C_{j_{[g]}} - \left(d_{j_{[g]}} - l_{j_{[g]}}\right)\right) = \omega_{j_{[g]}} l_{j_{[s]}} - \omega_{j_{[s]}}\left(C_{j_{[g]}} - d_{j_{[g]}} + l_{j_{[g]}}\right) \leq \omega_{j_{[g]}} l_{j_{[s]}} - \omega_{j_{[s]}} l_{j_{[g]}} \leq 0 \end{aligned}$$

because $p_{j_{[g]}} \leq p_{j_{[s]}} \; \forall s < g$.

(c) Suppose $d_{j_{[s]}} - C_{j_{[s]}} = 0$, $d_{j_{[s]}} \leq d_{j_{[g]}} - l_{j_{[g]}}$, and the task $j_{[g]}$ is tardy after the permutation, then

$$\begin{aligned} f - f' &= \omega_{j_{[g]}}\left(C_{j_{[g]}} - d_{j_{[g]}}\right) - \omega_{j_{[g]}}\left(C_{j_{[g]}} - l_{j_{[s]}} - d_{j_{[g]}}\right) - \omega_{j_{[s]}} \sum_{i=s+1}^{g} l_{j_{[i]}} \\ &= \omega_{j_{[g]}} l_{j_{[s]}} - \omega_{j_{[s]}} l_{j_{[g]}} - \omega_{j_{[s]}} \sum_{i=s+1}^{g-1} l_{j_{[i]}} \leq \omega_{j_{[g]}} l_{j_{[s]}} - \omega_{j_{[s]}} l_{j_{[g]}} \leq 0 \end{aligned}$$

because $p_{j_{[g]}} \leq p_{j_{[s]}} \; \forall s < g$.

(d) Suppose $d_{j_{[s]}} - C_{j_{[s]}} = 0$, $d_{j_{[s]}} \leq d_{j_{[g]}} - l_{j_{[g]}}$, and the task $j_{[g]}$ is not tardy after the permutation, i.e., $C_{j_{[g]}} - d_{j_{[g]}} \leq l_{j_{[s]}}$. Then

$$f - f' = \omega_{j_{[g]}}\left(C_{j_{[g]}} - d_{j_{[g]}}\right) - \omega_{j_{[s]}}\sum_{i=s+1}^{g} l_{j_{[i]}} \le \omega_{j_{[g]}} l_{j_{[s]}} - \omega_{j_{[s]}}\sum_{i=s+1}^{g} l_{j_{[i]}} \le 0,$$

similarly to the case (c).

Therefore, since there are no preceding tasks with time reserves for the tardy task $j_{[g]}$ within its interval of insertion, then there are no permutations leading to a decrease in the functional value. ∎

Statement 4.5 Suppose that the interval of insertion of the task $j_{[g]}$ in the sequence σ^{ord} is $I_{j_{[g]}} = \overline{p, g-1}$. Additional reserves within the interval of insertion for the task $j_{[g]}$ may only be created by the tasks $j_{[i]}$, $i \in \overline{1, p-1}$, for which

$$d_{j_{[i]}} > d_{j_{[g]}} - l_{j_{[g]}},\ d_{j_{[i]}} > C_{j_{[i]}},\ d_{j_{[i]}} > C_{j_{[p-1]}}. \tag{4.5}$$

Proof For all tasks $j_{[i]}$, $i \in \overline{1, p-1}$, we have $p_{j_{[i]}} \ge p_{j_{[k]}}$, $k \in I_{j_{[g]}}$. If conditions (4.5) are not satisfied, then the task $j_{[i]}$ will be tardy in the position p. Therefore, according to Statement 4.1, moving it into the interval of insertion of the task $j_{[g]}$ will increase the functional value. ∎

Statement 4.6 Suppose that a task $j_{[g]}$ is tardy in the sequence σ^{ord}. Moving the task $j_{[g]}$ into an earlier position as a result of permutations and insertions can decrease the functional value only if at least one of the following conditions is true:

1. $\exists j_{[i]}, p \le i \le g$, such that $R_{j_{[i]}} > 0$. There are tasks with time reserves within the interval of insertion of the task $j_{[g]}$ where p is the position in which the weighted tardiness of the task $j_{[g]}$ is minimal (or equals to zero).
2. $\exists j_{[q]}$, $q < g$, such that $d_{j_{[q]}} > C_{j_{[g]}}$. There is a task with a time reserve in the sequence σ^{ord} in position q that precedes the position g, and its permutation into the position after the task $j_{[g]}$ reduces the weighted tardiness of the task $j_{[g]}$. The task $j_{[q]}$ remains non-tardy.
3. $\exists j_{[q]}$, $q < g$, such that $C_{j_{[g]}} - l_{j_{[g]}} < d_{j_{[q]}} < C_{j_{[g]}}$. There is a non-tardy task $j_{[q]}$ which due date is greater than the start time of the task $j_{[g]}$ but less than the completion time of the task $j_{[g]}$. Also, the following must be satisfied:

$$\omega_{j_{[g]}} \cdot \min\left(C_{j_{[g]}} - d_{j_{[g]}}, l_{j_{[q]}}\right) - \omega_{j_{[q]}}\left(C_{j_{[g]}} - d_{j_{[q]}}\right) > 0.$$

   Consequently, a permutation of the task $j_{[q]}$ into a position after the task $j_{[g]}$ will decrease the functional value due to the use of reserve of the task $j_{[q]}$.
4. $\forall i, p \le i \le g$, $R_{j_{[i]}} \le 0$, but $\exists j_{[k]}$, $k < p$, such that $d_{j_{[k]}} > C_{j_{[p]}}$. We do not have reserves within the interval of insertion of the task $j_{[g]}$. But there is a task $j_{[k]}$, $k < p$, with the greater due date than $C_{j_{[p]}}$. Therefore, after permutation of the

task $j_{[k]}$ into the position p, a reserve is created within the interval of insertion of the task $j_{[g]}$.

5. The fourth condition is not fulfilled and $\forall i, p \le i \le g, R_{j_{[i]}} \le 0$, but $\exists j_{[k]}, k < p$, such that $d_{j_{[k]}} > C_{j_{[k]}}, d_{j_{[k]}} > C_{j_{[p]}} - l_{j_{[p]}}, d_{j_{[k]}} > d_{j_{[g]}} - l_{j_{[g]}}$. We do not have reserves within the interval of insertion of the task $j_{[g]}$. But there is a non-tardy task $j_{[k]}$, $k < p$, with the greater due date than the latest start time of the task $j_{[g]}$ that would make the task $j_{[g]}$ non-tardy. Consequently, there are permutations and insertions that can lead to the creation of reserves within the interval of insertion of the task $j_{[g]}$.

Proof According to Statement 4.3, a decrease in the functional value in consequence of the insertion of the task $j_{[g]}$ into an earlier position is possible only if there are reserves within the interval of insertion.

Fulfillment of one of the conditions 1–5 means that there are reserves within the interval of insertion of the task $j_{[g]}$ (conditions 1–3) or they will be created in consequence of permutations (conditions 4, 5). If conditions 4, 5 are not met, then permutation of tasks within positions $\overline{1, p-1}$ into the interval of insertion of the task $j_{[g]}$ will make them tardy. Since they have a greater priority than that of the task $j_{[g]}$, the insertion of the task $j_{[g]}$ will increase the functional value. Therefore, if none of the conditions 1–5 is fulfilled, then there are no insertions of the task $j_{[g]}$ into earlier positions that decrease the functional value. ∎

Corollary 4.3, SSO #4 Suppose that the number of tardy tasks in the sequence σ^{ord} is not zero. If for any tardy task $j_{[g]}$ there are no preceding tasks $j_{[s]}$, $s < g$, for which at least one of the conditions 1–5 of Statement 4.6 is true, then the sequence σ^{ord} corresponds to the optimal functional value.

Suppose that a tardy task $j_{[g]}$ in the sequence $\sigma^{ord} = \{j_{[1]}, j_{[2]}, \ldots, j_{[g]}\}$ has moved into a position $p = m$ in consequence of its insertion. Denote this task by $j^*_{[m]}$ ("$j_{[m]}$ marked with asterisk") and the obtained sequence by $\sigma(j_{[g]})$.

Statement 4.7 An insertion of a tardy task $j_{[k]}$, $k = \overline{m+1, g}$, into an earlier position in the sequence $\sigma(j_{[g]})$ will decrease the functional value only if at least one preceding task has a time reserve or (if there are no reserves) only in consequence of a permutation of the task $j^*_{[m]}$ into a position after the task $j_{[k]}$.

Proof The proof for the case when reserves exist is given in Statements 4.3 and 4.6. Let us consider the case when there are no reserves in the sequence $\sigma(j_{[g]})$.

It is obvious that among the tasks within positions $\overline{1, k-1}$ a permutation into position k is possible only for the task $j^*_{[m]}$ because $\forall i = \overline{1, k-1}\ d_{j_{[i]}} - C_{j_{[i]}} \le 0$ and $p_{j_{[i]}} \ge p_{j_{[k]}}$ for all unmarked tasks $i = \overline{1, k-1}$.

Perform permutation of the task $j^*_{[m]}$ into the position k. The value of the objective function before the permutation is

$$f = \sum_{i=m}^{k} \omega_{j_{[i]}} \cdot \max\left(0, C_{j_{[i]}} - d_{j_{[i]}}\right) = \sum_{i=m}^{k} \omega_{j_{[i]}} \left(C_{j_{[i]}} - d_{j_{[i]}}\right)$$

After the permutation of the task $j^*_{[m]}$ into the position k:

$$f' = \omega_{j_{[m]}} \left(C_{j_{[m]}} + \sum_{i=m+1}^{k} l_{j_{[i]}} - d_{j_{[m]}} \right) + \sum_{i=m+1}^{k} \omega_{j_{[i]}} \max\left(0,\ C_{j_{[i]}} - d_{j_{[i]}} - l_{j_{[m]}}\right)$$

where $C_{j_{[m]}}$ is the completion time of the task $j^*_{[m]}$ before the permutation.

Let us prove that $f - f'$ can be greater than zero.

$$\begin{aligned} f - f' &= \left[\omega_{j_{[m]}} \left(C_{j_{[m]}} - d_{j_{[m]}} \right) + \sum_{i=m+1}^{k} \omega_{j_{[i]}} \left(C_{j_{[i]}} - d_{j_{[i]}} \right) \right] \\ &\quad - \left[\omega_{j_{[m]}} \left(\tilde{N}_{j_{[m]}} + \sum_{i=m+1}^{k} l_{j_{[i]}} - d_{j_{[m]}} \right) + \sum_{i=m+1}^{k} \omega_{j_{[i]}} \cdot \max\left(0, C_{j_{[i]}} - d_{j_{[i]}} - l_{j_{[m]}}\right) \right] \\ &= \omega_{j_{[m]}} \left(C_{j_{[m]}} - d_{j_{[m]}} - C_{j_{[m]}} - \sum_{i=m+1}^{k} l_{j_{[i]}} + d_{j_{[m]}} \right) \\ &\quad + \sum_{i=m+1}^{k} \omega_{j_{[i]}} \left[\left(C_{j_{[i]}} - d_{j_{[i]}} \right) - \max\left(0, C_{j_{[i]}} - d_{j_{[i]}} - l_{j_{[m]}}\right) \right] \\ &= \sum_{i=m+1}^{k} \left[\omega_{j_{[i]}} \cdot \min\left(l_{j_{[m]}}, C_{j_{[i]}} - d_{j_{[i]}} \right) - \omega_{j_{[m]}} l_{j_{[i]}} \right]. \end{aligned} \tag{4.6}$$

Let us analyze Eq. (4.6). We break the sum (4.6) into two parts.

Denote by P the set of tasks within $\overline{m+1, k}$ that became non-tardy after the permutation of the task $j^*_{[m]}$, i.e., $P = \left\{ i : l_{j_{[m]}} \geq C_{j_{[i]}} - d_{j_{[i]}} \right\}$.

Denote by Q the set of tasks within $\overline{m+1, k}$ that remained tardy after the permutation of the task $j^*_{[m]}$, i.e., $Q = \left\{ i : l_{j_{[m]}} < C_{j_{[i]}} - d_{j_{[i]}} \right\}$.

Eq. (4.6) will take the form:

$$\sum_{j_{[i]} \in P} \left[\omega_{j_{[i]}} \left(\tilde{N}_{j_{[i]}} - d_{j_{[i]}} \right) - \omega_{j_{[m]}} l_{j_{[i]}} \right] + \sum_{j_{[i]} \in Q} \left(\omega_{j_{[i]}} l_{j_{[m]}} - \omega_{j_{[m]}} l_{j_{[i]}} \right) \tag{4.7}$$

We divide both sides of Eq. (4.7) by $l_{j_{[i]}} l_{j_{[m]}}$:

$$\sum_{j_{[i]} \in P} \left(\frac{\omega_{j_{[i]}}}{l_{j_{[i]}}} \xi_{j_{[i]}} - \frac{\omega_{j_{[m]}}}{l_{j_{[m]}}} \right) + \sum_{j_{[i]} \in Q} \left(\frac{\omega_{j_{[i]}}}{l_{j_{[i]}}} - \frac{\omega_{j_{[m]}}}{l_{j_{[m]}}} \right) \quad (4.8)$$

where $\xi_{j_{[i]}} = \frac{C_{j_{[i]}} - d_{j_{[i]}}}{l_{j_{[m]}}}$ and $\forall j_{[i]} : \xi_{j_{[i]}} \in [0, 1]$.

Tasks within positions $\overline{m+1, k}$ have a greater (or equal) priority than that of the task $j^*_{[m]}$. So, the value of Eq. (4.8) can be greater than zero. Hence, the permutation of the task $j^*_{[m]}$ into the position k can decrease the functional value. ∎

Example 4.1 (to Statement 4.7) Initial data for Example 4.1 is in Table 4.1.

σ^{ord} is the sequence of the tasks in non-increasing order of priorities (Table 4.2).

Sequence σ_1 (Table 4.3) is obtained as a result of optimization iteration for task 7. The position of insertion of task 7 equals to 1. The first position is occupied by task 1 for which $l_1 < l_7$, $d_1 < d_7$, $\omega_1 > \omega_7$. Therefore, the position of insertion of task 7 is 2. We mark task 7 with asterisk and arrange all the following tasks in non-increasing order of priorities.

Optimization iteration 2 for task 2. The position of insertion of task 2 is 2. Position 2 is occupied by task 7 marked with asterisk, for which $l_7 > l_2$. Since there are no tasks $j_{[i]}$ for which $d_{j_{[i]}} > d_{j_{[2]}} - l_{j_{[2]}}$ in the previous positions, we perform the procedure of releasing the reserve of the task marked with asterisk (Table 4.4).

Similar optimization procedures are performed for tasks 3, 4, 5, and 6 (Tables 4.5, 4.6, 4.7 and 4.8).

Table 4.1 Initial data for Example 4.1

j	1	2	3	4	5	6	7
l_j	25	30	45	50	67	71	80
ω_j	7	6	5	4	3	2	1
d_j	55	80	120	160	217	290	70

j the task number; l_j its processing time; d_j its due date; ω_j its weight

Table 4.2 Sequence σ^{ord}

j	p_j	l_j	ω_j	d_j	C_j	f_j
1	0.280	25	7	55	25	–
2	0.200	30	6	80	55	–
3	0.111	45	5	120	100	–
4	0.080	50	4	160	150	–
5	0.045	67	3	217	217	–
6	0.028	71	2	290	288	–
7	0.013	80	1	70	368	298

Here and further, $f_j = \omega_j \max(0, C_j - d_j)$. In this table, the functional value $f = 298$

Table 4.3 Iteration 1

j	p_j	l_j	ω_j	d_j	C_j	f_j
1	0.280	25	7	55	25	–
7*	0.013	80	1	70	105	35
2	0.200	30	6	80	135	330
3	0.111	45	5	120	180	300
4	0.080	50	4	160	230	280
5	0.045	67	3	217	297	240
6	0.028	71	2	290	368	156

The functional value $f = 1341$

Table 4.4 Iteration 2

j	p_j	l_j	ω_j	d_j	C_j	f_j
1	0.280	25	7	55	25	–
2	0.200	30	6	80	55	–
7**	0.013	80	1	70	135	65
3	0.111	45	5	120	180	300
4	0.080	50	4	160	230	280
5	0.045	67	3	217	297	240
6	0.028	71	2	290	368	156

The functional value $f = 1041$

This example was solved by the first polynomial component of the PSC-algorithm.

Statement 4.8 Suppose that we have performed an insertion of the task $j_{[g]}$ in the sequence σ^{ord}; tasks $j_{[l]}$, $l = \overline{g+1, n}$, are tardy; $l_{j_{[l]}} \geq l_{j_{[g]}}$ $\forall l = \overline{g+1, n}$; and for all preceding tasks $j_{[i]}$, $i = \overline{1, g}$, we have $R_{j_{[i]}} \leq 0$. In this case, tasks $j_{[l]}$ will remain on the occupied positions within the interval $\overline{g+1, n}$ in an optimal sequence.

Proof Validity of this statement is based on the fact that $\forall j_{[l]}$, $l = \overline{g+1, n}$, we have $l_{j_{[l]}} \geq l_{j_{[g]}}$. Reserves of the sequence σ^{ord} (initial sequence before inserting the task $j_{[g]}$ into the position m) are used in the sequence $\sigma(j_{[g]})$ by the shorter task $j_{[g]}$. Therefore, inserting tasks $j_{[l]}$, $l = \overline{g+1, n}$, into earlier positions will increase the total weighted tardiness of tasks moved into later positions. So, this will increase the functional value. ∎

Example 4.2 (to Statement 4.8) Initial data for Example 4.2 is in Table 4.9.

σ^{ord} is the sequence of the tasks in non-increasing order of priorities (Table 4.10).

Sequence σ_1 (Table 4.11) is obtained as a result of optimization iteration for task 4. The position of insertion of task 4 equals to 1. Since there are no reserves after iteration 1, then, according to Statement 4.8, we obtain an optimal sequence on the entire set of tasks.

Table 4.5 Iteration 3

j	p_j	l_j	ω_j	d_j	C_j	f_j
1	0.280	25	7	55	25	–
2	0.200	30	6	80	55	–
3	0.111	45	1	120	100	–
7**	0.013	80	5	70	180	110
4	0.080	50	4	160	230	280
5	0.045	67	3	217	297	240
6	0.028	71	2	290	368	156

The functional value $f = 786$

Table 4.6 Iteration 4

j	p_j	l_j	ω_j	d_j	C_j	f_j
1	0.280	25	7	55	25	–
2	0.200	30	6	80	55	–
3	0.111	45	1	120	100	–
4	0.080	50	5	160	150	–
7**	0.013	80	4	70	230	160
5	0.045	67	3	217	297	240
6	0.028	71	2	290	368	156

The functional value $f = 556$

Table 4.7 Iteration 5

j	p_j	l_j	ω_j	d_j	C_j	f_j
1	0.280	25	7	55	25	–
2	0.200	30	6	80	55	–
3	0.111	45	1	120	100	–
4	0.080	50	5	160	150	–
5	0.045	67	4	217	217	–
7**	0.013	80	3	70	297	227
6	0.028	71	2	290	368	156

The functional value $f = 383$

Table 4.8 Iteration 6

j	p_j	l_j	ω_j	d_j	C_j	f_j
1	0.280	25	7	55	25	–
2	0.200	30	6	80	55	–
3	0.111	45	1	120	100	–
4	0.080	50	5	160	150	–
5	0.045	67	4	217	217	–
6	0.028	71	3	290	288	–
7	0.013	80	2	70	368	298

The optimal functional value $f^{opt} = 298$

Table 4.9 Initial data for Example 4.2

j	1	2	3	4	5	6	7
l_j	25	30	45	50	67	71	80
ω_j	7	6	5	4	3	2	1
d_j	75	100	130	40	50	30	10

Table 4.10 Sequence σ^{ord}

j	p_j	l_j	ω_j	d_j	C_j	f_j
1	0.280	25	7	75	25	–
2	0.200	30	6	100	55	–
3	0.111	45	5	130	100	–
4	0.080	50	4	40	150	440
5	0.045	67	3	50	217	501
6	0.028	71	2	30	288	516
7	0.013	80	1	10	368	358

The functional value $f = 1815$

Table 4.11 Sequence σ_1

j	p_j	l_j	ω_j	d_j	C_j	f_j
4*	0.080	50	4	40	50	40
1	0.280	25	7	75	75	–
2	0.200	30	6	100	105	30
3	0.111	45	5	130	150	100
5	0.045	67	3	50	217	501
6	0.028	71	2	30	288	516
7	0.013	80	1	10	368	358

The optimal functional value $f^{opt} = 1545$

This example was solved by the first polynomial component of the PSC-algorithm.

Statement 4.9 Suppose that tasks $j_{[g]}$ and $j_{[k]}$, $g < k$, such that $p_{j_{[g]}} \geq p_{j_{[k]}}$, $l_{j_{[g]}} \leq l_{j_{[k]}}$, and $\max R(\sigma^{ord}) < l_{j_{[g]}}$, qualify for existing reserves in the sequence σ^{ord}. In this case, task $j_{[k]}$ can qualify for the reserves only if $d_{j_{[k]}} < d_{j_{[g]}}$.

Proof According to Eq. (4.4), the decrease in the weighted tardiness of a tardy task $j_{[g]}$ in consequence of its insertion into an earlier position is determined by its weight, its tardiness, and the length of the interval of insertion. The decrease does not depend on the processing time of the task itself. The increase in the weighted tardiness of tasks shifted into later positions in consequence of the insertion is determined by their weight, depends on whether the tasks within the interval of insertion have reserves or not and on the processing time of the inserted task. Since $\max R(\sigma^{ord}) < l_{j_{[g]}}$, then, in case of inserting into the same interval, the longer processing time of the inserted task and the lower its priority, the greater possible increase in the total weighted tardiness of tasks within the interval of insertion, and only if longer tasks have lower due dates, a decrease in the functional value is possible. ∎

4.2.2 *Sequence* σ^{fp}

Statement 4.10 Statements 4.1, 4.2, 4.4–4.6, 4.9 are valid for the sequence σ^{fp}.

Proof Let us show that differences between the sequences σ^{ord} and σ^{fp} do not break the proofs of the above statements. The main methods in the proof of Statements 4.1, 4.2, 4.4–4.6, 4.9 were permutations and insertions.

For permutations, proofs of the statements are similar if we use σ^{fp} instead of σ^{ord}, for the following reasons:

- we shift all tasks within the interval of permutation into earlier positions and, consequently, their total weighted tardiness is decreased. In this case, the tasks that break the order of the sequence σ^{ord} (moved in consequence of free permutations) are non-tardy, so we can eliminate them from consideration both in f and in f';
- for tardy tasks in the sequence σ^{fp}, the priority ordering is retained and, therefore, the priority of the task moved into a later position is greater than the priority of any task within the interval of permutation.

The procedure of insertion for the sequence σ^{fp} is similar to that for the sequence σ^{ord}, since the priority of an inserted task in σ^{fp} is always lower or equal to the priority of any task within the interval of insertion. This makes the presence of reserves of the tasks occupying earlier positions necessary for the insertion.

Thus, Statements 4.1, 4.2, 4.4–4.6, 4.9 are true also for the sequence σ^{fp}. ∎

There may be tasks within the interval of insertion in the sequence σ^{fp} that were moved in consequence of free permutations and break the priority ordering. Therefore, we must reorder all the tasks following the inserted task according to their priorities and, in turn, check for the tardy tasks in this sequence the possibility of moving them into earlier positions, due to the reserves of non-tardy tasks.

Statement 4.11, SSO #5 There are no permutations and insertions in the sequence σ^{fp} that decrease the functional value, and the sequence σ^{fp} is optimal on the entire set of tasks, if at least one of the following conditions holds:

1. $d_{j_{[i]}} - C_{j_{[i]}} \geq 0,\ \forall i = \overline{1, n}$;
2. $d_{j_{[i]}} - C_{j_{[i]}} \leq 0,\ \forall i = \overline{1, l},\ d_{j_{[i]}} - C_{j_{[i]}} > 0,\ \forall i = \overline{l+1, n}$;
3. $d_{j_{[i]}} - C_{j_{[i]}} \leq 0,\ \forall i = \overline{1, n}$;
4. Let R denote the set of tasks $j_{[i]}$: $d_{j_{[i]}} - C_{j_{[i]}} > 0$, Z denote the set of tasks $j_{[i]}$: $d_{j_{[i]}} - C_{j_{[i]}} \leq 0$. Then $d_{j_{[k]}} \leq d_{j_{[g]}} - l_{j_{[g]}} \forall k \in R, \forall g \in Z$.

Proof The first item does not require a proof, since $f = 0$. The proof of items 2–4 is similar to the proof of Statements 4.1, 4.3, and 4.4. ∎

Corollary 4.4 If at least one of the conditions formulated in Statement 4.11 holds, then the sequence σ^{fp} corresponds to an optimal functional value.

We introduce the following definitions.

Definition 4.11 A tardy task $j_{[g]}$ in the sequence σ^{fp} is called a *competing task* if there is at least one such preceding task $j_{[l]}$ in σ^{fp} that $d_{j_{[l]}} > d_{j_{[g]}} - l_{j_{[g]}}$ and $d_{j_{[l]}} - C_{j_{[l]}} > 0$.

Competing tasks in the sequence σ^{fp} are in non-increasing order of their priorities because σ^{fp} is obtained from the sequence σ^{ord} by free permutations of non-tardy tasks, and the permutations do not change the order of tardy tasks.

Definition 4.12 Tasks that were non-tardy in σ^{fp} but became tardy after insertions of tardy tasks into earlier positions, are called *originated tardy tasks*.

Remark 4.2 We determine the set of competing tardy tasks within the sequence σ^{fp}. The tardiness of originated tardy tasks in regard to their due dates appears during the insertion of competing tasks.

4.2.3 Iterations of Optimization

The algorithm consists of a series of uniform iterations. At each iteration, we check the possibility to use the time reserves of preceding tasks for the current competing task from the sequence σ^{fp} and construct an optimal schedule for tasks of the current subsequence. At the first iteration we construct an optimal schedule for tasks within positions $\overline{1, g_1}$ of σ^{fp} where $j_{[g_1]}$ is the first competing task.

According to Definition 4.11, we cannot reduce the weighted tardiness of non-competing tasks, so the sequence σ^{fp} is optimal within the interval $\overline{1, g_1 - 1}$.

At the second iteration, we consider a subsequence of tasks within interval $\overline{1, g_2}$. In it, the positions $\overline{1, g_1}$ occupies the optimal subsequence obtained at the first iteration. Positions $\overline{g_1 + 1, g_2}$ are occupied by the tasks that have these positions in σ^{fp}. $j_{[g_2]}$ is the next competing task in the sequence σ^{fp}.

Definition 4.13 At the current kth iteration of optimization we construct an optimal schedule on the *current subsequence* σ^k. The subsequence consists of:

- the optimal subsequence obtained at the previous iteration $k - 1$ executed for a competing task $j_{[g_{k-1}]}$;
- the current competing task $j_{[g_k]}$ (the next competing task from the sequence σ^{fp}); the task's priority does not exceed the priorities of tasks of the current subsequence, by construction of the sequence σ^{fp};
- the tasks that are in σ^{fp} between the competing tasks $j_{[g_{k-1}]}$ and $j_{[g_k]}$.

Remark 4.3 The tasks in positions between the competing tasks $j_{[g_{k-1}]}$ and $j_{[g_k]}$ may be non-tardy or non-competing tardy. A subsequence including these tasks is also optimal because a decrease in the weighted tardiness for non-competing tasks is impossible, according to Definition 4.11.

Remark 4.4 The number of competing tasks in the sequence σ^{fp} determines the number of iterations of optimization. We will show that we can eliminate some of the tasks from the set of competing tasks during the iteration of optimization. Thus, the total number of iterations is finite (the upper bound on the number of iterations is the total number of tasks) because, by construction of the algorithm, we eliminate the competing tasks, for which we executed the iteration of optimization, from the set of competing tasks and do not include them again into it. We can consider them as originated tardy tasks at the next iterations.

Suppose that we have already executed $k-1$ iterations and constructed an optimal subsequence σ^{k-1} on the set of tasks within positions $\overline{1, g_{k-1}}$ where $j_{[g_{k-1}]}$ is a competing task in sequence σ^{fp}. We proceed to the next competing task $j_{[g_k]}$ determined in the sequence σ^{fp}, according to the rules formulated in Definition 4.13. Then, under similar rules, we construct the subsequence σ^k that includes tasks within interval $\overline{1, g_k}$. We check the possibility of decreasing the weighted tardiness of the task $j_{[g_k]}$ due to the use of the existing time reserves or the reserves obtained as a result of permutations to the later positions of the tasks that earlier used these reserves at previous iterations. The functional value decreases or remains unchanged at each iteration.

Lemma 4.1 *Suppose that we have already executed* $k-1$ *iterations and constructed an optimal subsequence* σ^{k-1} *on the set of tasks within positions* $\overline{1, g-1}$. *Then, to obtain an optimal subsequence at the current iteration for the current competing task* $j_{[g]}$, *it is necessary and sufficient to obtain a subsequence in which we achieve the minimum possible increase in the total weighted tardiness in comparison with the optimal functional value at the previous iteration.*

Proof The proof follows from the fact that in a non-optimal subsequence within positions $\overline{1, g}$ the increase in the total weighted tardiness is not less than that in an optimal subsequence. ∎

Corollary 4.5 Thus, to obtain an optimal subsequence at the current iteration, it is necessary and sufficient to obtain a subsequence at this iteration in which the increase in the total weighted tardiness is the minimum possible in comparison with the optimal functional value obtained at the previous iteration.

In order to achieve the minimum possible increase in the total weighted tardiness, we insert the competing task $j_{[g]}$ into a position where its weighted tardiness is minimal or equal to zero and then perform a stepwise optimization for each tardy task within the intervals of insertion or permutation until we get the minimum possible increase in the total weighted tardiness.

4.2.4 The Current Subsequence

We use the following types of permutations and insertions in the current subsequence [24].

- *Permutations* (EFSR, Definition 4.3):
  - of tasks for which the time reserve is greater than zero;
  - of tasks that used reserves in consequence of permutations and insertions.

  We perform permutations if they decrease the functional value.
- *Insertions* (EBSR, Definition 4.5), they are used in these cases:
  - if there are reserves within the interval of insertion;
  - if reserves within the interval of insertion are created according to the conditions 4 and 5 of Statement 4.6.

We now give the rules for permutations and insertions performing in the subsequence σ^k.

Let us introduce rules for tasks marking in the subsequence σ^k when performing permutations and insertions.

In the subsequence σ^k, we mark with asterisk "*" the tasks that were tardy and in consequence of insertions moved into earlier positions using the time reserves of preceding tasks. We mark with two asterisks "**" the tasks that were earlier marked with asterisk "*" and in consequence of further permutations moved into later positions releasing part of the used reserves.

Suppose that we have already executed $k - 1$ iterations. Proceeding to the next competing task determines the beginning of a new iteration.

4.2.5 The Current Iteration

The following statements are true.

Statement 4.12 §1. Statements 4.1, 4.2, 4.4–4.6, 4.9 are also true for the current subsequence σ^k.

§2. If during the current iteration, after insertion of the task $j_{[g]}$, a tardy task $j_{[q]}$ in the current subsequence σ^k is preceded by an unmarked task $j_{[l]}$, $p_{j_{[l]}} < p_{j_{[q]}}$, and $p_{j_{[q]}} > p_{j_{[g]}}$, $q < g$, then a decrease in the functional value is possible in consequence of the move of the task $j_{[l]}$ into the position q.

Proof The proof of §1 is similar to the proof of Statement 4.10 where we proved that Statements 4.1, 4.2, 4.4–4.6, 4.9 are also true for the sequence σ^{fp}. In contrast to σ^{fp}, there are tasks in the current subsequence marked with "*" that were tardy and in consequence of insertions moved into earlier positions using the time reserves of preceding tasks. Also, there are tasks marked with "**" that during the iteration of optimization, in consequence of further permutations into the later positions, released part of the used reserves. This difference does not affect proofs of Statements 4.1, 4.2, 4.4–4.6, 4.9 that remain true for the subsequence σ^k.

§2. According to §1, Statement 4.6 is true for the subsequence σ^k. According to the rules for permutations performing during an optimal subsequence construction at the current iteration, there may exist tasks $j_{[q]}$ in the subsequence σ^k for which free permutations were performed at the previous steps of algorithm (condition 2 of Statement 4.6) and which became tardy in consequence of insertion of the current tardy task $j_{[g]}$, $g > q$. These tasks in σ^k are preceded by lower priority tasks $j_{[l]}$: $p_{j_{[l]}} < p_{j_{[q]}}$. According to the conditions of the statement, $p_{j_{[q]}} > p_{j_{[g]}}$. In this case, we adjust the interval of the free permutation of the task $j_{[q]}$ sequentially permuting the unmarked lower priority tasks after it, starting with the lowest priority task. The functional value decreases as a result of permutation of tasks $j_{[l]}$ into positions after the task $j_{[q]}$ if the task $j_{[q]}$ does not gain a reserve as a result of such permutations. ∎

Example 4.3 (to Statement 4.12) Initial data for Example 4.3 is in Table 4.12.

σ^{ord} is the sequence of the tasks in non-increasing order of priorities (Table 4.13).

As a result of free permutations, we obtain the sequence shown in Table 4.14.

Iteration 1: the first tardy task preceded by a task with time reserve is task 12. Determine for it the position of insertion, it is position 1. We perform its correction because the tasks in positions 1, …, 7 are tardy. We insert task 12 before task 11 (the task $j_{[q]}$) and mark task 12 with asterisk. Reorder all tasks following task 12* in non-increasing order of priorities. We get a sequence σ_1 (Table 4.15).

According to Statement 4.12, the weighted tardiness of task 11 can be reduced permuting unmarked tasks of lower priority into the position after it, starting with the minimum priority task. We find a task $j_{[l]}$ of a lower priority in the previous positions (task 2). Task 11 becomes non-tardy in consequence of the move of task 2 after it, and the functional value decreases (Table 4.16). Since tasks 7 and 5 are not tardy, we do not permute task 2 after them (since task 2 will gain a reserve).

An optimal sequence on the entire set of tasks is obtained. This example was solved by the first polynomial component of the PSC-algorithm. The use of Statement 4.12 is shown more deeply in Example 4.7 in Sect. 4.6 where the statement is satisfied for a number of tasks.

Statement 4.13 If at iteration k for a tardy task $j_{[g]}$ in the current subsequence σ^k exists a preceding non-tardy task $j_{[l]}$ such that

Table 4.12 Initial data for Example 4.3

j	1	2	3	4	5	6	7	8	9	10	11	12
l_j	5	27	20	45	23	8	26	53	4	3	14	57
ω_j	1	5	6	8	5	7	9	10	5	4	8	10
d_j	0	0	25	179	244	0	225	0	0	0	173	0

Table 4.13 Sequence σ^{ord}

j	p_j	l_j	ω_j	d_j	C_j	f_j
10	1.333	3	4	0	3	12
9	1.250	4	5	0	7	35
6	0.875	8	7	0	15	105
11	0.571	14	8	173	29	–
7	0.346	26	9	225	55	–
3	0.300	20	6	25	75	300
5	0.217	23	5	244	98	–
1	0.200	5	1	0	103	103
8	0.189	53	10	0	156	1560
2	0.185	27	5	0	183	915
4	0.178	45	8	179	228	392
12	0.175	57	10	0	285	2850

The functional value $f = 6272$

Table 4.14 Sequence σ^{fp}

j	p_j	l_j	ω_j	d_j	C_j	f_j
10	1.333	3	4	0	3	12
9	1.250	4	5	0	7	35
6	0.875	8	7	0	15	105
3	0.300	20	6	25	35	60
1	0.200	5	1	0	40	40
8	0.189	53	10	0	93	930
2	0.185	27	5	0	120	600
11	0.571	14	8	173	134	0
4	0.178	45	8	179	179	0
7	0.346	26	9	225	205	0
5	0.217	23	5	244	228	0
12	0.175	57	10	0	285	2850

The functional value $f = 4632$

Table 4.15 The first iteration. Sequence σ_1

j	p_j	l_j	ω_j	d_j	C_j	f_j
10	1.333	3	4	0	3	12
9	1.250	4	5	0	7	35
6	0.875	8	7	0	15	105
3	0.300	20	6	25	35	60
1	0.200	5	1	0	40	40
8	0.189	53	10	0	93	930
2	0.185	27	5	0	120	600
12*	0.175	57	10	0	177	1770
11	0.571	14	8	173	191	144
7	0.346	26	9	225	217	0
5	0.217	23	5	244	240	0
4	0.178	45	8	179	285	848

The functional value $f = 4544$

Table 4.16 Permutation of task 2. Optimal sequence

j	p_j	l_j	ω_j	d_j	C_j	f_j
10	1.333	3	4	0	3	12
9	1.250	4	5	0	7	35
6	0.875	8	7	0	15	105
3	0.300	20	6	25	35	60
1	0.200	5	1	0	40	40
8	0.189	53	10	0	93	930
12	0.175	57	10	0	150	1500
11	0.571	14	8	173	164	0
2	0.185	27	5	0	191	955
7	0.346	26	9	225	217	0
5	0.217	23	5	244	240	0
4	0.178	45	8	179	285	848

The optimal functional value $f^{opt} = 4485$

$$d_{j_{[l]}} > C_{j_{[g]}} - l_{j_{[g]}},\ \omega_{j_{[g]}} \min\left(l_{j_{[l]}}, C_{j_{[g]}} - d_{j_{[g]}}\right) > \omega_{j_{[l]}}\left(C_{j_{[g]}} - d_{j_{[l]}}\right), \tag{4.9}$$

then the permutation of the task $j_{[l]}$ into the position g decreases the functional value. The task $j_{[g]}$ after the permutation is marked with asterisk.

Proof According to §1 of Statement 4.12, condition 3 of Statement 4.6 is true for the current subsequence σ^k. Let us derive the conditions of a permutation that decreases the functional value, that is, $f - f' > 0$.

$$f = \sum_{i=l}^{g} \omega_{j_{[i]}} \max\left(0, C_{j_{[i]}} - d_{j_{[i]}}\right).$$

After the permutation of the task $j_{[l]}$ into the position g:

$$f' = \omega_{j_{[l]}} \max\left(0, C_{j_{[g]}} - d_{j_{[l]}}\right) + \sum_{i=l+1}^{g} \omega_{j_{[i]}} \max\left(0, C_{j_{[i]}} - l_{j_{[l]}} - d_{j_{[i]}}\right).$$

$$\begin{aligned} f - f' &= \sum_{i=l}^{g} \omega_{j_{[i]}} \max\left(0, C_{j_{[i]}} - d_{j_{[i]}}\right) - \omega_{j_{[l]}}\left(C_{j_{[g]}} - d_{j_{[l]}}\right) \\ &- \sum_{i=l+1}^{g} \omega_{j_{[i]}} \max\left(0, C_{j_{[i]}} - l_{j_{[l]}} - d_{j_{[i]}}\right) = -\omega_{j_{[l]}}\left(C_{j_{[g]}} - d_{j_{[l]}}\right) \\ &+ \sum_{i=l+1}^{g} \omega_{j_{[i]}} \max\left(0, C_{j_{[i]}} - d_{j_{[i]}}\right) - \sum_{i=l+1}^{g} \omega_{j[i]} \max\left(0, C_{j_{[i]}} - l_{j_{[l]}} - d_{j_{[i]}}\right). \end{aligned}$$

Suppose that all tasks within positions $i = \overline{l+1, g-1}$ are not tardy before the move, i.e., $C_{j_{[i]}} - d_{j_{[i]}} < 0$. Then

$$\begin{aligned} f - f' &= -\omega_{j_{[l]}}\left(C_{j_{[g]}} - d_{j_{[l]}}\right) + \omega_{j_{[g]}}\left(C_{j_{[g]}} - d_{j_{[g]}}\right) - \omega_{j_{[g]}} \max\left(0, C_{j_{[g]}} - l_{j_{[l]}} - d_{j_{[g]}}\right) \\ &= -\omega_{j_{[l]}}\left(C_{j_{[g]}} - d_{j_{[l]}}\right) + \omega_{j_{[g]}} \min\left(l_{j_{[l]}}, C_{j_{[g]}} - d_{j_{[g]}}\right) > 0. \end{aligned}$$

Thus, if

$$\omega_{j_{[g]}} \min\left(l_{j_{[l]}}, C_{j_{[g]}} - d_{j_{[g]}}\right) > \omega_{j_{[l]}}\left(C_{j_{[g]}} - d_{j_{[l]}}\right),$$

then the functional value decreases in consequence of the permutation of the task $j_{[l]}$ into the position g. ∎

Statement 4.14 If the first positions $i = 1, \ldots, k$ in the sequence σ^{ord} are occupied by the tasks with $R_{j_{[i]}} \le 0$, then we can decompose the sequences σ^{ord} and σ^{fp} into two subsequences: σ^1 includes the tasks $j_{[1]}, \ldots, j_{[k]}$, σ^2 contains $j_{[k+1]}, \ldots, j_{[n]}$. The subsequence σ^1 is optimal. The tasks that belong to σ^1 do not participate in permutations. We do the optimization only for the tasks of the subsequence σ^2.

Statement 4.15 Suppose that we have already executed $k-1$ iterations. We proceed to the next competing task $j_{[g]}$ and to iteration k, according to the rules formulated in Definition 4.13. The necessary conditions for inserting a competing task $j_{[g]}$ in the subsequence σ^k into the position p where the task's tardiness is minimal or equal to zero, are:

1. Tasks $j_{[i]} \in I_{j_{[g]}}$, $d_{j_{[i]}} > d_{j_{[g]}} - l_{j_{[g]}}$, have reserves within the interval of insertion of the task $j_{[g]}$.
2. If there are no reserves within positions $\overline{p, g-1}$, then tasks $j_{[i]}$ within positions $\overline{1, p-1}$ have reserves and the following conditions hold:

$$d_{j_{[i]}} > d_{j_{[g]}} - l_{j_{[g]}}, \quad d_{j_{[i]}} > C_{j_{[i]}}, \text{ and } \quad d_{j_{[i]}} > C_{j_{[p-1]}}. \tag{4.10}$$

3. If there are no reserves within positions $\overline{1, g-1}$, then there are tasks $j_{[i]}$ marked with "*" or "**" within those positions.

Proof According to §1 of Statement 4.12, conditions of Statements 4.1, 4.4, 4.5, and 4.6 are true for the current subsequence σ^k.

In case (1), the necessity of the condition $d_{j_{[i]}} > d_{j_{[g]}} - l_{j_{[g]}}$ is based on Statement 4.4.

In case (2), according to conditions (4.10), a task $j_{[i]}$ that can be moved to the interval of insertion to create a reserve, must meet the following requirements: (a) it must be non-tardy; (b) $d_{j_{[i]}} > d_{j_{[g]}} - l_{j_{[g]}}$; (c) it must have a reserve in the position $p-1$ and, therefore, within the interval of insertion of the task $j_{[g]}$. If these requirements are not met, then moving the task $j_{[i]}$ into the interval of insertion of the task $j_{[g]}$ will make the task $j_{[i]}$ tardy. According to Statement 4.1, this will increase the functional value.

In case (3), permutations of marked tasks into the later positions create reserves within the interval of insertion of the task $j_{[g]}$. Using of them by the task $j_{[g]}$ can lead to a decrease in the functional value (Statements 4.5 and 4.6).

Therefore, if the conditions of Statement 4.15 are not met, the insertion of the competing task $j_{[g]}$ into an earlier position increases the functional value. ∎

Remark 4.5 If, in case (3), the marked tasks are in positions greater than p, then we define a new position of insertion $p^{new} > p$ for the task $j_{[g]}$ and check the necessary conditions to insert the task $j_{[g]}$ into the position p^{new} in the same way.

Let us formulate the necessary conditions for move of a tardy task into an earlier position in the subsequence σ^k that decreases the functional value due to existing and released reserves.

Statement 4.16 Suppose that $j_{[g]}$ is a tardy (competing or originated) task in the current subsequence σ^k. A decrease in the functional value when moving the task $j_{[g]}$ into an earlier position is possible only if at least one of the following conditions is satisfied:

1. $\exists j_{[i]}$, $p \le i \le g$, such that $R_{j_{[i]}} > 0$, $d_{j_{[i]}} > d_{j_{[g]}} - l_{j_{[g]}}$ where p is the position in which the weighted tardiness of the task $j_{[g]}$ is minimal (or equals to zero). There are tasks $j_{[i]}$ with time reserves within the interval of insertion of the task $j_{[g]}$ and $d_{j_{[i]}} > d_{j_{[g]}} - l_{j_{[g]}}$ (Statement 4.4).

2. $\exists j_{[l]}$, $l < g$, such that $d_{j_{[l]}} > C_{j_{[g]}} - l_{j_{[g]}}$ and condition (4.9) is met. There is a non-tardy task $j_{[l]}$ which due date is greater than the start time of the task $j_{[g]}$, and condition (4.9) is satisfied (Statement 4.13).
3. $\forall i, p \le i \le g, R_{j_{[i]}} \le 0$, but $\exists j_{[l]}, l < p$, such that $d_{j_{[l]}} > C_{j_{[p]}}$. A free permutation of the task $j_{[l]}$ after the task $j_{[p]}$ is possible.
4. $\forall i, p \le i \le g, R_{j_{[i]}} \le 0$, but $\exists j_{[l]}, l < p$, such that $d_{j_{[l]}} > C_{j_{[l]}}$, $d_{j_{[l]}} > C_{j_{[p]}} - l_{j_{[p]}}$, $d_{j_{[l]}} > d_{j_{[g]}} - l_{j_{[g]}}$ (Statement 4.4).
5. $\forall i,\ i = \overline{1, g-1},\ R_{j_{[i]}} \le 0$, but $\exists j_{[l]}, j_{[r]}, l < r < g, p_{j_{[l]}} \le p_{j_{[r]}}$. Here, $j_{[l]}$ is an unmarked task, $j_{[r]}$ is an originated tardy task within the interval of insertion of the competing task $j_{[g]}$, and a free permutation was done for the task $j_{[r]}$ at the previous steps of the algorithm (§2 of Statement 4.12).
6. $\forall i,\ i = \overline{1, g-1},\ R_{j_{[i]}} \le 0$, but $\exists j^{*}_{[m]}$ ($j^{**}_{[m]}$), $m < g$ (Statement 4.7).

Proof According to §1 of Statement 4.12, Statement 4.6 holds for the current subsequence σ^k. Similar to Statement 4.6, fulfillment of one of the conditions 1–4 means that there are reserves within the interval of insertion of the task $j_{[g]}$ (conditions 1, 2) or they will be created in consequence of permutations (conditions 3, 4). Fulfillment of condition 5 or 6 means that the subsequence σ^k contains tasks that have used the reserves within the interval of insertion of the task $j_{[g]}$, and moving them into the later positions will lead to creation of reserves within this interval. If none of the conditions 1–6 is fulfilled, then a move of the task $j_{[g]}$ into an earlier position will increase the functional value. ∎

Corollary 4.6, SSO #6 Suppose that we have constructed an optimal subsequence σ^k on the set of tasks within positions $\overline{1, g}$ as a result of k iterations of the algorithm. If none of the conditions 1–6 of Statement 4.16 is satisfied for each of the tardy tasks within positions $\overline{g+1, n}$, then they remain in the positions they are, and the current sequence is optimal on the entire set of tasks.

Corollary 4.7, SSO #7 This follows from Corollary 4.6 when $k = 0$. If none of the conditions 1–6 of Statement 4.16 is satisfied for each tardy task in the sequence σ^{fp}, then the sequence σ^{fp} is optimal on the entire set of tasks.

Statement 4.17 The following procedures for tardy tasks are necessary during the iterations of optimization within the current subsequence:

1. Determination and extension of the interval of insertion (conditions 1–4 of Statement 4.16);
2. Determination of existence of time reserves within the extended interval of insertion (Definition 4.1);
3. If time reserves exist within the extended interval of insertion, the procedure of insertion for a tardy task into a position defined by the interval of insertion (Definition 4.6). Then we need to reorder all tasks that follow the inserted tardy

task according to the non-increasing order of their priorities (Statement 4.6) and to perform the optimization within the extended interval of insertion;
4. If the reserves are absent within the extended interval of insertion or we are unable to use them, the procedure of releasing the reserves by means of permutation of the tasks that earlier used them at the previous steps of the iteration (conditions 5, 6 of Statement 4.16, Statement 4.12);
5. Optimization within the interval of permutation due to the use of the reserves released by tardy tasks (conditions 5, 6 of Statement 4.16).

All procedures 1–5 are based on the conditions for a task permutations given in Statement 4.16, We perform them both for competing tasks and for tardy tasks within their intervals of insertion.

Proof The proof is obvious and is based on Statement 4.16. ∎

Statement 4.18 If there is such a task $j_{[k]}$ within the interval of insertion of a tardy task $j_{[g]}$ that $\omega_{j_{[k]}} \geq \omega_{j_{[g]}}$, $d_{j_{[k]}} \leq d_{j_{[g]}}$, $l_{j_{[k]}} \leq l_{j_{[g]}}$, then the interval of insertion for the task $j_{[g]}$ is determined by the positions $\overline{k+1, g-1}$.

Proof Validity of this statement is based on Statement 4.15 and on the following property [25]: if $\omega_{j_{[k]}} \geq \omega_{j_{[g]}}$, $l_{j_{[i]}} \leq l_{j_{[g]}}$, $d_{j_{[i]}} \leq d_{j_{[g]}}$, then the task $j_{[k]}$ precedes the task $j_{[g]}$ in an optimal schedule. Therefore, the position of insertion for the task $j_{[g]}$ is $p \leq k+1$. The optimization is done due to the reserves of non-tardy tasks within the interval of insertion and of tasks within positions $\overline{1, p-1}$ that meet the requirements of Statement 4.15 by moving them into the interval of insertion. ∎

Thus, the conditions formulated in Statement 4.18 allow us to reduce the interval of insertion for a current tardy task. Since we execute the optimization within this interval, we can correspondingly reduce the number of permutations.

Statement 4.19 Suppose that we have already executed $k-1$ iterations and constructed an optimal subsequence σ^{k-1} on the set of tasks within positions $\overline{1, g-1}$. The next competing task $j_{[g]}$ such that $p_{j_{[g]}} \leq p_{j_{[i]}}$, $i \in \overline{1, g-1}$, can take an earlier position at iteration k. This will lead to the minimum possible increase in the total weighted tardiness of the current subsequence in comparison with the functional value at iteration $k-1$ and to the construction of an optimal subsequence at iteration k only if at least one task $j_{[l]}$ within the interval $\overline{1, g-1}$ has a reserve and $d_{j_{[l]}} > d_{j_{[g]}} - l_{j_{[g]}}$ or (if reserves are absent) there are marked tasks within the interval $\overline{1, g-1}$.

Proof Validity of this statement is based on Statement 4.16. According to conditions 1–4 of Statement 4.16, we can reduce the functional value in the current subsequence only if $d_{j_{[l]}} > d_{j_{[g]}} - l_{j_{[g]}}$ for tasks with reserves. Otherwise, according to condition 6 of Statement 4.16, we can obtain additional reserves only by means of marked tasks that earlier used the reserves, permuting them into later positions. ∎

Statement 4.20 (Lawler [7]) Suppose that for competing tasks $j_{[k]}, j_{[r]} \in J$ in the sequence σ^{fp} we have $l_{j_{[k]}} \le l_{j_{[r]}}$, $\omega_{j_{[k]}} \ge \omega_{j_{[r]}}$, $d_{j_{[k]}} \le d_{j_{[r]}}$. Then the task $j_{[k]}$ will precede the task $j_{[r]}$ in an optimal schedule.

Statement 4.21 Non-competing tasks in the current subsequence σ^k cannot take positions earlier than those they occupied in σ^{fp}.

Proof The proof is obvious, according to Statement 4.4. ∎

Statement 4.22 Suppose that we have already executed $k - 1$ iterations and constructed an optimal subsequence σ^{k-1} on the set of tasks within positions $\overline{1, g-1}$. The competing task $j_{[g]}$ such that $p_{j_{[g]}} \le p_{j_{[i]}}$, $i \in \overline{1, g-1}$, cannot be moved into an earlier position at iteration k and remains in its position that corresponds to the minimum possible increase in the total weighted tardiness of the current subsequence in comparison with the functional value at iteration $k - 1$ and to the construction of an optimal subsequence at iteration k if there are no reserves within the interval $\overline{1, g-1}$ and for all marked tasks $j^*_{[m]}\left(j^{**}_{[m]}\right)$ within this interval we have: $l_{j^*_{[m]}}(l_{j^{**}_{[m]}}) \le l_{j_{[g]}}$, $\forall m = \overline{1, g-1}$.

Proof Since the subsequence σ^k is optimal on the set of tasks within the interval $\overline{1, g-1}$ and was constructed at the previous iteration, we base the proof on Statement 4.16. The competing task $j_{[g]}$ has the lowest priority in compare to all preceding unmarked tasks. Therefore, its move into an earlier position in the absence of reserves will increase the functional value. Since all marked tasks have shorter processing times, additional reserves released in consequence of permutation of the marked tasks into later positions are not enough to move the task $j_{[g]}$ into an earlier position. So, the task $j_{[g]}$ remains in its position, and the current subsequence is optimal within positions $\overline{1, g}$. ∎

Statement 4.23 Suppose that we have constructed an optimal subsequence σ^k on the set of tasks within positions $\overline{1, g-1}$ at the previous iteration. If the competing task $j_{[g]}$ such that $p_{j_{[g]}} \le p_{j_{[i]}}$, $i \in \overline{1, g-1}$, has not used the existing reserves during permutations and insertions, the minimum functional value corresponds to the position g occupied by this task, and there are no tasks marked with "*" ("**"), then we can eliminate the task $j_{[g]}$ from the set of competing tasks, and the current subsequence is optimal on the set of tasks within positions $\overline{1, g}$.

Proof The proof is based on Statement 4.16. According to Corollary 4.6, none of the conditions 1–6 is satisfied for the tardy task $j_{[g]}$, so this task remains in the occupied position. Since there are no marked tasks in the subsequence σ^k, then we can eliminate the task $j_{[g]}$ from the set of competing tasks, and the current subsequence is optimal on the set of tasks within positions $\overline{1, g}$. ∎

Statement 4.24 Suppose that we have already executed $k - 1$ iterations and constructed an optimal subsequence σ^{k-1} on the set of tasks within positions

$\overline{1, g-1}$. Let us proceed to the next competing task $j_{[g]}$ from the sequence σ^{fp}, $p_{j_{[g]}} \le p_{j_{[i]}}$, $i \in \overline{1, g-1}$, and formulate the conditions that allow us to shorten its interval of insertion (the interval where we achieve the minimum possible increase in the total weighted tardiness of the current subsequence in comparison with the functional value at iteration $k - 1$). If there is an unmarked tardy task or a task with zero reserve in the position of insertion p of the task $j_{[g]}$ and there are no tasks within positions $\overline{1, p-1}$ corresponding to conditions (4.10) and tasks marked with "*" ("**"), then we must insert the task $j_{[g]}$ into the position $p + 1$. If similar conditions are satisfied in the position $p + 1$, we must insert the task $j_{[g]}$ into the position $p + 2$, etc., up to the position g.

Proof The proof is based on Statement 4.16. Since conditions 3, 4, and 6 of Statement 4.16 are not satisfied, then we will execute the optimization within the interval of insertion determined by positions $\overline{p, g-1}$. The priority of the inserted competing task $j_{[g]}$ is less than or equal to the priorities of tasks within the interval of insertion and tasks within positions $\overline{1, p-1}$. There are no tasks within positions $\overline{1, p-1}$ that create reserves in consequence of permutations into later positions. The task occupying the position of insertion p has a higher or equal priority than that of the task $j_{[g]}$. Consequently, its insertion into the position p will only increase the functional value. Similarly, if there are tardy tasks or tasks with zero reserve in the positions $p + 1$, $p + 2$, …, then we must perform the insertion into the next position, and so on, up to the position g. ∎

The following statements allow us to determine and eliminate unpromising permutations and insertions.

Statement 4.25 Suppose that we have already executed $k - 1$ iterations and constructed an optimal subsequence σ^{k-1} on the set of tasks in positions $\overline{1, g-1}$. Let us proceed to the next competing task $j_{[g]}$ from the sequence σ^{fp}, $p_{j_{[g]}} \le p_{j_{[i]}}$, $i \in \overline{1, g-1}$. If during the iteration of optimization we either have $d_{j_{[i]}} \le d_{j_{[g]}} - l_{j_{[g]}}$ for all preceding non-tardy tasks $j_{[i]}$, or there are no reserves within positions $\overline{1, g-1}$ and $l_{j^*_{[m]}} \le l_{j_{[g]}}$ for marked tasks in these positions, then the task $j_{[g]}$ remains in its position that corresponds to the minimum possible increase in the total weighted tardiness of the current subsequence in comparison with the functional value on the iteration $k - 1$. This task cannot take a position less than g in an optimal schedule, and the current subsequence is optimal within positions $\overline{1, g}$.

Proof The proof is based on Statements 4.15 and 4.16. Necessary conditions for insertion of the task $j_{[g]}$ into an earlier position are not met. The task $j_{[g]}$ remains in its position because additional reserves released as a result of the permutations of marked tasks into later positions are not enough, and the reserves can not be created, since condition 6 of Statement 4.16 is not satisfied. The current subsequence is optimal on the set of tasks within positions $\overline{1, g}$. ∎

Statement 4.26 If during the insertion of a tardy task $j_{[g]}$ in an optimal subsequence σ^k there are no preceding tasks $j_{[i]}$ within positions $\overline{1, g-1}$ for which $d_{j_{[i]}} > d_{j_{[g]}} - l_{j_{[g]}}$ and $d_{j_{[i]}} > C_{j_{[i]}}$, but there are tasks $j^*_{[m]}$ ($j^{**}_{[m]}$), then the minimum possible increase in the total weighted tardiness of the current subsequence in comparison with the functional value at the previous iteration and an optimal subsequence at the current iteration are achieved only due to reserves released by the tasks $j^*_{[m]}$ ($j^{**}_{[m]}$) during their insertions into the position k corresponding to their priorities (if $p_{j^*_{[m]}}$ (or $p_{j^{**}_{[m]}}$) is less than $p_{j_{[g]}}$, then $k = g$). Also, the following is necessary:

$$\omega_{j^*_{[m]}}\left(C_{j_{[k]}} - d_{j^*_{[m]}}\right) < \sum_{i=m}^{g} \omega_{j_{[i]}} \max\left(0, C_{j_{[i]}} - d_{j_{[i]}}\right). \tag{4.11}$$

Proof The proof is based on Statement 4.16. Since there are no reserves for the insertion of the tardy task $j_{[g]}$ into earlier positions (conditions 1–4 of Statement 4.16 are not satisfied), then we execute the optimization due to the reserves released by the tasks $j^*_{[m]}$ ($j^{**}_{[m]}$) that earlier used them (condition 6 of Statement 4.16). In the Eq. (4.11), the left part is the weighted tardiness of the task $j^*_{[k]}$ after its insertion into the position k (the task $j^*_{[m]}$ after its move is $j^*_{[k]}$), and the sum on the right side of the condition (4.11) reflects the maximum possible decrease in the total weighted tardiness as a result of this move. If the condition (4.11) is not met, then we do not perform the permutations of the marked tasks. This significantly decreases the number of permutations. ∎

Statement 4.27 If a competing task $j_{[g]}$, $p_{j_{[g]}} \le p_{j_{[i]}}$, $i \in \overline{1, g-1}$, as a result of the iteration of its optimization that determines the possibility of using reserves for this task, took the position $k > p$, then the following competing tasks $j_{[l]}$, for which $d_{j_{[l]}} \ge d_{j_{[g]}}$, $l_{j_{[l]}} \ge l_{j_{[g]}}$, $p_{j_{[l]}} \le p_{j_{[g]}}$, can not take a position lower than $k + 1$ in an optimal schedule.

Proof The competing tasks $j_{[l]}$ within positions $\overline{g+1, n}$ are in non-increasing order of their priorities. Therefore, if $d_{j_{[l]}} \ge d_{j_{[g]}}$ and $l_{j_{[l]}} \ge l_{j_{[g]}}$, then the task $j_{[g]}$ precedes the task $j_{[l]}$ in an optimal schedule. Thus, the conditions formulated in Statement 4.27 allow to shorten the interval of insertion of the current tardy task. Since we perform the optimization procedure within this interval, this correspondingly decreases the number of permutations. ∎

Statement 4.28 (Potts and Van Wassenhove [25]) Suppose $l_{j_{[l]}} \le l_{j_{[r]}}$, $d_{j_{[l]}} \le d_{j_{[r]}}$, $\omega_{j_{[l]}} \ge \omega_{j_{[r]}}$ for some tardy tasks $j_{[l]}, j_{[r]} \in J$ in the current subsequence σ^k. Then the task $j_{[l]}$ will precede the task $j_{[r]}$ in an optimal schedule.

Statement 4.29 Suppose that we have already executed $k-1$ iterations and constructed an optimal subsequence σ^{k-1} on the set of tasks within positions $\overline{1, g-1}$. If the next competing task $j_{[g]}$, $p_{j_{[g]}} \le p_{j_{[i]}}$, $i \in \overline{1, g-1}$, after executing its optimization iteration, remains in its position that corresponds to the minimum possible increase in the total weighted tardiness of the current subsequence in comparison with the functional value at the previous iteration; there are no tasks marked with "*" ("**") within positions $\overline{1, g-1}$; and all tasks within positions $\overline{g+1, n}$ are tardy, then we can eliminate the task $j_{[g]}$ and tasks $j_{[i]}$ such that $d_{j_{[i]}} \geq d_{j_{[g]}}$, $l_{j_{[i]}} \geq l_{j_{[g]}}$, $i \in \overline{g+1, n}$, from the number of tasks competing for reserves. These tasks cannot take earlier positions in an optimal schedule, and the current subsequence is optimal within positions $\overline{1, g}$.

Proof The proof is based on Statement 4.16. Since the task $j_{[g]}$ after executing its optimization iteration remains in its original position, then existing reserves were not enough to move it into earlier positions (conditions 1–4 of Statement 4.16). Since tasks marked with "*" ("**") that earlier used reserves are absent (condition 6 of Statement 4.16 is not satisfied), then additional reserves cannot be created. The task $j_{[g]}$ remains in its position, we eliminate it from the set of competing tasks. Therefore, according to Statement 4.27, we can also eliminate tasks $j_{[i]}$, $i \in \overline{g+1, n}$, such that $d_{j_{[i]}} \geq d_{j_{[g]}}$, $l_{j_{[i]}} \geq l_{j_{[g]}}$. ∎

Statement 4.30 Suppose that we have already executed $k-1$ iterations and constructed an optimal subsequence σ^{k-1} on the set of tasks within positions $\overline{1, g-1}$. The tardy task $j_{[g]}$ at iteration k can take earlier position, which will lead to the minimum possible increase in the total weighted tardiness of the current subsequence in comparison with the functional value at the previous iteration and to the construction of an optimal subsequence at the current iteration, only if at least one task $j_{[l]}$ within positions $\overline{1, g-1}$ has a reserve and $d_{j_{[l]}} > d_{j_{[g]}} - l_{j_{[g]}}$ or (if there are no reserves) there are marked tasks within positions $\overline{1, g-1}$.

Proof The proof is obvious and is based on fulfillment of at least one of the conditions 1–6 of Statement 4.16. ∎

In Statements 4.31, 4.32, we formulate sufficient signs of optimality of a solution we have obtained.

Statement 4.31, SSO #8 Suppose that we have already executed $k-1$ iterations and constructed an optimal subsequence σ^{k-1} on the set of tasks within positions $\overline{1, g-1}$. If at iteration k in the subsequence σ^k the competing task $j_{[g]}$, $p_{j_{[g]}} \leq p_{j_{[i]}}$, $i \in \overline{1, g-1}$, as a result of the optimization iteration for it, has not used the existing reserves; the minimum possible increase in the total weighted tardiness of the current subsequence in comparison with the functional value at the previous iteration corresponds to the position g of this task; there are no tasks marked with "*" ("**") within positions $\overline{1, g-1}$ or the conditions of Statement 4.26 are not satisfied for the marked tasks; and $\forall j_{[r]}$, $r = \overline{g+1, n}$, $d_{j_{[r]}} \geq d_{j_{[g]}}$, $l_{j_{[r]}} \geq l_{j_{[g]}}$, and $C_{j_{[r]}} \geq d_{j_{[r]}}$, then we can eliminate the tasks $j_{[r]}$ from the set of competing tasks. These tasks remain in their positions, and the current subsequence is optimal on the entire set of tasks.

Proof The proof is based on Statement 4.16. The conditions 1–6 of Statement 4.16 are not satisfied for the current competing task $j_{[g]}$. Tasks $j_{[r]}$, $r = \overline{g+1,\ n}$, such that $d_{j_{[r]}} \geq d_{j_{[g]}}$, $l_{j_{[r]}} \geq l_{j_{[g]}}$, and $C_{j_{[r]}} \geq d_{j_{[r]}}$, are in non-increasing order of priorities and, according to Statement 4.27, must be processed after the task $j_{[g]}$. Therefore, they remain in their positions. ∎

We check the following SSO #9 immediately after appending the next competing task to the current subsequence.

Statement 4.32, SSO #9 Suppose that we have already executed $k - 1$ iterations and constructed an optimal subsequence σ^{k-1} on the set of tasks within positions $\overline{1, g-1}$. Suppose that at iteration k executed for the next competing task $j_{[g]}$, for all tasks $j_{[l]}$, $l = \overline{1, g-1}$, we have $p_{j_{[g]}} \leq p_{j_{[l]}}$, $d_{j_{[l]}} \leq C_{j_{[l]}}$; for all marked tasks $j^*_{[l]}$ $(j^{**}_{[l]})$ we have $l_{j^*_{[l]}} (l_{j^{**}_{[l]}}) \leq l_{j_{[i]}}$, $i = \overline{g, n}$; and there are no reserves within positions $\overline{g, n}$. Then we can eliminate tardy tasks within positions $\overline{g, n}$ from the set of competing tasks. They remain in their positions corresponding to the minimum possible increase in the total weighted tardiness of the current subsequence in comparison with the functional value at the previous iteration. The current sequence corresponds to the optimal functional value on the entire set of tasks.

Proof The proof is based on Statement 4.16. The priority of the task $j_{[g]}$ is less than or equal to the priorities of tasks within positions $\overline{1, g-1}$. All tasks in these positions have no reserves, and reserves cannot be created since the processing times of marked tasks are shorter than that of the task $j_{[g]}$. Therefore, the task $j_{[g]}$ cannot take an earlier position (the conditions 1–6 of Statement 4.16 are not satisfied for the task $j_{[g]}$). There are no reserves within positions $\overline{g, n}$, these tasks are in non-increasing order of their priorities. Therefore, permutations within this interval are eliminated. Since processing times of marked tasks are shorter than those of tasks within positions $\overline{g, n}$, then the tasks within this interval cannot be moved into earlier positions (conditions 1–6 of Statement 4.16 for the tasks within positions $\overline{g, n}$ are also not satisfied). So, the current sequence is optimal on the entire set of tasks. ∎

Corollary 4.8 In Statements 4.23, 4.29, 4.31, 4.32, we have given the conditions of eliminating one or several competing tasks from the set of competing tasks. We do not include them to the set again. Either all of the conditions of Statement 4.23, or all of the conditions of Statement 4.29, or all of the conditions of Statement 4.31, or all of the conditions of Statement 4.32 must be satisfied. For these tasks, we do not execute iterations of optimization, we can consider them as originated tardy tasks at the next iterations.

Statement 4.33 Suppose that we have already executed $k - 1$ iterations and constructed an optimal subsequence σ^{k-1} on the set of tasks within positions $\overline{1, g-1}$. Suppose that at iteration k for the next competing task $j_{[g]}$, $p_{j_{[g]}} \leq p_{j_{[i]}}$, $i \in \overline{1, g-1}$, there is a tardy task $j_{[r]}$ within the interval of insertion of the task $j_{[g]}$

such that $d_{j_{[r]}} < d_{j_{[g]}}$, $\omega_{j_{[r]}} > \omega_{j_{[g]}}$, and $l_{j_{[r]}} < l_{j_{[g]}}$. Then, we can decompose the current subsequence and skip consideration of tasks within positions $\overline{1, r}$ at this iteration of optimization. We execute the optimization (i.e., the minimization of the increase in the total weighted tardiness of the current subsequence in comparison with the functional value at the previous iteration and the construction of an optimal subsequence at the current iteration) within positions $\overline{r+1,\ g-1}$.

Proof The proof is based on Statement 4.16. According to [25], the task $j_{[r]}$ precedes the task $j_{[g]}$. Since the task $j_{[r]}$ is tardy, then the existing reserves are not enough to decrease the weighted tardiness for this task (the conditions 1–4 of Statement 4.16). And since $d_{j_{[r]}} < d_{j_{[g]}}$ and $l_{j_{[r]}} < l_{j_{[g]}}$, we can decompose the subsequence and make the above mentioned eliminations. This will significantly decrease the number of permutations performed. ∎

Statement 4.34 Suppose that we have already executed $k-1$ iterations and constructed an optimal subsequence σ^{k-1} on the set of tasks within positions $\overline{1, g-1}$. We execute iteration k for the next competing task $j_{[g]}$, $p_{j_{[g]}} \le p_{j_{[i]}}$, $i \in \overline{1, g-1}$. Suppose that the previous competing task in the subsequence σ^k is a task $j_{[q]}$, $q < g$, we have $l_{j_{[g]}} \ge l_{j_{[q]}}$, $d_{j_{[g]}} \le d_{j_{[q]}} < C_{j_{[p^{new}]}} - l_{j_{[p^{new}]}}$ where p^{new} is the position of insertion of the task $j_{[q]}$, and there are no reserves within positions $\overline{1, p^{new}}$. If the task $j_{[q]}$ took the position $s > p^{new}$ as a result of this iteration of optimization, then the position of the task $j_{[g]}$ in an optimal schedule (the position corresponding to the minimum increase in the total weighted tardiness of the current subsequence in comparison with the functional value at the previous iteration) cannot be less than $s + 1$.

Proof The proof is based on Statement 4.16. There are no reserves within positions $\overline{1, p^{new}}$ in the subsequence σ^k. The task $j_{[q]}$, $l_{j_{[g]}} \ge l_{j_{[q]}}$, in consequence of its iteration of optimization, took the position $s > p^{new}$ where it remains tardy (the conditions 3, 4 of Statement 4.16 for the task $j_{[g]}$ are not satisfied). Therefore, the task $j_{[g]}$ that has a lower priority and a longer processing time than the task $j_{[q]}$, cannot take a position lower than $s + 1$. We execute the optimization (the minimization of the increase in the total weighted tardiness of the current subsequence in comparison with the functional value at the previous iteration and the construction of an optimal subsequence at the current iteration) for the task $j_{[g]}$ within positions $\overline{s+1, g}$. This significantly reduces the number of permutations performed. ∎

In the following statements, we consider some cases when we achieve an optimal solution of the problem by the first polynomial component of the PSC-algorithm.

Statement 4.35, SSO #10 Suppose that $j_{[g]}$ is the first tardy task in the sequence σ^{ord}, tasks within positions $\overline{g+1,\ n}$ are tardy, and at least one of the following conditions holds:

1. $\max R(\sigma^{ord}) \le l_{j_{[g]}}$ and $\forall j_{[r]}$, $r = \overline{g+1,\ n}$, we have $l_{j_{[r]}} \ge l_{j_{[g]}}$;
2. $\forall l, r \in \overline{g,\ n}$, $l < r$, we have $l_{j_{[l]}} < l_{j_{[r]}}$, $d_{j_{[l]}} < d_{j_{[r]}}$.

In this case, an optimal solution is obtained by a polynomial subalgorithm with the complexity of each iteration determined by the function $O(n_{curr} \log n_{curr})$ where n_{curr} is the number of tasks optimized at the current iteration (Remark 4.6).

Proof In case (1), the procedure of insertion and further optimization due to the use of reserves of preceding tasks can be implemented only for one of the tardy tasks, according to the condition (4.4). At each iteration of the subalgorithm, we will need to find such a position within the interval of insertion for the current tardy task where we achieve the maximum possible decrease of the functional value in comparison with that at the beginning of the iteration. Since $\forall j_{[r]}$, $r = \overline{g+1,\ n}$, we have $l_{j_{[r]}} \ge l_{j_{[g]}}$, then the task $j_{[r]}$ can take an earlier position only if $d_{j_{[r]}} < d_{j_{[g]}}$ (Statement 4.9). Otherwise, this task remains in its position. To construct an optimal subsequence, we select a task for which we achieve the maximum decrease in the functional value in consequence of the task's insertion and further optimization. In case (2), according to Statement 4.27, $\forall l, r \in \overline{g,\ n}$, $l < r$, the task $j_{[l]}$ precedes the task $j_{[r]}$ in an optimal schedule. During the algorithm execution, we define such position of insertion for each tardy task where we achieve the maximum possible decrease of the functional value in a subsequence bounded with this tardy task. If some tardy task remains in its position in the consequence of its iteration, then the following tardy tasks will also remain in their positions in an optimal schedule (Statement 4.27). The complexity of each iteration of the subalgorithm is determined by the complexity of sorting, $O(n_{curr} \log n_{curr})$ where n_{curr} is the number of tasks optimized at the current iteration. ∎

Statement 4.36, SSO #11 Suppose that $j_{[g]}$ is the first competing task in the sequence σ^{fp}. If there are no reserves in this sequence within positions $\overline{1, p-1}$ where p is the position of insertion of the task $j_{[g]}$; the task $j_{[g]}$ remains tardy in position p; for each pair of tasks $j_{[s]}$, $j_{[t]}$, $s < t$, within positions $\overline{p, g-1}$, we have $l_{j_{[s]}} \le l_{j_{[t]}}$, $d_{j_{[s]}} \le d_{j_{[t]}}$, and $\omega_{j_{[s]}} \ge \omega_{j_{[t]}}$; and $\forall j_{[r]}$, $r = \overline{g+1, n}$, we have $l_{j_{[r]}} > l_{j_{[r-1]}}$, $d_{j_{[r]}} > d_{j_{[r-1]}}$, and $d_{j_{[r]}} < C_{j_{[r]}}$, then an optimal solution is obtained by a polynomial subalgorithm with the complexity of each iteration determined by the function $O(n_{curr})$ where n_{curr} is the number of tasks optimized at the current iteration (Remark 4.6).

Proof According to the statement, the task $j_{[g]}$ remains tardy in its position of insertion p; there are no reserves within positions $\overline{1, p-1}$ to extend the interval of insertion; tasks within positions $\overline{p, g-1}$ cannot change their order; and the task $j_{[r-1]}$ will precede the task $j_{[r]} \forall r = \overline{g+1, n}$ in an optimal schedule (Statement 4.28). In order to construct an optimal schedule on the entire set of tasks, it is sufficient to find such position of insertion for each next competing task within positions $\overline{g+1, n}$ where we achieve the minimum possible increase in the functional value in comparison with the functional value at the previous iteration of the subalgorithm. If, after that, the current competing task remains in its position, then the following competing tasks in an optimal schedule will also remain in their positions (Statement 4.27). The complexity of each iteration of the subalgorithm is determined by the function $O(n_{curr})$ where n_{curr} is the number of tasks optimized at the current iteration. ∎

Statement 4.37, SSO #12 Suppose that $j_{[g]}$ is the first competing task in the sequence σ^{fp}. If all tasks within positions $\overline{1, g-1}$ are in non-increasing order of their priorities; $\max R_i \le l_{j_{[g]}}$, $i = \overline{1, g-1}$; $\forall j_{[s]}$, $s = \overline{g+1, k}$, we have $d_{j_{[s]}} < C_{j_{[s]}}$, $l_{j_{[s]}} \ge l_{j_{[g]}}$; and $\forall j_{[s]}$, $s = \overline{k+1, n}$, we have $d_{j_{[s]}} \ge C_{j_{[s]}}$ where $k < n$, then an optimal solution is obtained by a polynomial subalgorithm with the complexity of each iteration determined by the function $O(n_{curr} \log n_{curr})$ where n_{curr} is the number of tasks optimized at the current iteration (Remark 4.6).

Proof If the conditions of this statement are satisfied, then the procedure of insertion and further optimization using the reserves of preceding tasks in accordance with the condition (4.4) can be implemented only for one of the tardy tasks. Similarly to Statement 4.35, to construct an optimal subsequence, we select such task that its insertion and further optimization yields the maximum decrease in the functional value. The complexity of each iteration of the subalgorithm is determined by the complexity of sorting, $O(n_{curr} \log n_{curr})$ where n_{curr} is the number of tasks optimized at the current iteration. ∎

Remark 4.6 If any of SSOs #10–12 is fulfilled, then an optimal solution is obtained by a corresponding polynomial subalgorithm. Each iteration of the subalgorithm has corresponding complexity: $O(n_k)$ or $O(n_k \log n_k)$ where n_k is the number of tasks optimized at iteration k of the subalgorithm. We propose to readers to describe the polynomial subalgorithms as an exercise. The complexity of the problem solving on the entire set of tasks is determined by the function $O(n^2)$ or $O(n^2 \log n)$.

Statement 4.38 (Sufficient Sign of Polynomial Solvability of the Current Iteration, SSPS) Suppose that we have already executed $k-1$ iterations and constructed an optimal subsequence σ^{k-1} on the set of tasks within positions $\overline{1, g-1}$. Suppose that all of the following conditions hold at the current iteration of optimization during the insertion of the current tardy (competing or originated) task $j_{[g]}$:

- actual position of insertion p^{new} determined during the algorithm execution is greater than the position of insertion of the task last inserted at the previous steps of the algorithm;
- there are no tasks that correspond to the conditions of Statement 4.12 within the interval of insertion of the task $j_{[g]}$;
- there are no tasks $j_{[l]}$ such that $d_{j_{[l]}} > d_{j_{[g]}} - l_{j_{[g]}}$, $d_{j_{[l]}} > C_{j_{[l]}}$, $d_{j_{[l]}} > C_{j_{[p^{new}-1]}}$ within positions $\overline{1, p^{new}-1}$;

- for tasks $j^*_{[m]}$ ($j^{**}_{[m]}$) within positions $\overline{1, p^{new} - 1}$, the following condition (that allows to move these tasks into later positions) does not hold:

$$\omega_{j^*_{[m]}}\left(C_{j_{[l]}} - d_{j^*_{[m]}}\right) < \sum_{i=m}^{p^{new}} \omega_{j_{[i]}} \max\left(0, C_{j_{[i]}} - d_{j_{[i]}}\right)$$

where the position l corresponds to the priority of $j^*_{[m]}$ ($j^{**}_{[m]}$) (if $p_{j^*_{[m]}}$ (or $p_{j^{**}_{[m]}}$) is less than $p_{j_{[p^{new}]}}$, then $l = p^{new}$). Then we achieve a solution at the current iteration by a polynomial subalgorithm with the complexity $O\left(n^2_{curr}\right)$ where n_{curr} is the number of tasks optimized at the current iteration.

Proof The proof is based on Statement 4.16. If the conditions of Statement 4.38 hold in the current subsequence σ^k, then reserves within the interval of insertion cannot be created by the interval extending or by the reserves releasing by tasks that earlier used them (conditions 2–6 of Statement 4.16). To construct an optimal schedule on the entire set of tasks, it is sufficient for each tardy task to find the position of insertion where we achieve the minimum increase in the total weighted tardiness of the current subsequence in comparison with the functional value at the previous iteration, and perform a further reordering within the interval of insertion. Hence, the complexity of the current iteration of the algorithm does not exceed $O\left(n^2_{curr}\right)$ where n_{curr} is the number of tasks optimized at the current iteration. The description of the current iteration of optimization is given in Sect. 4.3.1. We propose to readers to prove its turning into a polynomial subalgorithm as an exercise. ∎

Statement 4.39 The first polynomial component of the PSC-algorithm is implemented by:

1. The algorithmic procedures of verifying the SSOs #1–5 and #7 (Corollaries 4.1, 4.3, and 4.7, Statements 4.1, 4.4, and 4.11) that are executed in the general algorithm (Sect. 4.3.1) before the optimization iterations.
2. If item 1 is not fulfilled, then the algorithmic procedures of verifying the SSOs #10–12 that include the execution of subalgorithms with polynomial complexity, as shown above in Statements 4.35–4.37 and Remark 4.6.
3. If items 1 and 2 are not fulfilled, then during the algorithmic procedures that implement each current iteration of optimization (Sect. 4.3.1) we verify the following. If SSO #9 (Statement 4.32) is fulfilled at the beginning of the current iteration of optimization, then we have obtained an optimal solution on the entire set of tasks and we do not execute the current and subsequent iterations of optimization. If SSO #9 is not fulfilled, then fulfillment of SSPS (Statement 4.38) is required. If SSPS is true, then the current iteration of optimization is implemented by a polynomial algorithm with the complexity $O\left(n^2_{curr}\right)$ where n_{curr} is the number of tasks optimized at the current iteration. If SSO #6 or SSO #8 (Corollary 4.6, Statement 4.31) is fulfilled after the current iteration of

optimization, then we have obtained an optimal schedule on the entire set of tasks by the polynomial subalgorithm and do not execute the subsequent iterations of optimization.

Remark 4.7 At the first iteration of optimization, fulfillment of SSPS (Statement 4.38) is required, then we perform a verification of SSOs #6 and #8 (Corollary 4.6, Statement 4.31).

Remark 4.8 We consider Statements 4.35–4.37 as the part of the first polynomial component of the PSC-algorithm that verifies the SSOs. The rationale is the following. Statement 4.39 is about fulfillment of the first polynomial component of the PSC-algorithm. The second paragraph of the statement is included there because we check the conditions of Statements 4.35–4.37 by polynomial algorithms with $O(n^2)$ or $O(n^2 \log n)$ complexity. If the conditions are satisfied, we obtain an optimal solution without using the exact subalgorithm that has enumerative complexity. Thus, we consider a part of the first polynomial component all conditions and polynomial algorithms for their checkup that are executed without the use of iterative optimization procedures of the universal exact subalgorithm.

Statement 4.40 The following is true:

1. If SSO #9 (Statement 4.32) is fulfilled at the beginning of the current iteration of optimization, then we do not execute the current and subsequent iterations of optimization: we have obtained an optimal solution;
2. If item 1 is not fulfilled, then SSPS is verified during the current iteration of optimization. If SSPS is true, then the current iteration of optimization is implemented by a polynomial algorithm (Statement 4.38).
3. If, after the current iteration of optimization, regardless of SSPS fulfillment, SSO #6 or SSO #8 (Corollary 4.6, Statement 4.31) is fulfilled, then we have obtained an optimal schedule on the entire set of tasks and do not execute the subsequent iterations of optimization.
4. If SSPS is true at each iteration, then a sufficient condition is satisfied that turns the exact subalgorithm of the PSC-algorithm into a polynomial complexity algorithm.

Lemma 4.2 *Condition 4 of* Statement 4.40 *is satisfied for non-empty set of the problem instances.*

Proof The proof is based on Example 4.4 given below. ∎

Example 4.4 (to Lemma 4.2) Initial data for Example 4.4 is in Table 4.17.

The sequence σ^{ord} is in Table 4.18.

The sequence σ^{fp} (Table 4.19) is the result of free permutations. Tasks 17 and 5 in it are non-competing, in accordance with the definition.

Iteration 1. The first competing task is task 14. We determine the interval of insertion of task 14 into an earlier position: $d_{14} - l_{14} = 2032 - 92 = 1940$. Task 22 has the completion time 1935, and for the next task it is 2005, so we insert task 14

Table 4.17 Initial data for Example 4.4

j	l_j	ω_j	d_j	j	l_j	ω_j	d_j	j	l_j	ω_j	d_j	j	l_j	ω_j	d_j
1	81	7	1933	14	92	2	2032	27	61	6	1958	40	70	3	2337
2	61	3	2068	15	24	10	2116	28	98	10	2129	41	56	6	2101
3	83	4	2245	16	12	8	1878	29	18	4	2138	42	9	5	2320
4	1	5	1926	17	53	1	2421	30	61	10	2200	43	11	4	2059
5	91	1	2450	18	46	5	1888	31	39	2	2084	44	47	10	1908
6	68	10	1931	19	7	5	2122	32	41	9	2089	45	16	9	2307
7	70	5	2230	20	57	4	2002	33	53	1	2138	46	80	4	2311
8	36	7	2113	21	67	2	2405	34	52	3	2243	47	26	8	2039
9	67	9	2241	22	89	4	1995	35	35	10	2153	48	77	8	2263
10	95	9	2018	23	45	3	2108	36	15	9	2086	49	37	5	1989
11	42	10	2207	24	26	4	2024	37	86	4	2206	50	85	9	2074
12	4	1	2043	25	59	3	2020	38	58	2	2091				
13	40	4	1900	26	90	8	2027	39	64	8	2133				

after task 22 and mark it with asterisk. Arrange the subsequent tasks in σ^{fp} in non-increasing order of priorities. We obtain the sequence shown in Table 4.20.

The first competing task in σ_1 is task 38. In the same way, we determine the interval of insertion for task 38 into an earlier position. We insert it after task 14^*, then mark task 38 with asterisk and sort the subsequent tasks by priorities. We obtain the sequence σ_2 (Table 4.21).

Iteration 2. The next competing task in the sequence σ_2 is task 33. We determine the position of its insertion, it is the position after task 38^*. We perform the insertion, sort the subsequent tasks, and obtain the sequence σ_3 (Table 4.22).

Tasks 46, 40, and 21 in the sequence σ_3 are originated tardy tasks. We perform free permutation of task 42 into the position after task 46^- (condition 1 of Statement 4.16). Task 46 becomes non-tardy. We obtain the sequence σ_4 (Table 4.23).

Tardy tasks 40 and 21 in the sequence σ_4 cannot be moved into earlier positions. The reason is: reserves are absent within the intervals of their insertion and cannot be created, and all preceding tasks in the sequence σ_4 have a higher priority. Tasks 17 and 5 are not competing. Hence, the sequence σ_4 is optimal.

It is clear that all four conditions of Statement 4.38 (SSPS) are satisfied on both iterations of the algorithm. Consequently, the problem was solved by the first polynomial component of the PSC-algorithm with the complexity $O(n_1^2) + O(n_2^2)$ where $n_1 = 47$ and $n_2 = 48$ are the numbers of tasks optimized at iterations 1 and 2, respectively.

Table 4.18 Sequence σ^{ord}

j	p_j	l_j	ω_j	d_j	C_j	f_j	j	p_j	l_j	ω_j	d_j	C_j	f_j
4	5.000	1	5	1926	1	0	48	0.104	77	8	2263	931	0
19	0.714	7	5	2122	8	0	28	0.102	98	10	2129	1029	0
16	0.667	12	8	1878	20	0	13	0.100	40	4	1900	1069	0
36	0.600	15	9	2086	35	0	27	0.098	61	6	1958	1130	0
45	0.563	16	9	2307	51	0	10	0.095	95	9	2018	1225	0
42	0.556	9	5	2320	60	0	26	0.089	90	8	2027	1315	0
15	0.417	24	10	2116	84	0	1	0.086	81	7	1933	1396	0
43	0.364	11	4	2059	95	0	7	0.071	70	5	2230	1466	0
47	0.308	26	8	2039	121	0	20	0.070	57	4	2002	1523	0
35	0.286	35	10	2153	156	0	23	0.067	45	3	2108	1568	0
12	0.250	4	1	2043	160	0	34	0.058	52	3	2243	1620	0
11	0.238	42	10	2207	202	0	31	0.051	39	2	2084	1659	0
29	0.222	18	4	2138	220	0	25	0.051	59	3	2020	1718	0
32	0.220	41	9	2089	261	0	46	0.050	80	4	2311	1798	0
44	0.213	47	10	1908	308	0	2	0.049	61	3	2068	1859	0
8	0.194	36	7	2113	344	0	3	0.048	83	4	2245	1942	0
30	0.164	61	10	2200	405	0	37	0.047	86	4	2206	2028	0
24	0.154	26	4	2024	431	0	22	0.045	89	4	1995	2117	488
6	0.147	68	10	1931	499	0	40	0.043	70	3	2337	2187	0
49	0.135	37	5	1989	536	0	38	0.034	58	2	2091	2245	308
9	0.134	67	9	2241	603	0	21	0.030	67	2	2405	2312	0
39	0.125	64	8	2133	667	0	14	0.022	92	2	2032	2404	744
18	0.109	46	5	1888	713	0	33	0.019	53	1	2138	2457	319
41	0.107	56	6	2101	769	0	17	0.019	53	1	2421	2510	89
50	0.106	85	9	2074	854	0	5	0.011	91	1	2450	2601	151

The functional value $f = 2099$

4.3 The PSC-Algorithm

4.3.1 Description of the PSC-Algorithm

We propose the algorithm to solve the TWT problem that consists of two stages: the preliminary stage (Algorithm A0) and the optimization stage (Algorithm A1).

4.3.1.1 Algorithm A0 (The Preliminary Stage of the PSC-Algorithm)

1. Arrange tasks in non-decreasing order of their due dates d_j. If there are no tardy tasks in the obtained sequence (SSO #1, Corollary 4.1), then the sequence is optimal, the algorithm terminates. Otherwise, go to step 2.

Table 4.19 Sequence σ^{fp}

j	p_j	l_j	ω_j	d_j	C_j	f_j	j	p_j	l_j	ω_j	d_j	C_j	f_j
4	5.000	1	5	1926	1	0	27	0.098	61	6	1958	1028	0
19	0.714	7	5	2122	8	0	10	0.095	95	9	2018	1123	0
16	0.667	12	8	1878	20	0	26	0.089	90	8	2027	1213	0
36	0.600	15	9	2086	35	0	1	0.086	81	7	1933	1294	0
15	0.417	24	10	2116	59	0	7	0.071	70	5	2230	1364	0
43	0.364	11	4	2059	70	0	20	0.070	57	4	2002	1421	0
47	0.308	26	8	2039	96	0	23	0.067	45	3	2108	1466	0
35	0.286	35	10	2153	131	0	34	0.058	52	3	2243	1518	0
12	0.250	4	1	2043	135	0	31	0.051	39	2	2084	1557	0
11	0.238	42	10	2207	177	0	25	0.051	59	3	2020	1616	0
29	0.222	18	4	2138	195	0	2	0.049	61	3	2068	1677	0
32	0.220	41	9	2089	236	0	3	0.048	83	4	2245	1760	0
44	0.213	47	10	1908	283	0	37	0.047	86	4	2206	1846	0
8	0.194	36	7	2113	319	0	22	0.045	89	4	1995	1935	0
30	0.164	61	10	2200	380	0	40	0.043	70	3	2337	2005	0
24	0.154	26	4	2024	406	0	38	0.034	58	2	2091	2063	0
6	0.147	68	10	1931	474	0	21	0.030	67	2	2405	2130	0
49	0.135	37	5	1989	511	0	48	0.104	77	8	2263	2207	0
9	0.134	67	9	2241	578	0	46	0.050	80	4	2311	2287	0
39	0.125	64	8	2133	642	0	45	0.563	16	9	2307	2303	0
18	0.109	46	5	1888	688	0	42	0.556	9	5	2320	2312	0
41	0.107	56	6	2101	744	0	14	0.022	92	2	2032	2404	744
50	0.106	85	9	2074	829	0	33	0.019	53	1	2138	2457	319
28	0.102	98	10	2129	927	0	17	0.019	53	1	2421	2510	89
13	0.100	40	4	1900	967	0	5	0.011	91	1	2450	2601	151

The functional value $f = 1303$

2. Arrange tasks in non-increasing order of their priorities p_j. Denote the obtained sequence by σ^{ord}. Check the fulfillment of SSOs #2, 3, and 4 (Statements 4.1 and 4.4, Corollary 4.3). If any of SSOs is fulfilled, then the sequence σ^{ord} is optimal, the algorithm terminates. Otherwise, go to step 3.
3. *Determination of the set of competing tasks.* Perform free permutations in the sequence σ^{ord} starting from the task with the maximum due date:

   3.1. Determine a set of tasks W:

$$W = \{k : C_{j_{[q]}} \le d_{j_{[k]}} \le C_{j_{[q+1]}}, k, q = \overline{1, n},\ k < q, \exists j_{[i]} : d_{j_{[i]}} < C_{j_{[i]}}, i = \overline{k+1, q}\}.$$

Table 4.20 Sequence σ_1

j	p_j	l_j	ω_j	d_j	C_j	f_j	j	p_j	l_j	ω_j	d_j	C_j	f_j
4	5.000	1	5	1926	1	0	27	0.098	61	6	1958	1028	0
19	0.714	7	5	2122	8	0	10	0.095	95	9	2018	1123	0
16	0.667	12	8	1878	20	0	26	0.089	90	8	2027	1213	0
36	0.600	15	9	2086	35	0	1	0.086	81	7	1933	1294	0
15	0.417	24	10	2116	59	0	7	0.071	70	5	2230	1364	0
43	0.364	11	4	2059	70	0	20	0.070	57	4	2002	1421	0
47	0.308	26	8	2039	96	0	23	0.067	45	3	2108	1466	0
35	0.286	35	10	2153	131	0	34	0.058	52	3	2243	1518	0
12	0.250	4	1	2043	135	0	31	0.051	39	2	2084	1557	0
11	0.238	42	10	2207	177	0	25	0.051	59	3	2020	1616	0
29	0.222	18	4	2138	195	0	2	0.049	61	3	2068	1677	0
32	0.220	41	9	2089	236	0	3	0.048	83	4	2245	1760	0
44	0.213	47	10	1908	283	0	37	0.047	86	4	2206	1846	0
8	0.194	36	7	2113	319	0	22	0.045	89	4	1995	1935	0
30	0.164	61	10	2200	380	0	14*	0.022	92	2	2032	2027	0
24	0.154	26	4	2024	406	0	45	0.563	16	9	2307	2043	0
6	0.147	68	10	1931	474	0	42	0.556	9	5	2320	2052	0
49	0.135	37	5	1989	511	0	48	0.104	77	8	2263	2129	0
9	0.134	67	9	2241	578	0	46	0.050	80	4	2311	2209	0
39	0.125	64	8	2133	642	0	40	0.043	70	3	2337	2279	0
18	0.109	46	5	1888	688	0	38	0.034	58	2	2091	2337	492
41	0.107	56	6	2101	744	0	21	0.030	67	2	2405	2404	0
50	0.106	85	9	2074	829	0	33	0.019	53	1	2138	2457	319
28	0.102	98	10	2129	927	0	17	0.019	53	1	2421	2510	89
13	0.100	40	4	1900	967	0	5	0.011	91	1	2450	2601	151

The functional value $f = 1051$

3.2. Find a task $j_{[k']}$ with the maximum due date from the set W. If $W = \varnothing$, then we have obtained the sequence σ^{fp}, go to step 3.4. Otherwise, go to step 3.3.
3.3. Perform the permutation of the task $j_{[k']}$ to its corresponding position q' and eliminate $j_{[k']}$ from the set W. Go to step 3.2.
3.4. Check the fulfillment of SSOs #5 and #7 (Statement 4.11 and Corollary 4.7). If any of SSOs is fulfilled, then the sequence σ^{fp} is optimal, the algorithm terminates. Otherwise, determine the set of competing tasks:

$$K = \{j_{[g]} \in \sigma^{fp} : C_{j_{[g]}} - d_{j_{[g]}} > 0,\ \exists j_{[l]} : d_{j_{[l]}} > d_{j_{[g]}} - l_{j_{[g]}},\ d_{j_{[l]}} - C_{j_{[l]}} > 0\}.$$

Go to the optimization stage.

Table 4.21 Sequence σ_2

j	p_j	l_j	ω_j	d_j	C_j	f_j	j	p_j	l_j	ω_j	d_j	C_j	f_j
4	5.000	1	5	1926	1	0	27	0.098	61	6	1958	1028	0
19	0.714	7	5	2122	8	0	10	0.095	95	9	2018	1123	0
16	0.667	12	8	1878	20	0	26	0.089	90	8	2027	1213	0
36	0.600	15	9	2086	35	0	1	0.086	81	7	1933	1294	0
15	0.417	24	10	2116	59	0	7	0.071	70	5	2230	1364	0
43	0.364	11	4	2059	70	0	20	0.070	57	4	2002	1421	0
47	0.308	26	8	2039	96	0	23	0.067	45	3	2108	1466	0
35	0.286	35	10	2153	131	0	34	0.058	52	3	2243	1518	0
12	0.250	4	1	2043	135	0	31	0.051	39	2	2084	1557	0
11	0.238	42	10	2207	177	0	25	0.051	59	3	2020	1616	0
29	0.222	18	4	2138	195	0	2	0.049	61	3	2068	1677	0
32	0.220	41	9	2089	236	0	3	0.048	83	4	2245	1760	0
44	0.213	47	10	1908	283	0	37	0.047	86	4	2206	1846	0
8	0.194	36	7	2113	319	0	22	0.045	89	4	1995	1935	0
30	0.164	61	10	2200	380	0	14*	0.022	92	2	2032	2027	0
24	0.154	26	4	2024	406	0	38*	0.034	58	2	2091	2085	0
6	0.147	68	10	1931	474	0	45	0.563	16	9	2307	2101	0
49	0.135	37	5	1989	511	0	42	0.556	9	5	2320	2110	0
9	0.134	67	9	2241	578	0	48	0.104	77	8	2263	2187	0
39	0.125	64	8	2133	642	0	46	0.050	80	4	2311	2267	0
18	0.109	46	5	1888	688	0	40	0.043	70	3	2337	2337	0
41	0.107	56	6	2101	744	0	21	0.030	67	2	2405	2404	0
50	0.106	85	9	2074	829	0	33	0.019	53	1	2138	2457	319
28	0.102	98	10	2129	927	0	17	0.019	53	1	2421	2510	89
13	0.100	40	4	1900	967	0	5	0.011	91	1	2450	2601	151

The functional value $f = 559$

4.3.1.2 Algorithm A1 (The Optimization Stage of the PSC-Algorithm)

The algorithm consists of k uniform iterations where k is the number of competing tasks. We can eliminate some of them from the set of competing tasks K during the problem solving. The set of competing tasks is defined in the sequence σ^{fp} obtained as a result of free permutations in the sequence σ^{ord} (the sequence of tasks in non-increasing order of their priorities).

At each iteration, we check the possibility to use time reserves of non-tardy tasks for the current competing task and for new tardy tasks originated in consequence of the competing task's permutations. As a result of each iteration, the functional value

Table 4.22 Sequence σ_3

j	p_j	l_j	ω_j	d_j	C_j	f_j	j	p_j	l_j	ω_j	d_j	C_j	f_j
4	5.000	1	5	1926	1	0	27	0.098	61	6	1958	1028	0
19	0.714	7	5	2122	8	0	10	0.095	95	9	2018	1123	0
16	0.667	12	8	1878	20	0	26	0.089	90	8	2027	1213	0
36	0.600	15	9	2086	35	0	1	0.086	81	7	1933	1294	0
15	0.417	24	10	2116	59	0	7	0.071	70	5	2230	1364	0
43	0.364	11	4	2059	70	0	20	0.070	57	4	2002	1421	0
47	0.308	26	8	2039	96	0	23	0.067	45	3	2108	1466	0
35	0.286	35	10	2153	131	0	34	0.058	52	3	2243	1518	0
12	0.250	4	1	2043	135	0	31	0.051	39	2	2084	1557	0
11	0.238	42	10	2207	177	0	25	0.051	59	3	2020	1616	0
29	0.222	18	4	2138	195	0	2	0.049	61	3	2068	1677	0
32	0.220	41	9	2089	236	0	3	0.048	83	4	2245	1760	0
44	0.213	47	10	1908	283	0	37	0.047	86	4	2206	1846	0
8	0.194	36	7	2113	319	0	22	0.045	89	4	1995	1935	0
30	0.164	61	10	2200	380	0	14*	0.022	92	2	2032	2027	0
24	0.154	26	4	2024	406	0	38*	0.034	58	2	2091	2085	0
6	0.147	68	10	1931	474	0	33*	0.019	53	1	2138	2138	0
49	0.135	37	5	1989	511	0	45	0.563	16	9	2307	2154	0
9	0.134	67	9	2241	578	0	42	0.556	9	5	2320	2163	0
39	0.125	64	8	2133	642	0	48	0.104	77	8	2263	2240	0
18	0.109	46	5	1888	688	0	46	0.050	80	4	2311	2320	36
41	0.107	56	6	2101	744	0	40	0.043	70	3	2337	2390	159
50	0.106	85	9	2074	829	0	21	0.030	67	2	2405	2457	104
28	0.102	98	10	2129	927	0	17	0.019	53	1	2421	2510	89
13	0.100	40	4	1900	967	0	5	0.011	91	1	2450	2601	151

The functional value $f = 539$

decreases or remains unchanged. The optimization procedures are similar for competing and originated tardy tasks.

If SSO #10 (Statement 4.35) is fulfilled in the sequence σ^{ord}, or at least one of SSOs #11 or #12 (Statements 4.36, 4.37) is fulfilled in the sequence σ^{fp}, then we will obtain an optimal sequence on the entire set of tasks by a polynomial algorithm, according to Remark 4.6.

1. Suppose that we have already executed $k-1$ iterations. We consider the execution of iteration k for the next competing task $j_{[g]}$. Let g^{prev} ("g previous") is the position of the previous competing task for which we executed the previous iteration of optimization. $g^{prev} = 0$ at the first iteration. We start the next iteration searching the next competing task from position $g^{prev} + 1$.

Table 4.23 Free permutation of task 42. Sequence σ_4

j	p_j	l_j	ω_j	d_j	C_j	f_j	j	p_j	l_j	ω_j	d_j	C_j	f_j
4	5.000	1	5	1926	1	0	27	0.098	61	6	1958	1028	0
19	0.714	7	5	2122	8	0	10	0.095	95	9	2018	1123	0
16	0.667	12	8	1878	20	0	26	0.089	90	8	2027	1213	0
36	0.600	15	9	2086	35	0	1	0.086	81	7	1933	1294	0
15	0.417	24	10	2116	59	0	7	0.071	70	5	2230	1364	0
43	0.364	11	4	2059	70	0	20	0.070	57	4	2002	1421	0
47	0.308	26	8	2039	96	0	23	0.067	45	3	2108	1466	0
35	0.286	35	10	2153	131	0	34	0.058	52	3	2243	1518	0
12	0.250	4	1	2043	135	0	31	0.051	39	2	2084	1557	0
11	0.238	42	10	2207	177	0	25	0.051	59	3	2020	1616	0
29	0.222	18	4	2138	195	0	2	0.049	61	3	2068	1677	0
32	0.220	41	9	2089	236	0	3	0.048	83	4	2245	1760	0
44	0.213	47	10	1908	283	0	37	0.047	86	4	2206	1846	0
8	0.194	36	7	2113	319	0	22	0.045	89	4	1995	1935	0
30	0.164	61	10	2200	380	0	14*	0.022	92	2	2032	2027	0
24	0.154	26	4	2024	406	0	38*	0.034	58	2	2091	2085	0
6	0.147	68	10	1931	474	0	33*	0.019	53	1	2138	2138	0
49	0.135	37	5	1989	511	0	45	0.563	16	9	2307	2154	0
9	0.134	67	9	2241	578	0	48	0.104	77	8	2263	2231	0
39	0.125	64	8	2133	642	0	46	0.050	80	4	2311	2311	0
18	0.109	46	5	1888	688	0	42	0.556	9	5	2320	2320	0
41	0.107	56	6	2101	744	0	40	0.043	70	3	2337	2390	159
50	0.106	85	9	2074	829	0	21	0.030	67	2	2405	2457	104
28	0.102	98	10	2129	927	0	17	0.019	53	1	2421	2510	89
13	0.100	40	4	1900	967	0	5	0.011	91	1	2450	2601	151

The functional value $f = 503$

2. *The search of the next tardy task:*

   2.1. Find the next competing task $j_{[g]}$ within positions $\overline{g^{prev}+1,\ n}$. If a competing task is not found, the algorithm terminates, the current sequence is optimal. Otherwise, check the fulfillment of SSO #9 (Statement 4.32) in the current sequence. If SSO #9 is fulfilled, then the current sequence is optimal, the algorithm terminates. Otherwise, check if conditions of at least one of Statements 4.22, 4.25, 4.27 or 4.28 are satisfied. If it is true, then set $g^{prev} = g$; if $K = \varnothing$, the algorithm terminates, the current sequence is optimal, otherwise go to step 2.1 (begin new iteration). If conditions of Statements 4.22, 4.25, 4.27 and 4.28 are not satisfied, memorize the current sequence and the functional value, go to step 3.

2.2. Find the next tardy task $j_{[q]}$ within positions $\overline{p^{new}+1,\ g}$. If a tardy task is not found, set $g^{prev} = g$ and go to step 2.1 (begin new iteration). If it is found, go to step 3 considering the found task $j_{[q]}$ as the task $j_{[g]}$.

3. *Determination of the position of insertion:*

3.1. Find a position p where the tardiness of the task $j_{[g]}$ is minimal or equal to zero. Let $T^{nt}_{j_{[g]}} = d_{j_{[g]}} - l_{j_{[g]}}$ denote the maximum start time of the task $j_{[g]}$ when it remains non-tardy. If the task $j_{[g]}$ is marked with "**", then the position of insertion p is found from the expression: $C_{j_{[p-1]}} \le T^{nt}_{j^{**}_{[g]}} < C_{j_{[p]}}$ where $T^{nt}_{j^{**}_{[g]}}$ is the memorized earliest start time of the task $j^{**}_{[g]}$ at the previous iterations of the algorithm. Otherwise, the position p is defined from condition: $C_{j_{[p-1]}} \le T^{nt}_{j_{[g]}} < C_{j_{[p]}}$. The following specific cases are possible:

- $T^{nt}_{j_{[g]}} \le 0$. In this case, $p = 1$.
- $C_{j_{[i]}} > T^{nt}_{j_{[g]}}\ \forall i = \overline{1, g-1}$. In this case, $p = 1$.

3.2. Determine the position of insertion p^{new} for the task $j_{[g]}$ within the interval $\overline{p, g-1}$ as the maximum of two positions: p and $q + 1$ if position q is occupied by such a task $j_{[q]}$ that $l_{j_{[q]}} \le l_{j_{[g]}},\ d_{j_{[q]}} \le d_{j_{[g]}},\ \omega_{j_{[q]}} \ge \omega_{j_{[g]}}$.[2]

3.3. If $p^{new} = p$ ($p^{new} = q+1$) and this position is occupied by such an unmarked task $j_{[p^{new}]}$ that

$$d_{j_{[p^{new}]}} > C_{j_{[p^{new}]}},\ d_{j_{[p^{new}]}} > d_{j_{[g]}} - l_{j_{[g]}},\ p^{new} \ne g, \tag{4.12}$$

then insert the task $j_{[g]}$ into the position p^{new} and go to step 7, else go to step 4.1.

4. *Analysis of reserves of tasks within positions* $\overline{1,\ p^{new}-1}$:

4.1 Check if there are tasks $j_{[r]}$ within positions $\overline{1,\ p^{new}-1}$ that satisfy at least one of the following conditions:

- $d_{j_{[r]}} > d_{j_{[p^{new}]}} - l_{j_{[p^{new}]}},\ d_{j_{[r]}} - C_{j_{[r]}} > 0,\ d_{j_{[r]}} > C_{j_{[p^{new}-1]}};$ (4.13)

[2]If we would extend the problem statement to include tasks with negative due dates (when a due date is less than the start time of machine), the change of the algorithm would consist only in adjusting the position of insertion for these tasks. All other procedures would remain unchanged. According to [4], if $d_{j_{[g]}} < 0$, then the position of insertion of the task $j_{[g]}$ at step 3.2 is defined within the interval $\overline{p, g-1}$ as the maximum of three positions: p, $q + 1$ (defined at step 3.2), and $l + 1$ if position l is occupied by such a task $j_{[l]}$ that $d_{j_{[l]}} < 0,\ l_{j_{[l]}} < l_{j_{[g]}},\ \omega_{j_{[l]}} > \omega_{j_{[g]}}$.

- $p_{j_{[r]}} < p_{j_{[p^{new}]}}$; (4.14)
- the tasks $j_{[r]}$ are marked with "*" ("**").

If such tasks are found, then go to step 4.2, otherwise go to step 5.

4.2. If position p ($q + 1$) is occupied by a marked task, then insert the task $j_{[g]}$ to position $p + 1$ ($q + 2$). If this next position is occupied by a marked task, then go to the next position $p + 2$ ($q + 3$) and so on, up to position g. Finally, go to step 7. If the position p^{new} is occupied by a non-marked task, then insert the task $j_{[g]}$ into the position p^{new} and go to step 7.

5. *Adjustment of the position of insertion.* Execute this step if positions $\overline{1,\ p^{new} - 1}$ do not contain either any marked tasks, or tasks of lower priority than that of the task $j_{[p^{new}]}$, or reserves to extend the interval of its insertion (or if the existing reserves are not enough for the extension).

Find the position of insertion for the task $j_{[p^{new}]}$ within positions $\overline{p^{new},\ g}$, starting from the position p^{new} (which is the new position of the task $j_{[g]}$) and sequentially checking for all tasks $j_{[i]}$, $i = \overline{p^{new},\ g}$, the conditions: $d_{j_{[i]}} > C_{j_{[i]}}$ and $d_{j_{[i]}} > d_{j_{[p^{new}]}} - l_{j_{[p^{new}]}}$. If both conditions are true, then insert the task $j_{[p^{new}]}$ into the position i, set $p^{new} = i$, go to step 7. If at least one of the conditions is not satisfied within the entire interval $\overline{p^{new},\ g}$, then the task $j_{[p^{new}]}$ remains in the initial position g. In this case, set $p^{new} = g$, go to step 8.

6. *Optimization due to reserves of tasks within positions* $\overline{1,\ p^{new} - 1}$. Execute the following steps 6.1–6.4 starting with lower priority tasks, i.e., consider positions $p^{new} - 1$, $p^{new} - 2$, etc.

   6.1. Perform free permutations decreasing the tardiness of the task $j_{[p^{new}]}$. Determine the new position p^{new}. If the task $j_{[p^{new}]}$ became non-tardy, go to step 2.2, otherwise go to step 6.2.
   6.2. Find an unmarked task $j_{[l]}$, $l < p^{new}$, such that $p_{j_{[l]}} < p_{j_{[p^{new}]}}$. If such task is found, then go step 6.4, otherwise go to step 6.3.
   6.3. Find a task $j_{[l]}$, $l < p^{new}$, such that $d_{j_{[l]}} > C_{j_{[l]}}$, $d_{j_{[l]}} > C_{j_{[p^{new}]}} - l_{j_{[p^{new}]}}$, and

$$\omega_{j_{[p^{new}]}} \min\left(l_{j_{[l]}}, C_{j_{[p^{new}]}} - d_{j_{[p^{new}]}}\right) > \omega_{j_{[l]}}\left(C_{j_{[p^{new}]}} - d_{j_{[l]}}\right).$$

   If such task is found, then go step 6.4. If the conditions given in steps 6.1–6.3 are not satisfied during the optimization (the position p^{new} was not changed), then go to step 8.

   6.4. Perform a permutation of the task $j_{[l]}$ into the position p^{new} (after the task $j_{[p^{new}]}$). Set $p^{new} = p^{new} - 1$. If the task $j_{[p^{new}]}$ became non-tardy, then go to step 7, otherwise go to step 6.1 and repeat steps 6.1–6.4 again. If, the position p^{new} was changed after the steps 6.1–6.4 execution, then go to step 7.

7. *Reordering*. Mark the task $j_{[p^{new}]}$ with "*" (p^{new} is the new position of the task $j_{[g]}$). Arrange all tasks within positions $\overline{p^{new}+1,\ g}$ in non-increasing order of their priorities (g is the previous position of the competing task for which we execute the iteration). If we perform the reordering after the insertion of a generated tardy task into an intermediate position within the interval $\overline{1,\ g}$, then perform it within the entire interval of reordering $\overline{p^{new}+1,\ g}$. Remove asterisks "*" from marked tasks within this interval. If tasks marked with "*" are followed by tasks marked with "**" within the interval of reordering, then remove the sign "**" from the marked tasks. Otherwise, if the tasks marked with "**" are not preceded by tasks marked with "*", then keep the sign "**" and the memorized start time for these tasks. Perform free permutations within the interval of reordering. If the task $j_{[p^{new}]}$ is not tardy, then go to step 2.2. If the task $j_{[p^{new}]}$ is tardy, then if step 6 was executed for it, go to step 8, else go to step 6.
8. *Optimization due to releasing of reserves of tasks marked with "*" ("**"):*
   - 8.1. Find the next task $j^{*}_{[m]}$ ($j^{**}_{[m]}$) within the interval $\overline{1,\ p^{new}-1}$ where p^{new} is the new position of the task $j_{[g]}$ (if the task $j_{[g]}$ remains in position g after its insertion and execution of step 6, then $p^{new}=g$). First we analyze the task we marked with "*" ("**") last, then the task marked before last, etc. If a marked task is found, go to step 8.2. Otherwise: if $j_{[p^{new}]}$ is the competing task for which we execute the iteration of optimization, then check Statements 4.23, 4.29, and 4.31 to eliminate the task $j_{[g]}$ and some of the following tasks from the set of competing tasks. Check the fulfillment of SSPS (Statement 4.38); if it is fulfilled, then the current iteration was executed by a polynomial subalgorithm. If SSO #6 (Corollary 4.6) or SSO #8 (Statement 4.31) is satisfied, then the algorithm terminates, the current sequence is optimal. Otherwise, set $g^{prev}=p^{new}$, go to step 2.1. If $j_{[p^{new}]}$ is not the competing task for which we execute the optimization iteration, go to step 2.2.
   - 8.2. Check the condition:

$$\omega_{j^{*}_{[m]}}\left(C_{j_{[l]}}-d_{j^{*}_{[m]}}\right)<\sum_{i=m}^{p^{new}}\omega_{j_{[i]}}\max\left(0,C_{j_{[i]}}-d_{j_{[i]}}\right) \tag{4.15}$$

     where position l corresponds to the priority of the task $j^{*}_{[m]}$ ($j^{**}_{[m]}$) within the interval $\overline{m,\ p^{new}}$. If $p_{j^{*}_{[m]}}$ (or $p_{j^{**}_{[m]}}$) is less than $p_{j^{**}_{[p^{new}]}}$, then $l=p^{new}$. If Eq. (4.15) is true, then go to step 8.3. Otherwise, go to step 8.1.
   - 8.3. Memorize the current subsequence within positions $\overline{1,\ p^{new}}$. Denote it by σ^{prev} ("sigma previous").

8.4. Perform the permutation of the task $j^*_{[m]}$ ($j^{**}_{[m]}$) into the position l. Mark the task $j^*_{[l]}$ with "**" (if it is not yet marked with "**"). Memorize its new start time as its earliest start time. If the task $j^*_{[l]}$ ($j^{**}_{[l]}$) is not the competing task for which we execute the iteration of optimization, then go to step 8.5. Otherwise, if the task $j^*_{[l]}$ ($j^{**}_{[l]}$) is that competing task and took an intermediate position as a result of the permutation (it is followed by an unmarked task of a higher priority), then go to step 8.5. Otherwise, if the task $j^*_{[l]}$ ($j^{**}_{[l]}$) is the competing task for which we execute the iteration of optimization, and it took the initial position g as a result of the permutation according to its priority (it is followed by an unmarked task of a lower priority), then compare the functional values of the subsequence σ^{prev} and the subsequence memorized in step 2 of the previous iteration. Choose a subsequence with the lower functional value. Check Statements 4.23, 4.29, and 4.31 to eliminate the task $j_{[g]}$ and some of the following tasks from the set of competing tasks. Check the fulfillment of SSPS (Statement 4.38); if it is fulfilled, then the current iteration was executed by a polynomial subalgorithm. If SSO #6 (Corollary 4.6) or SSO #8 (Statement 4.31) is satisfied, then the algorithm terminates, the current sequence is optimal. If the task $j_{[g]}$ is followed by both tardy and non-tardy tasks, then memorize its start time $T^{nt}_{j_{[g]}}$ as its earliest start time. Set $g^{prev} = g$. If $K = \varnothing$, then the algorithm terminates. Otherwise, go to step 2 (begin new iteration).

8.5. Recursively execute procedures of Algorithm A1 (steps 3–8) for each tardy task $j_{[k]}$, $k = \overline{m,\ p^{new} - 1}$, considering the task $j_{[k]}$ as the task $j_{[g]}$. Here, m is the position of the task $j^*_{[l]}$ before its permutation. Use for the optimization reserves of tasks within the interval $\overline{1,\ k - 1}$. After considering all the tasks $j_{[k]}$, go to step 8.6.

8.6. Denote the subsequence within positions $\overline{1,\ p^{new}}$ by σ^{new}. If the functional value of the subsequence σ^{new} is not less than that of the subsequence σ^{prev}, then go to step 8.7. Otherwise, go to step 8.8. The algorithm will continue on a subsequence with a lower functional value.

8.7. Return to the subsequence σ^{prev} and go to step 8.1.

8.8. If the task $j^*_{[l]}$ ($j^{**}_{[l]}$) took the position corresponding to its priority (it is followed by unmarked tasks of the same or lower priorities), then remove the mark "*" ("**") from the task. Otherwise, mark this task with "**" and memorize its earliest start time. If there are another tasks marked with "**" within the interval of permutation of the task $j^*_{[l]}$ ($j^{**}_{[l]}$), then mark the tasks with "*" and remove their earliest start times. Go to step 8.9.

8.9. Find the next task $j^*_{[m]}$ ($j^{**}_{[m]}$) that satisfies (4.15) within the interval $\overline{1,\ l - 1}$. If such a task is found, then go to step 8.3 executing step 8 in turn for the task $j^*_{[l]}$ (denoting it by $j_{[p^{new}]}$ and setting $p^{new} = l$). If such a task is not found, or if the task $j_{[p^{new}]}$ became non-tardy, then if the task $j_{[p^{new}]}$ is the competing task for which we execute the iteration of optimization, then check the fulfillment of SSPS (Statement 4.38); if it is fulfilled, then the current iteration was executed by a polynomial subalgorithm. If SSO #6 (Corollary 4.6) or SSO #8 (Statement 4.31) is satisfied, then the algorithm terminates, the current sequence is optimal. Otherwise, set $g^{prev} = p^{new}$, go to step 2.1 (begin new iteration). If the task $j_{[p^{new}]}$ is not the competing task for which we execute the iteration of optimization, then go to step 2.2. We repeat steps 8.1–8.9 until all marked tasks are considered or until the task $j_{[p^{new}]}$ is not tardy.

We can represent the PSC-algorithm as a block diagram (Figs. 4.1 and 4.2). The block numbers correspond to the steps of Algorithm A1.

4.3.2 Properties of the First Polynomial Component of the PSC-Algorithm

As a result of investigating the problem's properties, we have strictly identified logical analytical conditions for fulfilling of the first polynomial component of the algorithm. If they are satisfied during a problem instance solving, then this subalgorithm yields an exact solution for this instance, and its complexity is a polynomial function of the problem's size. In this case, we achieve the exact solution executing only blocks of the algorithm that have polynomial complexity.

We can represent the fulfillment conditions of the first polynomial component of the algorithm in the form of a block diagram (Fig. 4.3).

At the block diagram we show the main stages of the problem solving:

- arranging tasks in non-increasing order of their priorities (the sequence σ^{ord});
- free permutations, determining the set of competing tasks (the sequence σ^{fp});
- iterations of optimization for each competing task to use the reserves of preceding tasks (the subsequences σ^{k}).

For each stage, we stated the conditions of reaching an optimal solution in polynomial time (Statements 4.35–4.39).

If the total tardiness is zero in the sequence of tasks arranged in non-decreasing order of due dates (SSO #1, Corollary 4.1), then the sequence is optimal on the entire set of tasks, and the solution is reached in polynomial time with the complexity $O(n \log n)$. If either SSO #2 or #3 or #4 (Statement 4.1, 4.4, or Corollary 4.3) is satisfied in the sequence σ^{ord}, then this sequence is optimal on the entire set of tasks, and the solution is reached in polynomial time with the complexity $O(n \log n)$. Otherwise, we perform free permutations and analyze the obtained sequence σ^{fp}. If SSO #5 or #7 (Statement 4.11 or Corollary 4.7) is satisfied, then the sequence σ^{fp} is optimal on the entire set of tasks, and the complexity of the algorithm is $O(n^2)$. If SSO #10 (Statement 4.35) is satisfied in the sequence σ^{ord},

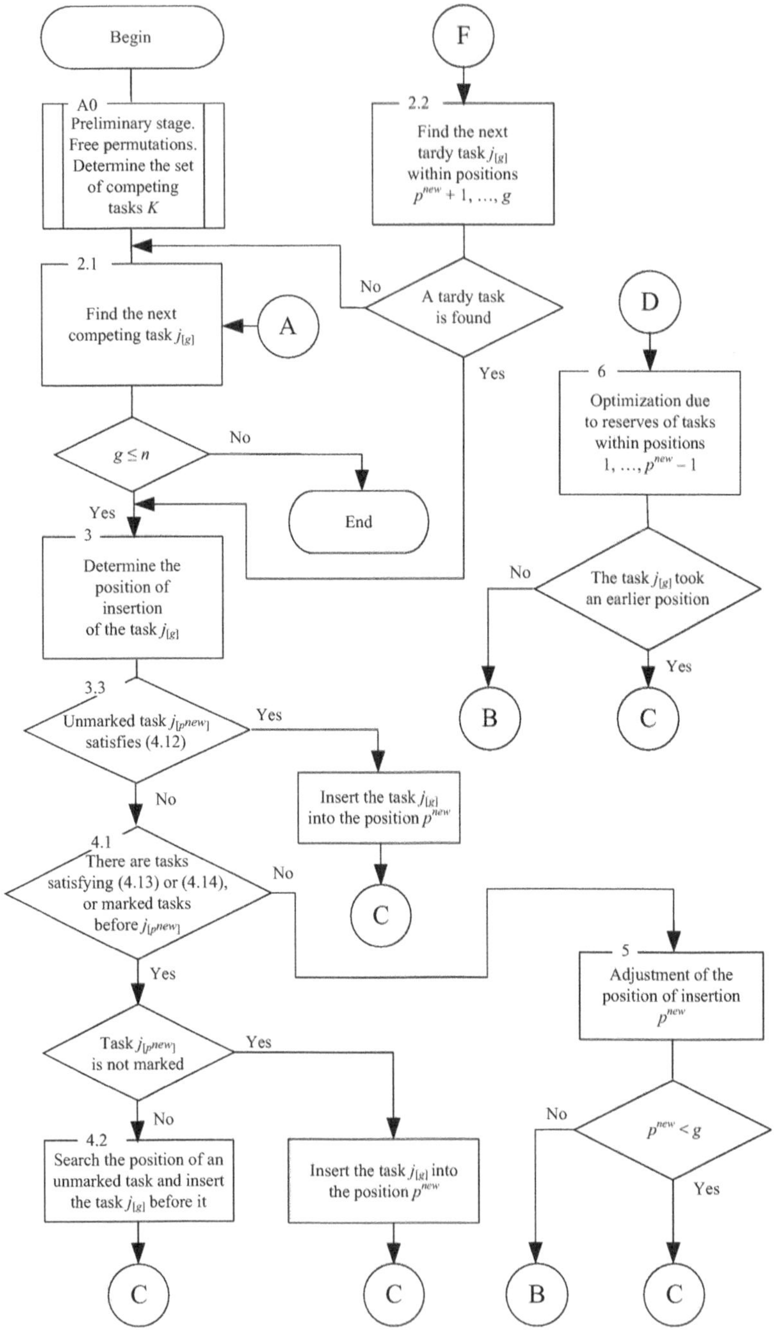

Fig. 4.1 Block diagram of the PSC-algorithm, part I

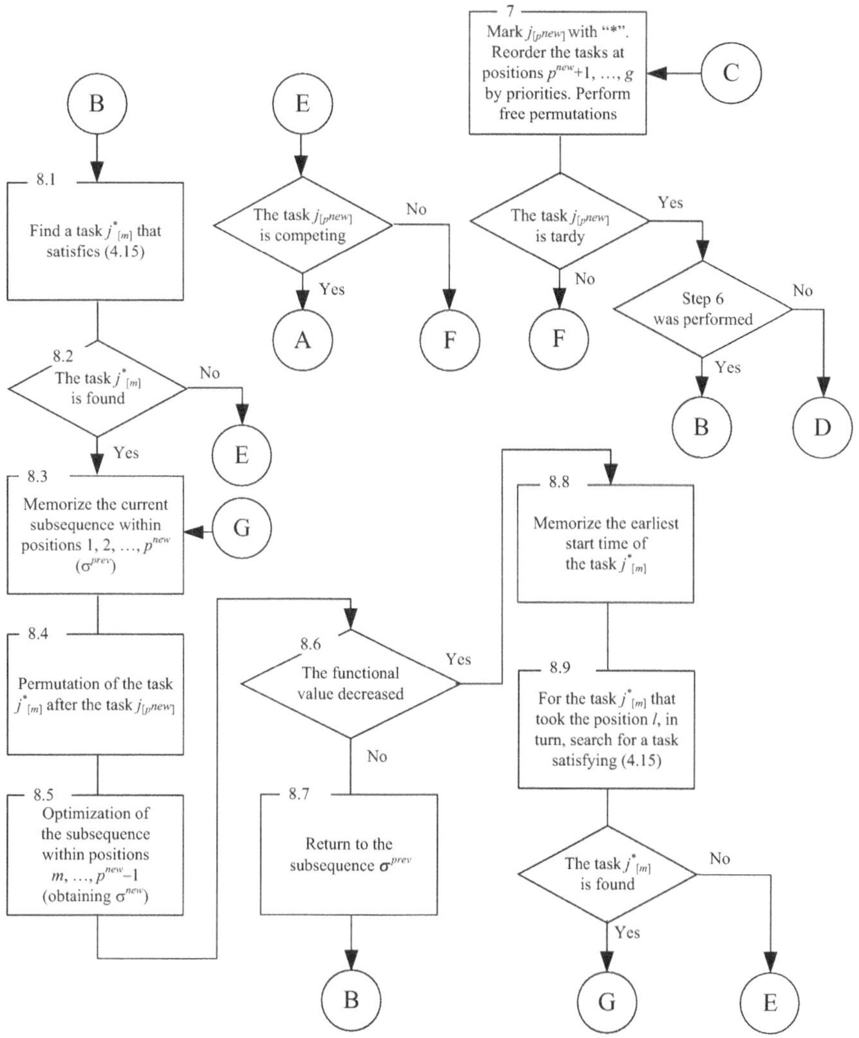

Fig. 4.2 Block diagram of the PSC-algorithm, part II

then we execute a polynomial time algorithm with the complexity $O(n^2 \log n)$. If SSO #11 or #12 (Statement 4.36 or 4.37) is satisfied in the sequence σ^{fp}, then we execute a polynomial time algorithm with the complexity $O(n^2)$ or $O(n^2 \log n)$, respectively. Otherwise, we execute iterations of optimization for competing tasks of the set K. If SSO #9 (Statement 4.32) is satisfied in the beginning of a current iteration, then the iteration is not executed, we have obtained an optimal solution on the entire set of tasks. Otherwise, we execute the current iteration and check the SSPS (Statement 4.38). If it is satisfied, then the current iteration is executed by a

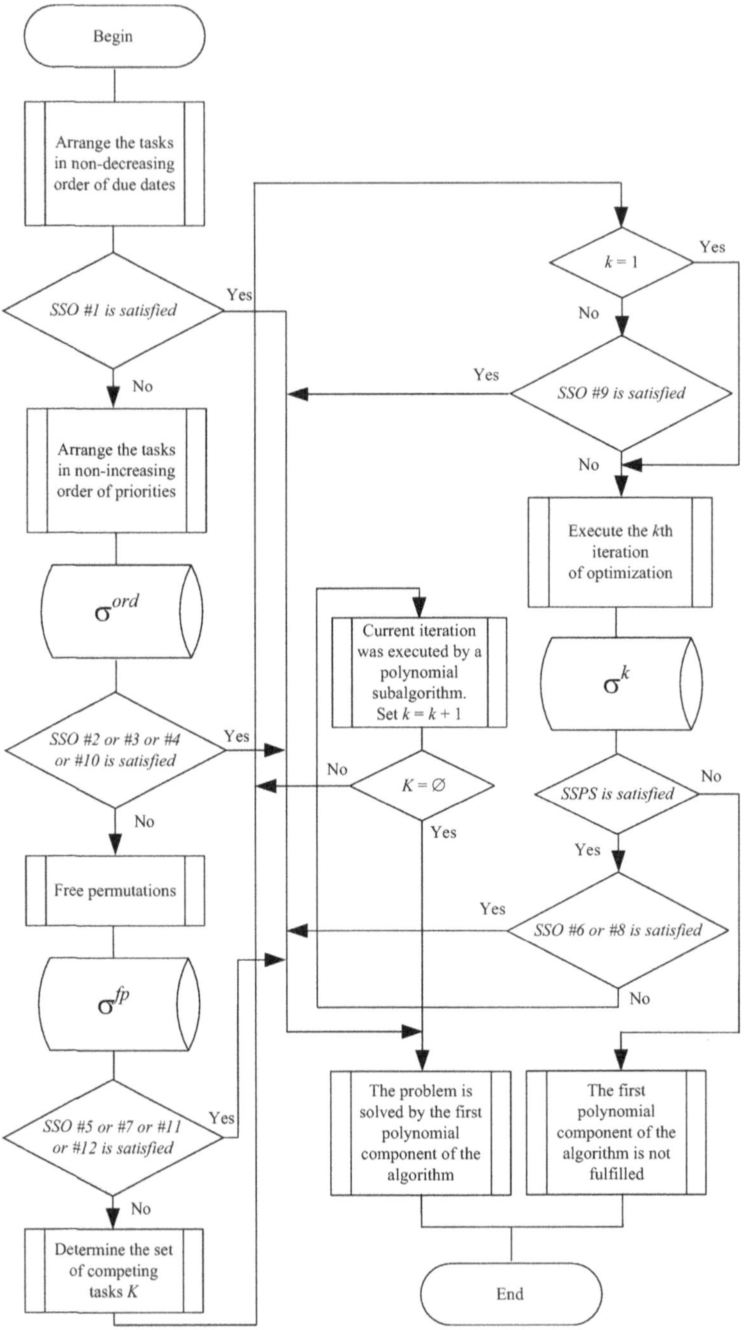

Fig. 4.3 Fulfillment conditions of the first polynomial component of the PSC-algorithm

polynomial subalgorithm with the complexity $O(n_k^2)$ where n_k is the number of tasks optimized at the current iteration. After the current iteration, we check SSOs #6 and #8 (Corollary 4.6, Statement 4.31). If any of them is satisfied, then we have obtained an optimal solution on the entire set of tasks by a polynomial subalgorithm. If none of the SSOs is satisfied at the current iteration, and SSPS is also false, then the first polynomial component of the algorithm is not fulfilled.

Therefore, the overall complexity of the first polynomial component of the PSC-algorithm does not exceed $O(n^3)$.

Experimental researches showed a statistical significance of the first polynomial component. Solving the problem instances, we had to execute iterations of optimization for much lesser number of tardy tasks than their entire number. This significantly decreased the complexity of the algorithm.

4.3.3 Description of the Logic of the PSC-Algorithm Construction for the Case When None of the SSOs Is Satisfied

The PSC-algorithm is based on permutations aimed to use the reserves of preceding non-tardy tasks by tardy tasks.

The algorithm consists of k uniform iterations where k is the number of competing tasks in the sequence σ^{fp}. During the solving, we can eliminate some tasks from the set of competing tasks following the determined rules. This reduces the number of iterations.

At the first iteration, we build an optimal schedule for the tasks within positions $\overline{1, g_1}$ in the sequence σ^{fp} where $j_{[g_1]}$ is the first competing task. At the next iteration, we consider the subsequence of tasks within positions $\overline{1, g_2}$ where the positions $\overline{1, g_1}$ are occupied by the optimal subsequence obtained at the first iteration; positions $\overline{g_1 + 1, g_2}$ are occupied by the tasks that are in these positions in the sequence σ^{fp}. $j_{[g_2]}$ is the next competing task in the sequence σ^{fp}. And so on, until we execute iterations of optimization for all competing tasks.

Let us consider the first iteration of optimization. We insert the first competing task $j_{[g_1]}$ into a position p where it will have minimal or zero weighted tardiness. Then we reorder the tasks within the interval of insertion $\overline{p, g_1 - 1}$ in non-increasing order of their priorities and perform free permutations. Originated tardy tasks may appear within the interval of insertion. We need to construct an optimal schedule for a subsequence within positions $\overline{1, g_1}$ by redistributing the reserves of non-tardy tasks between the inserted task and the originated tardy tasks. Therefore, we execute step by step optimization within the current subsequence for each tardy task. We use the time reserves of non-tardy tasks within the interval of insertion $\overline{p, g_1 - 1}$ as well as of the tasks within positions $\overline{1, p - 1}$ that have reserves within the interval of insertion of the tardy task under consideration. The iteration is completed when we executed the optimization for all tardy tasks in the current subsequence and constructed an optimal subsequence on the set of tasks within positions $\overline{1, g_1}$.

At each next iteration k, we insert the next competing task $j_{[g_k]}$ into an earlier position in the optimal subsequence obtained at the previous iteration. Then we execute the step by step optimization for each next tardy task within the current subsequence and redistribution of reserves within positions $\overline{1, g_k}$, similarly to the first iteration.

To simplify the description of the PSC-algorithm, it is convenient to distinguish several logical procedures in it. We will describe them below. For a specific problem instance, we will have to execute not all procedures but only a necessary part of them depending on the initial data of the instance.

Procedure A Arrange the tasks of the given subsequence in non-increasing order of their priorities p_j (step 2 of Algorithm A0, see above).

According to Definition 4.8, we build the sequence σ^{ord} by executing Procedure A on the entire set of tasks.

Procedure B Perform free permutations in the given subsequence of tasks according to Definition 4.9 (step 3 of Algorithm A0).

According to Definition 4.10, we build the sequence σ^{fp} by applying Procedure B to the sequence σ^{ord}.

Procedure C Determine the set of competing tasks in the sequence σ^{fp} (Definition 4.11, step 3.4 of Algorithm A0).

We execute the following procedures D–I for each competing task $j_{[g]}$ and for originated tardy tasks that became tardy as a result of insertion of the task $j_{[g]}$.

Procedure D Determination of the interval of insertion for the current tardy task $j_{[g]}$ in the current sequence: it is the position p where the weighted tardiness of the task is minimal or equal to zero (Definition 4.6, step 3 of Algorithm A1). We determine the actual position of insertion p^{new} within positions $\overline{p, g-1}$ as the maximum of two positions: p and $q+1$ if the task $j_{[q]}$ in position q is such that $l_{j_{[q]}} \le l_{j_{[g]}}$, $d_{j_{[q]}} \le d_{j_{[g]}}$, $\omega_{j_{[q]}} \ge \omega_{j_{[g]}}$ (Statement 4.18). If the task in the position p^{new} is not tardy, not marked and has reserves, the position of insertion is p^{new}. But if the position p^{new} is occupied by a tardy task of a higher priority, a marked task or a task with zero reserve, then we insert the task $j_{[g]}$ after it. If the task $j_{[g]}$ will be tardy after insertion into the position p^{new}, then we execute Procedure E (starting optimization with this task). Otherwise, we insert the task $j_{[g]}$ into the position p^{new}, mark it with asterisk and execute Procedure H.

Remark 4.9 We always mark the task $j_{[g]}$ with asterisk after its insertion and execute Procedure H (which includes reordering by priorities and free permutations, that is, procedures A and B) for all tasks following it within the interval of insertion $\overline{p^{new}+1,\ g}$. Then we execute optimization procedures D–I for each tardy task within the interval of insertion. If there are no tardy tasks, we proceed to the next competing task from the sequence σ^{fp} and to the next iteration.

Procedure E Analysis of reserves of tasks within positions $\overline{1, p^{new}-1}$ for the current tardy (competing or originated) task $j_{[p^{new}]}$ (step 4.1 of Algorithm A1): check if there are tasks $j_{[r]}$ within positions $\overline{1, p^{new}-1}$ that satisfy at least one of the following conditions:

- $d_{j_{[r]}} > d_{j_{[p^{new}]}} - l_{j_{[p^{new}]}}$, $d_{j_{[r]}} - C_{j_{[r]}} > 0$, $d_{j_{[r]}} > C_{j_{[p^{new}-1]}}$;
- $p_{j_{[r]}} < p_{j_{[p^{new}]}}$;
- the tasks $j_{[r]}$ are marked with "*" ("**").

If such tasks are found, then execute Procedure F, else execute Procedure G.

Remark 4.10 During the optimization for each next tardy (competing or originated) task, we determine its own position of insertion p^{new} and its corresponding interval $\overline{1, p^{new}-1}$. In this case, if it turns out that the interval of insertion is empty, then $p^{new} = g$. The value of p^{new} is changed during the algorithm execution. It indicates the current position of the tardy (competing or originated) task.

Procedure F Optimization (decreasing the weighted tardiness of the current tardy task) due to the use of reserves within the extended interval of insertion (conditions 3, 4 of Statement 4.16, Statements 4.12 and 4.13, step 6 of Algorithm A1). We execute the optimization permuting the tasks from interval $\overline{1, p^{new}-1}$ to the interval of insertion of the current tardy task to obtain additional time reserves (Definition 4.1). We do this by means of:

- free permutations (Procedure B);
- permutations of tasks $j_{[i]}$ that satisfy the conditions $d_{j_{[i]}} - C_{j_{[i]}} > 0$, $d_{j_{[i]}} > d_{j_{[p^{new}]}} - l_{j_{[p^{new}]}}$ (Statement 4.4). There are no reserves within the interval of insertion of the current tardy task for tasks that do not satisfy these conditions, so their permutation increases the functional value;
- permutations of unmarked lower priority tasks $j_{[l]}$ that satisfy Statement 4.12;
- permutations of tasks $j_{[l]}$, $d_{j_{[l]}} > C_{j_{[l]}}$, that satisfy Statement 4.13:

$$d_{j_{[l]}} > C_{j_{[p^{new}]}} - l_{j_{[p^{new}]}},\ \omega_{j_{[p^{new}]}} \min\left(l_{j_{[l]}}, C_{j_{[p^{new}]}} - d_{j_{[p^{new}]}}\right) > \omega_{j_{[l]}}\left(C_{j_{[p^{new}]}} - d_{j_{[l]}}\right).$$

Permutations of such tasks decrease the functional value. After Procedure F, we execute Procedure H.

Procedure G Adjustment of the position of insertion to reduce the interval of insertion (step 5 of Algorithm A1): find the position of insertion for the task $j_{[p^{new}]}$

within the interval $\overline{p^{new}+1,\ g}$, starting from the position p^{new} and sequentially checking two conditions for all tasks $j_{[i]}$ within this interval: $d_{j_{[i]}} > C_{j_{[i]}}$ and $d_{j_{[i]}} > d_{j_{[p^{new}]}} - l_{j_{[p^{new}]}}$. If both conditions are true in a position i, then insert the task $j_{[p^{new}]}$ into the position i. If at least one of the conditions is not satisfied within the entire interval $\overline{p^{new}+1,\ g}$, then the task $j_{[p^{new}]}$ remains in the initial position g.

Procedure H Execute Procedures A and B (priority ordering and free permutations) within the extended interval of insertion $\overline{p^{new}+1,\ g}$ of the current tardy task (step 7 of Algorithm A1).

Procedure I Optimization within the interval of permutation of marked tasks into the later positions. The reserves released as a result of the permutation are used by tardy tasks (conditions 5 and 6 of Statement 4.16, step 8 of Algorithm A1). Find a task $j^*_{[m]}$ ($j^{**}_{[m]}$) within positions $\overline{1,\ p^{new}-1}$ such that Eq. (4.11) holds:

$$\omega_{j^*_{[m]}}\left(C_{j_{[l]}} - d_{j^*_{[m]}}\right) < \sum_{i=m}^{p^{new}} \omega_{j_{[i]}} \max\left(0, C_{j_{[i]}} - d_{j_{[i]}}\right)$$

where l is the position within $\overline{m,\ p^{new}}$ corresponding to the priority of $j^*_{[m]}$ ($j^{**}_{[m]}$). If, in this case, $p_{j^*_{[m]}}$ (or $p_{j^{**}_{[m]}}$) is less than $p^{\mathrm{H}}_{j_{[p]}}$, then $l = p^{new}$. Memorize the current sequence and the functional value. Permute the task $j^*_{[m]}$ ($j^{**}_{[m]}$) into the position l. Mark the task $j^*_{[l]}$ with "**" (if it is not yet marked with "**") and memorize its new start time as its earliest start time. Recursively execute the optimization (Procedures D–I) for the subsequence within positions $\overline{1,\ p^{new}-1}$ starting from the position m which was occupied by the task $j^*_{[l]}$ ($j^{**}_{[l]}$) before the permutation. If, as a result of the optimization, the functional value decreases, then continue with the obtained sequence and call Procedure I for the current task $j^*_{[l]}$ ($j^{**}_{[l]}$). Otherwise, return to the memorized sequence and continue Procedure I for the next marked task that satisfies the above condition. Procedure I terminates if such task is not found or the task $j_{[p^{new}]}$ becomes non-tardy.

Procedure I involves a recursive call of procedures D–I for a subsequence of tasks within the interval of permutation of the marked task. We can analyze several such subsequences in parallel, that is, the use of parallel computing systems is possible.

The algorithm terminates when the optimization is done for all competing tasks.

Substantiation of Procedures D–I Let us analyze the properties of sequences σ^{ord}, σ^{fp}, and σ^{curr}.

All tasks in the sequence σ^{ord} are in non-increasing order of their priorities. We formulated the properties of the sequence σ^{ord} in Statements 4.1–4.9.

To obtain the sequence σ^{fp}, we performed free permutations that break the priority ordering of tasks in the sequence σ^{ord}. Therefore, we reorder the tasks in accordance with their priorities after insertions. We formulated the properties of the sequence σ^{fp} in Statement 4.10.

Then we execute a step by step optimization for each subsequent tardy task in the current subsequence σ^k, as follows. We insert a current tardy (competing or originated) task into the position where its weighted tardiness becomes minimal or equal to zero (Procedure D). During the optimization (Procedure I) this position can be changed. Then we mark the inserted task with asterisk (this means that it used the time reserves of preceding tasks as much as possible) and perform priority ordering and free permutations (Procedure H). If the task remains tardy in the position of insertion, then we execute optimization (decreasing the weighted tardiness of this task) within the extended interval of insertion (Procedures D and E) and Procedures D–I for optimization of each subsequent tardy task within the interval of insertion. Procedures D–H decrease the weighted tardiness of the tardy task due to existing reserves. Procedure I decreases it due to reserves obtained as a result of moving marked tasks (i.e., the tasks that earlier used the reserves) into the later positions.

Remark 4.11 New originated tardy tasks may appear after the insertion of the current tardy task. We optimize them, in turn, using Procedures D–I. After their insertions we reorder tasks according to priorities. Thus, we always execute the optimization procedures within a priority-ordered subsequence. Consequently, Statements 4.1, 4.2, 4.4–4.6, 4.9 that are valid for the sequence σ^{ord}, are as well valid for the sequence σ^{fp} and for the current sequence.

The substantiation of Procedures D–I is based on the above analysis of properties of the sequences σ^{ord}, σ^{fp}, and the current subsequence, and also on the following facts.

It follows from Statement 4.1 that if there are no reserves in the sequence σ^{ord}, then this sequence is optimal on the entire set of tasks. A decrease in the functional value is only possible if there are reserves of non-tardy tasks in the sequence σ^{ord} that are used by tardy tasks.

According to Statement 4.2, the insertion of a tardy task into position $f < p$ where p is the position where its weighted tardiness becomes minimal or equal to zero, cannot lead to a decrease in the functional value.

In Statement 4.4, we determine two conditions for the sequence σ^{ord} Their fulfillment guaranties that there are no permutations or insertions that decreases the functional value for any of the tardy tasks, even if reserves exist (the existing reserves cannot be used because any move will only increase the functional value).

In Statement 4.5, we determine three conditions for the sequence σ^{ord}. Their fulfillment allows to obtain additional reserves in consequence of tasks permutation into the interval of insertion.

If at least one of the conditions of Statement 4.6 is satisfied for the sequence σ^{ord}, then a decrease in the functional value is possible after a move of the determined tardy tasks into an earlier position.

In Statement 4.9, we formulated the conditions to create additional reserves in consequence of marked tasks permutations into later positions. A permutation of marked tasks that do not meet those conditions will increase the functional value.

In Statements 4.12 and 4.13, we showed several conditions for the current subsequence. If they are satisfied, then the permutation of the determined tasks into the position after the current tardy task decreases the functional value. A permutation of tasks that do not meet those conditions will increase the functional value.

If at least one of the necessary conditions we formulated in Statements 4.15 and 4.16 is satisfied for the subsequence σ^k, then a decrease in the functional value is possible after a move of the determined tardy tasks into an earlier position.

In Statement 4.17, we formalized the necessary optimization procedures that lead to a decrease in the functional value.

On the basis of the following statements, we can decrease the number of procedures executed during the problem solving.

Statements 4.18, 4.20, 4.24, 4.27, 4.28, 4.30, 4.33, and 4.34 formulate the conditions to shorten the interval of insertion. Statement 4.21 sets the conditions to decrease the number of unpromising feasible schedules. Statements 4.22 and 4.25 determine the conditions to decrease the number of unpromising insertions. Statement 4.26 specifies those for unpromising permutations. Statement 4.33 sets the conditions for the current subsequence decomposition.

The conditions of Statements 4.23, 4.29, 4.31, and 4.32 allow to eliminate some tasks from the set of competing tasks.

Also, we eliminate the following evidently inefficient permutations during the algorithm execution:

- if the sequence σ^{fp} contains subsequences with only *tardy* tasks, then permutations of tasks within such subsequences will increase the functional value;
- if the sequence σ^{fp} contains subsequences with only *non-tardy* tasks, then permutations of tasks within such subsequences will not decrease the functional value.

4.3.3.1 Outline of the Algorithm

1. Build the sequence σ^{ord} (execute Procedure A on the entire set of tasks).
2. Build the sequence σ^{fp} (execute Procedure B for the sequence σ^{ord}).
3. Determine the set of competing tasks K in the sequence σ^{fp} (Procedure C).
4. Suppose that we have already executed $k-1$ iterations. We consider the execution of iteration k for the next competing task $j_{[g]}$. Let g^{prev} ("g previous") is the position of the previous tardy (competing or originated) task for which we executed the optimization. $g^{prev}=0$ at the first iteration. In the beginning of each other iteration, g^{prev} is the position of the previous competing task for

which we executed the previous iteration of optimization. We start the next iteration searching the next tardy task from position $g^{prev} + 1$.

4.1. Find the next tardy task $j_{[g]}$ within positions $\overline{g^{prev} + 1,\ n}$. If a tardy task is not found, the algorithm terminates, the current sequence is optimal. Otherwise: if $j_{[g]} \notin K$, then go to step 4.2, else check if conditions of at least one of Statements 4.22, 4.25, 4.27, or 4.28 are satisfied. If it is true, set $g^{prev} = g$; if $K = \varnothing$, then the algorithm terminates, the current sequence is optimal, otherwise go to step 4.1 (begin new iteration). If conditions of Statements 4.22, 4.25, 4.27 and 4.28 are not satisfied, memorize the current sequence and the functional value, go to step 4.2.
4.2. Determine the position of insertion (Procedure D). If the task $j_{[g]}$ is tardy after its insertion into the position p^{new}, then go to step 4.3 (the optimization starts from this task). Otherwise, insert the task $j_{[g]}$ into the position p^{new}, mark it with asterisk and go to step 4.6.
4.3. Analyze the reserves of tasks within positions $\overline{1, p^{new} - 1}$ (Procedure E). If the tasks needed to extend the interval of insertion are found, go to step 4.4. Otherwise, go to step 4.5.
4.4. Insert the task $j_{[g]}$ into the position p^{new}, mark it with asterisk and execute Procedure H. Execute the optimization (decreasing the weighted tardiness of the task $j_{[p^{new}]}$) due to the use of reserves within the extended interval of insertion (Procedure F). Go to step 4.6.
4.5. Adjust the position of insertion (Procedure G). Go to step 4.6.
4.6. Reorder the subsequence within the extended interval of insertion (Procedure H). If the task $j_{[p^{new}]}$ is not tardy, set $g^{prev} = p^{new}$ and go to step 4.2 executing, in turn, the optimization procedures D–I for each tardy task within positions $\overline{p^{new} + 1, g}$. Otherwise (if the task $j_{[p^{new}]}$ is tardy and we executed Procedure F), go to step 4.7. If there are no tardy tasks, proceed to the next competing task (set $g^{prev} = g$; if $K = \varnothing$, then the algorithm terminates) and to step 4.1, the next iteration begins.
4.7. Find the next marked task $j^{*}_{[m]}$ ($j^{**}_{[m]}$) within positions $\overline{1, p^{new} - 1}$ that satisfies the condition (4.11). First we analyze the task we marked with "*" ("**") last, then the task marked before last, etc. If a marked task is found, then memorize the current subsequence within positions $\overline{1,\ p^{new}}$. Denote it by σ^{act}, go to step 4.8. Otherwise: set $g^{prev} = p^{new}$. If $p^{new} = g$, then check Statements 4.23, 4.29, and 4.31 to eliminate the task $j_{[g]}$ and some of the following tasks from the set of competing tasks. Go to step 4.1, the next iteration begins.
4.8. Execute the optimization due to releasing of reserves of marked tasks (Procedure I). Permute the task $j^{*}_{[m]}$ ($j^{**}_{[m]}$) into the position l and memorize its start time. If the task $j^{*}_{[l]}$ ($j^{**}_{[l]}$) is not the competing task for which we execute the iteration of optimization, or if it is the competing task but is followed by a non-marked task of greater priority, then set $g^{prev} = m - 1$ where m is the

position of the task $j^*_{[l]}$ ($j^{**}_{[l]}$) before the permutation, go to step 4.2 executing Procedures D–I, in turn, for each tardy task within positions $\overline{1, p^{new} - 1}$ (we obtain the sequence σ^{new}). We execute steps 8.4–8.9 of Algorithm A1 here until at least one of the following conditions is true:

- the task $j_{[p^{new}]}$ becomes non-tardy;
- $p^{new} = g$;
- all tasks $j^*_{[m]}$ ($j^{**}_{[m]}$) were considered.

Then go to step 4.9.

4.9. Proceed to the next competing task (set $g^{prev} = g$; if $K = \varnothing$, then the algorithm terminates) and to step 4.1, the next iteration begins. If all competing tasks were considered, the algorithm terminates.

Remark 4.12 This outline of the algorithm is a description of the optimization procedures and their execution sequence aimed to facilitate their understanding. A more complete description of the algorithm is given in Sect. 4.3.1.

Statement 4.41 Procedures F and I implement all types of permutations based on conditions 1–6 of Statement 4.16 that decrease the functional value in the current subsequence. Also, the procedures define the rules for the permutations.

Proof Fulfillment of the necessary conditions we formulated in Statement 4.16 makes the decrease in the functional value possible when moving tardy tasks into earlier positions. If neither of these conditions is met, then the current sequence is optimal on the entire set of tasks, and there are no permutations and insertions that decrease the functional value.

Let us analyze the conditions of Statement 4.16.

Fulfillment of the condition 1 means that there are reserves within the interval of insertion of a tardy task $j_{[g]}$. In this case, we do the insertion (Procedure H).

The condition 2 means that there is a non-tardy task $j_{[l]}$, $l < p^{new}$, which due date is greater than the start time of the task $j_{[p^{new}]}$ (or of the task $j_{[g]}$ if the tardy task is in position g) and condition (4.9) is met. We check the condition 2 in step 8.3 of the Algorithm A1 and in Procedure F.

The condition 3: There is a task $j_{[l]}$, $l < p^{new}$, such that $d_{j_{[l]}} > C_{j_{[p^{new}]}}$. A free permutation of the task $j_{[l]}$ after the task $j_{[p^{new}]}$ is possible (step 6.1 of the Algorithm A1 and Procedure F).

The condition 4: There is a task $j_{[l]}$, $l < p^{new}$, which due date is greater than the start time of the task $j_{[p^{new}]}$ (or $j_{[g]}$). Tasks that do not satisfy the condition 4 do not participate in permutations. We check the condition 4 in step 8.4 of the Algorithm A1 and in Procedure F.

The condition 5 means that the task $j_{[p^{new}]}$ (or $j_{[g]}$) is preceded by an unmarked lower priority task $j_{[l]}$. We check the condition 5 in step 6.2 of the Algorithm A1 and in Procedure F.

The condition 6 of Statement 4.16 is to release the reserves of tasks that earlier used them (those tasks are marked with asterisks). We perform the permutation if condition (4.15) is met. We check the condition 6 in step 6.2 of the Algorithm A1 and in Procedure I.

Hence, the considered conditions 1–6 of Statement 4.16 define all types of permutations implemented in Procedures F and I. ∎

Statement 4.42 The current iteration of optimization, based on directional permutations and implemented through a finite number of operations of optimization procedures D–I executed for the current competing task in the current subsequence and recursively executed to reorder tasks within the intervals of insertions and intervals of permutations, is finite.

Proof Finiteness of the algorithm is based on the following facts:

1. Permutations are directional, they are based on the use of reserves of the preceding non-tardy tasks for tardy tasks. As a result of this, the number of considered feasible sequences is decreased.
2. Formalization of necessary conditions and optimization procedures D–I allows to build an optimal schedule on the entire set of tasks at the current iteration.
3. The types of permutations and insertions that decrease the functional value (Statement 4.16).
4. The step by step optimization within the interval of insertion of the current competing task and, similarly, within the intervals of permutations. It allows to proceed sequentially to the next tardy task within this interval and to construct an optimal schedule by Procedures D–I for a subsequence bounded with the first task of the current sequence and the position of the competing task in the sequence σ^{fp}. The number of iterations is finite, since the number of competing tasks decreases after each iteration of optimization (Remark 4.4 and Corollary 4.8). As a result of the insertion of a competing task, originated tardy tasks may appear. Their number is limited with the number of non-tardy tasks in the current subsequence. This also determines the finiteness of the optimization operations number. At the iteration for the next competing task $j_{[g]}$ we solve the following problem: to insert the task $j_{[g]}$ into the optimal subsequence obtained at the previous iteration of the algorithm. We do it optimally redistributing reserves within the interval $\overline{1, g}$ using Procedures D–I. For a specific problem instance, we will have to execute not all types of optimization procedures but only a necessary part of them depending on the initial data of the instance. Usually we perform only certain types of permutations for some tasks. We do not consider part of feasible subsequences that obviously do not lead to an optimal solution.
5. Absence of looping in the algorithm. Indeed, looping is possible only if two or more tasks compete for the same position during tardy tasks insertions (step 7 of Algorithm A1). We mark the inserted task with asterisk. The algorithm is built so that if there is already a task with asterisk in the position of insertion of the current task $j_{[g]}$, then we insert the task $j_{[g]}$ into the next position (step 4.2).

During the execution of step 8 (optimization due to releasing of reserves of tasks marked with asterisks), we consider all possible combinations of moves of the marked tasks into the later positions with further optimization of subsequences. We perform only permutations of the marked tasks that decrease the functional value. This guarantees the finiteness of the procedures we execute and, as a consequence, the absence of looping in the algorithm.

6. Directional construction of non-repeating schedules, except for cases when we return to the subsequences that were initially optimal, after the execution of optimization procedures in the subsequences of tasks within the intervals of insertions and permutations. The number of such subsequences does not exceed the number of tasks in the current subsequence. ∎

The following statement is an obvious corollary of the above given logic of the algorithm construction (for the case when none of the SSOs and the SSPS condition are satisfied during the algorithm execution) and also of Statements 4.2, 4.3, 4.5–4.10, 4.13–4.30, 4.33, 4.34, 4.41, 4.42, and Lemma 4.1.

Statement 4.43 The algorithm yields an optimal schedule.

Proof Suppose we have obtained an optimal subsequence at the previous iteration on the set of maximum priority tasks. The priority of the current competing task does not exceed the priority of any of the tasks in the optimal subsequence obtained at the previous iteration. So, it follows in an obvious way from the logic of the algorithm construction and from Statements 4.2, 4.2, 4.3, 4.5–4.10, 4.13–4.30, 4.33, 4.34, 4.41, 4.42 that the optimization procedures during the current iteration yield a subsequence in which we achieve the minimum possible increase in the total weighted tardiness compared to the optimal functional value at the previous iteration. Then it follows from Lemma 4.1 that this subsequence is optimal. ∎

Corollary 4.9 We need to arrange the competing tasks in non-increasing order of their priorities to guarantee that the priority of the current competing task at each iteration does not exceed the priority of any of the tasks in the optimal subsequence obtained at the previous iteration.

Corollary 4.10 As a result of the algorithm execution, we obtain one optimal schedule. To obtain the entire set of optimal schedules, it is sufficient to perform permutations of non-tardy tasks among themselves so that they remain non-tardy, without changing the position of another tasks.

4.3.4 *Properties of the Exact Subalgorithm for the Problem Solving*

Let us investigate the efficiency of the exact subalgorithm for the problem solving on the basis of both the results of theoretical studies and a qualitative analysis of its execution logic.

1. The upper bound for the computational complexity of the exact subalgorithm is $O((n_R + n_K)!) + O(n^2)$. Here, $O(n^2)$ is the complexity of the sequence σ^{fp} construction; n_K the number of competing tasks in the sequence σ^{fp} (the set K); n_R the number of tasks with time reserves in σ^{fp} (the set R).

This upper bound is a real estimate of the computational complexity of the PSC-algorithm if for all competing tasks in σ^{fp}:

(a)
$$\forall i < g,\ j_{[i]},\ j_{[g]} \in K,\ d_{j_{[i]}} > d_{j_{[g]}}; \tag{4.16}$$

(b) priorities, due dates, and processing times of competing tasks differ insignificantly;

(c) there are tasks $j_r \in R$ in σ^{fp}, and for some of them we have performed free permutations (see Definition 4.9);

(d) the condition (4.16) is satisfied also for tasks with reserves:

$$\forall i < g,\ j_{[i]}, j_{[g]} \in R,\ d_{j_{[i]}} > d_{j_{[g]}}. \tag{4.17}$$

Substantiation The less the processing times, the due dates, and priorities of tasks differ, the more we consider different combinations of permutations. We exhaust the reserves slower and eliminate the less number of unpromising permutations during F and I optimization procedures execution. The reason is: the parameters of all tasks are practically the same, so we repeat similar permutations for each competing task (Statements 4.4, 4.12, 4.13, 4.16, 4.26).

The presence of tasks $j_r \in R$ in σ^{fp} for which we performed a free permutation causes the combinatorial search within the intervals of insertion after their insertion and subsequent ordering of tasks by priorities. These tasks again will gain the reserves which will be used for each tardy task at the optimization iteration. And since the due dates differ insignificantly, this combinatorial search will be performed on almost the same subsequences. The less the difference in due dates, the more number of subsequences upper-bounded by the due date are considered.

If (4.16) is true, the position of insertion of the task $j_{[g]}$ will be less than that of the task $j_{[i]}$ (because $d_{j_{[i]}} > d_{j_{[g]}}$). The interval of insertion $I_{j_{[g]}}$ increases in comparison with $I_{j_{[i]}}$ by the value of the interval between the positions of their insertion. Most of the originated tardy tasks belonging to the interval $I_{j_{[g]}}$ will also belong to $I_{j_{[i]}}$ (since their processing times differ insignificantly). Because of this, optimization procedures F and I will be executed again for them after the ordering procedure H (first at the iteration for the task $j_{[g]}$ and then for the task $j_{[i]}$).

If, on the other hand, (4.17) is true, we extend the interval of insertion (including to it the additional tasks for which we meet the permutation conditions) for each competing task. As a result, we will have to perform all optimization procedures on the entire set of tasks.

Thus, when the above conditions 1 (a–d) are true, we will statistically significantly face the problem instances which real complexity is commensurate with an exhaustive search. ∎

Remark 4.13 If only conditions 1 (a–c) are true and condition 1 (d) is false, then the complexity of the current optimization iteration is also quite high.

We check the conditions 1 (a–d) when we analyze the sequence σ^{fp}. If we meet these conditions, the above upper bound becomes a real estimate for the computational complexity of the PSC-algorithm.

2. Overall efficiency of the exact subalgorithm is determined by the following factors:

   2.1. At the optimization stage, we decompose the sequence σ^{fp} to a subsequences of a smaller size. We execute the optimization on the subsequence bounded with the position of insertion of the current competing task and the position of this task in the sequence σ^{fp}. During the solving, we can include to this subsequence only tasks that create the reserves at the interval of insertion of the current competing task. In this way we decompose the problem to subproblems of a smaller size.

   2.2. During the problem solving we check the conditions of Statements 4.23, 4.29, 4.31, and 4.32 that allow to eliminate some tasks from the set of competing tasks. In Statements 4.2, 4.4, 4.6, 4.16, 4.18–4.30, 4.33, 4.34, we formulate the rules for eliminating the permutations that do not lead to an optimal solution. In Statement 4.40 we show the conditions for significant reduce in the complexity of the exact subalgorithm. We will have to execute the exhaustive search for some subsequences only in special cases of the problem (if conditions 1 (a–d) are true).

   2.3. The simplest problem instances (the conditions under which the combinatorial search is significantly reduced) are:

      (a) there are no tasks for extending the interval of insertion for all competing tasks in the sequence σ^{fp} (see Statements 4.5 and 4.10) and the following is true for all tasks with reserves within intervals of their insertion:

$$\max_{j_i \in R}\left(d_{j_i} - C_{j_i}\right) < \min_{j_k \in K} l_{j_k}. \tag{4.18}$$

Substantiation Suppose that we insert a competing task j_k into the position belonging to a task $j_r \in R$, and there are no other tasks between the tasks j_k and j_r. In this case, Eq. (4.3) transforms to this:

$$\omega_{j_k} \min\left(C_{j_k} - d_{j_k}, l_{j_r}\right) < \omega_{j_r}\left(l_{j_k} - R_{j_r}\right) \ \forall j_k \in K, j_r \in R \tag{4.19}$$

If condition (4.18) is true, then only one of the competing tasks can take an earlier position. It follows from (4.19) that after insertion of any competing task all tasks with reserves within its interval of insertion will become tardy. The problem reduces to the selection of a competing task and the search for such its position within its interval of insertion that the subsequent optimization minimally increases the functional value. Suppose that a task j_q has the $\min\limits_{j_k \in K} l_{j_k}$. Tasks $j_i \in K$: $l_{j_i} > l_{j_q}$ can take an earlier position only if $d_{j_i} < d_{j_q}$. As a result, the algorithm will not include an enumeration of marked competing tasks in Procedure I. An optimal solution will be achieved by a subalgorithm with complexity $O(n_K n^2)$. We can statistically significantly solve instances of any practically possible dimension. ∎

We check the condition when we analyze the sequence σ^{fp}.

(b) the following is true for all competing tasks:

$$\forall j_i \in K,\ \omega_{j_1} \geq \omega_{j_2} \geq \cdots \geq \omega_{j_{n_K}},\ l_{j_1} \leq l_{j_2} \leq \cdots \leq l_{j_{n_K}},\ d_{j_1} \leq d_{j_2} \leq \cdots \leq d_{j_{n_K}} \tag{4.20}$$

Substantiation In this case, competing tasks are arranged in the order $j_1, j_2, \ldots, j_{n_K}$ in an optimal schedule ([25], Statement 4.28). This significantly reduces the complexity of the algorithm execution because we do not consider competing tasks that took up earlier positions to release reserves during the optimization in Procedure I. The longer the processing time of each next competing task, the faster we exhaust the reserves (Eq. 4.3), and the number of permutations are reduced. ∎

We check the condition when we analyze the sequence σ^{fp}.

(c) The number of tasks with productive reserves (satisfying $d_{j_{[i]}} > d_{j_{[k]}} - l_{j_{[k]}}$, $j_i \in R$, $j_{[k]} \in K$) is minimal.

Substantiation We exclude tasks with unproductive reserves from consideration during the problem solving (Statements 4.4 and 4.10). If the number of tasks with productive reserves is minimal, then we may have a small number of competing tasks for each of which there are productive reserves. Only for such competing tasks we will execute the current optimization iteration. This reduces the number of iterations. ∎

We check the condition when we analyze the sequence σ^{fp}.

We must meet conditions 2.3 (a–c) for all or for the majority of competing tasks and tasks with reserves in order to obtain a significant efficient reduction in the number of permutations performed.

The conditions presented in sub-item 2.3 (a) can be considered a heuristic condition that turns the exact subalgorithm into a polynomial one. These conditions are the subject of further theoretical research with an aim to make them theoretically grounded.

4.4 Approximation Algorithm for the TWT Problem Solving

Recommendations for Using the Approximation Algorithm If analysis of the sequence σ^{fp} shows that conditions 2.3 (a–c) (Sect. 4.3.4) are met, then we recommend to solve the problem for an arbitrary n by the exact algorithm.

Due to the presence of factors 2.1 and 2.2 (Sect. 4.3.4), other conditions are possible which drastically reduce the computational complexity of the exact subalgorithm. We taken into account this possibility in the approximation algorithm by introducing a restriction on the computation time to execute the maximum possible number of iterations of the exact subalgorithm. We obtain the restriction from the introduced polynomial restriction on the number of operations taking into account the characteristics of the computing facilities.

If analysis of the sequence σ^{fp} shows that the conditions causing a time-consuming search (conditions 1 (a–d) or 1 (a–c), Sect. 4.3.4) are satisfied, then we necessarily recommend to solve the problem by the approximation algorithm.

4.4.1 Heuristics that Turn the Implementation of the Current Iteration of the Exact Subalgorithm into a Polynomial Subalgorithm

Consider the properties of the developed PSC-algorithm. The PSC-algorithm consists of a series of uniform iterations. At each iteration, we check the possibility to use the time reserves ($d_j - C_j > 0$) of preceding tasks for the current competing task from the sequence σ^{fp} and construct an optimal schedule for tasks of the current subsequence. As a result of each iteration, the functional value decreases or remains unchanged. Such property allows researchers to develop efficient heuristic and approximation algorithms based on our PSC-algorithm. A result of all the iterations of optimization is an optimal sequence on the entire set of tasks.

The PSC-algorithm executes the following basic procedures at each optimization iteration:

1. Determination of the interval of insertion for the current tardy task in the current sequence. We search the position p, where the tardiness of the task will be minimal (Procedure D).
2. Analysis of reserves within positions $\overline{1, p-1}$ for the current tardy task (extension of the interval of insertion, Procedure E), adjustment of the position of insertion (Procedure G), insertion and priority ordering within the interval of insertion (Procedure H).
3. Step-by-step optimization (decreasing the weighted tardiness of the current tardy task) due to the use of reserves within the extended interval of insertion (Procedure F).
4. Decreasing the weighted tardiness of tardy tasks due to the reserves released by the tasks that previously used them (marked tasks), by permuting them into later positions (Procedure I).

Remark 4.13 We fully describe the procedures of PSC-algorithm in Sect. 4.3.3.

We now highlight the procedures implementing the current iteration of the exact subalgorithm which are related to a non-polynomial complexity.

1. As a result of insertion and further step-by-step optimization within the interval of insertion, we reorder the tasks within this interval and rebuild the schedule. If the position of insertion for the current tardy task $j_{[k]}$ is smaller than that for the previous one for which we performed the insertion procedure, then we reorder all tasks, including the tasks within the previous interval of insertion. Then, we restart optimization procedures within this extended interval, which may lead to a non-polynomial complexity.
2. We perform Procedure I if the task remains tardy after Procedure F. Procedure I includes recursive call of Procedures D–H for a subsequence of tasks within the interval of permutation of a marked task. In this case, we search, in turn, another marked task to insert it after the first marked task and decrease the functional value. Therefore, we do an enumeration of marked tasks with further optimization of a subsequence. This may be related to a non-polynomial complexity. Procedures D–H have polynomial complexity.
3. We considered in Sect. 4.3.4 the most time-consuming case in the exact algorithm of searching the reserves for competing tasks.

It is clear that all other procedures have polynomial complexity.

We now present and substantiate heuristics that allow to eliminate the time-consuming search and construct a polynomial approximation algorithm for the TWT problem solving based on the exact PSC-algorithm.

Heuristic 1 We eliminate the procedure of priority ordering at the interval of insertion (Procedure H). We performed this procedure after insertion of current competing task of the sequence σ^{fp}. Now we only shift the tasks within the interval of insertion into later positions and perform for each tardy task within the interval of insertion the optimization Procedure F and the modified Procedure I (see Heuristic 2).

Substantiation In consequence of Heuristic 1 execution, we do not reconstruct the schedule within the interval of insertion of current competing task but adjust it by optimization procedures (Procedure F and modified Procedure I). Thus, we eliminate repeated (in the general case, multiple) insertion and optimization for a number of tasks within the interval of insertion, which could lead to a non-polynomial complexity. Competing tasks are in non-increasing order of their priorities in the sequence σ^{fp}. So, we sequentially perform the insertion procedures starting from higher priority tasks. It allows the higher priority tasks to use the existing reserves the most efficiently and, therefore, to decrease the functional value (since an insertion of a higher priority task in the case of equal due dates into the same interval decreases the functional value more). ∎

Heuristic 2 (Modified Procedure I) We mark with asterisk only competing tasks and perform Procedure I for the current tardy task as follows. We select the most recently marked task for permutation into a later position. Then we check the condition (4.11). If it holds, we perform the permutation and step-by-step optimization within the interval of permutation by Procedure F (free permutations and the permutations according to Statements 4.12 and 4.13). Otherwise, we proceed to the next most recently marked task and similarly check the condition (4.11) and perform step-by-step optimization. The modified Procedure I terminates if the task for which we perform Procedure I becomes non-tardy or all marked tasks are considered.

Substantiation Heuristic 2 allows to eliminate the search of various combinations of permutations of the marked tasks (we exclude the recursive execution of Procedures D–I). We at first consider the task marked last to release previously used reserves of non-tardy tasks. Competing tasks are in non-increasing order of their priorities in the sequence σ^{fp}. Thus, the task we consider has the lowest priority. This allows us to achieve the maximum possible decrease in the functional value when using reserves by higher priority tasks. ∎

4.4.2 The Approximation Algorithm AA for TWT Problem Solving

The second polynomial component of the PSC-algorithm for the TWT problem is the approximation algorithm AA. It consists of the preliminary stage and the main stage. The main stage is based on the optimization stage A1 of the exact algorithm (Sect. 4.3.1.2). It differs from algorithm Al in that we have changed Procedures F and I on the basis of Heuristics 1 and 2 (we perform insertions only for competing tasks; do not perform orderings within the intervals of insertion; perform only permutations improving the functional value within the intervals of permutation and insertion). Also, we do not check SSOs for the current subsequences.

Preliminary Stage has the following three steps.

1. Specify a polynomial of the problem size restriction on the execution time of the algorithm taking into account the available computing facilities.
2. Execute Algorithm A0 (Sect. 4.3.1.1).
3. Execute the maximum number of iterations of the exact subalgorithm A1 (Sect. 4.3.1.2) satisfying the imposed restriction on the execution time of the algorithm. If all competing tasks are considered within the allotted time, then we have obtained an exact solution, the algorithm terminates. Otherwise, go to the main stage of Algorithm AA.

Main Stage has the following seven steps.

1. Suppose that we have already executed $k - 1$ iterations and built an optimal subsequence σ^{curr} within the interval $\overline{1,\ g-1}$. Remove asterisks from originated tardy tasks in the sequence σ^{curr}. Replace two asterisk marks "**" for competing tasks with the single asterisk "*". We now consider the execution of iteration k by the approximation algorithm for the next competing task $j_{[g]}$ on the current subsequence σ^{curr}. $j_{[g]}$ is the task for which we did not execute the optimization.

2. *The search of the next tardy task:*

   2.1. Find the next competing task $j_{[g]}$ within positions $\overline{g,\ n}$ in the sequence σ^{curr}. If it is not found, the algorithm terminates. Otherwise, check the conditions of Statements 4.22, 4.25, 4.27 and 4.28. If at least one of them is satisfied, then set $g = g + 1$, return to step 2.1, new iteration begins. Otherwise, go to step 3.

   2.2. Find the next tardy task $j_{[p^{new}]}$ within positions $\overline{p^{new}+1,\ g}$ in the sequence σ^{curr}. If it is not found, set $g = g + 1$, go to step 2.1, new iteration begins. If a tardy task is found, go to step 6.

3. *Determination of the position of insertion:*

   3.1. Find a position p where the tardiness of the task $j_{[g]}$ is minimal or equal to zero. Let $T^{nt}_{j_{[g]}} = d_{j_{[g]}} - l_{j_{[g]}}$ denote the maximum start time of the task $j_{[g]}$ when it remains non-tardy. Define the position p from condition: $C_{j_{[p-1]}} \le T^{nt}_{j_{[g]}} < C_{j_{[p]}}$. The following specific cases are possible:

   - $T^{nt}_{j_{[g]}} \le 0$. In this case, $p = 1$.
   - $C_{j_{[i]}} > T^{nt}_{j_{[g]}}\ \forall i = \overline{1, g-1}$. In this case, $p = 1$.

   3.2. Determine the position of insertion p^{new} for the task $j_{[g]}$ within the interval $\overline{p, g-1}$ as the maximum of two positions: p and $q + 1$ if position q is occupied by such a task $j_{[q]}$ that $l_{j_{[q]}} \le l_{j_{[g]}}$, $d_{j_{[q]}} \le d_{j_{[g]}}$, $\omega_{j_{[q]}} \ge \omega_{j_{[g]}}$. Go to step 4.

   3.3. If $p^{new} = p$ ($p^{new} = q + 1$) and this position is occupied by such an unmarked task $j_{[p^{new}]}$ that

$$d_{j_{[p^{new}]}} > C_{j_{[p^{new}]}},\ d_{j_{[p^{new}]}} > d_{j_{[g]}} - l_{j_{[g]}},\ p^{new} \neq g, \tag{4.21}$$

then insert the task $j_{[g]}$ into the position p^{new}, mark it with asterisk, and go to step 6, else go to step 4.1.

4. *Analysis of reserves of tasks within positions* $\overline{1, p^{new} - 1}$:

   4.1. Check if there are tasks $j_{[r]}$ within positions $\overline{1, p^{new} - 1}$ that satisfy at least one of the following conditions:

   - $d_{j_{[r]}} > d_{j_{[p^{new}]}} - l_{j_{[p^{new}]}},\ d_{j_{[r]}} - C_{j_{[r]}} > 0,\ d_{j_{[r]}} > C_{j_{[p^{new}-1]}};$ (4.22)
   - $p_{j_{[r]}} < p_{j_{[p^{new}]}};$ (4.23)
   - the tasks $j_{[r]}$ are marked with "*".

   If such tasks are found, then go to step 4.2, otherwise go to step 5.

   4.2. If position p $(q + 1)$ is occupied by a marked task, then insert the task $j_{[g]}$ to position $p + 1$ $(q + 2)$. If this next position is occupied by a marked task, then go to the next position $p + 2$ $(q + 3)$ and so on, up to position g. Mark the inserted task with asterisk and go to step 6. If the position p^{new} is occupied by a non-marked task, then insert the task $j_{[g]}$ into the position p^{new}, mark it with asterisk, and go to step 6.

5. *Adjustment of the position of insertion.* Execute this step if positions $\overline{1, p^{new} - 1}$ do not contain either any marked tasks, or tasks of lower priority than that of the task $j_{[p^{new}]}$, or reserves to extend the interval of its insertion (or if the existing reserves are not enough for the extension).
   Find the position of insertion for the task $j_{[p^{new}]}$ within positions $\overline{p^{new}, g}$, starting from the position p^{new} (which is the new position of the task $j_{[g]}$) and sequentially checking for all tasks $j_{[i]}$, $i = \overline{p^{new}, g}$, the conditions: $d_{j_{[i]}} > C_{j_{[i]}}$ and $d_{j_{[i]}} > d_{j_{[p^{new}]}} - l_{j_{[p^{new}]}}$. If both conditions are true, then insert the task $j_{[p^{new}]}$ into the position i, mark it with asterisk, set $p^{new} = i$, go to step 6. If at least one of the conditions is not satisfied within the entire interval $\overline{p^{new}, g}$, then the task $j_{[p^{new}]}$ remains in the initial position g. In this case, set $p^{new} = g$, go to step 7.
6. *Optimization due to reserves of tasks within positions* $\overline{1, p^{new} - 1}$. Execute the following steps 6.1–6.4 sequentially, starting with lower priority tasks, i.e., consider positions $p^{new} - 1$, $p^{new} - 2$, etc. Perform each step as many times as there are tasks corresponding to the conditions of this step.

6.1. Perform free permutations decreasing the tardiness of the task $j_{[p^{new}]}$. Determine the new position p^{new}. If the task $j_{[p^{new}]}$ remains tardy, go to step 6.2. If $j_{[p^{new}]}$ is not tardy, go to step 2.2.

6.2. Find an unmarked task $j_{[l]}$, $l < p^{new}$, such that $p_{j_{[l]}} < p_{j_{[p^{new}]}}$. If such task is not found, then go step 6.3. Otherwise, permute the task $j_{[l]}$ into the position p^{new} (after the task $j_{[p^{new}]}$), set $p^{new} = p^{new} - 1$. If the task $j_{[p^{new}]}$ remains tardy, go to step 6.2, else go to step 6.4.

6.3. Find a task $j_{[l]}$, $l < p^{new}$, such that $d_{j_{[l]}} > C_{j_{[l]}}$, $d_{j_{[l]}} > C_{j_{[p^{new}]}} - l_{j_{[p^{new}]}}$, and

$$\omega_{j_{[p^{new}]}} \min\left(l_{j_{[l]}}, C_{j_{[p^{new}]}} - d_{j_{[p^{new}]}}\right) > \omega_{j_{[l]}}\left(C_{j_{[p^{new}]}} - d_{j_{[l]}}\right).$$

If such task is not found, then go step 6.4. Otherwise, permute the task $j_{[l]}$ into the position p^{new} (after the task $j_{[p^{new}]}$), set $p^{new} = p^{new} - 1$. If the task $j_{[p^{new}]}$ remains tardy, go to step 6.3, else go to step 6.4.

6.4. If, the task $j_{[p^{new}]}$ has remained tardy, or the conditions formulated in steps 6.1–6.3 did not satisfy during the optimization (the position p^{new} has remained the same and equals to g), then go to step 7. If $j_{[p^{new}]}$ is not tardy, go to step 2.2.

7. *Optimization due to releasing of reserves of tasks marked with "*":*

7.1. Find the next unconsidered marked task $j^*_{[m]}$ within the interval $\overline{1, p^{new} - 1}$ where p^{new} is the new position of the task $j_{[g]}$ (if the task $j_{[g]}$ remains in position g after its insertion and execution of step 6, then $p^{new} = g$). First we analyze the task we marked with "*" last, then the task marked before last, etc. If a marked task is found, go to step 7.2. If a marked task is not found and $j_{[p^{new}]}$ is the competing task for which we execute the optimization iteration, then check Statements 4.23, 4.29, and 4.31 to eliminate the task $j_{[g]}$ and some of the following tasks from the set of competing tasks. Set $g = g + 1$ and go to step 2.1, new iteration begins. If a marked task is not found and $j_{[p^{new}]}$ is not the competing task for which we execute the optimization iteration, then go to step 2.2.

7.2. Check the condition:

$$\omega_{j^*_{[m]}}\left(C_{j_{[l]}} - d_{j^*_{[m]}}\right) < \sum_{i=m}^{p^{new}} \omega_{j_{[i]}} \max\left(0, C_{j_{[i]}} - d_{j_{[i]}}\right) \quad (4.24)$$

where position l corresponds to the priority of the task $j^*_{[m]}$ within the interval $\overline{m, p^{new}}$ (if $p_{j^*_{[m]}} < p_{j_{[p^{new}]}}$, then $l = p^{new}$). If Eq. (4.24) is true, then go to step 7.3, else go to step 7.1.

7.3. Memorize the current subsequence within positions $\overline{1, p^{new}}$. Denote it by σ^{prev} ("sigma previous").

7.4. Perform the permutation of the task $j^*_{[m]}$ into the position l, go to step 7.5.

7.5. Perform improving the functional value permutations (free permutations and permutations of tasks satisfying the conditions of Statements 4.12 and 4.13) for each tardy task $j_{[k]}$, $k = \overline{m, p^{new} - 1}$. Here, m is the position of the task $j^*_{[l]}$ before the permutation. Use for the optimization reserves of tasks within the interval $\overline{1, k - 1}$. Go to step 7.6.

7.6. Denote the subsequence within positions $\overline{1, p^{new}}$ by σ^{new}. If the functional value of the subsequence σ^{new} is not less than that of the subsequence σ^{prev}, then go to step 7.7. Otherwise, go to step 7.8. The algorithm will continue on a subsequence with a lower functional value.

7.7. Return to the subsequence σ^{prev} and go to step 7.1 to search for the next unconsidered marked task.

7.8. If the task $j^*_{[l]}$ took the position corresponding to its priority (it is followed by unmarked tasks of the same or lower priorities), then remove the mark "*" from this task. Go to step 7.9.

7.9. Find the next task $j^*_{[m]}$ that satisfies (4.24) within the interval $\overline{1, l - 1}$. If such a task is found, then go to step 7.3 executing step 7 in turn for the task $j^*_{[l]}$ (denoting it by $j_{[p^{new}]}$ and setting $p^{new} = l$). If such a task is not found or the task $j_{[p^{new}]}$ became non-tardy, then if the task $j_{[p^{new}]}$ is the competing task for which we execute the iteration of optimization, set $g^{prev} = p^{new}$, go to step 2.1 (begin new iteration). If the task $j_{[p^{new}]}$ is not the competing task for which we execute the iteration of optimization, then go to step 2.2. We repeat steps 7.1–7.9 until all marked tasks are considered or until the task $j_{[p^{new}]}$ is not tardy. We also recursively execute step 7 for each tardy task within the interval of insertion of the task $j_{[g]}$, i.e., $\overline{p^{new}, g}$.

Approximation Algorithm AA is the PSC-algorithm of the second class. Indeed, due to Heuristics 1 and 2 and the polynomial restriction on the amount of computations for the iterations of optimization of the exact algorithm, it has a polynomial complexity. We check the SSOs #1–5, 7, 10–12 before the optimization iterations (see Corollaries 4.1 and 4.7, Statements 4.1, 4.4, 4.6, 4.11, 4.35, 4.36, 4.37). If at least one of these SSOs is satisfied during the algorithm AA execution, then an obtained feasible solution is optimal. A distinctive feature of the algorithm is the following fact. If the solution is obtained only by the iterations of optimization of the exact algorithm, then the approximation algorithm guarantees an optimal solution obtaining. This situation may occur quite often, since in the general case the total number of competing tasks may decrease by more than one task after each iteration of optimization.

Average statistical estimate of deviation of an approximate solution from the optimum does not exceed 5% of it.

Remark 4.15 Paper [26] proposes a special case of this approximation algorithm for the problem with equal weights, based on the sequence σ^{fp} construction. The authors have constructed an estimate of deviation from an optimal functional value for the solutions obtained for any problem instance. On the basis of this estimate, they build an estimate of deviation from the optimum for any heuristic algorithms and study the dependence between the deviation of the approximate solution from the optimum and the number of competing tasks in a sequence σ^{fp}. The research showed that for problems of dimension $n \leq 1000$ when the number of competing tasks is more than 35% of the entire number of tasks, we statistically significantly obtain a solution within 1–2% deviation from the optimum by the approximation algorithm with $O(n^2)$ complexity. According to the proposed heuristic, the same dependence is true for real practical dimensions instances ($n > 1000$) that cannot be solved by any of the existing methods. This allows us to obtain an efficient solution of the problem of almost any dimension by means of the proposed approximation algorithm.

Remark 4.16 One can also use Algorithm AA directly on the sequence σ^{fp}, without executing the optimization iterations of the exact algorithm.

Example 4.5 (to the Approximation Algorithm) Initial data are the test problem #6 from twelve test problems [27]. The sequence σ^{ord} is in Table 4.24.

The sequence σ^{fp} (Table 4.25) is the result of free permutations.

For clarity, we perform the approximation algorithm AA starting from the sequence σ^{fp} (assuming that the number of optimization iterations of the exact algorithm is zero). Define the set of competing tasks. According to Definition 4.11, tasks 15 and 20 are not competing.

Iteration 1. The first competing task is task 13. Its position of insertion is 4 (after task 5). But this position occupies task 3 with $l_3 < l_{13}$, $d_3 < d_{13}$, $\omega_3 > \omega_{13}$. In accordance with Statement 4.28, we insert task 13 after task 3 and mark it with asterisk. We obtain the sequence σ_1 shown in Table 4.26.

We perform improving permutations in the sequence σ_1. We insert task 9 after task 2 (the free permutation) and task 7 after task 10 according to Statement 4.13. We obtain the sequence σ_2 (Table 4.27).

Iteration 2. The next competing task in the sequence σ_2 is task 14. Its position of insertion is 3, after task 6. But tasks 5, 3, and 13*, according to Statement 4.28,

Table 4.24 Sequence σ^{ord}

j	p_j	l_j	ω_j	d_j	C_j	f_j	j	p_j	l_j	ω_j	d_j	C_j	f_j
1	0.160	86	14	544	86	0	11	0.110	104	11	932	1076	1584
2	0.150	100	15	873	186	0	12	0.100	102	10	1185	1178	0
3	0.140	97	14	407	283	0	13	0.100	103	10	465	1281	8160
4	0.140	102	14	236	385	2086	14	0.090	110	10	364	1391	10,270
5	0.130	99	13	318	484	2158	15	0.090	93	8	1304	1484	1440
6	0.130	100	13	220	584	4732	16	0.080	104	8	691	1588	7176
7	0.130	96	12	1207	680	0	17	0.070	113	8	774	1701	7416
8	0.120	98	12	721	778	684	18	0.070	101	7	223	1802	11,053
9	0.120	101	12	889	879	0	19	0.060	94	6	940	1896	5736
10	0.120	93	11	1193	972	0	20	0.050	103	5	1392	1999	3035

The functional value $f = 65{,}530$

Table 4.25 Sequence σ^{fp}

j	p_j	l_j	ω_j	d_j	C_j	f_j	j	p_j	l_j	ω_j	d_j	C_j	f_j
4	0.140	102	14	236	102	0	7	0.130	96	12	1207	1085	0
6	0.130	100	13	220	202	0	10	0.120	93	11	1193	1178	0
5	0.130	99	13	318	301	0	13	0.100	103	10	465	1281	8160
3	0.140	97	14	407	398	0	14	0.090	110	10	364	1391	10,270
1	0.160	86	14	544	484	0	15	0.090	93	8	1304	1484	1440
8	0.120	98	12	721	582	0	16	0.080	104	8	691	1588	7176
9	0.120	101	12	889	683	0	17	0.070	113	8	774	1701	7416
2	0.150	100	15	873	783	0	18	0.070	101	7	223	1802	11,053
11	0.110	104	11	932	887	0	19	0.060	94	6	940	1896	5736
12	0.100	102	10	1185	989	0	20	0.050	103	5	1392	1999	3035

The functional value f = 54,286

Table 4.26 Sequence σ_1

j	p_j	l_j	ω_j	d_j	C_j	f_j	j	p_j	l_j	ω_j	d_j	C_j	f_j
4	0.140	102	14	236	102	0	12	0.100	102	10	1185	1092	0
6	0.130	100	13	220	202	0	7	0.130	96	12	1207	1188	0
5	0.130	99	13	318	301	0	10	0.120	93	11	1193	1281	968
3	0.140	97	14	407	398	0	14	0.090	110	10	364	1391	10,270
13*	0.100	103	10	465	501	360	15	0.090	93	8	1304	1484	1440
1	0.160	86	14	544	587	602	16	0.080	104	8	691	1588	7176
8	0.120	98	12	721	685	0	17	0.070	113	8	774	1701	7416
9	0.120	101	12	889	786	0	18	0.070	101	7	223	1802	11,053
2	0.150	100	15	873	886	195	19	0.060	94	6	940	1896	5736
11	0.110	104	11	932	990	638	20	0.050	103	5	1392	1999	3035

The functional value f = 48,889

Table 4.27 Sequence σ_2

j	p_j	l_j	ω_j	d_j	C_j	f_j	j	p_j	l_j	ω_j	d_j	C_j	f_j
4	0.140	102	14	236	102	0	12	0.100	102	10	1185	1092	0
6	0.130	100	13	220	202	0	10	0.120	93	11	1193	1185	0
5	0.130	99	13	318	301	0	7	0.130	96	12	1207	1281	888
3	0.140	97	14	407	398	0	14	0.090	110	10	364	1391	10,270
13*	0.100	103	10	465	501	360	15	0.090	93	8	1304	1484	1440
1	0.160	86	14	544	587	602	16	0.080	104	8	691	1588	7176
8	0.120	98	12	721	685	0	17	0.070	113	8	774	1701	7416
2	0.150	100	15	873	785	0	18	0.070	101	7	223	1802	11,053
9	0.120	101	12	889	886	0	19	0.060	94	6	940	1896	5736
11	0.110	104	11	932	990	638	20	0.050	103	5	1392	1999	3035

The functional value f = 48,614

Table 4.28 Sequence σ_3

j	p_j	l_j	ω_j	d_j	C_j	f_j	j	p_j	l_j	ω_j	d_j	C_j	f_j
4	0.140	102	14	236	102	0	11	0.110	104	11	932	1100	1848
6	0.130	100	13	220	202	0	12	0.100	102	10	1185	1202	170
5	0.130	99	13	318	301	0	10	0.120	93	11	1193	1295	1122
3	0.140	97	14	407	398	0	7	0.130	96	12	1207	1391	2208
13*	0.100	103	10	465	501	360	15	0.090	93	8	1304	1484	1440
14*	0.090	110	10	364	611	2470	16	0.080	104	8	691	1588	7176
1	0.160	86	14	544	697	2142	17	0.070	113	8	774	1701	7416
8	0.120	98	12	721	795	888	18	0.070	101	7	223	1802	11,053
2	0.150	100	15	873	895	330	19	0.060	94	6	940	1896	5736
9	0.120	101	12	889	996	1284	20	0.050	103	5	1392	1999	3035

The functional value f = 48,678

must precede task 14 in an optimal schedule. Therefore, we insert task 14 after task 13*, mark it with asterisk, and obtain the sequence σ_3 (Table 4.28).

We do not perform Procedure F because we lack reserves for task 1. We perform for it improving permutations (Procedure I). We permute task 14* into the position after task 1 and mark it with two asterisks. The functional value decreases (sequence σ_4, Table 4.29).

Conditions of procedures F and I are not met for tasks 8, 2, 9, and 11, there are no insertions and permutations. Tasks 10 and 7 are preceded by task 12 of a lower priority. According to Statement 4.12, we permute task 12 into the position after tasks 10 and 7. We obtain the sequence σ_5 (Table 4.30).

Task 10 in the sequence σ_5 cannot be inserted after task 7 because the latter one gains a reserve as a result of such permutation, and the functional value increases. Tasks 15 and 20 are not competing. Tasks 16, 17, and 19 cannot be moved into earlier positions because reserves are absent within the intervals of their insertion.

Table 4.29 Sequence σ_4

j	p_j	l_j	ω_j	d_j	C_j	f_j	j	p_j	l_j	ω_j	d_j	C_j	f_j
4	0.140	102	14	236	102	0	11	0.110	104	11	932	1100	1848
6	0.130	100	13	220	202	0	12	0.100	102	10	1185	1202	170
5	0.130	99	13	318	301	0	10	0.120	93	11	1193	1295	1122
3	0.140	97	14	407	398	0	7	0.130	96	12	1207	1391	2208
13*	0.100	103	10	465	501	360	15	0.090	93	8	1304	1484	1440
1	0.160	86	14	544	587	602	16	0.080	104	8	691	1588	7176
14**	0.090	110	10	364	697	3330	17	0.070	113	8	774	1701	7416
8	0.120	98	12	721	795	888	18	0.070	101	7	223	1802	11,053
2	0.150	100	15	873	895	330	19	0.060	94	6	940	1896	5736
9	0.120	101	12	889	996	1284	20	0.050	103	5	1392	1999	3035

The functional value f = 47,998

Table 4.30 Sequence σ_5

j	p_j	l_j	ω_j	d_j	C_j	f_j	j	p_j	l_j	ω_j	d_j	C_j	f_j
4	0.140	102	14	236	102	0	11	0.110	104	11	932	1100	1848
6	0.130	100	13	220	202	0	10	0.120	93	11	1193	1193	0
5	0.130	99	13	318	301	0	7	0.130	96	12	1207	1289	984
3	0.140	97	14	407	398	0	12	0.100	102	10	1185	1391	2060
13*	0.100	103	10	465	501	360	15	0.090	93	8	1304	1484	1440
1	0.160	86	14	544	587	602	16	0.080	104	8	691	1588	7176
14**	0.090	110	10	364	697	3330	17	0.070	113	8	774	1701	7416
8	0.120	98	12	721	795	888	18	0.070	101	7	223	1802	11,053
2	0.150	100	15	873	895	330	19	0.060	94	6	940	1896	5736
9	0.120	101	12	889	996	1284	20	0.050	103	5	1392	1999	3035

The functional value $f = 47{,}542$

Existing reserves are not enough for task 18. Conditions of Procedure I are not satisfied for tasks 16–19, therefore they remain on the occupied positions. The sequence σ_5 is final. It coincides with the optimal solution obtained by the exact algorithm.

4.5 Illustration to the Problem Solving by the PSC-Algorithm Using Procedures A–I

Example 4.6 Initial data for Example 4.6 is in Table 4.31.

Execute Procedure A and obtain the sequence σ^{ord} (Table 4.32).

Perform free permutations (Procedure B). The result is in Table 4.33.

Determine the set of competing tasks (Procedure C): $K = \{12, 13\}$.

Iteration 1: the first competing task $j_{[g]}$ is task 12 (its position $g = 12$). Determine the position of its insertion (Procedure D), this is position 1. $p^{new} = k + 1 = 7$. Since tardy task 2 occupies the position 7, we perform the analysis of reserves within positions 1, …, 6 (Procedure E). There are no tasks in these positions that allow to extend the interval of insertion for task 12 (Procedure F), therefore, we perform adjustment of the position of insertion (Procedure G). As a result, we insert task 12 into position 8 before task 11. Mark task 12 with asterisk and execute the

Table 4.31 Initial data for Example 4.6

j	1	2	3	4	5	6	7	8	9	10	11	12	13
l_j	5	27	20	45	23	8	26	53	4	3	14	57	55
ω_j	1	5	6	8	5	7	9	10	5	4	8	10	6
d_j	0	0	25	179	244	0	225	0	0	0	173	0	200

Table 4.32 Sequence σ^{ord}

Position	j	p_j	l_j	ω_j	d_j	C_j	f_j
1	10	1.333	3	4	0	3	12
2	9	1.250	4	5	0	7	35
3	6	0.875	8	7	0	15	105
4	11	0.571	14	8	173	29	–
5	7	0.346	26	9	225	55	–
6	3	0.300	20	6	25	75	300
7	5	0.217	23	5	244	98	–
8	1	0.200	5	1	0	103	103
9	8	0.189	53	10	0	156	1560
10	2	0.185	27	5	0	183	915
11	4	0.178	45	8	179	228	392
12	12	0.175	57	10	0	285	2850
13	13	0.109	55	6	200	340	840

The functional value $f = 7112$

Table 4.33 Sequence σ^{fp}

Position	j	p_j	l_j	ω_j	d_j	C_j	f_j
1	10	1.333	3	4	0	3	12
2	9	1.250	4	5	0	7	35
3	6	0.875	8	7	0	15	105
4	3	0.300	20	6	25	35	60
5	1	0.200	5	1	0	40	40
6	8	0.189	53	10	0	93	930
7	2	0.185	27	5	0	120	600
8	11	0.571	14	8	173	134	0
9	4	0.178	45	8	179	179	0
10	7	0.346	26	9	225	205	0
11	5	0.217	23	5	244	228	0
12	12	0.175	57	10	0	285	2850
13	13	0.109	55	6	200	340	840

The functional value $f = 5472$

reordering (Procedure H), that is, Procedures A and B for tasks within positions 9, …, 12. We obtain a sequence σ_1 (Table 4.34).

For each tardy task within the interval of insertion 9, …, 12 we perform step by step optimization (Procedures D–I). Therefore, consider task 11. The position of insertion of task 11 is 8 (Procedure D). But position 8 is occupied by tardy task 12^*. So, the position of insertion is 9, we perform the analysis of reserves (Procedure E). Since a marked task 12^* and tasks of a lower priority than that of task 11 are present in positions 1, …, 8, we execute the procedure of optimization due to the use of reserves within the extended interval of insertion (Procedure E). According to

Table 4.34 The first iteration. Sequence σ_1

Position	j	p_j	l_j	ω_j	d_j	C_j	f_j
1	10	1.333	3	4	0	3	12
2	9	1.250	4	5	0	7	35
3	6	0.875	8	7	0	15	105
4	3	0.300	20	6	25	35	60
5	1	0.200	5	1	0	40	40
6	8	0.189	53	10	0	93	930
7	2	0.185	27	5	0	120	600
8	12^*	0.175	57	10	0	177	1770
9	11	0.571	14	8	173	191	144
10	7	0.346	26	9	225	217	0
11	5	0.217	23	5	244	240	0
12	4	0.178	45	8	179	285	848
13	13	0.109	55	6	200	340	840

The functional value $f = 5384$

Statement 4.12, we can decrease the weighted tardiness of task 11 putting after it unmarked lower priority tasks, starting with the minimum priority task. Find the minimum priority task within positions 1, …, 8, it is task 2. Task 11 becomes non-tardy in consequence of the permutation of task 2 after it, and the functional value decreases. We execute the reordering (Procedure H) and proceed to the next tardy task 4 within the interval of insertion of the task 12^*. The current sequence is given in Table 4.35.

Table 4.35 The permutation of task 2

Position	j	p_j	l_j	ω_j	d_j	C_j	f_j
1	10	1.333	3	4	0	3	12
2	9	1.250	4	5	0	7	35
3	6	0.875	8	7	0	15	105
4	3	0.300	20	6	25	35	60
5	1	0.200	5	1	0	40	40
6	8	0.189	53	10	0	93	930
7	12^*	0.175	57	10	0	150	1500
8	11	0.571	14	8	173	164	0
9	2	0.185	27	5	0	191	955
10	7	0.346	26	9	225	217	0
11	5	0.217	23	5	244	240	0
12	4	0.178	45	8	179	285	848
13	13	0.109	55	6	200	340	840

The functional value $f = 5325$

The position of insertion for task 4 is 7. But there is marked task 12* in position 7. Position 8 is occupied by task 11 that should precede task 4 in an optimal schedule, according to Statement 4.28. A tardy task of a higher priority is in position 9. Therefore, the position of insertion is 10. Task 4 will be tardy in position 10, so we perform the analysis of reserves within positions 1, …, 9 (Procedure E). Since the marked task 12* exists within those positions, we execute the procedure of optimization due to the use of reserves within the extended interval of insertion (Procedure F). There are no free permutations and tasks that satisfy the conditions of Statements 4.12 and 4.13 within the interval of insertion. So, we insert task 4 into the position after task 2 and execute the reordering (Procedure H). The current sequence is in Table 4.36.

Since task 4* is tardy, we execute the optimization due to releasing of reserves of marked tasks (Procedure I). To do this, we find the task marked with asterisk last within positions 1, …, 9. This is task 12*, it should be inserted after the task 4*, but the functional value increases with this move. So, we proceed to originated tardy tasks.

The originated tardy tasks for task 4* are tasks 7 and 5. The position of insertion for task 7 is occupied by task 4*. Therefore, task 7 remains in its position. We execute the optimization due to releasing of reserves (Procedure I). To do this, we find the task marked with asterisk last within positions 1, …, 10. This is task 4*. We memorize the current sequence and the functional value and insert task 4* into position 11 because its priority is less than that of task 7. The current sequence is in Table 4.37.

Task 7 became non-tardy. The situation for task 5 will be the same as for task 7. Task 4 returns into its position. Thus, we return to the sequence shown in

Table 4.36 The insertion of task 4

Position	j	p_j	l_j	ω_j	d_j	C_j	f_j
1	10	1.333	3	4	0	3	12
2	9	1.250	4	5	0	7	35
3	6	0.875	8	7	0	15	105
4	3	0.300	20	6	25	35	60
5	1	0.200	5	1	0	40	40
6	8	0.189	53	10	0	93	930
7	12*	0.175	57	10	0	150	1500
8	11	0.571	14	8	173	164	0
9	2	0.185	27	5	0	191	955
10	4*	0.178	45	8	179	236	456
11	7	0.346	26	9	225	262	333
12	5	0.217	23	5	244	285	205
13	13	0.109	55	6	200	340	840

The functional value $f = 5471$

Table 4.37 The insertion of task 4*

Position	j	p_j	l_j	ω_j	d_j	C_j	f_j
1	10	1.333	3	4	0	3	12
2	9	1.250	4	5	0	7	35
3	6	0.875	8	7	0	15	105
4	3	0.300	20	6	25	35	60
5	1	0.200	5	1	0	40	40
6	8	0.189	53	10	0	93	930
7	12*	0.175	57	10	0	150	1500
8	11	0.571	14	8	173	164	0
9	2	0.185	27	5	0	191	955
10	7	0.346	26	9	225	217	0
11	4*	0.178	45	8	179	262	664
12	5	0.217	23	5	244	285	205
13	13	0.109	55	6	200	340	840

The functional value $f = 5346$

Table 4.35. The last unconsidered competing task 13 is excluded from consideration, according to Statement 4.28.

It is clear that we have obtained an optimal schedule on the entire set of tasks (Table 4.35) and that there are no permutations that decrease the functional value.

Example 4.7 Initial data for Example 4.7 is in Table 4.38.

Execute Procedure A and obtain the sequence σ^{ord} (Table 4.39).

Perform free permutations (Procedure B). The result is in Table 4.40.

Determine the set of competing tasks (Procedure C): $K = \{8, 9, 10\}$.

Iteration 1 for the first competing task 8. Procedure D: the position of insertion for the task 8 is 3. Procedure H: we insert the task 8 into position 3, mark it with asterisk, execute priority ordering within the interval of insertion 3, …, 7 and free permutations. We obtain a sequence σ_1 (Table 4.41).

Optimization (Procedures D–I to decrease the weighted tardiness) for the originated tardy task 7. Procedure D: the position of insertion is 3. But within the interval of insertion we have a marked task 8^*. Therefore, in order to avoid looping, according to the algorithm, we insert the task 7 not after task 6 but after the task 8^*. Then we perform priority ordering and free permutations (Table 4.42).

Table 4.38 Initial data for Example 4.7

j	1	2	3	4	5	6	7	8	9	10
l_j	1	4	8	37	35	29	55	82	21	70
ω_j	10	9	5	10	8	6	4	5	1	2
d_j	250	150	200	155	187	150	100	150	150	250

Table 4.39 Sequence σ^{ord}

Position	j	p_j	l_j	ω_j	d_j	C_j	f_j
1	1	10.000	1	10	250	1	0
2	2	2.250	4	9	150	5	0
3	3	0.625	8	5	200	13	0
4	4	0.270	37	10	155	50	0
5	5	0.229	35	8	187	85	0
6	6	0.207	29	6	150	114	0
7	7	0.073	55	4	100	169	276
8	8	0.061	82	5	150	251	505
9	9	0.048	21	1	150	272	122
10	10	0.029	70	2	250	342	184

The functional value $f = 1087$

Table 4.40 Sequence σ^{fp}

Position	j	p_j	l_j	ω_j	d_j	C_j	f_j
1	2	2.250	4	9	150	4	0
2	6	0.207	29	6	150	33	0
3	7	0.073	55	4	100	88	0
4	4	0.270	37	10	155	125	0
5	5	0.229	35	8	187	160	0
6	3	0.625	8	5	200	168	0
7	1	10.000	1	10	250	169	0
8	8	0.061	82	5	150	251	505
9	9	0.048	21	1	150	272	122
10	10	0.029	70	2	250	342	184

The functional value $f = 811$

Table 4.41 The first iteration. Sequence σ_1

Position	j	p_j	l_j	ω_j	d_j	C_j	f_j
1	2	2.250	4	9	150	4	0
2	6	0.207	29	6	150	33	0
3	8^*	0.061	82	5	150	115	0
4	4	0.270	37	10	155	152	0
5	5	0.229	35	8	187	187	0
6	1	10.000	1	10	250	188	0
7	3	0.625	8	5	200	196	0
8	7	0.073	55	4	100	251	604
9	9	0.048	21	1	150	272	122
10	10	0.029	70	2	250	342	184

The functional value $f = 910$

Table 4.42 Sequence σ_2

Position	j	p_j	l_j	ω_j	d_j	C_j	f_j
1	2	2.250	4	9	150	4	0
2	6	0.207	29	6	150	33	0
3	8^*	0.061	82	5	150	115	0
4	7^*	0.073	55	4	100	170	280
5	3	0.625	8	5	200	178	0
6	4	0.270	37	10	155	215	600
7	1	10.000	1	10	250	216	0
8	5	0.229	35	8	187	251	512
9	9	0.048	21	1	150	272	122
10	10	0.029	70	2	250	342	184

The functional value $f = 1698$

Since the task 7^* remains tardy after the insertion, we need to extend the interval of its insertion. Existing reserves within positions 1, …, 3 are not enough to extend the interval of insertion, so we execute the optimization due to releasing of reserves (Procedure I). Doing this, we determine the possibility to move task 7 into earlier positions due to the reserves released by task 8^*. We memorize the sequence, permute the task 8^* into the position after task 7 (since task 7 has greater priority than task 8^*), mark the task 8^* with two asterisks (since it is shifted into a later position and has given out a part of its reserves). We obtain the sequence shown in Table 4.43.

Procedures D–I for the next originated tardy task 4. The condition of Statement 4.13 is satisfied for tasks 3 and 4 (Procedure E). Therefore, we insert task 3 into the position after task 4. The current sequence is in Table 4.44.

Similarly, task 8^{**} successively releases reserves for tasks 4, 3, and 5 (we execute the recursive call of Procedure I for task 8^{**}). The condition to permute task

Table 4.43 Sequence σ_3

Position	j	p_j	l_j	ω_j	d_j	C_j	f_j
1	2	2.250	4	9	150	4	0
2	6	0.207	29	6	150	33	0
3	7^*	0.073	55	4	100	88	0
4	8^{**}	0.061	82	5	150	170	100
5	3	0.625	8	5	200	178	0
6	4	0.270	37	10	155	215	600
7	1	10.000	1	10	250	216	0
8	5	0.229	35	8	187	251	512
9	9	0.048	21	1	150	272	122
10	10	0.029	70	2	250	342	184

The functional value $f = 1518$

Table 4.44 Sequence σ_4

Position	j	p_j	l_j	ω_j	d_j	C_j	f_j
1	2	2.250	4	9	150	4	0
2	6	0.207	29	6	150	33	0
3	7^*	0.073	55	4	100	88	0
4	8^{**}	0.061	82	5	150	170	100
5	4	0.270	37	10	155	207	520
6	3	0.625	8	5	200	215	75
7	1	10.000	1	10	250	216	0
8	5	0.229	35	8	187	251	512
9	9	0.048	21	1	150	272	122
10	10	0.029	70	2	250	342	184

The functional value $f = 1513$

7^* into a position after either task 4, 3, 5 or 8 (Statement 4.26) does not hold. Thus, task 8 returns into its initial position, we remove two asterisks from it and proceed to the next competing task. The current schedule is in Table 4.45.

Iteration 2: Procedures D–I for competing task 9. We insert the task 9 into position 5 after task 4, execute priority ordering and free permutations. We obtain the schedule shown in Table 4.46.

We have two tardy tasks (8 and 10) in the current sequence. But task 8 cannot use the reserves of the preceding tasks anymore, except for the reserves used by task 9. Therefore, we memorize the current sequence, insert the task 9^* into the position after task 8. The functional value increases to 811, so we return to the memorized sequence with the functional value 794.

Iteration 3: Procedures D–I for competing task 10. The position of insertion for the task 10 is 5 (before task 5). There is task 1 within the interval of insertion of task 10 such that $l_1 < l_{10}$, $d_1 = d_{10}$, $\omega_1 > \omega_{10}$. So, task 1 precedes task 10 in an optimal

Table 4.45 Sequence σ_5

Position	j	p_j	l_j	ω_j	d_j	C_j	f_j
1	2	2.250	4	9	150	4	0
2	6	0.207	29	6	150	33	0
3	7^*	0.073	55	4	100	88	0
4	4	0.270	37	10	155	125	0
5	3	0.625	8	5	200	133	0
6	1	10.000	1	10	250	134	0
7	5	0.229	35	8	187	169	0
8	8	0.061	82	5	150	251	505
9	9	0.048	21	1	150	272	122
10	10	0.029	70	2	250	342	184

The functional value $f = 811$

Table 4.46 Sequence σ_6

Position	j	p_j	l_j	ω_j	d_j	C_j	f_j
1	2	2.250	4	9	150	4	0
2	6	0.207	29	6	150	33	0
3	7*	0.073	55	4	100	88	0
4	4	0.270	37	10	155	125	0
5	9*	0.048	21	1	150	146	0
6	5	0.229	35	8	187	181	0
7	3	0.625	8	5	200	189	0
8	1	10.000	1	10	250	190	0
9	8	0.061	82	5	150	272	610
10	10	0.029	70	2	250	342	184

The functional value $f = 794$

schedule (Statement 4.28). Therefore, the position of insertion of the task 10 is 9 (before task 8). But this position is occupied by tardy task 8, so we fall one position below, job 10 remains in its position (and there are no preceding tasks that can create reserves for task 10). All competing tasks are considered. The sequence shown in Table 4.46 is optimal.

4.6 Illustrative Examples

In this section, we show the solving process for three problem instances taken from OR-Library [28]. Instances of greater dimensions than those from OR-Library are given in [29].

Example 4.8 Instance #22 from file wt50.txt in OR-Library [28]. Initial data is in Table 4.47.

The sequence σ^{ord} is in Table 4.48.

The sequence σ^{fp} (Table 4.49) is the result of free permutations.

Iteration 1. The first tardy task in σ^{fp} preceded by a task with reserve is task 19. We insert it into the position after task 7, mark it with asterisk and arrange tasks that follow it within the interval of insertion in non-increasing order of priorities. We obtain the sequence shown in Table 4.50.

According to Statement 4.12, we insert task 7 into the position after task 46, then after task 20. Perform free permutation of task 46 and put it after task 20. The current sequence is in Table 4.51.

The functional value is without change after iteration 1.

Iteration 2 for competing task 11. We insert it after task 19*, mark it with asterisk and arrange tasks that follow it within the interval of insertion in non-increa-sing order of priorities. We obtain the sequence shown in Table 4.52.

Table 4.47 Initial data for Example 4.8

j	l_j	ω_j	d_j	j	l_j	ω_j	d_j	j	l_j	ω_j	d_j	j	l_j	ω_j	d_j
1	22	1	212	14	18	8	43	27	14	1	208	40	54	3	0
2	3	7	0	15	44	2	0	28	79	5	72	41	64	4	181
3	84	4	85	16	14	7	0	29	40	2	0	42	81	10	0
4	69	4	206	17	13	2	0	30	57	1	0	43	14	4	63
5	7	7	0	18	83	7	9	31	29	2	0	44	20	2	52
6	31	2	13	19	35	8	0	32	63	2	0	45	68	9	0
7	35	8	91	20	12	3	136	33	55	6	0	46	9	7	148
8	28	6	0	21	78	5	0	34	94	7	120	47	33	1	0
9	39	4	0	22	68	8	119	35	88	6	204	48	56	4	28
10	8	5	0	23	81	10	115	36	44	5	0	49	6	10	0
11	44	10	34	24	46	4	114	37	64	6	0	50	42	8	0
12	80	6	60	25	37	6	0	38	12	7	0				
13	7	1	112	26	78	2	0	39	19	2	3				

According to Statement 4.12, we insert marked task 19^* after task 46 and then after task 20. Since the task 19^* takes the position corresponding to its priority, we remove the asterisk from it. The current sequence is in Table 4.53.

We perform free permutation of task 46 into the position after task 20. The current sequence is in Table 4.54.

Optimal functional value at iteration 1 in subsequence within positions 1, …, 12 was equal to 2813. At iteration 2, we appended the competing task 11 with weighted tardiness 1830. After iteration 2 in subsequence within positions 1, …, 13, the functional value became equal to 4611, that is, increased by 1798. This is the minimum possible increase in the total weighted tardiness in comparison with the optimal functional value at the previous iteration (Lemma 4.1 and Statement 4.43). The functional value on the entire set of tasks decreased by 32, as a result of optimization for task 11, and became equal to 150,800.

Competing task 19 has not used the reserves of preceding tasks during its iteration of optimization and returned into its initial position. Therefore, the tasks $j_{[i]}$ that satisfy the conditions $l_{j_{[i]}} \geq l_{19}$, $d_{j_{[i]}} \geq d_{19}$, and $\omega_{j_{[i]}} \leq \omega_{19}$ are eliminated from the set of competing tasks and remain in their initial positions in an optimal schedule (Statement 4.28). These are the tasks: 50, 25, 22, 36, 33, 9, 37, 24, 18, 12, 34, 48, 35, 21, 28, 41, 4, 40, 29, 3, 15, 32, 26, 30. After the iteration of optimization, task 8 remained in its initial position. Therefore, we eliminated tasks 31 and 6 from the set of competing tasks, they remained in the initial positions. Similarly, tasks 17 (together with tasks 39, 44, 47) and 13 (together with tasks 27 and 1) remain in their initial positions. Iterations for tasks 45, 23, and 42 led to an increase in the functional value, so these tasks also remained in their initial positions. Consequently, the obtained sequence σ_5 is optimal.

Example 4.9 Instance #11 from file wt50.txt [28]. Initial data is in Table 4.55.

Table 4.48 Sequence σ^{ord}

j	p_j	l_j	ω_j	d_j	C_j	f_j	j	p_j	l_j	ω_j	d_j	C_j	f_j
2	2.333	3	7	0	3	21	9	0.103	39	4	0	799	3196
49	1.667	6	10	0	9	90	44	0.100	20	2	52	819	1534
5	1.000	7	7	0	16	112	37	0.094	64	6	0	883	5298
46	0.778	9	7	148	25	0	24	0.087	46	4	114	929	3260
10	0.625	8	5	0	33	165	18	0.084	83	7	9	1012	7021
38	0.583	12	7	0	45	315	12	0.075	80	6	60	1092	6192
16	0.500	14	7	0	59	413	34	0.074	94	7	120	1186	7462
14	0.444	18	8	43	77	272	27	0.071	14	1	208	1200	992
43	0.286	14	4	63	91	112	48	0.071	56	4	28	1256	4912
20	0.250	12	3	136	103	0	31	0.069	29	2	0	1285	2570
7	0.229	35	8	91	138	376	35	0.068	88	6	204	1373	7014
19	0.229	35	8	0	173	1384	6	0.065	31	2	13	1404	2782
11	0.227	44	10	34	217	1830	21	0.064	78	5	0	1482	7410
8	0.214	28	6	0	245	1470	28	0.063	79	5	72	1561	7445
50	0.190	42	8	0	287	2296	41	0.063	64	4	181	1625	5776
25	0.162	37	6	0	324	1944	4	0.058	69	4	206	1694	5952
17	0.154	13	2	0	337	674	40	0.056	54	3	0	1748	5244
13	0.143	7	1	112	344	232	29	0.050	40	2	0	1788	3576
45	0.132	68	9	0	412	3708	3	0.048	84	4	85	1872	7148
23	0.123	81	10	115	493	3780	1	0.045	22	1	212	1894	1682
42	0.123	81	10	0	574	5740	15	0.045	44	2	0	1938	3876
22	0.118	68	8	119	642	4184	32	0.032	63	2	0	2001	4002
36	0.114	44	5	0	686	3430	47	0.030	33	1	0	2034	2034
33	0.109	55	6	0	741	4446	26	0.026	78	2	0	2112	4224
39	0.105	19	2	3	760	1514	30	0.018	57	1	0	2169	2169

The functional value f = 151,279

The sequence σ^{ord} is in Table 4.56.

The sequence σ^{fp} (Table 4.57) is the result of free permutations.

Iteration 1 for competing task 32. We determine the position of insertion: 1199 − 65 = 1134. Insert task 32 into the position after task 10. Since the interval of insertion consists of one task 16, and the functional value increases in the consequence of the insertion, then task 32 returns to its initial position. The resulting sequence σ_1 coincides with σ^{fp}. The functional value did not change at iteration 1.

Iteration 2 for competing task 31. We insert task 31 into the position after task 30 and mark it with asterisk. We arrange tasks within the interval of insertion in non-increasing order of priorities. The resulting sequence is in Table 4.58.

In a similar way, we insert the first tardy in this sequence (task 29) into the position after task 31. Then we similarly insert task 25 into the position after task 28

Table 4.49 Sequence σ^{fp}

j	p_j	l_j	ω_j	d_j	C_j	f_j	j	p_j	l_j	ω_j	d_j	C_j	f_j
2	2.333	3	7	0	3	21	9	0.103	39	4	0	799	3196
49	1.667	6	10	0	9	90	44	0.100	20	2	52	819	1534
5	1.000	7	7	0	16	112	37	0.094	64	6	0	883	5298
10	0.625	8	5	0	24	120	24	0.087	46	4	114	929	3260
38	0.583	12	7	0	36	252	18	0.084	83	7	9	1012	7021
16	0.500	14	7	0	50	350	12	0.075	80	6	60	1092	6192
14	0.444	18	8	43	68	200	34	0.074	94	7	120	1186	7462
43	0.286	14	4	63	82	76	27	0.071	14	1	208	1200	992
7	0.229	35	8	91	117	208	48	0.071	56	4	28	1256	4912
20	0.250	12	3	136	129	0	31	0.069	29	2	0	1285	2570
46	0.778	9	7	148	138	0	35	0.068	88	6	204	1373	7014
19	0.229	35	8	0	173	1384	6	0.065	31	2	13	1404	2782
11	0.227	44	10	34	217	1830	21	0.064	78	5	0	1482	7410
8	0.214	28	6	0	245	1470	28	0.063	79	5	72	1561	7445
50	0.190	42	8	0	287	2296	41	0.063	64	4	181	1625	5776
25	0.162	37	6	0	324	1944	4	0.058	69	4	206	1694	5952
17	0.154	13	2	0	337	674	40	0.056	54	3	0	1748	5244
13	0.143	7	1	112	344	232	29	0.050	40	2	0	1788	3576
45	0.132	68	9	0	412	3708	3	0.048	84	4	85	1872	7148
23	0.123	81	10	115	493	3780	1	0.045	22	1	212	1894	1682
42	0.123	81	10	0	574	5740	15	0.045	44	2	0	1938	3876
22	0.118	68	8	119	642	4184	32	0.032	63	2	0	2001	4002
36	0.114	44	5	0	686	3430	47	0.030	33	1	0	2034	2034
33	0.109	55	6	0	741	4446	26	0.026	78	2	0	2112	4224
39	0.105	19	2	3	760	1514	30	0.018	57	1	0	2169	2169

The functional value f = 150,832

and task 3 (that became tardy after the insertion of task 25) after task 25* because task 25 is marked with asterisk to avoid cycling. We perform free permutation for task 10 and insert it after task 3. Tardy task 16 is inserted after task 43 on the basis of Statement 4.13. Denote the obtained sequence by σ_3 (Table 4.59):

The minimum possible increase in the total weighted tardiness at iteration 2 in comparison with the optimal functional value at iteration 1 is 272 (Lemma 4.1 and Statement 4.43).

Iteration 3 for competing task 23. According to Statement 4.28, we insert the task 23 into the position after task 29*. Reorder the tasks within the interval of insertion by priority. We obtain a sequence σ_4 (Table 4.60).

We perform free permutations within the interval of insertion and obtain the sequence shown in Table 4.61.

Table 4.50 Sequence σ_1

j	p_j	l_j	ω_j	d_j	C_j	f_j	j	p_j	l_j	ω_j	d_j	C_j	f_j
2	2.333	3	7	0	3	21	9	0.103	39	4	0	799	3196
49	1.667	6	10	0	9	90	44	0.100	20	2	52	819	1534
5	1.000	7	7	0	16	112	37	0.094	64	6	0	883	5298
10	0.625	8	5	0	24	120	24	0.087	46	4	114	929	3260
38	0.583	12	7	0	36	252	18	0.084	83	7	9	1012	7021
16	0.500	14	7	0	50	350	12	0.075	80	6	60	1092	6192
14	0.444	18	8	43	68	200	34	0.074	94	7	120	1186	7462
43	0.286	14	4	63	82	76	27	0.071	14	1	208	1200	992
7	0.229	35	8	91	117	208	48	0.071	56	4	28	1256	4912
19*	0.229	35	8	0	152	1216	31	0.069	29	2	0	1285	2570
46	0.778	9	7	148	161	91	35	0.068	88	6	204	1373	7014
20	0.250	12	3	136	173	111	6	0.065	31	2	13	1404	2782
11	0.227	44	10	34	217	1830	21	0.064	78	5	0	1482	7410
8	0.214	28	6	0	245	1470	28	0.063	79	5	72	1561	7445
50	0.190	42	8	0	287	2296	41	0.063	64	4	181	1625	5776
25	0.162	37	6	0	324	1944	4	0.058	69	4	206	1694	5952
17	0.154	13	2	0	337	674	40	0.056	54	3	0	1748	5244
13	0.143	7	1	112	344	232	29	0.050	40	2	0	1788	3576
45	0.132	68	9	0	412	3708	3	0.048	84	4	85	1872	7148
23	0.123	81	10	115	493	3780	1	0.045	22	1	212	1894	1682
42	0.123	81	10	0	574	5740	15	0.045	44	2	0	1938	3876
22	0.118	68	8	119	642	4184	32	0.032	63	2	0	2001	4002
36	0.114	44	5	0	686	3430	47	0.030	33	1	0	2034	2034
33	0.109	55	6	0	741	4446	26	0.026	78	2	0	2112	4224
39	0.105	19	2	3	760	1514	30	0.018	57	1	0	2169	2169

The functional value $f = 150{,}866$

Optimization for tardy task 16: we insert the task 16 into the position after task 10 and mark it with asterisk. According to Statement 4.7, task 16* is then inserted after task 43 and marked with "**". The current sequence is in Table 4.62.

Optimization for tardy task 25: we insert the task 25 after task 28 and mark it with asterisk. Then we arrange the tasks within the interval of insertion in non-increasing order of priorities and perform free permutations. We obtain the sequence shown in Table 4.63.

Similarly, we perform insertions, priority orderings and free permutations for tardy tasks 3 and 4. As a result, we obtain the sequence shown in Table 4.64.

The minimum possible increase in the total weighted tardiness at iteration 3 in comparison with the optimal functional value at iteration 2 is 2009 (Lemma 4.1 and Statement 4.43). Last sequence is optimal because we could not decrease the functional value at the iterations of optimization for the remaining competing tasks (from task 47 to the end of the sequence), and all these tasks remained in their

Table 4.51 Optimization in accordance with Statement 4.12. Sequence σ_2

j	p_j	l_j	ω_j	d_j	C_j	f_j	j	p_j	l_j	ω_j	d_j	C_j	f_j
2	2.333	3	7	0	3	21	9	0.103	39	4	0	799	3196
49	1.667	6	10	0	9	90	44	0.100	20	2	52	819	1534
5	1.000	7	7	0	16	112	37	0.094	64	6	0	883	5298
10	0.625	8	5	0	24	120	24	0.087	46	4	114	929	3260
38	0.583	12	7	0	36	252	18	0.084	83	7	9	1012	7021
16	0.500	14	7	0	50	350	12	0.075	80	6	60	1092	6192
14	0.444	18	8	43	68	200	34	0.074	94	7	120	1186	7462
43	0.286	14	4	63	82	76	27	0.071	14	1	208	1200	992
19*	0.229	35	8	0	117	936	48	0.071	56	4	28	1256	4912
20	0.250	12	3	136	129	0	31	0.069	29	2	0	1285	2570
46	0.778	9	7	148	138	0	35	0.068	88	6	204	1373	7014
7	0.229	35	8	91	173	656	6	0.065	31	2	13	1404	2782
11	0.227	44	10	34	217	1830	21	0.064	78	5	0	1482	7410
8	0.214	28	6	0	245	1470	28	0.063	79	5	72	1561	7445
50	0.190	42	8	0	287	2296	41	0.063	64	4	181	1625	5776
25	0.162	37	6	0	324	1944	4	0.058	69	4	206	1694	5952
17	0.154	13	2	0	337	674	40	0.056	54	3	0	1748	5244
13	0.143	7	1	112	344	232	29	0.050	40	2	0	1788	3576
45	0.132	68	9	0	412	3708	3	0.048	84	4	85	1872	7148
23	0.123	81	10	115	493	3780	1	0.045	22	1	212	1894	1682
42	0.123	81	10	0	574	5740	15	0.045	44	2	0	1938	3876
22	0.118	68	8	119	642	4184	32	0.032	63	2	0	2001	4002
36	0.114	44	5	0	686	3430	47	0.030	33	1	0	2034	2034
33	0.109	55	6	0	741	4446	26	0.026	78	2	0	2112	4224
39	0.105	19	2	3	760	1514	30	0.018	57	1	0	2169	2169

The functional value f = 150,832

positions. Since task 2 remains in its position, tasks 27, 48, 33, and 19 are eliminated from the number of competing tasks. Similarly, after the iteration for task 44, tasks 12 and 5 are eliminated from the number of competing tasks, and task 18 is eliminated after the iteration for task 7.

Example 4.10 Instance #2 from file wt50.txt [28]. Initial data is in Table 4.65.

The sequence σ^{ord} is in Table 4.66.

The sequence σ^{fp} (Table 4.67) is the result of free permutations.

Iteration 1. The first competing task is task 14. We determine the interval of insertion of task 14 into an earlier position: $d_{14} - l_{14} = 2125 - 92 = 2033$. Task 40 has the completion time 2005, and for the next task it is 2063, so we insert task 14 after task 40. Arrange the subsequent tasks in non-increasing order of priorities. Mark the task 14 with asterisk. The current sequence is in Table 4.68.

Table 4.52 Iteration 2. Sequence σ_3

j	p_j	l_j	ω_j	d_j	C_j	f_j	j	p_j	l_j	ω_j	d_j	C_j	f_j
2	2.333	3	7	0	3	21	9	0.103	39	4	0	799	3196
49	1.667	6	10	0	9	90	44	0.100	20	2	52	819	1534
5	1.000	7	7	0	16	112	37	0.094	64	6	0	883	5298
10	0.625	8	5	0	24	120	24	0.087	46	4	114	929	3260
38	0.583	12	7	0	36	252	18	0.084	83	7	9	1012	7021
16	0.500	14	7	0	50	350	12	0.075	80	6	60	1092	6192
14	0.444	18	8	43	68	200	34	0.074	94	7	120	1186	7462
43	0.286	14	4	63	82	76	27	0.071	14	1	208	1200	992
19*	0.229	35	8	0	117	936	48	0.071	56	4	28	1256	4912
11*	0.227	44	10	34	161	1270	31	0.069	29	2	0	1285	2570
46	0.778	9	7	148	170	154	35	0.068	88	6	204	1373	7014
20	0.250	12	3	136	182	138	6	0.065	31	2	13	1404	2782
7	0.229	35	8	91	217	1008	21	0.064	78	5	0	1482	7410
8	0.214	28	6	0	245	1470	28	0.063	79	5	72	1561	7445
50	0.190	42	8	0	287	2296	41	0.063	64	4	181	1625	5776
25	0.162	37	6	0	324	1944	4	0.058	69	4	206	1694	5952
17	0.154	13	2	0	337	674	40	0.056	54	3	0	1748	5244
13	0.143	7	1	112	344	232	29	0.050	40	2	0	1788	3576
45	0.132	68	9	0	412	3708	3	0.048	84	4	85	1872	7148
23	0.123	81	10	115	493	3780	1	0.045	22	1	212	1894	1682
42	0.123	81	10	0	574	5740	15	0.045	44	2	0	1938	3876
22	0.118	68	8	119	642	4184	32	0.032	63	2	0	2001	4002
36	0.114	44	5	0	686	3430	47	0.030	33	1	0	2034	2034
33	0.109	55	6	0	741	4446	26	0.026	78	2	0	2112	4224
39	0.105	19	2	3	760	1514	30	0.018	57	1	0	2169	2169

The functional value f = 150,916

The next competing task is task 38. In the same way, we determine the interval of insertion for task 38 into an earlier position. We insert it after task 14* because if the interval of insertion contains a task marked with asterisk, then we insert the competing task after it. Then we reorder the subsequent tasks according to priorities. The current sequence is in Table 4.69.

Task 38* is tardy, so we perform free permutation of task 40 into the position after task 38* (condition 1 of Statement 4.16). Tasks 38* and 40 become non-tardy. We obtain the sequence shown in Table 4.70.

The next tardy task is task 46. According to condition 2 of Statement 4.16, task 42 permutes into the position after task 46 (Table 4.71).

The minimum possible increase in the total weighted tardiness at iteration 1 is 367 (Lemma 4.1 and Statement 4.43).

Iteration 2. The next competing task is task 33. We similarly insert it into the position after task 38* and reorder the subsequent tasks by priorities (Table 4.72).

Table 4.53 Releasing of reserves by task 19. Sequence σ_4

j	p_j	l_j	ω_j	d_j	C_j	f_j	j	p_j	l_j	ω_j	d_j	C_j	f_j
2	2.333	3	7	0	3	21	9	0.103	39	4	0	799	3196
49	1.667	6	10	0	9	90	44	0.100	20	2	52	819	1534
5	1.000	7	7	0	16	112	37	0.094	64	6	0	883	5298
10	0.625	8	5	0	24	120	24	0.087	46	4	114	929	3260
38	0.583	12	7	0	36	252	18	0.084	83	7	9	1012	7021
16	0.500	14	7	0	50	350	12	0.075	80	6	60	1092	6192
14	0.444	18	8	43	68	200	34	0.074	94	7	120	1186	7462
43	0.286	14	4	63	82	76	27	0.071	14	1	208	1200	992
11*	0.227	44	10	34	126	920	48	0.071	56	4	28	1256	4912
46	0.778	9	7	148	135	0	31	0.069	29	2	0	1285	2570
20	0.250	12	3	136	147	33	35	0.068	88	6	204	1373	7014
19	0.229	35	8	0	182	1456	6	0.065	31	2	13	1404	2782
7	0.229	35	8	91	217	1008	21	0.064	78	5	0	1482	7410
8	0.214	28	6	0	245	1470	28	0.063	79	5	72	1561	7445
50	0.190	42	8	0	287	2296	41	0.063	64	4	181	1625	5776
25	0.162	37	6	0	324	1944	4	0.058	69	4	206	1694	5952
17	0.154	13	2	0	337	674	40	0.056	54	3	0	1748	5244
13	0.143	7	1	112	344	232	29	0.050	40	2	0	1788	3576
45	0.132	68	9	0	412	3708	3	0.048	84	4	85	1872	7148
23	0.123	81	10	115	493	3780	1	0.045	22	1	212	1894	1682
42	0.123	81	10	0	574	5740	15	0.045	44	2	0	1938	3876
22	0.118	68	8	119	642	4184	32	0.032	63	2	0	2001	4002
36	0.114	44	5	0	686	3430	47	0.030	33	1	0	2034	2034
33	0.109	55	6	0	741	4446	26	0.026	78	2	0	2112	4224
39	0.105	19	2	3	760	1514	30	0.018	57	1	0	2169	2169

The functional value f = 150,827

Conditions 1 and 2 of Statement 4.16 are true. So, we successively insert tasks 14^*, 38^*, and 37 into the position after task 33^*. Then, in accordance with condition 2 of Statement 4.16, we insert tasks 45 and 42 into the position after task 46. The resulting sequence (Table 4.73) is optimal because the execution of iterations for the remaining competing tasks does not decrease the functional value, and all these tasks remain in their positions. The minimum possible increase in the total weighted tardiness at iteration 2 comparing to the optimal functional value at iteration 1 is 430 (Lemma 4.1 and Statement 4.43).

Table 4.54 Free permutation of task 46. Sequence σ_5

j	p_j	l_j	ω_j	d_j	C_j	f_j	j	p_j	l_j	ω_j	d_j	C_j	f_j
2	2.333	3	7	0	3	21	9	0.103	39	4	0	799	3196
49	1.667	6	10	0	9	90	44	0.100	20	2	52	819	1534
5	1.000	7	7	0	16	112	37	0.094	64	6	0	883	5298
10	0.625	8	5	0	24	120	24	0.087	46	4	114	929	3260
38	0.583	12	7	0	36	252	18	0.084	83	7	9	1012	7021
16	0.500	14	7	0	50	350	12	0.075	80	6	60	1092	6192
14	0.444	18	8	43	68	200	34	0.074	94	7	120	1186	7462
43	0.286	14	4	63	82	76	27	0.071	14	1	208	1200	992
11*	0.227	44	10	34	126	920	48	0.071	56	4	28	1256	4912
20	0.250	12	3	136	138	6	31	0.069	29	2	0	1285	2570
46	0.778	9	7	148	147	0	35	0.068	88	6	204	1373	7014
19	0.229	35	8	0	182	1456	6	0.065	31	2	13	1404	2782
7	0.229	35	8	91	217	1008	21	0.064	78	5	0	1482	7410
8	0.214	28	6	0	245	1470	28	0.063	79	5	72	1561	7445
50	0.190	42	8	0	287	2296	41	0.063	64	4	181	1625	5776
25	0.162	37	6	0	324	1944	4	0.058	69	4	206	1694	5952
17	0.154	13	2	0	337	674	40	0.056	54	3	0	1748	5244
13	0.143	7	1	112	344	232	29	0.050	40	2	0	1788	3576
45	0.132	68	9	0	412	3708	3	0.048	84	4	85	1872	7148
23	0.123	81	10	115	493	3780	1	0.045	22	1	212	1894	1682
42	0.123	81	10	0	574	5740	15	0.045	44	2	0	1938	3876
22	0.118	68	8	119	642	4184	32	0.032	63	2	0	2001	4002
36	0.114	44	5	0	686	3430	47	0.030	33	1	0	2034	2034
33	0.109	55	6	0	741	4446	26	0.026	78	2	0	2112	4224
39	0.105	19	2	3	760	1514	30	0.018	57	1	0	2169	2169

The functional value $f = 150{,}800$

Table 4.55 Initial data for Example 4.9

j	l_j	ω_j	d_j	j	l_j	ω_j	d_j	j	l_j	ω_j	d_j	j	l_j	ω_j	d_j
1	21	9	1098	14	18	10	1016	27	66	3	1012	40	3	8	1134
2	61	3	871	15	34	10	828	28	23	7	1149	41	9	4	1041
3	84	7	1082	16	86	7	1214	29	59	6	867	42	33	5	1102
4	72	8	1002	17	66	8	933	30	87	9	870	43	40	6	1170
5	97	2	1233	18	34	1	982	31	47	3	808	44	83	3	794
6	9	7	1005	19	66	1	933	32	65	5	1199	45	45	2	1050
7	27	1	807	20	78	4	1167	33	77	2	1248	46	2	10	795
8	31	6	929	21	43	8	801	34	13	9	873	47	49	3	1203
9	61	8	794	22	88	5	1125	35	57	9	1032	48	67	3	1246
10	38	8	1124	23	98	6	875	36	89	4	970	49	59	3	1024
11	6	4	961	24	20	2	1220	37	5	6	795	50	52	2	813
12	97	3	843	25	76	6	1013	38	35	2	1181				
13	43	5	1110	26	94	10	884	39	55	8	1081				

Table 4.56 Sequence σ^{ord}

j	p_j	l_j	ω_j	d_j	C_j	f_j	j	p_j	l_j	ω_j	d_j	C_j	f_j
46	5.0000	2	10	795	2	0	24	0.1000	20	2	1220	942	0
40	2.6667	3	8	1134	5	0	3	0.0833	84	7	1082	1026	0
37	1.2000	5	6	795	10	0	16	0.0814	86	7	1214	1112	0
6	0.7778	9	7	1005	19	0	25	0.0789	76	6	1013	1188	1050
34	0.6923	13	9	873	32	0	32	0.0769	65	5	1199	1253	270
11	0.6667	6	4	961	38	0	31	0.0638	47	3	808	1300	1476
14	0.5556	18	10	1016	56	0	23	0.0612	98	6	875	1398	3138
41	0.4444	9	4	1041	65	0	47	0.0612	49	3	1203	1447	732
1	0.4286	21	9	1098	86	0	38	0.0571	35	2	1181	1482	602
28	0.3043	23	7	1149	109	0	22	0.0568	88	5	1125	1570	2225
15	0.2941	34	10	828	143	0	20	0.0513	78	4	1167	1648	1924
10	0.2105	38	8	1124	181	0	49	0.0508	59	3	1024	1707	2049
8	0.1935	31	6	929	212	0	2	0.0492	61	3	871	1768	2691
21	0.1860	43	8	801	255	0	27	0.0455	66	3	1012	1834	2466
35	0.1579	57	9	1032	312	0	36	0.0449	89	4	970	1923	3812
42	0.1515	33	5	1102	345	0	48	0.0448	67	3	1246	1990	2232
43	0.1500	40	6	1170	385	0	45	0.0444	45	2	1050	2035	1970
39	0.1455	55	8	1081	440	0	50	0.0385	52	2	813	2087	2548
9	0.1311	61	8	794	501	0	7	0.0370	27	1	807	2114	1307
17	0.1212	66	8	933	567	0	44	0.0361	83	3	794	2197	4209
13	0.1163	43	5	1110	610	0	12	0.0309	97	3	843	2294	4353
4	0.1111	72	8	1002	682	0	18	0.0294	34	1	982	2328	1346
26	0.1064	94	10	884	776	0	33	0.0260	77	2	1248	2405	2314
30	0.1034	87	9	870	863	0	5	0.0206	97	2	1233	2502	2538
29	0.1017	59	6	867	922	330	19	0.0152	66	1	933	2568	1635

The functional value $f = 47{,}217$

Table 4.57 Sequence σ^{fp}

j	p_j	l_j	ω_j	d_j	C_j	f_j	j	p_j	l_j	ω_j	d_j	C_j	f_j
46	5.0000	2	10	795	2	0	28	0.3043	23	7	1149	1061	0
37	1.2000	5	6	795	7	0	40	2.6667	3	8	1134	1064	0
6	0.7778	9	7	1005	16	0	10	0.2105	38	8	1124	1102	0
34	0.6923	13	9	873	29	0	16	0.0814	86	7	1214	1188	0
11	0.6667	6	4	961	35	0	32	0.0769	65	5	1199	1253	270
14	0.5556	18	10	1016	53	0	31	0.0638	47	3	808	1300	1476
41	0.4444	9	4	1041	62	0	23	0.0612	98	6	875	1398	3138
1	0.4286	21	9	1098	83	0	47	0.0612	49	3	1203	1447	732
15	0.2941	34	10	828	117	0	38	0.0571	35	2	1181	1482	602
8	0.1935	31	6	929	148	0	22	0.0568	88	5	1125	1570	2225
21	0.1860	43	8	801	191	0	20	0.0513	78	4	1167	1648	1924
35	0.1579	57	9	1032	248	0	49	0.0508	59	3	1024	1707	2049
42	0.1515	33	5	1102	281	0	2	0.0492	61	3	871	1768	2691
39	0.1455	55	8	1081	336	0	27	0.0455	66	3	1012	1834	2466
9	0.1311	61	8	794	397	0	36	0.0449	89	4	970	1923	3812
17	0.1212	66	8	933	463	0	48	0.0448	67	3	1246	1990	2232
13	0.1163	43	5	1110	506	0	45	0.0444	45	2	1050	2035	1970
4	0.1111	72	8	1002	578	0	50	0.0385	52	2	813	2087	2548
26	0.1064	94	10	884	672	0	7	0.0370	27	1	807	2114	1307
30	0.1034	87	9	870	759	0	44	0.0361	83	3	794	2197	4209
29	0.1017	59	6	867	818	0	12	0.0309	97	3	843	2294	4353
3	0.0833	84	7	1082	902	0	18	0.0294	34	1	982	2328	1346
25	0.0789	76	6	1013	978	0	33	0.0260	77	2	1248	2405	2314
24	0.1000	20	2	1220	998	0	5	0.0206	97	2	1233	2502	2538
43	0.1500	40	6	1170	1,038	0	19	0.0152	66	1	933	2568	1635

The functional value f = 45,837

Table 4.58 Sequence σ_2

j	p_j	l_j	ω_j	d_j	C_j	f_j	j	p_j	l_j	ω_j	d_j	C_j	f_j
46	5.0000	2	10	795	2	0	29	0.1017	59	6	867	969	612
37	1.2000	5	6	795	7	0	24	0.1000	20	2	1220	989	0
6	0.7778	9	7	1005	16	0	3	0.0833	84	7	1082	1073	0
34	0.6923	13	9	873	29	0	16	0.0814	86	7	1214	1159	0
11	0.6667	6	4	961	35	0	25	0.0789	76	6	1013	1235	1332
14	0.5556	18	10	1016	53	0	32	0.0769	65	5	1199	1300	505
41	0.4444	9	4	1041	62	0	23	0.0612	98	6	875	1398	3138
1	0.4286	21	9	1098	83	0	47	0.0612	49	3	1203	1447	732
15	0.2941	34	10	828	117	0	38	0.0571	35	2	1181	1482	602
8	0.1935	31	6	929	148	0	22	0.0568	88	5	1125	1570	2225
21	0.1860	43	8	801	191	0	20	0.0513	78	4	1167	1648	1924
35	0.1579	57	9	1032	248	0	49	0.0508	59	3	1024	1707	2049
42	0.1515	33	5	1102	281	0	2	0.0492	61	3	871	1768	2691
39	0.1455	55	8	1081	336	0	27	0.0455	66	3	1012	1834	2466
9	0.1311	61	8	794	397	0	36	0.0449	89	4	970	1923	3812
17	0.1212	66	8	933	463	0	48	0.0448	67	3	1246	1990	2232
13	0.1163	43	5	1110	506	0	45	0.0444	45	2	1050	2035	1970
4	0.1111	72	8	1002	578	0	50	0.0385	52	2	813	2087	2548
26	0.1064	94	10	884	672	0	7	0.0370	27	1	807	2114	1307
30	0.1034	87	9	870	759	0	44	0.0361	83	3	794	2197	4209
31*	0.0638	47	3	808	806	0	12	0.0309	97	3	843	2294	4353
40	2.6667	3	8	1134	809	0	18	0.0294	34	1	982	2328	1346
28	0.3043	23	7	1149	832	0	33	0.0260	77	2	1248	2405	2314
10	0.2105	38	8	1124	870	0	5	0.0206	97	2	1233	2502	2538
43	0.1500	40	6	1170	910	0	19	0.0152	66	1	933	2568	1635

The functional value f = 46,540

Table 4.59 The result of iteration 2. Sequence σ_3

j	p_j	l_j	ω_j	d_j	C_j	f_j	j	p_j	l_j	ω_j	d_j	C_j	f_j
46	5.0000	2	10	795	2	0	3*	0.0833	84	7	1082	1051	0
37	1.2000	5	6	795	7	0	10	0.2105	38	8	1124	1089	0
6	0.7778	9	7	1005	16	0	43	0.1500	40	6	1170	1129	0
34	0.6923	13	9	873	29	0	16	0.0814	86	7	1214	1215	7
11	0.6667	6	4	961	35	0	24	0.1000	20	2	1220	1235	30
14	0.5556	18	10	1016	53	0	32	0.0769	65	5	1199	1300	505
41	0.4444	9	4	1041	62	0	23	0.0612	98	6	875	1398	3138
1	0.4286	21	9	1098	83	0	47	0.0612	49	3	1203	1447	732
15	0.2941	34	10	828	117	0	38	0.0571	35	2	1181	1482	602
8	0.1935	31	6	929	148	0	22	0.0568	88	5	1125	1570	2225
21	0.1860	43	8	801	191	0	20	0.0513	78	4	1167	1648	1924
35	0.1579	57	9	1032	248	0	49	0.0508	59	3	1024	1707	2049
42	0.1515	33	5	1102	281	0	2	0.0492	61	3	871	1768	2691
39	0.1455	55	8	1081	336	0	27	0.0455	66	3	1012	1834	2466
9	0.1311	61	8	794	397	0	36	0.0449	89	4	970	1923	3812
17	0.1212	66	8	933	463	0	48	0.0448	67	3	1246	1990	2232
13	0.1163	43	5	1110	506	0	45	0.0444	45	2	1050	2035	1970
4	0.1111	72	8	1002	578	0	50	0.0385	52	2	813	2087	2548
26	0.1064	94	10	884	672	0	7	0.0370	27	1	807	2114	1307
30	0.1034	87	9	870	759	0	44	0.0361	83	3	794	2197	4209
31*	0.0638	47	3	808	806	0	12	0.0309	97	3	843	2294	4353
29*	0.1017	59	6	867	865	0	18	0.0294	34	1	982	2328	1346
40	2.6667	3	8	1134	868	0	33	0.0260	77	2	1248	2405	2314
28	0.3043	23	7	1149	891	0	5	0.0206	97	2	1233	2502	2538
25*	0.0789	76	6	1013	967	0	19	0.0152	66	1	933	2568	1635

The functional value $f = 44{,}633$

Table 4.60 Sequence σ_4

j	p_j	l_j	ω_j	d_j	C_j	f_j	j	p_j	l_j	ω_j	d_j	C_j	f_j
46	5.0000	2	10	795	2	0	10	0.2105	38	8	1124	1027	0
37	1.2000	5	6	795	7	0	43	0.1500	40	6	1170	1067	0
6	0.7778	9	7	1005	16	0	24	0.1000	20	2	1220	1087	0
34	0.6923	13	9	873	29	0	3	0.0833	84	7	1082	1171	623
11	0.6667	6	4	961	35	0	16	0.0814	86	7	1214	1257	301
14	0.5556	18	10	1016	53	0	25	0.0789	76	6	1013	1333	1920
41	0.4444	9	4	1041	62	0	32	0.0769	65	5	1199	1398	995
1	0.4286	21	9	1098	83	0	47	0.0612	49	3	1203	1447	732
15	0.2941	34	10	828	117	0	38	0.0571	35	2	1181	1482	602
8	0.1935	31	6	929	148	0	22	0.0568	88	5	1125	1570	2225
21	0.1860	43	8	801	191	0	20	0.0513	78	4	1167	1648	1924
35	0.1579	57	9	1032	248	0	49	0.0508	59	3	1024	1707	2049
42	0.1515	33	5	1102	281	0	2	0.0492	61	3	871	1768	2691
39	0.1455	55	8	1081	336	0	27	0.0455	66	3	1012	1834	2466
9	0.1311	61	8	794	397	0	36	0.0449	89	4	970	1923	3812
17	0.1212	66	8	933	463	0	48	0.0448	67	3	1246	1990	2232
13	0.1163	43	5	1110	506	0	45	0.0444	45	2	1050	2035	1970
4	0.1111	72	8	1002	578	0	50	0.0385	52	2	813	2087	2548
26	0.1064	94	10	884	672	0	7	0.0370	27	1	807	2114	1307
30	0.1034	87	9	870	759	0	44	0.0361	83	3	794	2197	4209
31*	0.0638	47	3	808	806	0	12	0.0309	97	3	843	2294	4353
29*	0.1017	59	6	867	865	0	18	0.0294	34	1	982	2328	1346
23*	0.0612	98	6	875	963	528	33	0.0260	77	2	1248	2405	2314
40	2.6667	3	8	1134	966	0	5	0.0206	97	2	1233	2502	2538
28	0.3043	23	7	1149	989	0	19	0.0152	66	1	933	2568	1635

The functional value $f = 45{,}320$

Table 4.61 Sequence σ_5

j	p_j	l_j	ω_j	d_j	C_j	f_j	j	p_j	l_j	ω_j	d_j	C_j	f_j
46	5.0000	2	10	795	2	0	3	0.0833	84	7	1082	1073	0
37	1.2000	5	6	795	7	0	10	0.2105	38	8	1124	1111	0
6	0.7778	9	7	1005	16	0	43	0.1500	40	6	1170	1151	0
34	0.6923	13	9	873	29	0	24	0.1000	20	2	1220	1171	0
11	0.6667	6	4	961	35	0	16	0.0814	86	7	1214	1257	301
14	0.5556	18	10	1016	53	0	25	0.0789	76	6	1013	1333	1920
41	0.4444	9	4	1041	62	0	32	0.0769	65	5	1199	1398	995
1	0.4286	21	9	1098	83	0	47	0.0612	49	3	1203	1447	732
15	0.2941	34	10	828	117	0	38	0.0571	35	2	1181	1482	602
8	0.1935	31	6	929	148	0	22	0.0568	88	5	1125	1570	2225
21	0.1860	43	8	801	191	0	20	0.0513	78	4	1167	1648	1924
35	0.1579	57	9	1032	248	0	49	0.0508	59	3	1024	1707	2049
42	0.1515	33	5	1102	281	0	2	0.0492	61	3	871	1768	2691
39	0.1455	55	8	1081	336	0	27	0.0455	66	3	1012	1834	2466
9	0.1311	61	8	794	397	0	36	0.0449	89	4	970	1923	3812
17	0.1212	66	8	933	463	0	48	0.0448	67	3	1246	1990	2232
26	0.1064	94	10	884	557	0	45	0.0444	45	2	1050	2035	1970
30	0.1034	87	9	870	644	0	50	0.0385	52	2	813	2087	2548
31*	0.0638	47	3	808	691	0	7	0.0370	27	1	807	2114	1307
29*	0.1017	59	6	867	750	0	44	0.0361	83	3	794	2197	4209
23*	0.0612	98	6	875	848	0	12	0.0309	97	3	843	2294	4353
40	2.6667	3	8	1134	851	0	18	0.0294	34	1	982	2328	1346
28	0.3043	23	7	1149	874	0	33	0.0260	77	2	1248	2405	2314
4	0.1111	72	8	1002	946	0	5	0.0206	97	2	1233	2502	2538
13	0.1163	43	5	1110	989	0	19	0.0152	66	1	933	2568	1635

The functional value f = 44,169

Table 4.62 Sequence σ_6

j	p_j	l_j	ω_j	d_j	C_j	f_j	j	p_j	l_j	ω_j	d_j	C_j	f_j
46	5.0000	2	10	795	2	0	3	0.0833	84	7	1082	1073	0
37	1.2000	5	6	795	7	0	10	0.2105	38	8	1124	1111	0
6	0.7778	9	7	1005	16	0	43	0.1500	40	6	1170	1151	0
34	0.6923	13	9	873	29	0	16**	0.0814	86	7	1214	1237	161
11	0.6667	6	4	961	35	0	24	0.1000	20	2	1220	1257	74
14	0.5556	18	10	1016	53	0	25	0.0789	76	6	1013	1333	1920
41	0.4444	9	4	1041	62	0	32	0.0769	65	5	1199	1398	995
1	0.4286	21	9	1098	83	0	47	0.0612	49	3	1203	1447	732
15	0.2941	34	10	828	117	0	38	0.0571	35	2	1181	1482	602
8	0.1935	31	6	929	148	0	22	0.0568	88	5	1125	1570	2225
21	0.1860	43	8	801	191	0	20	0.0513	78	4	1167	1648	1924
35	0.1579	57	9	1032	248	0	49	0.0508	59	3	1024	1707	2049
42	0.1515	33	5	1102	281	0	2	0.0492	61	3	871	1768	2691
39	0.1455	55	8	1081	336	0	27	0.0455	66	3	1012	1834	2466
9	0.1311	61	8	794	397	0	36	0.0449	89	4	970	1923	3812
17	0.1212	66	8	933	463	0	48	0.0448	67	3	1246	1990	2232
26	0.1064	94	10	884	557	0	45	0.0444	45	2	1050	2035	1970
30	0.1034	87	9	870	644	0	50	0.0385	52	2	813	2087	2548
31*	0.0638	47	3	808	691	0	7	0.0370	27	1	807	2114	1307
29*	0.1017	59	6	867	750	0	44	0.0361	83	3	794	2197	4209
23*	0.0612	98	6	875	848	0	12	0.0309	97	3	843	2294	4353
40	2.6667	3	8	1134	851	0	18	0.0294	34	1	982	2328	1346
28	0.3043	23	7	1149	874	0	33	0.0260	77	2	1248	2405	2314
4	0.1111	72	8	1002	946	0	5	0.0206	97	2	1233	2502	2538
13	0.1163	43	5	1110	989	0	19	0.0152	66	1	933	2568	1635

The functional value $f = 44{,}103$

Table 4.63 Sequence σ_7

j	p_j	l_j	ω_j	d_j	C_j	f_j	j	p_j	l_j	ω_j	d_j	C_j	f_j
46	5.0000	2	10	795	2	0	10	0.2105	38	8	1124	1080	0
37	1.2000	5	6	795	7	0	28	0.3043	23	7	1149	1103	0
6	0.7778	9	7	1005	16	0	43	0.1500	40	6	1170	1143	0
34	0.6923	13	9	873	29	0	24	0.1000	20	2	1220	1163	0
11	0.6667	6	4	961	35	0	3	0.0833	84	7	1082	1247	1155
14	0.5556	18	10	1016	53	0	16	0.0814	86	7	1214	1333	833
41	0.4444	9	4	1041	62	0	32	0.0769	65	5	1199	1398	995
1	0.4286	21	9	1098	83	0	47	0.0612	49	3	1203	1447	732
15	0.2941	34	10	828	117	0	38	0.0571	35	2	1181	1482	602
8	0.1935	31	6	929	148	0	22	0.0568	88	5	1125	1570	2225
21	0.1860	43	8	801	191	0	20	0.0513	78	4	1167	1648	1924
35	0.1579	57	9	1032	248	0	49	0.0508	59	3	1024	1707	2049
42	0.1515	33	5	1102	281	0	2	0.0492	61	3	871	1768	2691
39	0.1455	55	8	1081	336	0	27	0.0455	66	3	1012	1834	2466
9	0.1311	61	8	794	397	0	36	0.0449	89	4	970	1923	3812
17	0.1212	66	8	933	463	0	48	0.0448	67	3	1246	1990	2232
26	0.1064	94	10	884	557	0	45	0.0444	45	2	1050	2035	1970
30	0.1034	87	9	870	644	0	50	0.0385	52	2	813	2087	2548
31*	0.0638	47	3	808	691	0	7	0.0370	27	1	807	2114	1307
29*	0.1017	59	6	867	750	0	44	0.0361	83	3	794	2197	4209
23*	0.0612	98	6	875	848	0	12	0.0309	97	3	843	2294	4353
40	2.6667	3	8	1134	851	0	18	0.0294	34	1	982	2328	1346
25*	0.0789	76	6	1013	927	0	33	0.0260	77	2	1248	2405	2314
4	0.1111	72	8	1002	999	0	5	0.0206	97	2	1233	2502	2538
13	0.1163	43	5	1110	1,042	0	19	0.0152	66	1	933	2568	1635

The functional value $f = 43{,}936$

Table 4.64 Sequence σ_8

j	p_j	l_j	ω_j	d_j	C_j	f_j	j	p_j	l_j	ω_j	d_j	C_j	f_j
46	5.0000	2	10	795	2	0	10	0.2105	38	8	1124	1121	0
37	1.2000	5	6	795	7	0	28	0.3043	23	7	1149	1144	0
6	0.7778	9	7	1005	16	0	43	0.1500	40	6	1170	1184	84
34	0.6923	13	9	873	29	0	13	0.1163	43	5	1110	1227	585
11	0.6667	6	4	961	35	0	24	0.1000	20	2	1220	1247	54
14	0.5556	18	10	1016	53	0	16	0.0814	86	7	1214	1333	833
41	0.4444	9	4	1041	62	0	32	0.0769	65	5	1199	1398	995
1	0.4286	21	9	1098	83	0	47	0.0612	49	3	1203	1447	732
15	0.2941	34	10	828	117	0	38	0.0571	35	2	1181	1482	602
8	0.1935	31	6	929	148	0	22	0.0568	88	5	1125	1570	2225
21	0.1860	43	8	801	191	0	20	0.0513	78	4	1167	1648	1924
35	0.1579	57	9	1032	248	0	49	0.0508	59	3	1024	1707	2049
42	0.1515	33	5	1102	281	0	2	0.0492	61	3	871	1768	2691
39	0.1455	55	8	1081	336	0	27	0.0455	66	3	1012	1834	2466
9	0.1311	61	8	794	397	0	36	0.0449	89	4	970	1923	3812
17	0.1212	66	8	933	463	0	48	0.0448	67	3	1246	1990	2232
26	0.1064	94	10	884	557	0	45	0.0444	45	2	1050	2035	1970
30	0.1034	87	9	870	644	0	50	0.0385	52	2	813	2087	2548
31*	0.0638	47	3	808	691	0	7	0.0370	27	1	807	2114	1307
29*	0.1017	59	6	867	750	0	44	0.0361	83	3	794	2197	4209
23*	0.0612	98	6	875	848	0	12	0.0309	97	3	843	2294	4353
25*	0.0789	76	6	1013	924	0	18	0.0294	34	1	982	2328	1346
4*	0.1111	72	8	1002	996	0	33	0.0260	77	2	1248	2405	2314
3*	0.0833	84	7	1082	1,080	0	5	0.0206	97	2	1233	2502	2538
40	2.6667	3	8	1134	1,083	0	19	0.0152	66	1	933	2568	1635

The functional value f = 43,504

Table 4.65 Initial data for Example 4.10

j	l_j	ω_j	d_j	j	l_j	ω_j	d_j	j	l_j	ω_j	d_j	j	l_j	ω_j	d_j
1	81	7	1933	14	92	2	2125	27	61	6	1958	40	70	3	2247
2	61	3	2068	15	24	10	2116	28	98	10	2129	41	56	6	2101
3	83	4	2245	16	12	8	1878	29	18	4	2138	42	9	5	2332
4	1	5	1926	17	53	1	2077	30	61	10	2200	43	11	4	2059
5	91	1	1835	18	46	5	1888	31	39	2	2084	44	47	10	1908
6	68	10	1931	19	7	5	2122	32	41	9	2089	45	16	9	2307
7	70	5	2230	20	57	4	2002	33	53	1	1954	46	80	4	2299
8	36	7	2113	21	67	2	2291	34	52	3	2243	47	26	8	2039
9	67	9	2241	22	89	4	1995	35	35	10	2153	48	77	8	2263
10	95	9	2018	23	45	3	2108	36	15	9	2086	49	37	5	1989
11	42	10	2207	24	26	4	2024	37	86	4	2206	50	85	9	2074
12	4	1	2043	25	59	3	2020	38	58	2	2091				
13	40	4	1900	26	90	8	2027	39	64	8	2133				

Table 4.66 Sequence σ^{ord}

j	p_j	l_j	ω_j	d_j	C_j	f_j	j	p_j	l_j	ω_j	d_j	C_j	f_j
4	5.000	1	5	1926	1	0	48	0.104	77	8	2263	931	0
19	0.714	7	5	2122	8	0	28	0.102	98	10	2129	1029	0
16	0.667	12	8	1878	20	0	13	0.100	40	4	1900	1069	0
36	0.600	15	9	2086	35	0	27	0.098	61	6	1958	1130	0
45	0.563	16	9	2307	51	0	10	0.095	95	9	2018	1225	0
42	0.556	9	5	2332	60	0	26	0.089	90	8	2027	1315	0
15	0.417	24	10	2116	84	0	1	0.086	81	7	1933	1396	0
43	0.364	11	4	2059	95	0	7	0.071	70	5	2230	1466	0
47	0.308	26	8	2039	121	0	20	0.070	57	4	2002	1523	0
35	0.286	35	10	2153	156	0	23	0.067	45	3	2108	1568	0
12	0.250	4	1	2043	160	0	34	0.058	52	3	2243	1620	0
11	0.238	42	10	2207	202	0	31	0.051	39	2	2084	1659	0
29	0.222	18	4	2138	220	0	25	0.051	59	3	2020	1718	0
32	0.220	41	9	2089	261	0	46	0.050	80	4	2299	798	0
44	0.213	47	10	1908	308	0	2	0.049	61	3	2068	1859	0
8	0.194	36	7	2113	344	0	3	0.048	83	4	2245	1942	0
30	0.164	61	10	2200	405	0	37	0.047	86	4	2206	2028	0
24	0.154	26	4	2024	431	0	22	0.045	89	4	1995	2117	488
6	0.147	68	10	1931	499	0	40	0.043	70	3	2247	2187	0
49	0.135	37	5	1989	536	0	38	0.034	58	2	2091	2245	308
9	0.134	67	9	2241	603	0	21	0.030	67	2	2291	2312	42
39	0.125	64	8	2133	667	0	14	0.022	92	2	2125	2,404	558
18	0.109	46	5	1888	713	0	33	0.019	53	1	1954	2457	503
41	0.107	56	6	2101	769	0	17	0.019	53	1	2077	2510	433
50	0.106	85	9	2074	854	0	5	0.011	91	1	1835	2601	766

The functional value $f = 3098$

Table 4.67 Sequence σ^{fp}

j	p_j	l_j	ω_j	d_j	C_j	f_j	j	p_j	l_j	ω_j	d_j	C_j	f_j
4	5.000	1	5	1926	1	0	27	0.098	61	6	1958	1028	0
19	0.714	7	5	2122	8	0	10	0.095	95	9	2018	1123	0
16	0.667	12	8	1878	20	0	26	0.089	90	8	2027	1213	0
36	0.600	15	9	2086	35	0	1	0.086	81	7	1933	1294	0
15	0.417	24	10	2116	59	0	7	0.071	70	5	2230	1364	0
43	0.364	11	4	2059	70	0	20	0.070	57	4	2002	1421	0
47	0.308	26	8	2039	96	0	23	0.067	45	3	2108	1466	0
35	0.286	35	10	2153	131	0	34	0.058	52	3	2243	1518	0
12	0.250	4	1	2043	135	0	31	0.051	39	2	2084	1557	0
11	0.238	42	10	2207	177	0	25	0.051	59	3	2020	1616	0
29	0.222	18	4	2138	195	0	2	0.049	61	3	2068	1677	0
32	0.220	41	9	2089	236	0	3	0.048	83	4	2245	1760	0
44	0.213	47	10	1908	283	0	37	0.047	86	4	2206	1846	0
8	0.194	36	7	2113	319	0	22	0.045	89	4	1995	1935	0
30	0.164	61	10	2200	380	0	40	0.043	70	3	2247	005	0
24	0.154	26	4	2024	406	0	38	0.034	58	2	2091	2063	0
6	0.147	68	10	1931	474	0	21	0.030	67	2	2291	2130	0
49	0.135	37	5	1989	511	0	48	0.104	77	8	2,263	2207	0
9	0.134	67	9	2241	578	0	46	0.050	80	4	2299	2287	0
39	0.125	64	8	2133	642	0	45	0.563	16	9	2307	2303	0
18	0.109	46	5	1888	688	0	42	0.556	9	5	2332	2312	0
41	0.107	56	6	2101	744	0	14	0.022	92	2	2125	2404	558
50	0.106	85	9	2074	829	0	33	0.019	53	1	1954	2457	503
28	0.102	98	10	2129	927	0	17	0.019	53	1	2077	2510	433
13	0.100	40	4	1900	967	0	5	0.011	91	1	1835	2601	766

The functional value $f = 2260$

Table 4.68 Sequence σ_1

j	p_j	l_j	ω_j	d_j	C_j	f_j	j	p_j	l_j	ω_j	d_j	C_j	f_j
4	5.000	1	5	1926	1	0	27	0.098	61	6	1958	1028	0
19	0.714	7	5	2122	8	0	10	0.095	95	9	2018	1123	0
16	0.667	12	8	1878	20	0	26	0.089	90	8	2027	1213	0
36	0.600	15	9	2086	35	0	1	0.086	81	7	1933	1294	0
15	0.417	24	10	2116	59	0	7	0.071	70	5	2230	1364	0
43	0.364	11	4	2059	70	0	20	0.070	57	4	2002	1421	0
47	0.308	26	8	2039	96	0	23	0.067	45	3	2108	1466	0
35	0.286	35	10	2153	131	0	34	0.058	52	3	2243	1518	0
12	0.250	4	1	2043	135	0	31	0.051	39	2	2084	1557	0
11	0.238	42	10	2207	177	0	25	0.051	59	3	2020	1616	0
29	0.222	18	4	2138	195	0	2	0.049	61	3	2068	1677	0
32	0.220	41	9	2089	236	0	3	0.048	83	4	2245	1760	0
44	0.213	47	10	1908	283	0	37	0.047	86	4	2206	1846	0
8	0.194	36	7	2113	319	0	22	0.045	89	4	1995	1935	0
30	0.164	61	10	2200	380	0	40	0.043	70	3	2247	2005	0
24	0.154	26	4	2024	406	0	14*	0.022	92	2	2125	2097	0
6	0.147	68	10	1931	474	0	45	0.563	16	9	2307	2113	0
49	0.135	37	5	1989	511	0	42	0.556	9	5	2332	2122	0
9	0.134	67	9	2241	578	0	48	0.104	77	8	2263	2199	0
39	0.125	64	8	2133	642	0	46	0.050	80	4	2299	2279	0
18	0.109	46	5	1888	688	0	38	0.034	58	2	2091	2337	492
41	0.107	56	6	2101	744	0	21	0.030	67	2	2291	2404	226
50	0.106	85	9	2074	829	0	33	0.019	53	1	1954	2457	503
28	0.102	98	10	2129	927	0	17	0.019	53	1	2077	2510	433
13	0.100	40	4	1900	967	0	5	0.011	91	1	1835	2601	766

The functional value $f = 2420$

Table 4.69 Sequence σ_2

j	p_j	l_j	ω_j	d_j	C_j	f_j	j	p_j	l_j	ω_j	d_j	C_j	f_j
4	5.000	1	5	1926	1	0	27	0.098	61	6	1958	1028	0
19	0.714	7	5	2122	8	0	10	0.095	95	9	2018	1123	0
16	0.667	12	8	1878	20	0	26	0.089	90	8	2027	1213	0
36	0.600	15	9	2086	35	0	1	0.086	81	7	1933	1294	0
15	0.417	24	10	2116	59	0	7	0.071	70	5	2230	1364	0
43	0.364	11	4	2059	70	0	20	0.070	57	4	2002	1421	0
47	0.308	26	8	2039	96	0	23	0.067	45	3	2108	1466	0
35	0.286	35	10	2153	131	0	34	0.058	52	3	2243	1518	0
12	0.250	4	1	2043	135	0	31	0.051	39	2	2084	1557	0
11	0.238	42	10	2207	177	0	25	0.051	59	3	2020	1616	0
29	0.222	18	4	2138	195	0	2	0.049	61	3	2068	1677	0
32	0.220	41	9	2089	236	0	3	0.048	83	4	2245	1760	0
44	0.213	47	10	1908	283	0	37	0.047	86	4	2206	1846	0
8	0.194	36	7	2113	319	0	22	0.045	89	4	1995	1935	0
30	0.164	61	10	2200	380	0	40	0.043	70	3	2247	2005	0
24	0.154	26	4	2024	406	0	14*	0.022	92	2	2125	2097	0
6	0.147	68	10	1931	474	0	38*	0.034	58	2	2091	2155	128
49	0.135	37	5	1989	511	0	45	0.563	16	9	2307	2171	0
9	0.134	67	9	2241	578	0	42	0.556	9	5	2332	2180	0
39	0.125	64	8	2133	642	0	48	0.104	77	8	2263	2257	0
18	0.109	46	5	1888	688	0	46	0.050	80	4	2299	2337	152
41	0.107	56	6	2101	744	0	21	0.030	67	2	2291	2404	226
50	0.106	85	9	2074	829	0	33	0.019	53	1	1954	2457	503
28	0.102	98	10	2129	927	0	17	0.019	53	1	2077	2510	433
13	0.100	40	4	1900	967	0	5	0.011	91	1	1835	2601	766

The functional value $f = 2208$

Table 4.70 Free permutation of task 40. Sequence σ_3

j	p_j	l_j	ω_j	d_j	C_j	f_j	j	p_j	l_j	ω_j	d_j	C_j	f_j
4	5.000	1	5	1926	1	0	27	0.098	61	6	1958	1028	0
19	0.714	7	5	2122	8	0	10	0.095	95	9	2018	1123	0
16	0.667	12	8	1878	20	0	26	0.089	90	8	2027	1213	0
36	0.600	15	9	2086	35	0	1	0.086	81	7	1933	1294	0
15	0.417	24	10	2116	59	0	7	0.071	70	5	2230	1364	0
43	0.364	11	4	2059	70	0	20	0.070	57	4	2002	1421	0
47	0.308	26	8	2039	96	0	23	0.067	45	3	2108	1466	0
35	0.286	35	10	2153	131	0	34	0.058	52	3	2243	1518	0
12	0.250	4	1	2043	135	0	31	0.051	39	2	2084	1557	0
11	0.238	42	10	2207	177	0	25	0.051	59	3	2020	1616	0
29	0.222	18	4	2138	195	0	2	0.049	61	3	2068	1677	0
32	0.220	41	9	2089	236	0	3	0.048	83	4	2245	1760	0
44	0.213	47	10	1908	283	0	37	0.047	86	4	2206	1846	0
8	0.194	36	7	2113	319	0	22	0.045	89	4	1995	1935	0
30	0.164	61	10	2200	380	0	14*	0.022	92	2	2125	2027	0
24	0.154	26	4	2024	406	0	38*	0.034	58	2	2091	2085	0
6	0.147	68	10	1931	474	0	40	0.043	70	3	2247	2155	0
49	0.135	37	5	1989	511	0	45	0.563	16	9	2307	2171	0
9	0.134	67	9	2241	578	0	42	0.556	9	5	2332	2180	0
39	0.125	64	8	2133	642	0	48	0.104	77	8	2263	2257	0
18	0.109	46	5	1888	688	0	46	0.050	80	4	2299	2337	152
41	0.107	56	6	2101	744	0	21	0.030	67	2	2291	2404	226
50	0.106	85	9	2074	829	0	33	0.019	53	1	1954	2457	503
28	0.102	98	10	2129	927	0	17	0.019	53	1	2077	2510	433
13	0.100	40	4	1900	967	0	5	0.011	91	1	1835	2601	766

The functional value $f = 2080$

Table 4.71 Permutation of task 42. Sequence σ_4

j	p_j	l_j	ω_j	d_j	C_j	f_j	j	p_j	l_j	ω_j	d_j	C_j	f_j
4	5.000	1	5	1926	1	0	27	0.098	61	6	1958	1028	0
19	0.714	7	5	2122	8	0	10	0.095	95	9	2018	1123	0
16	0.667	12	8	1878	20	0	26	0.089	90	8	2027	1213	0
36	0.600	15	9	2086	35	0	1	0.086	81	7	1933	1294	0
15	0.417	24	10	2116	59	0	7	0.071	70	5	2230	1364	0
43	0.364	11	4	2059	70	0	20	0.070	57	4	2002	1421	0
47	0.308	26	8	2039	96	0	23	0.067	45	3	2108	1466	0
35	0.286	35	10	2153	131	0	34	0.058	52	3	2243	1518	0
12	0.250	4	1	2043	135	0	31	0.051	39	2	2084	1557	0
11	0.238	42	10	2207	177	0	25	0.051	59	3	2020	1616	0
29	0.222	18	4	2138	195	0	2	0.049	61	3	2068	1677	0
32	0.220	41	9	2089	236	0	3	0.048	83	4	2245	1760	0
44	0.213	47	10	1908	283	0	37	0.047	86	4	2206	1846	0
8	0.194	36	7	2113	319	0	22	0.045	89	4	1995	1935	0
30	0.164	61	10	2200	380	0	14*	0.022	92	2	2125	2027	0
24	0.154	26	4	2024	406	0	38*	0.034	58	2	2091	2085	0
6	0.147	68	10	1931	474	0	40	0.043	70	3	2247	2155	0
49	0.135	37	5	1989	511	0	45	0.563	16	9	2307	2171	0
9	0.134	67	9	2241	578	0	48	0.104	77	8	2263	2248	0
39	0.125	64	8	2133	642	0	46	0.050	80	4	2299	2328	116
18	0.109	46	5	1888	688	0	42	0.556	9	5	2332	2337	25
41	0.107	56	6	2101	744	0	21	0.030	67	2	2291	2404	226
50	0.106	85	9	2074	829	0	33	0.019	53	1	1954	2457	503
28	0.102	98	10	2129	927	0	17	0.019	53	1	2077	2510	433
13	0.100	40	4	1900	967	0	5	0.011	91	1	1835	2601	766

The functional value $f = 2069$

Table 4.72 Sequence σ_5

j	p_j	l_j	ω_j	d_j	C_j	f_j	j	p_j	l_j	ω_j	d_j	C_j	f_j
4	5.000	1	5	1926	1	0	27	0.098	61	6	1958	1028	0
19	0.714	7	5	2122	8	0	10	0.095	95	9	2018	1123	0
16	0.667	12	8	1878	20	0	26	0.089	90	8	2027	1213	0
36	0.600	15	9	2086	35	0	1	0.086	81	7	1933	1294	0
15	0.417	24	10	2116	59	0	7	0.071	70	5	2230	1364	0
43	0.364	11	4	2059	70	0	20	0.070	57	4	2002	1421	0
47	0.308	26	8	2039	96	0	23	0.067	45	3	2108	1466	0
35	0.286	35	10	2153	131	0	34	0.058	52	3	2243	1518	0
12	0.250	4	1	2043	135	0	31	0.051	39	2	2084	1557	0
11	0.238	42	10	2207	177	0	25	0.051	59	3	2020	1616	0
29	0.222	18	4	2138	195	0	2	0.049	61	3	2068	1677	0
32	0.220	41	9	2089	236	0	3	0.048	83	4	2245	1760	0
44	0.213	47	10	1908	283	0	37	0.047	86	4	2206	1846	0
8	0.194	36	7	2113	319	0	22	0.045	89	4	1995	1935	0
30	0.164	61	10	2200	380	0	14*	0.022	92	2	2125	2027	0
24	0.154	26	4	2024	406	0	38*	0.034	58	2	2091	2085	0
6	0.147	68	10	1931	474	0	33*	0.019	53	1	1954	2138	184
49	0.135	37	5	1989	511	0	45	0.563	16	9	2307	2154	0
9	0.134	67	9	2241	578	0	42	0.556	9	5	2332	2163	0
39	0.125	64	8	2133	642	0	48	0.104	77	8	2263	2240	0
18	0.109	46	5	1888	688	0	46	0.050	80	4	2299	2320	84
41	0.107	56	6	2101	744	0	40	0.043	70	3	2247	2390	429
50	0.106	85	9	2074	829	0	21	0.030	67	2	2291	2457	332
28	0.102	98	10	2129	927	0	17	0.019	53	1	2077	2510	433
13	0.100	40	4	1900	967	0	5	0.011	91	1	1835	2601	766

The functional value $f = 2228$

Table 4.73 Optimal sequence

j	p_j	l_j	ω_j	d_j	C_j	f_j	j	p_j	l_j	ω_j	d_j	C_j	f_j
4	5.000	1	5	1926	1	0	27	0.098	61	6	1958	1028	0
19	0.714	7	5	2122	8	0	10	0.095	95	9	2018	1123	0
16	0.667	12	8	1878	20	0	26	0.089	90	8	2027	1213	0
36	0.600	15	9	2086	35	0	1	0.086	81	7	1933	1294	0
15	0.417	24	10	2116	59	0	7	0.071	70	5	2230	1364	0
43	0.364	11	4	2059	70	0	20	0.070	57	4	2002	1421	0
47	0.308	26	8	2039	96	0	23	0.067	45	3	2108	1466	0
35	0.286	35	10	2153	131	0	34	0.058	52	3	2243	1518	0
12	0.250	4	1	2043	135	0	31	0.051	39	2	2084	1557	0
11	0.238	42	10	2207	177	0	25	0.051	59	3	2020	1616	0
29	0.222	18	4	2138	195	0	2	0.049	61	3	2068	1677	0
32	0.220	41	9	2089	236	0	3	0.048	83	4	2245	1760	0
44	0.213	47	10	1908	283	0	22	0.045	89	4	1995	1849	0
8	0.194	36	7	2113	319	0	33*	0.019	53	1	1954	1902	0
30	0.164	61	10	2200	380	0	14*	0.022	92	2	2125	1994	0
24	0.154	26	4	2024	406	0	38*	0.034	58	2	2091	2052	0
6	0.147	68	10	1931	474	0	37	0.047	86	4	2206	2138	0
49	0.135	37	5	1989	511	0	48	0.104	77	8	2263	2215	0
9	0.134	67	9	2241	578	0	46	0.050	80	4	2299	2295	0
39	0.125	64	8	2133	642	0	45	0.563	16	9	2307	2311	36
18	0.109	46	5	1888	688	0	42	0.556	9	5	2332	2320	0
41	0.107	56	6	2101	744	0	40	0.043	70	3	2247	2390	429
50	0.106	85	9	2074	829	0	21	0.030	67	2	2291	2457	332
28	0.102	98	10	2129	927	0	17	0.019	53	1	2077	2510	433
13	0.100	40	4	1900	967	0	5	0.011	91	1	1835	2601	766

The functional value $f = 1996$

References

1. Zgurovsky, M.Z., Pavlov, A.A.: Prinyatie Resheniy v Setevyh Sistemah s Ogranichennymi Resursami (Принятие решений в сетевых системах с ограниченными ресурсами; Decision Making in Network Systems with Limited Resources). Naukova dumka, Kyiv (2010) (in Russian)
2. Pavlov, A.A., Misura, E.B.: Metodologicheskie i teoreticheskie osnovy PDS-algoritma resheniya zadachi minimizacii summarnogo vzveshennogo zapazdyvaniya (Методологические и теоретические основы ПДС-алгоритма решения задачи минимизации суммарного взвешенного запаздывания; Methodological and theoretical basics of PSC-algorithm to solve the total weighted tardiness minimization problem). Visnyk NTUU KPI Inform. Oper. Comput. Sci. **63**, 93–99 (2015) (in Russian)
3. Pavlov, A.A., Misura, E.B.: Novyi podhod k resheniyu zadachi "Minimizaciya summarnogo vzveshennogo opozdaniya pri vypolnenii nezavisimyh zadanii s direktivnymi srokami odnim priborom" (Новый подход к решению задачи "Минимизация суммарного взвешенного опоздания при выполне-нии независимых заданий с директивными сроками одним прибором"; A new approach for solving the problem: "Minimization of total weighted tardiness when fulfilling independent tasks in due times at one mashine"). Syst. Res. Inform. Technol. **2002**(2), 7–32 (2002) (in Russian)
4. Pavlov, A.A., Misura, E.B., Kostik, D.Y.: Minimizaciya summarnogo zapazdyvaniya pri nalichii zadanii s otricatel'nymi znacheniyami direktivnyh srokov (Минимизация суммарного запаздывания при наличии заданий с отрицательными значениями директивных сроков; Minimizing total tasks tardiness in the presence of negative deadline values). Visnyk NTUU KPI Inform. Oper. Comput. Sci. **53**, 3–5 (2011) (in Russian)
5. Pavlov, A.A., Misura, E.B., Lisetsky, T.N.: Realizaciya zadachi summarnogo vzveshennogo momenta okonchaniya vypolneniya zadanii v sisteme ierarhicheskogo planirovaniya (Реализация задачи суммарного взвешенного момента окончания выполнения заданий в системе иерархического планирования; Application of the total weighted completion times of tasks problem in an hierarchical planning system). Paper presented at the 1st international conference Informaciini tehnologiї yak innovaciynyi shlyah rozvitku Ukrainy u XXI stolitti, Transcarpathian State University, Uzhhorod, 6–8 Dec 2012 (in Russian)
6. Pavlov, A.A., Misura, E.B., Shevchenko, K.Y.: Pobudova PDS-algorytmu rozv'jazannya zadachi minimizacyi sumarnoho zvajenoho zapiznennya vykonannya robit na odnomu pryladi (Побудова ПДС-алгоритму розв'язання задачі мінімізації сумарного зваженого запізнення виконання робіт на одному приладі; Construction of a PDC-algorithm for solving the single machine total weighted tardiness problem). Visnyk NTUU KPI Inform. Oper. Comput. Sci. **56**, 58–70 (2012) (in Ukrainian)
7. Lawler, E.L.: A "pseudopolynomial" algorithm for sequencing jobs to minimize total tardiness. Ann. Discr. Math. **1**, 331–342 (1977). https://doi.org/10.1016/S0167-5060(08)70742-8
8. Akturk, M.S., Ozdemir, D.: A new dominance rule to minimize total weighted tardiness with unequal release dates. Eur. J. Oper. Res. **135**(2), 394–412 (2001). https://doi.org/10.1016/s0377-2217(00)00319-2
9. Emmons, H.: One-machine sequencing to minimize certain functions of job tardiness. Oper. Res. **17**(4), 701–715 (1969). https://doi.org/10.1287/opre.17.4.701
10. Fisher, M.L.: A dual algorithm for the one-machine scheduling problem. Math. Progr. **11**(1), 229–251 (1976). https://doi.org/10.1007/bf01580393
11. Potts, C.N., Van Wassenhove, L.N.: A branch and bound algorithm for the total weighted tardiness problem. Oper. Res. **33**(2), 363–377 (1985). https://doi.org/10.1287/opre.33.2.363
12. Potts, C.N., Van Wassenhove, L.N.: Dynamic programming and decomposition approaches for the single machine total tardiness problem. Eur. J. Oper. Res. **32**(3), 405–414 (1987). https://doi.org/10.1016/s0377-2217(87)80008-5

13. Rinnooy Kan, A.H.G., Lageweg, B.J., Lenstra, J.K.: Minimizing total costs in one-machine scheduling. Oper. Res. **23**(5), 908–927 (1975). https://doi.org/10.1287/opre.23.5.908
14. Rachamadugu, R.M.V.: A note on the weighted tardiness problem. Oper. Res. **35**(3), 450–452 (1987). https://doi.org/10.1287/opre.35.3.450
15. Sen, T., Sulek, J.M., Dileepan, P.: Static scheduling research to minimize weighted and unweighted tardiness: a state-of-the-art survey. Int. J. Prod. Econ. **83**(1), 1–12 (2003). https://doi.org/10.1016/s0925-5273(02)00265-7
16. Chambers, R.J., Carraway, R.L., Lowe, T.J., et al.: Dominance and decomposition heuristics for single machine scheduling. Oper. Res. **39**(4), 639–647 (1991). https://doi.org/10.1287/opre.39.4.639
17. Akturk, M.S., Yildirim, M.B.: A new lower bounding scheme for the total weighted tardiness problem. Comp. Oper. Res. **25**(4), 265–278 (1998). https://doi.org/10.1016/s0305-0548(97)00073-7
18. Abdul-Razaq, T.S., Potts, C.N., Van Wassenhove, L.N.: A survey of algorithms for the single-machine total weighted tardiness scheduling problem. Discr. Appl. Math. **26**(2–3), 235–253 (1990). https://doi.org/10.1016/0166-218x(90)90103-j
19. Hoogeveen, J.A., Van de Velde, S.L.: Stronger Lagrangian bounds by use of slack variables: applications to machine scheduling problems. Math. Prog. **70**(1–3), 173–190 (1995). https://doi.org/10.1007/bf01585935
20. Szwarc, W., Liu, J.J.: Weighted tardiness single machine scheduling with proportional weights. Manag. Sci. **39**(5), 626–632 (1993). https://doi.org/10.1287/mnsc.39.5.626
21. Tanaka, S., Fujikuma, S., Araki, M.: An exact algorithm for single-machine scheduling without machine idle time. J. Sched. **12**(6), 575–593 (2009). https://doi.org/10.1007/s10951-008-0093-5
22. Karakostas, G., Kolliopoulos, S.G., Wang, J.: An FPTAS for the minimum total weighted tardiness problem with a fixed number of distinct due dates. ACM Trans. Algor., **8**(4), 40:1–40:16 (2012) https://doi.org/10.1145/2344422.2344430
23. Tanaev, V.S., Shkurba, V.V.: Vvedenie v Teoriju Raspisaniy (Введение в теорию расписаний; Introduction to Scheduling Theory). Nauka, Moscow (1975). (in Russian)
24. Grosso, A., Della Croce, F., Tadei, R.: An enhanced dynasearch neighborhood for the single-machine total weighted tardiness scheduling problem. Oper. Res. Lett. **32**, 68–72 (2004). https://doi.org/10.1016/S0167-6377(03)00064-6
25. Potts, C.N., Van Wassenhove, L.N.: A decomposition algorithm for the single machine total tardiness problem. Oper. Res. Lett. **1**, 177–181 (1982). https://doi.org/10.1016/0167-6377(82)90035-9
26. Pavlov, A.A., Misura, E.B., Melnikov, O.V., Mukha, I.P.: NP-Hard scheduling problems in planning process automation in discrete systems of certain classes. In: Hu, Z., Petoukhov, S., Dychka, I., He, M. (eds) Advances in Computer Science for Engineering and Education. ICCSEEA 2018. Advances in Intelligent Systems and Computing, vol. 754, pp. 429–436. Springer, Cham (2019). https://doi.org/10.1007/978-3-319-91008-6_43
27. Baker, K.R., Trietsch, D.: Principles of Sequencing and Scheduling. Wiley, New York (2009). https://doi.org/10.1002/9780470451793
28. Beasley, J.E.: OR-library: weighted tardiness. http://people.brunel.ac.uk/~mastjjb/jeb/orlib/wtinfo.html (1998). Accessed 23 Jan 2018
29. Tanaka, S., Fujikuma, S., Araki, M.: OR-library: weighted tardiness. https://sites.google.com/site/shunjitanaka/sips/benchmark-results-sips (2013). Accessed 23 Jan 2018

Chapter 5
The Total Earliness/Tardiness Minimization on a Single Machine with Arbitrary Due Dates

Abstract We solve a single machine problem of constructing a schedule of tasks with arbitrary due dates on a single machine that minimizes the total earliness/tardiness in relation to their due dates. This problem is solved in three different formulations: (1) the start time of the machine is fixed. In this case the problem is NP-hard; (2) the start time of the machine belongs to a specified time segment. The problem is intractable because there is no exact polynomial algorithm for its solving; (3) the start time of the machine is arbitrary. The problem is intractable because there is no exact polynomial algorithm for its solving. For the first two problems we give PSC-algorithms, each of them contains sufficient signs of a feasible solution optimality and is based on the optimal solution for the single machine problem to minimize the total tardiness of tasks in relation to their various due dates (with equal weights). We show that the PSC-algorithm of its solving is a simplified modification of the PSC-algorithm presented in Chap. 4. For the third problem solving we build an efficient approximation algorithm.

5.1 The Formulation of E/T 1, E/T 2, E/T 3 Problems

This chapter contains our new original results based on our earlier works [1–5].

The Problems Statement A set of independent tasks $J = \{j_1, j_2, \ldots, j_n\}$ is to be processed without interruptions on a single machine. The machine can process only one task at a time. We assume that all tasks and the machine are constantly available from the time point zero. Idle times of the machine are not allowed. A task j, where $j = 1, 2, \ldots, n$, requires the processing time l_j and should be in the best case completed exactly in its due date d_j. For any given schedule, we can define the earliness E_j and the tardiness T_j of a task j by expressions:

$$E_j = \max\left(0, d_j - C_j\right) = \left(d_j - C_j\right)^+, \tag{5.1}$$

M. Z. Zgurovsky and A. A. Pavlov, *Combinatorial Optimization Problems in Planning and Decision Making*, Studies in Systems, Decision and Control 173,
https://doi.org/10.1007/978-3-319-98977-8_5

$$T_j = \max(0, C_j - d_j) = (C_j - d_j)^+ \tag{5.2}$$

where C_j is the completion time of task j.

Our goal is to find a schedule that minimizes the total earliness and tardiness (further called earliness/tardiness or E/T) of all tasks:

$$f = \sum_{j=1}^{n} (E_j + T_j). \tag{5.3}$$

We consider the following three cases of the problem:

- E/T 1: The start time of tasks execution s is fixed and equal to zero;
- E/T 2: The start time of tasks is in the time segment $[t_1, t_k]$, $t_k > t_1 > 0$;
- E/T 3: The start time of tasks s is not fixed.

The problem belongs to the class of NP-hard problems. We include it to the mathematics of the four-level model of planning (including operative planning) and decision making (Chap. 9). At the third level of the model, we solve the problem during the operational plan construction. At the fourth level, we use it during the operative adjustment of the constructed plan in case of its partial violation.

Scheduling against due dates has been receiving considerable attention in the literature [6]. One reason for this phenomenon is the growing pressure of competition on the international markets: firms have to offer a great variety of different and individual products while customers expect ordered goods to be delivered on time. To meet these kinds of requirements, principles and philosophies like lean management, simultaneous engineering, just-in-time production, etc. have been developed. For example, the just-in-time principle states that the right amount of goods should be produced or delivered at exactly the right time. Thus jobs are expected to be punctual, since late as well as early delivery is seen negatively.

Many authors have studied the E/T problem [7, 8] which was first introduced by Kanet [9]. Using the notation of Lawler et al. [10], the static, single-machine earliness/tardiness (E/T) problem can be described as $1||\sum(E_j + T_j)$. A useful review of early/tardy scheduling is provided in the book [11]. Baker and Scudder [12] review the open literature on the E/T scheduling problems, describing various E/T models and different penalties for task earliness and tardiness [13, 14]. When task due dates are predetermined but distinct, E/T problems are NP-complete [15]. Yano and Kim [16] develop a heuristic for the special case of weighted penalties wherein the weight is proportional to the processing time of the task [13, 14]. They use a pair-wise interchange procedure to improve upon initial heuristic solutions.

Davis and Kanet's TIMETABLER [17] can be embedded within an existing heuristic solution to improve 1-E/T schedule quality at low computational costs [13, 14]. Ow and Morton [18] investigate a priority-based function and a variation of filtered beam search for E/T problems, concluding that solution quality degrades significantly if task earliness costs are ignored in favor of considering only tardiness cost. Szwarc and Mukhopadhyay [19] developed a clustering approach for determining the optimal starting time for tasks in a 1-E/T problem instance, assuming the job sequence is known a priori. The clustering approach produces optimal task timings for 500 job instances in a few seconds of computation time. Sridharan and Zhou [20] use a dispatching heuristic for 1-E/T problems that permits inserted machine idle time when appropriate.

An algorithm of Wan and Yen [21] is based on tabu search [8]. Algorithms for optimal job sequencing are relatively less studied. Szwarc [22] proposed a branch-and-bound algorithm based on some adjacent ordering conditions for tasks with distinct penalty weights. Lee and Choi [23] proposed a genetic algorithm and Yano and Kim [16] studied several dominance properties for sequencing jobs with penalty weights proportional to processing times.

The most important results for unrestrictive and restrictive scheduling E/T problems are presented in the papers [6, 24–27].

Hoogeven and Van de Velde [28] proposed an algorithm based on a branch-and-bound method. Because of the exponentially growing computation time, this algorithm can only be used to solve instances where the number of jobs is no larger than 20. Therefore, in practice, almost always the approximate algorithms are used.

The most efficient of known exact algorithms for weighted earliness/tardiness problem is the dynamic programming algorithm of Tanaka et al. [29]. It allows to solve instances with up to 300 tasks within 1 h on a computer with a 2.4 GHz Pentium 4 processor.

We present in this chapter efficient PSC-algorithms to solve E/T 1 and E/T 2 problems. A PSC-algorithm to solve E/T 3 problem is a part of the PSC-algorithm to solve E/T 2 problem.

Let us give a brief plan for solving the problems we formulated.

E/T 1 The earliness/tardiness problem with fixed start time of tasks.

1.1. The first polynomial component of the PSC-algorithm for the problem solving.

1.1.1. Formulation of sufficient sign of optimality (SSO) of a feasible[1] solution.

1.1.2. The polynomial algorithm that checks the SSO. The algorithm is based on our PSC-algorithm to solve the problem of minimizing the total tardiness of tasks on a single machine (TT). It includes its first polynomial component, its approximation algorithm, and its exact subalgorithm with additional restriction imposed on the number of iterations to be executed.

1.2. The second polynomial component: the approximation algorithm for E/T 1 problem with qualitatively justified heuristic SSOs.

1.3. An example to the problem solving.

E/T 2 The start time of tasks is in the time segment $[t_1, t_k]$.

[1]All schedules obtained during the considered problem solving are feasible including those with tardy tasks.

2.1. The first polynomial component of PSC-algorithm for E/T 2 problem solving converges to that for E/T 1 problem.
2.1.1. Formulation of SSO.
2.1.2. The polynomial algorithm that checks the SSO, based on the PSC-algorithm for TT problem solving.
2.2. The approximation algorithm for E/T 2 problem.
2.2.1. Approximation algorithm for E/T 3 problem.
2.2.2. Approximation algorithm for E/T 2 problem based on the algorithm for E/T 3.

Our algorithms for E/T 1 and E/T 2 problems are based on our PSC-algorithm to solve the TT problem. Since the algorithmic solution for the two problems includes the algorithm to solve the TT problem, we give below its formulation, main theoretical propositions, SSOs, and the PSC-algorithm for the problem solving.

5.2 PSC-Algorithm to Solve the Problem of Minimizing the Total Tardiness of Independent Tasks on a Single Machine

The combinatorial optimization problem "Minimizing the total tardiness of independent tasks on a single machine" (TT) is a particular case of the TWT problem presented in Chap. 4 provided that the weights of all tasks equal to one. According to Du and Leung [30], the problem is NP-hard in the ordinary sense, in contrast to the TWT problem.

The PSC-algorithm for the TT problem solving is an obvious simplified version of the PSC-algorithm for the TWT problem because the weights of all tasks are the same. Let us show how we can obtain the PSC-algorithm for the TT problem from that for the TWT problem.

The algorithm for the TWT problem solving (Chap. 4) consists of a series of uniform iterations. The number of competing tasks in the sequence σ^{fp} determines the number of iterations. At each iteration, we add the next competing task from the sequence σ^{fp} and check the possibility to use the time reserves of preceding tasks for it. By this we construct an optimal schedule for tasks of the current subsequence ended with the added competing task. The functional value decreases or remains unchanged at each iteration.

Competing tasks in the sequence σ^{fp} of the TT problem are arranged in non-decreasing order of their processing times. This is the main distinguishing property of the sequence σ^{fp} from that in the TWT problem. This has allowed us to simplify the rules for eliminating unpromising permutations and insertions and to reduce their number. Also, we have formulated and justified the rules for eliminating some tasks from the set of competing tasks.

Let us consider the differences in the theoretical properties of the TT problem from the TWT problem in more detail.

5.2.1 Main Theoretical Propositions

Let us introduce the following notation and definitions [1]. Let i or j_i be the number of task corresponding to the indexation specified by the functional; $j_{[i]}$ be the number of a task in position i in a feasible schedule. Let $\overline{a, b}$ denote the interval of integer numbers from a to b, that is, $\overline{a, b} = \mathbb{Z} \cap [a, b] = a,\ a + 1,\ \ldots,\ b$.

All the definitions necessary to present the problem solving method coincide with those given in Chap. 4 except for the given below.

Definition 5.1 *Sequence* σ^{ord} (sigma ordered) is a sequence of tasks j, $j = \overline{1, n}$, of the set J that are in non-decreasing order of processing times l_j, i.e., $\forall j, i, j < i$: $l_j \le l_i$, and if $l_j = l_i$, then $d_j \le d_i$.

Definition 5.2 A tardy task $j_{[g]}$ in the sequence σ^{fp} is called a *competing task* if there is at least one such preceding task $j_{[l]}$ in σ^{fp} that $d_{j_{[l]}} > d_{j_{[g]}}$ and $d_{j_{[l]}} > C_{j_{[l]}}$.

Definition 5.3 At the current kth iteration of optimization we construct an optimal schedule on the *current subsequence* σ^k. The subsequence consists of:

- the optimal subsequence obtained at the previous iteration $k - 1$ executed for a competing task $j_{[g_{k-1}]}$;
- the current competing task $j_{[g_k]}$ (the next competing task from the sequence σ^{fp}); the task's processing time is not less than the processing times of tasks of the current subsequence, by construction of the sequence σ^{fp};
- the tasks that are in σ^{fp} between the competing tasks $j_{[g_{k-1}]}$ and $j_{[g_k]}$.

Theorem 5.1, Lemma 5.2, Statements 5.1–5.3, 5.7, 5.10, 5.14, 5.17, 5.21, 5.39–5.43, and their proofs completely coincide with the Theorem 4.1, Lemma 4.2, Statements 4.1–4.3, 4.7, 4.10, 4.14, 4.17, 4.21, 4.39, 4.40 (Sect. 4.2), 4.41–4.43 (Sect. 4.3.3.1) and their proofs, respectively. Lemma 5.1 and Statements 5.4–5.6, 5.8, 5.9, 5.11–5.13, 5.15, 5.16, 5.18–5.20, 5.22–5.38 are, respectively, the modifications of Lemma 4.1 and Statements 4.4–4.6, 4.8, 4.9, 4.11–4.13, 4.15, 4.16, 4.18–4.20, 4.22–4.38. The modifications essentially simplify the algorithm implementation to solve the TT problem. Their proofs are similar to the proofs of the corresponding statements given in Chap. 4. We used the above mentioned distinguishing property of the TT problem to modify all those statements. All tasks in the sequence σ^{fp} of the TT problem are in non-decreasing order of their processing times (the shorter the processing time, the higher the task's priority). Also, we used the equality of the weights of all tasks to one. We have used only these properties of the problem for modifications in Statements 5.4, 5.8, 5.9, 5.12, 5.13, 5.20, 5.22, 5.23, 5.26, 5.28, 5.35–5.37 and Lemma 5.1.

Theorem 5.1, Sufficient Sign of Optimality of a Feasible Solution (SSO) #1 ([31, Chap. 3]; coincides with Theorem 4.1.) *The total tardiness is zero in an optimal schedule if and only if it is zero in the schedule in which the tasks are in non-decreasing order of their due dates.*

Corollary 5.1 If there are tardy tasks in the schedule in which the tasks are in non-decreasing order of their due dates, then it is not possible to build a schedule without tardy tasks.

Statement 5.1, SSO #2 (coincides with Statement 4.1.) If tasks with a time reserve do not precede the tardy tasks in the sequence σ^{ord}, then there are no moves (permutations or insertions) of tasks that lead to a decrease in the functional value, and the sequence σ^{ord} is optimal on the entire set of tasks.

Statement 5.2 (coincides with Statement 4.2.) Moving a tardy task $j_{[g]}$ into a position $f < p$, where p is the position of its insertion, cannot decrease the functional value.

Statement 5.3 (coincides with Statement 4.3.) Suppose that we have $R_{j_{[i]}} \leq 0 \; \forall j_{[i]}$, $i = \overline{1, p-1}$, in the sequence σ^{ord}. Here, $R_{j_{[i]}} = d_{j_{[i]}} - C_{j_{[i]}}$ is the time reserve of a task $j_{[i]}$. Denote such sequence by σ^{ord1}. A tardy task $j_{[g]}$ can take an earlier position in the sequence σ^{ord1} in consequence of its insertion, which will decrease the functional value, only if at least one of the tasks within the interval of insertion $I_{j_{[g]}} = \overline{p, g-1}$ in the sequence σ^{ord1} has a positive time reserve.

Statement 5.4, SSO #3 (modification of Statement 4.4.) If for any tardy task $j_{[g]}$ in the sequence σ^{ord} there are no preceding tasks $j_{[s]}$, $s = \overline{1, g-1}$, for which $d_{j_{[s]}} - C_{j_{[s]}} > 0$ and $d_{j_{[s]}} > d_{j_{[g]}}$, then there are no permutations or insertions of tasks that decrease the functional value, and the sequence σ^{ord} is optimal on the entire set of tasks.

Statement 5.4 is basic for the proof of a number of statements following below. Therefore, we now give its proof completely [1].

Proof 1. Insertions Insert the task $j_{[g]}$ into the position of insertion p (Definition 4.6). Considering the condition of the statement that there are no preceding tasks $j_{[s]}$, $s = \overline{1, g-1}$, for which $d_{j_{[s]}} - C_{j_{[s]}} > 0$ and $d_{j_{[s]}} > d_{j_{[g]}}$, we find that the task $j_{[g]}$ should be inserted into the interval $\overline{p, g-1}$ where we have $C_{j_{[i]}} \geq d_{j_{[i]}}$ for any task $j_{[i]}$, $i = \overline{p, g-1}$.

Suppose $C_{j_{[i]}} - d_{j_{[i]}} = 0$, $i = \overline{p, g-1}$. Then,

$$f = C_{j_{[g]}} - d_{j_{[g]}},\; f' = \sum_{i=p}^{g-1} \left(C_{j_{[i]}} + l_{j_{[g]}} - d_{j_{[i]}} \right),$$

$$f - f' = C_{j_{[g]}} - d_{j_{[g]}} - \sum_{i=p}^{g-1} \left(C_{j_{[i]}} + l_{j_{[g]}} - d_{j_{[i]}} \right) \leq \sum_{i=p}^{g-1} \left(l_{j_{[i]}} - l_{j_{[g]}} \right) \leq 0$$

because $l_{j_{[i]}} \leq l_{j_{[g]}} \; \forall i < g$.

Proof for the case $C_{j_{[i]}} - d_{j_{[i]}} > 0$ is similar.

2. *Permutations.* Permutations of tasks with zero time reserve and with a non-zero reserve are possible. Let us consider all possible cases.

(a) Suppose $d_{j_{[s]}} - C_{j_{[s]}} > 0$, $d_{j_{[s]}} \le d_{j_{[g]}}$, and the task $j_{[g]}$ is tardy after the permutation, i.e., $C_{j_{[g]}} - l_{j_{[s]}} > d_{j_{[g]}}$. In this case, $f = C_{j_{[g]}} - d_{j_{[g]}}$,

$$f' = \left(C_{j_{[g]}} - l_{j_{[s]}} - d_{j_{[g]}}\right) + \left(C_{j_{[g]}} - d_{j_{[s]}}\right).$$

$$\begin{aligned} f - f' &= \left(C_{j_{[g]}} - d_{j_{[g]}}\right) - \left(C_{j_{[g]}} - l_{j_{[s]}} - d_{j_{[g]}}\right) - \left(C_{j_{[g]}} - d_{j_{[s]}}\right) \\ &= l_{j_{[s]}} - \left(C_{j_{[g]}} - d_{j_s}\right) \le l_{j_{[s]}} - \left(C_{j_g} - d_{j_g}\right) \le 0 \end{aligned}$$

because $C_{j_{[g]}} - d_{j_{[g]}} > l_{j_{[s]}}$.

(b) Suppose $d_{j_{[s]}} - C_{j_{[s]}} > 0$, $d_{j_{[s]}} \le d_{j_{[g]}}$, and the task $j_{[g]}$ is not tardy after the permutation, i.e., $C_{j_{[g]}} - d_{j_{[g]}} \le l_{j_{[s]}}$. Then, $f' = C_{j_{[g]}} - d_{j_{[s]}}$,

$$f - f' = \left(C_{j_{[g]}} - d_{j_{[g]}}\right) - \left(C_{j_{[g]}} - d_{j_{[s]}}\right) \le 0$$

because $d_{j_{[s]}} \le d_{j_{[g]}}$.

(c) Suppose $d_{j_{[s]}} - C_{j_{[s]}} = 0$, $d_{j_{[s]}} \le d_{j_{[g]}}$, and the task $j_{[g]}$ is tardy after the permutation, then

$$\begin{aligned} f - f' &= C_{j_{[g]}} - d_{j_{[g]}} - \left(C_{j_{[g]}} - l_{j_{[s]}} - d_{j_{[g]}}\right) - \sum_{i=s+1}^{g} l_{j_{[i]}} \\ &= l_{j_{[s]}} - \sum_{i=s+1}^{g} l_{j_{[i]}} = l_{j_{[s]}} - l_{j_{[g]}} - \sum_{i=s+1}^{g-1} l_{j_{[i]}} < l_{j_{[s]}} - l_{j_{[g]}} \le 0 \end{aligned}$$

because $l_{j_{[s]}} \le l_{j_{[g]}}\ \forall s < g$.

(d) Suppose $d_{j_{[s]}} - C_{j_{[s]}} = 0$, $d_{j_{[s]}} \le d_{j_{[g]}}$, and the task $j_{[g]}$ is not tardy after the permutation, i.e., $C_{j_{[g]}} - d_{j_{[g]}} \le l_{j_{[s]}}$. Then,

$$f - f' = C_{j_{[g]}} - d_{j_{[g]}} - \sum_{i=s+1}^{g} l_{j_{[i]}} \le l_{j_{[s]}} - \sum_{i=s+1}^{g} l_{j_{[i]}} \le 0,$$

similarly to the case (c).

Therefore, since there are no preceding tasks with time reserves for the tardy task $j_{[g]}$ within its interval of insertion, then there are no permutations leading to a decrease in the functional value. ∎

Thus, ordering of tasks in the sequence σ^{ord} by non-decreasing processing times and equality of their weights to one have allowed us to simplify the conditions of Statement 4.4.

Corollary 5.2 Suppose that there are no reserves in the sequence σ^{ord} within the interval of insertion $\overline{p, g-1}$ of a tardy task $j_{[g]}$. Then, the task $j_{[g]}$ can take an earlier position, which will decrease the functional value, only if there are such tasks $j_{[k]}$ in the sequence σ^{ord} within the interval $\overline{1, p-1}$ that $d_{j_{[k]}} - C_{j_{[k]}} > 0$, $d_{j_{[k]}} > d_{j_{[g]}}$. In this case, we can create reserves within the interval of insertion of the task $j_{[g]}$ permuting the tasks $j_{[k]}$ into later positions.

Corollary 5.3 If $l_{j_{[k]}} \le l_{j_{[g]}}$ and $d_{j_{[k]}} \le d_{j_{[g]}}$ for each task with reserves $j_{[k]}$ and each tardy task $j_{[g]}$ in the sequence σ^{ord}, then there are no permutations and insertions that decrease the functional value.

Statement 5.5 (modification of Statement 4.5.) Suppose that the interval of insertion of the task $j_{[g]}$ in the sequence σ^{ord} is $I_{j_{[g]}} = \overline{p, g-1}$. Additional reserves within the interval of insertion for the task $j_{[g]}$ may only be created by the tasks $j_{[i]}$, $i \in \overline{1, p-1}$, for which

$$d_{j_{[i]}} > d_{j_{[g]}},\ d_{j_{[i]}} > C_{j_{[i]}},\ d_{j_{[i]}} > C_{j_{[p-1]}}. \tag{5.4}$$

The modification in Statement 5.5 is based on Statement 5.4 which was used for the modification.

Statement 5.6 (modification of Statement 4.6.) Suppose that a task $j_{[g]}$ is tardy in the sequence σ^{ord}. Moving the task $j_{[g]}$ into an earlier position as a result of permutations and insertions can decrease the functional value only if at least one of the following conditions is true:

1. $\exists j_{[i]}, p \le i \le g$, such that $R_{j_{[i]}} > 0, d_{j_{[i]}} > d_{j_{[g]}}$. There are tasks $j_{[i]}$ with time reserves within the interval of insertion of the task $j_{[g]}$ for which $d_{j_{[i]}} > d_{j_{[g]}}$. Here, p is the position in which the tardiness of the task $j_{[g]}$ is minimal (or equals to zero).
2. $\exists j_{[q]},\ q < g$, such that $d_{j_{[q]}} > C_{j_{[g]}}$. There is a task with a time reserve in the sequence σ^{ord} in position q that precedes the position g, and its permutation into the position after the task $j_{[g]}$ reduces the tardiness of the task $j_{[g]}$. The task $j_{[q]}$ remains non-tardy.
3. $\exists j_{[q]},\ q < g$, such that $C_{j_{[g]}} - l_{j_{[g]}} < d_{j_{[q]}} < C_{j_{[g]}}$. There is a non-tardy task $j_{[q]}$ which due date is greater than the start time of the task $j_{[g]}$ but less than the completion time of the task $j_{[g]}$. Also, the following must be satisfied:

   $$\min\left(C_{j_{[g]}} - d_{j_{[g]}}, l_{j_{[q]}}\right) - \left(C_{j_{[g]}} - d_{j_{[q]}}\right) > 0.$$

   Consequently, a permutation of the task $j_{[q]}$ into a position after the task $j_{[g]}$ will decrease the functional value due to the use of reserve of the task $j_{[q]}$.
4. $\forall i,\ p \le i \le g,\ R_{j_{[i]}} \le 0$, but $\exists j_{[k]},\ k < p$, such that $d_{j_{[k]}} > d_{j_{[g]}},\ R_{j_{[k]}} > 0$, and $d_{j_{[k]}} > C_{j_{[p-1]}}$. We do not have reserves within the interval of insertion of the task

$j_{[g]}$. But there is a task $j_{[k]}$, $k < p$, with the due date greater than $d_{j_{[g]}}$ and a reserve greater than zero.

The modification in Statement 5.6 is based on Statements 5.1, 5.3, 5.4, and also on the fact of equality of the task weights to one.

Corollary 5.4, SSO #4 Suppose that the number of tardy tasks in the sequence σ^{ord} is greater than one. If the sequence σ^{ord} for any tardy task $j_{[g]}$ has no preceding tasks $j_{[s]}$, $s < g$, for which at least one of the conditions 1–4 of Statement 5.6 is true, then the sequence σ^{ord} corresponds to the optimal functional value.

Suppose that a tardy task $j_{[g]}$ in the sequence $\sigma^{ord} = \{j_{[1]}, j_{[2]}, \ldots, j_{[g]}\}$ has moved into a position $p = m$ in consequence of its insertion. Denote this task by $j^*_{[m]}$ ("$j_{[m]}$ marked with asterisk") and the obtained sequence by $\sigma(j_{[g]})$.

Statement 5.7 (coincides with Statement 4.7.) An insertion of a tardy task $j_{[k]}$, $k = \overline{m+1, g}$, into an earlier position in the sequence $\sigma(j_{[g]})$ will decrease the functional value only if at least one preceding task has a time reserve or (if there are no reserves) only in consequence of a permutation of the task $j^*_{[m]}$ into a position after the task $j_{[k]}$.

Statement 5.8 (modification of Statement 4.8.) Suppose that we have performed an insertion of the task $j_{[g]}$ in the sequence σ^{ord}; tasks $j_{[l]}$, $l = \overline{g+1, n}$, are tardy; and for all preceding tasks $j_{[i]}$, $i = \overline{1, g}$, we have $R_{j_{[i]}} \le 0$. In this case, tasks $j_{[l]}$ will remain on the occupied positions within the interval $\overline{g+1, n}$ in an optimal sequence.

We used ordering of tasks in the sequence σ^{ord} by non-decreasing processing times and equality of their weights in modification of Statement 5.8.

Statement 5.9 (modification of Statement 4.9.) Suppose that tasks $j_{[g]}$ and $j_{[k]}$, $g < k$, such that $l_{j_{[g]}} \le l_{j_{[k]}}$ and $\max R(\sigma^{ord}) < l_{j_{[g]}}$, qualify for existing reserves in the sequence σ^{ord}. In this case, task $j_{[k]}$ can qualify for the reserves only if $d_{j_{[k]}} < d_{j_{[g]}}$.

We used ordering of tasks in the sequence σ^{ord} by non-decreasing processing times and equality of their weights in modification of Statement 5.9.

Statement 5.10 (coincides with Statement 4.10.) Statements 5.1, 5.2, 5.4–5.6, 5.9 are valid for the sequence σ^{fp}.

Statement 5.11, SSO #5 (modification of Statement 4.11.) There are no permutations and insertions in the sequence σ^{fp} that decrease the functional value, and the sequence σ^{fp} is optimal on the entire set of tasks, if at least one of the following conditions holds:

1. $d_{j_{[i]}} - C_{j_{[i]}} \ge 0,\ \forall i = \overline{1, n}$;
2. $d_{j_{[i]}} - C_{j_{[i]}} \le 0,\ \forall i = \overline{1, l},\ d_{j_{[i]}} - C_{j_{[i]}} \ge 0,\ \forall i = \overline{l+1, n}$;

3. $d_{j_{[i]}} - C_{j_{[i]}} \le 0$, $\forall i = \overline{1, n}$;
4. Let R denote the set of tasks $j_{[i]}$: $d_{j_{[i]}} - C_{j_{[i]}} > 0$, Z denote the set of tasks $j_{[i]}$: $d_{j_{[i]}} - C_{j_{[i]}} \le 0$. Then, $d_{j_{[k]}} \le d_{j_{[g]}}$, $l_{j_{[k]}} \le l_{j_{[g]}} \forall j_{[k]} \in R$, $j_{[g]} \in Z$.

The modification in Statement 5.11 is based on Statement 5.4.

Corollary 5.5 If at least one of the conditions formulated in Statement 5.11 holds, then the sequence σ^{fp} corresponds to an optimal functional value.

Statement 5.12 (modification of Statement 4.12.) §1. Statements 5.1, 5.2, 5.4–5.6, 5.9 are also true for the current subsequence σ^k.

§2. If during the current iteration, after insertion of the task $j_{[g]}$, a tardy task $j_{[q]}$ in the current subsequence σ^k is preceded by an unmarked task $j_{[l]}$, $l_{j_{[l]}} > l_{j_{[q]}}$, and $l_{j_{[q]}} < l_{j_{[g]}}$, $q < g$, then a decrease in the functional value is possible in consequence of the move of the task $j_{[l]}$ into the position q.

We used ordering of tasks in the sequence σ^{ord} by non-decreasing processing times and equality of their weights in modification of Statement 5.12.

Statement 5.13 (modification of Statement 4.13.) If at iteration k for a tardy task $j_{[g]}$ in the current subsequence σ^k exists a preceding non-tardy task $j_{[l]}$ such that

$$d_{j_{[l]}} > d_{j_{[g]}},\ \min\left(l_{j_{[l]}}, C_{j_{[g]}} - d_{j_{[g]}}\right) > \left(C_{j_{[g]}} - d_{j_{[l]}}\right), \tag{5.5}$$

then the permutation of the task $j_{[l]}$ into the position g decreases the functional value. The task $j_{[g]}$ after the permutation is marked with asterisk.

We used ordering of tasks in the sequence σ^{ord} by non-decreasing processing times and equality of their weights in modification of Statement 5.13.

Statement 5.14 (coincides with Statement 4.14.) If the first positions $i = 1, \ldots, k$ in the sequence σ^{ord} are occupied by the tasks with $R_{j_{[i]}} \le 0$, then we can decompose the sequences σ^{ord} and σ^{fp} into two subsequences: σ^1 includes the tasks $j_{[1]}, \ldots, j_{[k]}$, σ^2 contains $j_{[k+1]}, \ldots, j_{[n]}$. The subsequence σ^1 is optimal. The tasks that belong to σ^1 do not participate in permutations. We do the optimization only for the tasks of the subsequence σ^2.

Lemma 5.1 (modification of Lemma 4.1.) Suppose that we have already executed k – 1 iterations and constructed an optimal subsequence σ^{k-1} on the set of tasks within positions $\overline{1, g-1}$. Then, to obtain an optimal subsequence at the current iteration for the current competing task $j_{[g]}$, it is necessary and sufficient to obtain a subsequence in which we achieve the minimum possible increase in the total tardiness in comparison with the optimal functional value at the previous iteration.

Proof The proof follows from the fact that in a non-optimal subsequence within positions $\overline{1, g}$ the increase in the total tardiness is not less than that in an optimal subsequence. ∎

We used the equality of the tasks' weights to one in modification of Lemma 5.1.

Corollary 5.6 Thus, to obtain an optimal subsequence at the current iteration, it is necessary and sufficient to obtain a subsequence at this iteration in which the increase in the total tardiness is the minimum possible in comparison with the optimal functional value obtained at the previous iteration.

Statement 5.15 (modification of Statement 4.15.) Suppose that we have already executed $k-1$ iterations. We proceed to the next competing task $j_{[g]}$ and to the iteration k, according to the rules formulated in Definition 5.3. The necessary conditions for inserting a competing task $j_{[g]}$ in the subsequence σ^k into the position p where the task's tardiness is minimal or equal to zero, are:

1. Tasks $j_{[i]} \in I_{j_{[g]}}$, $d_{j_{[i]}} > d_{j_{[g]}}$, have reserves within the interval of insertion of the task $j_{[g]}$.
2. If there are no reserves within positions $\overline{p, g-1}$, then tasks $j_{[i]}$ within positions $\overline{1, p-1}$ have reserves and the following conditions hold:

$$d_{j_{[i]}} > d_{j_{[g]}},\ d_{j_{[i]}} > C_{j_{[i]}} \text{ and } d_{j_{[i]}} > C_{j_{[p-1]}}. \tag{5.6}$$

3. If there are no reserves within positions $\overline{1, g-1}$, then there are tasks $j_{[i]}$ marked with "*" or "**" within those positions.

The modification in Statement 5.15 is based on Statements 5.4, 5.7, 5.13.

Statement 5.16 (modification of Statement 4.16.) Suppose that $j_{[g]}$ is a tardy (competing or originated) task in the current subsequence σ^k. A decrease in the functional value when moving the task $j_{[g]}$ into an earlier position is possible only if at least one of the following conditions is satisfied:

1. $\exists j_{[i]}, p \le i \le g$, such that $R_{j_{[i]}} > 0$, $d_{j_{[i]}} > d_{j_{[g]}}$ where p is the position in which the tardiness of the task $j_{[g]}$ is minimal (or equals to zero) (Statement 5.4).
2. $\exists j_{[l]}, l < g$, such that $d_{j_{[l]}} > C_{j_{[g]}}$. A free permutation of the task $j_{[l]}$ after the task $j_{[g]}$ is possible.
3. $\exists j_{[l]}, l < g$, such that $d_{j_{[l]}} > C_{j_{[g]}} - l_{j_{[g]}}$, $\min\left(l_{j_{[l]}}, C_{j_{[g]}} - d_{j_{[g]}}\right) > C_{j_{[g]}} - d_{j_{[l]}}$ (Statement 5.13).
4. $\forall i,\ p \le i \le g$, $R_{j_{[i]}} \le 0$, but $\exists j_{[l]}, l < p$, such that $d_{j_{[l]}} > d_{j_{[g]}}$, $d_{j_{[l]}} > C_{j_{[l]}}$, $d_{j_{[l]}} > C_{j_{[p-1]}}$ (Statement 5.4).
5. $\forall i, i = \overline{1, g-1}$, $R_{j_{[i]}} \le 0$, but $\exists j_{[l]}, j_{[r]}, l < r < g$, $l_{j_{[l]}} \ge l_{j_{[r]}}$, $l_{j_{[r]}} < l_{j_{[g]}}$. Here, $j_{[l]}$ is an unmarked task, $j_{[r]}$ is an originated tardy task within the interval of insertion of the competing task $j_{[g]}$, and a free permutation was done for the task $j_{[r]}$ at the previous steps of the algorithm (§2 of Statement 5.12).
6. $\forall i, i = \overline{1, g-1}$, $R_{j_{[i]}} \le 0$, but $\exists j^*_{[m]}$ ($j^{**}_{[m]}$), $m < g$ (Statement 5.7).

The modification in Statement 5.16 is based on Statements 5.4, 5.7, 5.12, 5.13.

Corollary 5.7, SSO #6 Suppose that we have constructed an optimal subsequence σ^k on the set of tasks within positions $\overline{1,g}$ as a result of k iterations of the algorithm. If none of the conditions 1–6 of Statement 5.16 is satisfied for each of the tardy tasks within positions $\overline{g+1,n}$, then they remain in the positions they are, and the current sequence is optimal on the entire set of tasks.

Corollary 5.8, SSO #7 This follows from Corollary 5.7 when k = 0. If none of the conditions 1–6 of Statement 5.16 is satisfied for each tardy task in the sequence σ^{fp}, then the sequence σ^{fp} is optimal on the entire set of tasks.

Statement 5.17 (coincides with Statement 4.17.) The following procedures for tardy tasks are necessary during the iterations of optimization within the current subsequence:

1. Determination and extension of the interval of insertion (conditions 1–4 of Statement 5.16);
2. Determination of existence of time reserves within the extended interval of insertion;
3. If time reserves exist within the extended interval of insertion, the procedure of insertion for a tardy task into a position defined by the interval of insertion. Then we need to reorder all tasks that follow the inserted tardy task according to the non-decreasing order of their processing times (Statement 5.6) and to perform the optimization within the extended interval of insertion;
4. If the reserves are absent within the extended interval of insertion or we are unable to use them, the procedure of releasing the reserves by means of permutation of the tasks that earlier used them at the previous steps of the iteration (condition 5 of Statement 5.16);
5. Optimization within the interval of permutation due to the use of the reserves released by tardy tasks (condition 6 of Statement 5.16).

All procedures 1–5 are based on the conditions for a task permutations given in Statement 5.16. We perform them both for competing tasks and for tardy tasks within their intervals of insertion.

Statement 5.18 (modification of Statement 4.18.) If there is such a task $j_{[k]}$ within the interval of insertion of a tardy task $j_{[g]}$ that $d_{j_{[k]}} \le d_{j_{[g]}}$, $l_{j_{[k]}} \le l_{j_{[g]}}$, then the interval of insertion for the task $j_{[g]}$ is determined by the positions $\overline{k+1, g-1}$.

The modification in Statement 5.18 is based on known statement from [32].

Statement 5.19 (modification of Statement 4.19.) Suppose that we have already executed $k-1$ iterations and constructed an optimal subsequence on the set of tasks within positions $\overline{1, g-1}$. The next competing task $j_{[g]}$ such that $l_{j_{[g]}} \ge l_{j_{[i]}}$, $i \in \overline{1, g-1}$, can take an earlier position at iteration k. This will lead to the minimum possible increase in the total tardiness of the current subsequence in comparison with the functional value at iteration $k - 1$ and to the construction of an optimal subsequence at iteration k only if at least one task $j_{[l]}$ within the interval $\overline{1, g-1}$ has a reserve and $d_{j_{[l]}} > d_{j_{[g]}}$ or (if reserves are absent) there are marked tasks within the interval $\overline{1, g-1}$.

The modification in Statement 5.19 is based on Statements 5.4 and 5.7.

Statement 5.20 (Lawler [32]; modification of Statement 4.20.) Suppose that for competing tasks $j_{[k]}, j_{[r]} \in J$ in the sequence σ^{fp} we have $l_{j_{[k]}} \le l_{j_{[r]}}$, $d_{j_{[k]}} \le d_{j_{[r]}}$. Then the task $j_{[k]}$ will precede the task $j_{[r]}$ in an optimal schedule.

We used the equality of the tasks' weights in modification of Statement 5.20.

Statement 5.21 (coincides with Statement 4.21.) Non-competing tasks in the current subsequence σ^k cannot take positions earlier than those they occupied in σ^{fp}.

Statement 5.22 (modification of Statement 4.22.) Suppose that we have already executed $k - 1$ iterations and constructed an optimal subsequence σ^{k-1} on the set of tasks within positions $\overline{1, g-1}$. The competing task $j_{[g]}$ such that $l_{j_{[g]}} \ge l_{j_{[i]}}$, $i \in \overline{1, g-1}$, cannot be moved into an earlier position at iteration k and remains in its position that corresponds to the minimum possible increase in the total tardiness of the current subsequence in comparison with the functional value at iteration $k - 1$ and to the construction of an optimal subsequence at iteration k if there are no reserves within the interval $\overline{1, g-1}$ and for all marked tasks $j^*_{[m]}\left(j^{**}_{[m]}\right)$ within this interval we have: $l_{j^*_{[m]}}(l_{j^{**}_{[m]}}) \le l_{j_{[g]}}$, $\forall m = \overline{1, g-1}$.

We used the equality of the tasks' weights in modification of Statement 5.22.

Statement 5.23 (modification of Statement 4.23.) Suppose that we have constructed an optimal subsequence σ^k on the set of tasks within positions $\overline{1, g-1}$ at the previous iteration. If the competing task $j_{[g]}$ such that $l_{j_{[g]}} \ge l_{j_{[i]}}$, $i \in \overline{1, g-1}$, has not used the existing reserves during permutations and insertions, the minimum functional value corresponds to the position g occupied by this task, and there are no tasks marked with "*" ("**"), then we can eliminate the task $j_{[g]}$ from the set of competing tasks, and the current subsequence is optimal on the set of tasks within positions $\overline{1, g}$.

We used ordering of tasks in the sequence σ^{ord} by non-decreasing processing times and equality of their weights in modification of Statement 5.23.

Statement 5.24 (modification of Statement 4.24.) Suppose that we have already executed $k - 1$ iterations and constructed an optimal subsequence σ^{k-1} on the set of tasks within positions $\overline{1, g-1}$. Let us proceed to the next competing task $j_{[g]}$ from the sequence σ^{fp}, $l_{j_{[g]}} \ge l_{j_{[i]}}$, $i \in \overline{1, g-1}$, and formulate the conditions that allow us to shorten its interval of insertion (the interval where we achieve the minimum possible increase in the total tardiness of the current subsequence in comparison with the functional value at iteration $k - 1$). If there is an unmarked tardy task or a task with zero reserve in the position of insertion p of the task $j_{[g]}$ and there are no tasks within positions $\overline{1, p-1}$ corresponding to conditions (5.6) and tasks marked with "*" ("**"), then we must insert the task $j_{[g]}$ into the position $p + 1$. If similar conditions are satisfied in the position $p + 1$, we must insert the task $j_{[g]}$ into the position $p + 2$, etc., up to the position g.

The modification in Statement 5.24 is based on Statements 5.15 and 5.16.

Statement 5.25 (modification of Statement 4.25.) Suppose that we have already executed $k - 1$ iterations and constructed an optimal subsequence σ^{k-1} on the set of tasks in positions $\overline{1, g-1}$. Let us proceed to the next competing task $j_{[g]}$ from the sequence σ^{fp}, $l_{j_{[g]}} \geq l_{j_{[i]}}$, $i \in \overline{1, g-1}$. If during the iteration of optimization we either have $d_{j_{[i]}} \leq d_{j_{[g]}}$ for all preceding non-tardy tasks $j_{[i]}$, or there are no reserves within positions $\overline{1, g-1}$, then the task $j_{[g]}$ remains in its position that corresponds to the minimum possible increase in the total tardiness of the current subsequence in comparison with the functional value on the iteration $k - 1$. This task cannot take position less than g in an optimal schedule, and the current subsequence is optimal within positions $\overline{1, g}$.

The modification in Statement 5.25 is based on Statement 5.4.

Statement 5.26 (modification of Statement 4.26.) If during the insertion of a tardy task $j_{[g]}$ in an optimal subsequence σ^k there are no preceding tasks $j_{[i]}$ within positions $\overline{1, g-1}$ for which $d_{j_{[i]}} > d_{j_{[g]}}$ and $d_{j_{[i]}} > C_{j_{[i]}}$, but there are such tasks $j^*_{[m]}$ ($j^{**}_{[m]}$) that $l_{j^*_{[m]}}$ ($l_{j^{**}_{[m]}}$) $\geq l_{j_{[g]}}$, then the minimum possible increase in the total tardiness of the current subsequence in comparison with the functional value at the previous iteration and an optimal subsequence at the current iteration are achieved only due to reserves released by the tasks $j^*_{[m]}$ ($j^{**}_{[m]}$) during their insertions into the position after the task $j_{[g]}$. Also, the following is necessary:

$$C_{j_{[g]}} - d_{j^*_{[m]}} < \sum_{i=m}^{g} \max\left(0, C_{j_{[i]}} - d_{j_{[i]}}\right). \tag{5.7}$$

We used ordering of tasks in the sequence σ^{ord} by non-decreasing processing times and equality of their weights in modification of Statement 5.26.

Statement 5.27 (modification of Statement 4.27.) If a competing task $j_{[g]}$, $l_{j_{[g]}} \geq l_{j_{[i]}}$, $i \in \overline{1, g-1}$, as a result of the iteration of its optimization that determines the possibility of using reserves for this task, took the position $k > p$, then the following competing tasks $j_{[l]}$, for which $d_{j_{[l]}} \geq d_{j_{[g]}}$, $l_{j_{[l]}} \geq l_{j_{[g]}}$, can not take a position lower than $k + 1$ in an optimal schedule.

The modification in Statement 5.27 is based on known statement from [32]. We used ordering of tasks in the sequence σ^{ord} by non-decreasing processing times and equality of their weights in the modification.

Statement 5.28 (Potts and Van Wassenhove [33]; modification of Statement 4.28.) Suppose $l_{j_{[l]}} \le l_{j_{[r]}}$, $d_{j_{[l]}} \le d_{j_{[r]}}$ for some tardy tasks $j_{[l]}, j_{[r]} \in J$ in the current subsequence σ^k. Then the task $j_{[l]}$ will precede the task $j_{[r]}$ in an optimal schedule.

We used ordering of tasks in the sequence σ^{ord} by non-decreasing processing times and equality of their weights in the modification of Statement 5.28.

Statement 5.29 (modification of Statement 4.29.) Suppose that we have already executed $k-1$ iterations and constructed an optimal subsequence σ^{k-1} on the set of tasks within positions $\overline{1, g-1}$. If the next competing task $j_{[g]}$, $l_{j_{[g]}} \ge l_{j_{[i]}}$, $i \in \overline{1, g-1}$, after executing its optimization iteration, remains in its position that corresponds to the minimum possible increase in the total tardiness of the current subsequence in comparison with the functional value at the previous iteration; there are no tasks marked with "*" ("**") within positions $\overline{1, g-1}$; and all tasks within positions $\overline{g+1, n}$ are tardy, then we can eliminate the task $j_{[g]}$ and tasks $j_{[i]}$ such that $d_{j_{[i]}} \ge d_{j_{[g]}}$, $i \in \overline{g+1, n}$, from the number of tasks competing for reserves. These tasks cannot take earlier positions in an optimal schedule, and the current subsequence is optimal within positions $\overline{1, g}$.

The modification in Statement 5.29 is based on known statement from [32].

Statement 5.30 (modification of Statement 4.30.) Suppose that we have already executed $k-1$ iterations and constructed an optimal subsequence σ^{k-1} on the set of tasks within positions $\overline{1, g-1}$. The tardy task $j_{[g]}$ at iteration k can take earlier position, which will lead to the minimum possible increase in the total tardiness of the current subsequence in comparison with the functional value at the previous iteration and to the construction of an optimal subsequence at the current iteration, only if at least one task $j_{[l]}$ within positions $\overline{1, g-1}$ has a reserve and $d_{j_{[l]}} > d_{j_{[g]}}$ or (if there are no reserves) there are marked tasks within positions $\overline{1, g-1}$.

The modification in Statement 5.30 is based on Statement 5.16.

Statement 5.31, SSO #8 (modification of Statement 4.31.) Suppose that we have already executed $k-1$ iterations and constructed an optimal subsequence σ^{k-1} on the set of tasks within positions $\overline{1, g-1}$. If at iteration k in the subsequence σ^k the competing task $j_{[g]}$, $l_{j_{[g]}} \ge l_{j_{[i]}}$, $i \in \overline{1, g-1}$, as a result of the optimization iteration for it, has not used the existing reserves; the minimum possible increase in the total tardiness of the current subsequence in comparison with the functional value at the previous iteration corresponds to the position g of this task; there are no tasks marked with "*" ("**") within positions $\overline{1, g-1}$ or the conditions of Statement 5.26 are not satisfied for the marked tasks; and $\forall j_{[r]}$, $r = \overline{g+1, n}$, $d_{j_{[r]}} \ge d_{j_{[g]}}$, and $C_{j_{[r]}} \ge d_{j_{[r]}}$, then we can eliminate the tasks $j_{[r]}$ from the set of competing tasks. These tasks remain in their positions, and the current subsequence is optimal on the entire set of tasks.

The modification in Statement 5.31 is based on Statement 5.16.

We check the following SSO #9 immediately after the next competing task appending to the current subsequence.

Statement 5.32, SSO #9 (modification of Statement 4.32.) Suppose that we have already executed $k - 1$ iterations and constructed an optimal subsequence σ^{k-1} on the set of tasks within positions $\overline{1, g-1}$. Suppose that at iteration k executed for the next competing task $j_{[g]}$, for all tasks $j_{[l]}$, $l = \overline{1, g-1}$, we have $l_{j_{[g]}} \geq l_{j_{[l]}}$, $d_{j_{[l]}} \leq C_{j_{[l]}}$; for all marked tasks $j^*_{[l]}$ ($j^{**}_{[l]}$) we have $l_{j^*_{[l]}}$ ($l_{j^{**}_{[l]}}$) $\leq l_{j_{[g]}}$; and there are no reserves within positions $\overline{g, n}$. Then we can eliminate tardy tasks within positions $\overline{g, n}$ from the set of competing tasks. They remain in their positions corresponding to the minimum possible increase in the total tardiness of the current subsequence in comparison with the functional value at the previous iteration. The current sequence corresponds to the optimal functional value on the entire task set.

The modification in Statement 5.32 is based on Statement 5.16.

Corollary 5.9 In Statements 5.29, 5.31, 5.32, we have given the conditions of eliminating one or several competing tasks from the set of competing tasks. We do not include them to the set again. Either all of the conditions of Statement 5.29, or all of the conditions of Statement 5.31, or all of the conditions of Statement 5.32 must be satisfied. For these tasks, we do not execute iterations of optimization, we can consider them as originated tardy tasks during the next iterations.

Statement 5.33 (modification of Statement 4.33.) Suppose that we have already executed $k - 1$ iterations and constructed an optimal subsequence σ^{k-1} on the set of tasks within positions $\overline{1, g-1}$. Suppose that at iteration k for the next competing task $j_{[g]}$, $l_{j_{[g]}} \geq l_{j_{[i]}}$, $i \in \overline{1, g-1}$, there is a tardy task $j_{[r]}$ within the interval of insertion of the task $j_{[g]}$ such that $d_{j_{[r]}} < d_{j_{[g]}}$, $l_{j_{[r]}} < l_{j_{[g]}}$. Then, we can decompose the current subsequence and skip consideration of tasks within positions $\overline{1, r}$ at this iteration of optimization. We execute the optimization (i.e., the minimization of the increase in the total tardiness of the current subsequence in comparison with the functional value at the previous iteration and the construction of an optimal subsequence at the current iteration) within positions $\overline{r+1,\ g-1}$.

The modification in Statement 5.33 is based on Statement 5.16.

Statement 5.34 (modification of Statement 4.34.) Suppose that we have already executed $k - 1$ iterations and constructed an optimal subsequence σ^{k-1} on the set of tasks within positions $\overline{1, g-1}$. We execute iteration k for the next competing task $j_{[g]}$, $l_{j_{[g]}} \geq l_{j_{[i]}}$, $i \in \overline{1, g-1}$. Suppose that the previous competing task in the subsequence σ^k is a task $j_{[q]}$, $q < g$, we have $d_{j_{[g]}} < d_{j_{[q]}} < C_{j_{[p^{new}]}} - l_{j_{[p^{new}]}}$ where p^{new} is the position of insertion of the task $j_{[q]}$, and there are no reserves within positions $\overline{1, p^{new}}$. If the task $j_{[q]}$ took the position $s > p^{new}$ as a result of this iteration of optimization, then the position of the task $j_{[g]}$ in an optimal schedule (the position corresponding to the minimum increase in the total tardiness of the current subsequence in comparison with the functional value at the previous iteration) cannot be less than $s + 1$.

The modification in Statement 5.34 is based on Statement 5.16.

Statement 5.35, SSO #10 (modification of Statement 4.35.) Suppose that $j_{[g]}$ is the first tardy task in the sequence σ^{ord}, tasks within positions $\overline{g+1,\ n}$ are tardy, and at least one of the following conditions holds:

1. $\max R\left(\sigma^{ord}\right) \le l_{j_{[g]}}$;
2. $\forall l, r \in \overline{g,\ n}$, $l < r$, we have $d_{j_{[l]}} < d_{j_{[r]}}$.

In this case, an optimal solution is obtained by a polynomial subalgorithm with the complexity of each iteration determined by the function $O(n_{curr} \log n_{curr})$ where n_{curr} is the number of tasks optimized at the current iteration (Remark 5.1 below).

We used ordering of tasks in the sequence σ^{ord} by non-decreasing processing times and equality of their weights in the modification of Statement 5.35.

Statement 5.36, SSO #11 (modification of Statement 4.36.) Suppose that $j_{[g]}$ is the first competing task in the sequence σ^{fp}. If there are no reserves in this sequence within positions $\overline{1, p-1}$ where p is the position of insertion of the task $j_{[g]}$; the task $j_{[g]}$ remains tardy in position p; for each pair of tasks $j_{[s]}, j_{[t]}$, $s < t$, within positions $\overline{p, g-1}$, we have $l_{j_{[s]}} \le l_{j_{[t]}}$, $d_{j_{[s]}} \le d_{j_{[t]}}$; and $\forall j_{[r]}$, $r = \overline{g+1, n}$, we have $d_{j_{[r]}} \le C_{j_{[r]}}$, $d_{j_{[r]}} \ge d_{j_{[r-1]}}$, then an optimal solution is obtained by a polynomial subalgorithm with the complexity of each iteration determined by the function $O(n_{curr})$ where n_{curr} is the number of tasks optimized at the current iteration (Remark 5.1 below).

We used ordering of tasks in the sequence σ^{ord} by non-decreasing processing times and equality of their weights in the modification of Statement 5.36.

Statement 5.37, SSO #12 (modification of Statement 4.37.) Suppose that $j_{[g]}$ is the first competing task in the sequence σ^{fp}. If all tasks within positions $\overline{1, g-1}$ are in non-increasing order of their processing times; $\max R_i \le l_{j_{[g]}}$, $i = \overline{1, g-1}$; $\forall j_{[s]}$, $s = \overline{g+1, k}$, we have $d_{j_{[s]}} < C_{j_{[s]}}$, $l_{j_{[s]}} \ge l_{j_{[g]}}$; and $\forall j_{[s]}$, $s = \overline{k+1, n}$, we have $d_{j_{[s]}} \ge C_{j_{[s]}}$ where $k < n$, then an optimal solution is obtained by a polynomial subalgorithm with the complexity of each iteration determined by the function $O(n_{curr} \log n_{curr})$ where n_{curr} is the number of tasks optimized at the current iteration (Remark 5.1 below).

We used ordering of tasks in the sequence σ^{ord} by non-decreasing processing times and equality of their weights in the modification of Statement 5.37.

Remark 5.1 If any of SSOs #10–12 is fulfilled, then an optimal solution is obtained by a corresponding polynomial subalgorithm. Each iteration of the subalgorithm has corresponding complexity: $O(n_k)$ or $O(n_k \log n_k)$ where n_k is the number of tasks optimized at iteration k of the subalgorithm. We propose to readers to describe the polynomial subalgorithms as an exercise. The complexity of the problem solving on the entire set of tasks is determined by the function $O(n^2)$ or $O(n^2 \log n)$.

Statement 5.38 (Sufficient Sign of Polynomial Solvability of the Current Iteration, SSPS) (modification of Statement 4.38.) Suppose that we have already executed $k - 1$ iterations and constructed an optimal subsequence σ^{k-1} on the set of tasks within positions $\overline{1, g - 1}$. Suppose that all of the following conditions hold at the current iteration of optimization during the insertion of the current tardy (competing or originated) task $j_{[g]}$:

- actual position of insertion p^{new} determined during the algorithm execution is greater than the position of insertion of the task last inserted at the previous steps of the algorithm;
- there are no tasks that correspond to the conditions of Statement 5.12 within the interval of insertion of the task $j_{[g]}$;
- there are no tasks $j_{[l]}$ such that $d_{j_{[l]}} > d_{j_{[g]}}$, $d_{j_{[l]}} > C_{j_{[l]}}$, $d_{j_{[l]}} > C_{j_{[p^{new}-1]}}$ within positions $\overline{1, p^{new} - 1}$;
- for tasks $j^*_{[m]}$ ($j^{**}_{[m]}$) within positions $\overline{1, p^{new} - 1}$, the following condition (that allows to move these tasks into later positions) does not hold:

$$C_{j_{[l]}} - d_{j^*_{[m]}} < \sum_{i=m}^{p^{new}} \max\left(0, C_{j_{[i]}} - d_{j_{[i]}}\right)$$

where the position l corresponds to the processing time of $j^*_{[m]}$ ($j^{**}_{[m]}$) (if $l_{j^*_{[m]}}$ (or $l_{j^{**}_{[m]}}$) is greater than $l_{j_{[p^{new}]}}$, then $l = p^{new}$). Then we achieve a solution at the current iteration by a polynomial subalgorithm with the complexity $O\left(n^2_{curr}\right)$ where n_{curr} is the number of tasks optimized at the current iteration.

The modification in Statement 5.38 is based on Statement 5.16.

Statement 5.39 (coincides with Statement 4.39.) The first polynomial component of the PSC-algorithm is implemented by:

1. The algorithmic procedures of verifying the SSOs #1–5 and #7 (Theorem 5.1, Corollaries 5.4 and 5.8, Statements 5.1, 5.4, and 5.11) that are executed in the general algorithm before the optimization iterations.
2. If item 1 is not fulfilled, then the algorithmic procedures of verifying the SSOs #10–12 that include the execution of subalgorithms with polynomial complexity, as shown above in Statements 5.35–5.37 and Remark 5.1.
3. If items 1 and 2 are not fulfilled, then during the algorithmic procedures that implement each current iteration of optimization we verify the following. If SSO #9 (Statement 5.32) is fulfilled at the beginning of the current iteration of optimization, then we have obtained an optimal solution on the entire set of tasks and we do not execute the current and subsequent iterations of optimization. If SSO #9 is not fulfilled, then fulfillment of SSPS (Statement 5.38) is required. If SSPS is true, then the current iteration of optimization is implemented by a polynomial algorithm with the complexity $O\left(n^2_{curr}\right)$ where n_{curr} is

the number of tasks optimized at the current iteration. If SSO #6 or SSO #8 (Corollary 5.7, Statement 5.31) is fulfilled after the current iteration of optimization, then we have obtained an optimal schedule on the entire set of tasks by the polynomial subalgorithm and do not execute the subsequent iterations of optimization.

Remark 5.2 At the first iteration of optimization, fulfillment of SSPS (Statement 5.38) is required, then we perform a verification of SSOs #6 and #8 (Corollary 5.7, Statement 5.31).

Remark 5.3 We consider Statements 5.35–5.37 as the part of the first polynomial component of the PSC-algorithm that verifies the SSOs. The rationale is the following. Statement 5.39 is about fulfillment of the first polynomial component of the PSC-algorithm. The second paragraph of the statement is included there because we check the conditions of Statements 5.35–5.37 by polynomial algorithms with $O(n^2)$ or $O(n^2 \log n)$ complexity. If the conditions are satisfied, we obtain an optimal solution without using the exact subalgorithm that has enumerative complexity. Thus, we consider a part of the first polynomial component all conditions and polynomial algorithms for their checkup that are executed without the use of iterative optimization procedures of the universal exact subalgorithm.

Statement 5.40 (coincides with Statement 4.40.) The following is true:

1. If SSO #9 (Statement 5.32) is fulfilled at the beginning of the current iteration of optimization, then we do not execute the current and subsequent iterations of optimization: we have obtained an optimal solution;
2. If item 1 is not fulfilled, then SSPS is verified during the current iteration of optimization. If SSPS is true, then the current iteration of optimization is implemented by a polynomial algorithm (Statement 5.38).
3. If, after the current iteration of optimization, regardless of SSPS fulfillment, SSO #6 or SSO #8 (Corollary 5.7, Statement 5.31) is fulfilled, then we have obtained an optimal schedule on the entire set of tasks and do not execute the subsequent iterations of optimization.
4. If SSPS is true at each iteration, then a sufficient condition is satisfied that turns the exact subalgorithm of the PSC-algorithm into a polynomial complexity algorithm.

Lemma 5.2 (coincides with Lemma 4.2.) *Condition 4 of* Statement 5.40 *is satisfied for non-empty set of the problem instances.*

Statement 5.41 (coincides with Statement 4.41.) Procedures F and I implement all types of permutations based on conditions 1–6 of Statement 5.16 that decrease the functional value in the current subsequence. Also, the procedures define the rules for the permutations.

Statement 5.42 (coincides with Statement 4.42.) The current iteration of optimization, based on directional permutations and implemented through a finite number of operations of optimization procedures D–I executed for the current competing task in the current subsequence and recursively executed to reorder tasks within the intervals of insertions and intervals of permutations, is finite.

The following statement is an obvious corollary of the above given logic of the algorithm construction (for the case when none of the SSOs and the SSPS condition are satisfied during the algorithm execution) and also of Statements 5.2, 5.3, 5.5–5.10, 5.13–5.19, 5.21, 5.22, 5.24–5.29, 5.33, 5.34, 5.41, 5.42, and Lemma 5.1.

Statement 5.43 (coincides with Statement 4.43.) The algorithm yields an optimal schedule.

Corollary 5.10 As a result of the algorithm execution, we obtain one optimal schedule. To obtain the entire set of optimal schedules, it is sufficient to perform permutations of non-tardy tasks among themselves so that they remain non-tardy, without changing the position of another tasks.

5.2.2 Modification of the PSC-Algorithm and Procedures A–I of the TWT Problem for the TT Problem

As a result of the modification, we have changed the implementation of the optimization unit (Procedure F) due to changes in conditions of Statements 4.4–4.6, 4.8, 4.9, 4.11–4.13, 4.15, 4.16, 4.18–4.20, 4.22–4.38. The changes have simplified implementation of the algorithm for the case of equal weights.

The algorithm for TT problem solving corresponds to that given in Sect. 4.3.1 for the TWT problem (algorithms A0 and A1) with the following modifications:

- assume the weights of all tasks to be equal to one;
- use the value $1/l_j$ as the priority of a task j;
- check the conditions of statements of this chapter corresponding to those mentioned in the algorithm in Chap. 4;
- in Algorithm A1:
  - use $d_{j_{[p^{new}]}} > d_{j_{[g]}}$ instead of $d_{j_{[p^{new}]}} > d_{j_{[g]}} - l_{j_{[g]}}$ in Eq. (4.12) at step 3.3;
  - use $d_{j_{[r]}} > d_{j_{[p^{new}]}}$ instead of $d_{j_{[r]}} > d_{j_{[p^{new}]}} - l_{j_{[p^{new}]}}$ in Eq. (4.13) at step 4.1;
  - use $l_{j_{[r]}} > l_{j_{[p^{new}]}}$ instead of $p_{j_{[r]}} < p_{j_{[p^{new}]}}$ in Eq. (4.14) at step 4.1;
  - at step 5, check the condition $d_{j[i]} > d_{j_{[p^{new}]}}$ instead of $d_{j[i]} > d_{j_{[p^{new}]}} - l_{j_{[p^{new}]}}$ for all tasks $j_{[i]}$;
  - use $d_{j_{[l]}} > d_{j_{[p^{new}]}}$ instead of $d_{j_{[l]}} > d_{j_{[p^{new}]}} - l_{j_{[p^{new}]}}$ at step 6.3.

Procedures A–I executed during the TT problem solving functionally coincide with those for the TWT problem solving (Sect. 4.3.3). We have simplified their implementation (the conditions for their execution) taking into account modifications in the corresponding statements.

In Procedure A (ordering, step 2 of Algorithm A0), we arrange the tasks of the given subsequence in non-increasing order of their processing times, according to Definition 5.1.

Procedure B (free permutations, step 3 of Algorithm A0) is unchanged, according to Definitions 4.9, 4.10.

Procedure C (determination of the set of competing tasks in the sequence σ^{fp}, step 3.4 of Algorithm A0) works in accordance with Definition 5.2.

Procedure D (determination of the interval of insertion for the current tardy task in the current sequence, Definition 4.6, step 3 of Algorithm A1, modified Statement 5.18): we determine the interval of insertion as the position at which the tardiness of the task will be minimum or equal to zero. The implementation of the procedure is unchanged accounting only that all tasks' weights equal to one.

Procedure E (analysis of reserves at the interval of insertion for the current tardy (competing or originated) task, step 4 of Algorithm A1) is unchanged.

Procedure F (optimization, i.e., decreasing the tardiness of the current tardy task due to the use of reserves within the extended interval of insertion, step 6 of Algorithm A1): we have changed the conditions for extending the interval of insertion and for optimization within the extended interval of insertion according to the conditions 2, 4, 5 of the modified Statement 5.16 and modified Statements 5.4, 5.12, 5.13.

Procedure G (adjustment of the position of insertion to reduce the interval of insertion, step 5 of Algorithm A1): we check conditions $d_{j[i]} > C_{j[i]}$ and $d_{j[i]} > d_{j_{[p^{new}]}}$ for all tasks $j_{[i]}$ within the interval $\overline{p^{new}+1,\ g}$.

Procedure H (ordering and free permutations within the extended interval of insertion, step 7 of Algorithm A1) is unchanged.

Procedure I (optimization within the interval of permutation of marked tasks into the later positions; the reserves released as a result of the permutation are used by tardy tasks; conditions 5 and 6 of modified Statement 5.16, step 8 of Algorithm A1): the implementation of the procedure is unchanged accounting only that all task weights equal to one.

The block diagram of the PSC-algorithm coincides with Figs. 4.1–4.2 taking into account that all task weights equal to one.

We propose to readers to describe the corresponding algorithm as an exercise.

5.2.3 Properties of the First Polynomial Component of the PSC-Algorithm for the TT Problem Solving

The first polynomial component of the PSC-algorithm for the TT problem solving (Sect. 4.3.2) and the block diagram of its description (Fig. 4.3) functionally coincide with the first polynomial component of the PSC-algorithm for the TWT problem solving. But we have simplified its implementation due to the simplification of Statements 5.35–5.39. Satisfying the conditions of these statements, we can achieve an optimal solution of the problem in a polynomial time.

5.2.4 Properties of the Exact Subalgorithm for the TT Problem Solving

Let us investigate the efficiency of the exact subalgorithm for the problem solving on the basis of both the results of theoretical studies and a qualitative analysis of its execution logic.

1. The upper bound for the computational complexity of the exact subalgorithm is $O((n_R+n_K)!)+O(n^2)$. Here, $O(n^2)$ is the complexity of the sequence σ^{fp} construction; n_K the number of competing tasks in the sequence σ^{fp} (the set K); n_R the number of tasks with time reserves in σ^{fp} (the set R).

This upper bound is a real estimate of the computational complexity of the PSC-algorithm if for all competing tasks in σ^{fp}:

(a) $\forall i<g,\ j_{[i]},\ j_{[g]}\in K,\ d_{j_{[i]}}>d_{j_{[g]}};$ (5.8)

(b) processing times and due dates of competing tasks differ insignificantly;
(c) there are tasks $j_r\in R$ in σ^{fp}, and for some of them we have performed free permutations (see Definition 4.9);
(d) the condition (5.8) is satisfied also for tasks with reserves:

$$\forall i<g, j_{[i]}, j_{[g]}\in R, d_{j_{[i]}}>d_{j_{[g]}}. \quad (5.9)$$

Substantiation The less the processing times and the due dates of tasks differ, the more we consider different combinations of permutations. We exhaust the reserves slower and eliminate the less number of unpromising permutations during F and I optimization procedures execution. The reason is: the parameters of all tasks are practically the same, so we repeat similar permutations for each competing task (Statements 5.4, 5.13, 5.16, 5.26).

The presence of tasks $j_r\in R$ in σ^{fp} for which we performed a free permutation causes the combinatorial search within the intervals of insertion after their insertion and subsequent ordering of tasks by the processing times. These tasks again will gain the reserves which will be used for each tardy task at the optimization iteration. And since the due dates differ insignificantly, this combinatorial search will be performed on almost the same subsequences. The less the difference in due dates, the more number of subsequences upper-bounded by the due date are considered.

If (5.8) is true, the position of insertion of the task $j_{[g]}$ will be less than that of the task $j_{[i]}$ (because $d_{j_{[i]}}>d_{j_{[g]}}$). The interval of insertion $I_{j_{[g]}}$ increases in comparison with $I_{j_{[i]}}$ by the value of the interval between the positions of their insertion. Most of the originated tardy tasks belonging to the interval $I_{j_{[g]}}$ will also belong to $I_{j_{[i]}}$ (since their processing times differ insignificantly). Because of this, optimization procedures F and I will be executed again for them after the ordering procedure H (first at the iteration for the task $j_{[g]}$ and then for the task $j_{[i]}$).

If, on the other hand, (5.9) is true, we extend the interval of insertion (including to it the additional tasks for which we meet the permutation conditions) for each competing task. As a result, we will have to perform all optimization procedures on the entire set of tasks.

Thus, when the above conditions 1 (a–d) are true, we will statistically significantly face the problem instances which real complexity is commensurate with an exhaustive search. ∎

Remark 5.4 If only conditions 1 (a–c) are true and condition 1 (d) is false, then the complexity of the current optimization iteration is also quite high.

The main difference between the total tardiness and total weighted tardiness problems is that the competing tasks are in non-decreasing order of their processing times in the TT problem. In Statements 5.4, 5.6, 5.16, 5.18, 5.19, 5.22, 5.24–5.29, 5.33, 5.34 we check the less number of conditions than for the weighted tardiness problem. As a result, the conditions of these statements fulfill more often. This leads to a reduction in the number of permutations performed and an increase in the number of their eliminations.

We check the conditions 1 (a–d) when we analyze the sequence σ^{fp}. If we meet these conditions, the above upper bound becomes a real estimate for the computational complexity of the PSC-algorithm.

2. Overall efficiency of the exact subalgorithm is determined by the following factors:

2.1. At the optimization stage, we decompose the sequence σ^{fp} to a subsequences of a smaller size. We execute the optimization on the subsequence bounded with the position of insertion of the current competing task and the position of this task in the sequence σ^{fp}. During the solving, we can include to this subsequence only tasks that create the reserves at the interval of insertion of the current competing task. In this way we decompose the problem to subproblems of a smaller size.

2.2. During the problem solving we check the conditions of Statements 5.29, 5.31, and 5.32 that allow to eliminate some tasks from the set of competing tasks. In Statements 5.2, 5.4, 5.6, 5.16, 5.18, 5.19, 5.21, 5.22, 5.24–5.29, 5.33, 5.34, we formulate the rules for eliminating the permutations that do not lead to an optimal solution. In Statement 5.40 we show the conditions for significant reduce in the complexity of the exact subalgorithm.

2.3. Experimental studies as a whole have shown that if conditions 1 (a–d) are false, then the search is insignificant during solving the problem instances. We will have to execute the exhaustive search for some subsequences only in special cases of the problem (if conditions 1 (a–d) are true).

2.4. The simplest problem instances (the conditions under which the combinatorial search is significantly reduced) are:

(a) there are no tasks for extending the interval of insertion for all competing tasks in the sequence σ^{fp} (see Statements 5.5 and 5.10) and the following is true for all tasks with reserves within intervals of their insertion:

$$\max_{j_i \in R}\left(d_{j_i} - C_{j_i}\right) < \min_{j_k \in K} l_{j_k} \tag{5.10}$$

Substantiation Suppose that we insert a competing task j_k into the position belonging to a task $j_r \in R$, and there are no other tasks between the tasks j_k and j_r. In this case, it follows from Eq. (4.3) with all weights equal to one that:

$$\min\left(C_{j_k} - d_{j_k}, l_{j_r}\right) < l_{j_k} - R_{j_r}\ \forall j_k \in K, j_r \in R. \tag{5.11}$$

If condition (5.10) is true, then only one of the competing tasks can take an earlier position. It follows from (5.11) that after insertion of any competing task all tasks with reserves within its interval of insertion will become tardy. The problem reduces to the selection of a competing task and the search for such its position within its interval of insertion that the subsequent optimization minimally increases the functional value. Suppose that a task j_q has the $\min_{j_k \in K} l_{j_k}$. Tasks $j_i \in K$: $l_{j_i} > l_{j_q}$ can take an earlier position only if $d_{j_i} < d_{j_q}$. As a result, the algorithm will not include an enumeration of marked competing tasks in Procedure I. An optimal solution will be achieved by a subalgorithm with complexity $O(n_K n^2)$. We can statistically significantly solve instances of any practically possible dimension. ∎

We check the condition when we analyze the sequence σ^{fp}.

(b) the following is true for all competing tasks:

$$\forall j_i \in K,\ l_{j_1} \le l_{j_2} \le \cdots \le l_{j_{n_K}},\ d_{j_1} \le d_{j_2} \le \cdots \le d_{j_{n_K}}. \tag{5.12}$$

Substantiation In this case, competing tasks are arranged in the order $j_1, j_2, \ldots, j_{n_K}$ in an optimal schedule ([33], Statement 5.16). This significantly reduces the complexity of the algorithm execution because we do not consider competing tasks that took up earlier positions to release reserves during the optimization in Procedure I. The longer the processing time of each next competing task, the faster we exhaust the reserves (Eq. (4.3) with all weights equal to one), and the number of permutations are reduced. ∎

We check the condition when we analyze the sequence σ^{fp}.

(c) The number of tasks with productive reserves (satisfying $d_{j_{[i]}} > d_{j_{[k]}}$, $j_i \in R$, $j_{[k]} \in K$) is minimal.

Substantiation We exclude tasks with unproductive reserves from consideration during the problem solving (Statements 5.4 and 5.10). If the number of tasks with productive reserves is minimal, then we may have a small number of competing tasks for each of which there are productive reserves. Only for such competing tasks we will execute the current optimization iteration. This reduces the number of iterations. ∎

We check the condition when we analyze the sequence σ^{fp}.

We must meet conditions 2.4 (a–c) for all or for the majority of competing tasks and tasks with reserves in order to obtain a significant efficient reduction in the number of permutations performed.

The conditions presented in sub-item 2.4 (a) can be considered a heuristic condition that turns the exact subalgorithm into a polynomial one. These conditions are the subject of further theoretical research with an aim to make them theoretically grounded.

5.2.5 Recommendations for Using the Approximation Algorithm

Heuristics 1 and 2 given in Sect. 4.4.1 are applied to the TT problem unchanged.

Approximation Algorithm AA is the PSC-algorithm of the second class. Indeed, due to Heuristics 1 and 2 and the polynomial restriction on the amount of computations for the iterations of optimization of the exact algorithm, it has a polynomial complexity. We check the SSOs #1–5, 7, 10–12 before the optimization iterations (see Theorem 5.1, Corollaries 5.4 and 5.8, Statements 5.1, 5.4, 5.11, 5.35, 5.36, 5.37). If at least one of these SSOs is satisfied during the algorithm AA execution, then an obtained feasible solution is optimal. A distinctive feature of the algorithm is the following fact. If the solution is obtained only by the iterations of optimization of the exact algorithm, then the approximation algorithm guarantees an optimal solution obtaining. This situation may occur quite often, since in the general case the total number of competing tasks may decrease by more than one task after each iteration of optimization.

If analysis of the sequence σ^{fp} shows that conditions 2.4 (a–c) (Sect. 5.2.4) are met, then we recommend to solve the problem for an arbitrary n by the exact algorithm.

Due to the presence of factors 2.1 and 2.2 (Sect. 5.2.4), other conditions are possible which drastically reduce the computational complexity of the exact subalgorithm. We taken into account this possibility in the approximation algorithm by introducing a restriction on the computation time to execute the maximum possible number of iterations of the exact subalgorithm. We obtain the restriction from the introduced polynomial restriction on the number of operations taking into account the characteristics of the computing facilities.

If analysis of the sequence σ^{fp} shows that the conditions causing a time-consuming search (conditions 1 (a–d) or 1 (a–c), Sect. 5.2.4) are satisfied, then we necessarily recommend to solve the problem by the approximation algorithm.

5.2.6 *The Approximation Algorithm Modification*

The approximation algorithm for TT problem solving corresponds to that given in Sect. 4.4.2 for the TWT problem with the following modifications:

- assume the weights of all tasks to be equal to one;
- use the value $1/l_j$ as the priority of a task j;
- check the conditions of statements of this chapter corresponding to those mentioned in the algorithm in Chap. 4;
- at the main stage of the algorithm:
  - use $d_{j_{[p^{new}]}} > d_{j_{[g]}}$ instead of $d_{j_{[p^{new}]}} > d_{j_{[g]}} - l_{j_{[g]}}$ in Eq. (4.21) at step 3.3;
  - use $d_{j_{[r]}} > d_{j_{[p^{new}]}}$ instead of $d_{j_{[r]}} > d_{j_{[p^{new}]}} - l_{j_{[p^{new}]}}$ in Eq. (4.22) at step 4.1;
  - use $l_{j_{[r]}} > l_{j_{[p^{new}]}}$ instead of $p_{j_{[r]}} < p_{j_{[p^{new}]}}$ in Eq. (4.23) at step 4.1;
  - at step 5, check the condition $d_{j[i]} > d_{j[p^{new}]}$ instead of $d_{j[i]} > d_{j_{[p^{new}]}} - l_{j_{[p^{new}]}}$ for all tasks $j_{[i]}$;
  - use $d_{j_{[l]}} > d_{j_{[p^{new}]}}$ instead of $d_{j_{[l]}} > C_{j_{[p^{new}]}} - l_{j_{[p^{new}]}}$ at step 6.3.

We propose to readers to describe the corresponding algorithm as an exercise.

5.3 PSC-Algorithm for E/T 1 Problem Solving. Fixed Start Time of the Tasks Execution ($s = 0$)

5.3.1 *The First Polynomial Component of PSC-Algorithm for the Problem Solving*

SSO Formulation A sufficient sign of optimality of a feasible solution for E/T 1 problem is the absence of early tasks in an optimal solution obtained as a result of the corresponding TT problem solving.

Justification A schedule that minimizes only the total tardiness yields the lower bound of the minimal total tardiness that can be achieved. If the total earliness is zero in this schedule, then it is optimal by the E/T criterion (5.3).

Polynomial Algorithm That Checks the SSO The algorithm is based on the PSC-algorithm to solve the TT problem. The first polynomial component of the PSC-algorithm for E/T 1 is implemented by the PSC-algorithm for the TT problem. We impose a predetermined (by a polynomial function of the problem dimension) restriction on the amount of calculations to the exact subalgorithm of both PSC-algorithms. In the case of the restriction violation, the first polynomial component of the PSC-algorithm for E/T 1 problem is not fulfilled because we did not obtain an optimal solution for the TT problem.

5.3.2 The Approximation Algorithm for E/T 1 Problem Solving (Algorithm A1)

If the first polynomial component of the PSC-algorithm is not fulfilled, then we solve the E/T 1 problem with the approximation algorithm which is based on fundamentals of the exact subalgorithm constructing for the TT problem.

Algorithm A1 has the following two steps.

Step 1 is the construction of an optimal schedule σ^{fp} using the PSC-algorithm for the TT problem solving. If the total earliness is zero in the obtained schedule, then it is optimal by the E/T criterion.

Step 2 is the schedule σ^{fp} optimization by the E/T criterion. Step 2 consists of uniform iterations. At each iteration we check the possibility of decreasing the functional value due to longer tasks insertions into the positions of early tasks.

2.1. Find a next early task $j_{[p]}$ in the current schedule where it is in position p. Determine the set of longer tasks within interval $\overline{p+1,n}$:

$$L = \{i \in \overline{p+1,n} : l_{j_{[i]}} > l_{j_{[p]}}\}$$

2.2. Find for each task $j_{[k]} \in L$ the total functional value which the tasks within the interval of insertion of the task $j_{[k]}$ will have after its insertion into the position p:

$$\left| d_{j_{[k]}} - \left(C_{j_{[k]}} - \sum_{i=p}^{k-1} l_{j_{[i]}} \right) \right| + \sum_{i=p}^{k-1} \left| d_{j_{[i]}} - \left(C_{j_{[i]}} + l_{j_{[k]}} \right) \right|. \tag{5.13}$$

2.3. Select the task $j_{[k]} \in L$ corresponding to the minimum value of Eq. (5.13), so that the functional value maximally decreases after its insertion into the position p.
2.4. Move the task $j_{[k]}$ into the position p. Search for the next early task within the interval $\overline{p+1,n}$ in the obtained schedule. Find the new set of longer tasks that are candidates for the insertion and perform a similar insertion procedure. We do this until we consider all early tasks.
2.5. Reorder the tasks within the intervals of early and tardy tasks in the resulting schedule in accordance with Lemmas 5.3 and 5.4 [18] (see below) if this decreases the functional value.
2.6. If all early tasks were considered, then the algorithm terminates, else go to step 2.1.

If no permutation has been done during the algorithm execution, then the obtained schedule is "statistically optimal" by the E/T criterion (see the heuristic SSO #1 below). Next, we check the heuristic SSO #2 given below. If it is fulfilled, then the obtained schedule is also "statistically optimal" by the E/T criterion. The end of the algorithm.

Remark 5.5 Schedules are called "statistically optimal" if they completely coincide with an optimal solutions obtained by an exhaustive enumeration method.

Justification of the Algorithm A1 There are following known properties of the E/T minimization problem.

Lemma 5.3 (Ow and Morton [18]) *If tasks in a sequence are in non-decreasing order of processing times, and there are no early tasks, then such a sequence is optimal by the E/T criterion.*

Lemma 5.4 (Ow and Morton [18]) *If tasks in a sequence are in non-increasing order of processing times, and there are no tardy tasks, then such a sequence is optimal by the E/T criterion.*

The algorithm is based on the following heuristic.

Heuristic 1 We search for a solution with a fixed start time in the neighborhood formed by the set of sequences constructed as a result of directional permutations in an optimal sequence by the TT criterion. The permutations of longer tasks into positions of early tasks decrease the functional value (5.3). The permutations into positions of non-early tasks will increase tardiness. Permutations of tasks that shorter than early ones will certainly increase the functional value, according to Lemma 5.3.

Heuristic SSO #1 If there are no permutations of longer tasks into earlier positions occupied by early tasks that decrease the functional value (5.3) in a schedule optimal by the TT criterion, then this schedule is optimal by the E/T criterion.

Justification If none of the permutations reducing the functional value according to Eq. (5.13) was applied during the Algorithm A1 execution, then the schedule optimal by the TT criterion is also optimal by the E/T criterion.

Heuristic SSO #2 If tardy tasks hold positions 1, …, k in a schedule optimal by the TT criterion and early tasks arranged in non-increasing order of processing times hold positions $k + 1$, …, n, then this schedule is optimal by the E/T criterion.

Justification Permutations of tardy tasks into later positions will increase the tardiness of the tardy tasks and the earliness of the early tasks (since they will be moved into earlier positions in consequence of the permutations). Similarly, moves of early tasks into earlier positions will increase their earliness and the tardiness of tardy tasks (since they will be moved into later positions).

The complexity of Algorithm A1 is determined by the complexity of the algorithm for the TT problem.

5.3.3 Example for E/T 1 Problem Solving by Algorithm A1

Example 5.1 Initial data for Example 5.1 is in Table 5.1.

We solve this problem by the algorithm for the TT criterion with fixed start time of the tasks execution ($s = 0$) and obtain an optimal schedule (Table 5.2).

For each early task in this schedule we do the following. We find the set of longer tasks that qualify for a move. Then we select from them the task corresponding to the minimum value of Eq. (5.13). Finally, we insert the selected task into the position of the early task. We obtain the schedule shown in Table 5.3.

This schedule coincides with a schedule obtained by an exact algorithm.

Table 5.1 Initial data for Example 5.1

j	1	2	3	4	5	6	7	8
l_j	65	81	92	96	116	117	132	133
d_j	180	334	303	724	499	399	650	691

j the task number; l_j its processing time; d_j its due date

Table 5.2 A schedule optimal by the TT criterion at zero start time

j	l_j	d_j	C_j	T_j
1	65	180	65	
2	81	334	146	
3	92	303	238	
6	117	399	355	
5	116	499	471	
7	132	650	603	
8	96	724	699	141

The functional value $f = 141$

Table 5.3 A schedule optimal by the E/T criterion for Example 5.1

j	l_j	d_j	C_j	$\lvert C_j - d_j \rvert$
6	117	399	117	282
1	65	180	182	2
3	92	303	274	29
2	81	334	355	21
5	116	499	471	28
7	132	650	603	47
8	133	691	736	45
4	96	724	832	108

The functional value $f = 562$

5.4 PSC-Algorithm for E/T 2 Problem Solving. Start Time of the Tasks Execution Is in the Time Segment $[t_1, t_k]$

5.4.1 *The First Polynomial Component of PSC-Algorithm for E/T 2 Problem Solving*

Similarly to E/T 1 problem, the *SSO* for E/T 2 problem is the absence of early tasks in an optimal solution obtained by PSC-algorithm for the corresponding TT problem with the start time of tasks execution equal to t_1.

Justification Suppose we have an optimal by the E/T criterion schedule for an arbitrary start time of tasks execution $s > t_1$, $s \in [t_1, t_k]$. Shift the start time to t_1 and solve the problem by PSC-algorithm for the TT problem. The total tardiness in the obtained schedule cannot be greater than that for any $s > t_1$, and the total earliness is still zero. Therefore, this schedule is optimal by the E/T criterion for start time of tasks equal to t_1.

Hence, *the first polynomial component of PSC-algorithm for E/T 2 problem* converges to that for E/T 1 problem.

If the schedule obtained by the PSC-algorithm for the TT problem at $s = t_1$ does not satisfy the SSO, then we execute the *approximation algorithm* that includes Algorithms A1–A3. *Approximation Algorithm A3* to determine the actual start time s_a of the tasks execution in a given time interval is based on *Approximation Algorithm A2* for *E/T 3* problem solving (with not fixed start time of tasks execution).

5.4.2 *The Approximation Algorithm for E/T 3 Problem Solving (Algorithm A2: Determining the Latest Start Time of the Tasks Execution at Which the Minimum Functional Value Is Reached)*

Let $E_{\min} = \min_j E_j$, N_R be the number of early tasks, and N_T be the number of tardy tasks and the tasks that complete exactly in due date.

Algorithm A2 has the following three steps:

1. Build a schedule σ_0 by PSC-algorithm for the TT problem at the start time of tasks execution $s_0 = 0$ (similarly to E/T 1 and E/T 2 problems, we add a restriction on the amount of calculations into the exact subalgorithm for the TT problem). Increase the start time of tasks execution: $s_{i+1} = s_i + E_{\min}$.
2. Build a schedule optimal by the TT criterion for $s_{i+1} > 0$.
3. Optimize the schedule with Algorithm A1.

Step 1

1.1. Build an optimal by the TT criterion schedule σ_0 for $s_0 = 0$, set $i = 0$. If there are no early tasks in the schedule σ_0, then it is optimal by the E/T criterion, $s_{opt} = 0$, the algorithm terminates. Otherwise, go to step 1.2.
1.2. Determine values N_R, N_T, and $E_{\min}$ for the schedule σ_0.
1.3. If $N_R > N_T$ (or $N_R - N_T > 1$ if the number of tasks is odd), then go to step 1.4, else go to step 3.
1.4. $s_{i+1} = s_i + E_{\min}$; $i = i + 1$. Go to step 2.

Step 2 Use the PSC-algorithm for the TT problem to decrease the total tardiness in the schedule σ_0. If there are no early tasks in the obtained schedule σ_i, then it is optimal by the E/T criterion, $s_{opt} = s_i$, the algorithm terminates. Otherwise, if $N_R > N_T$ (or $N_R - N_T > 1$ if the number of tasks is odd) in the obtained schedule σ_i, then go to step 1.4, else go to step 3.
Step 3 Use Algorithm A1 to optimize the schedule σ_i at the obtained optimal value of $s_i = s_{opt}$. We obtain a schedule optimal by the E/T criterion. Then we check heuristic SSOs for Algorithm A1. If $N_R = N_T$ (or $N_R - N_T = 1$ if the number of tasks is odd) and $E_{\min} - k > 0$ in the obtained schedule, then we can increase the start time of the tasks execution by k, that is, we have a set of optimal start times with the same functional value: $M = \{s_{opt}, s_{opt} + 1, \ldots, s_{opt} + k\}$. The end of the algorithm.

Justification of Algorithm A2 The following statements are true.

Statement 5.44 If $N_R > N_T$ in some schedule, then increasing the start time of the tasks execution by $E_{\min}$ decreases the E/T functional value by $E_{\min}(N_R - N_T)$.

Proof Indeed, the total earliness is decreased by $E_{\min}N_R$ in this case, while the total tardiness is increased by $E_{\min}N_T$. Therefore, the functional value is decreased by $E_{\min}(N_R - N_T)$. ∎

Statement 5.45 Suppose $N_R > N_T$ in an optimal schedule obtained by PSC-algorithm for the TT problem. If there are no early tasks in the schedule obtained after increasing the start time of the tasks execution by $E_{\min}$ and optimization by the TT criterion, then such schedule is optimal by the E/T criterion for the current start time of the tasks execution.

Statement 5.46 The functional value decreases with the increasing of the start time of the tasks execution while $N_R > N_T$ is true. If $N_T > N_R$, then the functional value increases.

Proof is based on Statement 5.44. ∎

Statement 5.47 Suppose that $E_{\min} = k$ and $N_R = N_T$ (or $N_R - N_T = 1$ if the number of tasks is odd) in an optimal by the E/T criterion schedule. Then we can increase the start time of the tasks execution by k without a change in the functional value.

Proof This is true because the values of N_R and N_T do not change in this case. ∎

Statement 5.48 For any given start time of the tasks execution s_i we can build an optimal by the E/T criterion schedule on the basis of an optimal by the TT criterion schedule with the start time s_i.

5.4.3 *The Approximation Algorithm for E/T 2 Problem Solving*

Algorithm A2 is based on the following heuristic.

Heuristic 2 We do the search of the optimal start time of the tasks execution and an optimal by the E/T criterion schedule this way. We build an optimal by the TT criterion schedule σ_0 for the initial start time $s_0 = 0$ (step 1.1 of Algorithm A2). If $N_R \le N_T$ in σ_0, then the optimal start time of the tasks execution is $s_{opt} = 0$. If there are no early tasks in σ_0, then it is optimal by the E/T criterion. Otherwise, if $N_R > N_T$, we increase the start time of the tasks execution by $E_{\min}$ and optimize the obtained schedule by the TT criterion: we obtain a schedule σ_{new} (steps 1.3, 1.4 and 2 of Algorithm A2, Statements 5.44, 5.45, and 5.48). We do such increasing of the start time by $E_{\min}$ with the subsequent optimization by the TT criterion while $N_R > N_T$ (steps 1.3 and 1.4 of Algorithm A2, Statement 5.46). If $N_R \le N_T$ in σ_{new} (for an even number of tasks) or $N_R - N_T \le 1$ (for an odd number of tasks), then the obtained start time of the tasks execution is optimal and corresponds to an optimal by the E/T criterion schedule. $N_R < N_T$ can be true only if several tasks had the minimal earliness $E_{\min}$ before the last increase in the start time. We optimize the schedule σ_{new} by the E/T criterion using Algorithm A1 and check heuristic SSOs for Algorithm A1. If $N_R = N_T$ (or $N_R - N_T = 1$ if the number of tasks is odd) and $E_{\min} = k > 0$ in an obtained E/T schedule, then we can increase the start time of the tasks execution by k (step 3 of Algorithm A2, Statement 5.47).

Hence, a *heuristic SSO* for the latest start time of the tasks execution s_{opt} that corresponds to a minimal functional value (5.3) is the following condition: either $s_{opt} = 0$ and $N_R \le N_T$, or $s_{opt} > 0$ and $N_R \le N_T$ (for an even number of tasks) or $N_R - N_T \le 1$ (for an odd number of tasks) in an optimal by the TT criterion schedule obtained by Heuristic 2 (Example 5.4 below).

Remark 5.6 If $N_R = N_T$ in some schedule and increasing the start time of the tasks execution by $E_{\min}$ also yields a schedule with $N_R = N_T$, then we again increase the start time of the tasks execution by $E_{\min}$ (Example 5.3 below).

The complexity of Algorithm A2 is determined by the complexity of the algorithm for the TT problem.

Since we did not find other authors' results for comparison on the problems under consideration, we performed the following studies to evaluate the efficiency of Algorithm A2. For different start times of the tasks execution (with a given step of change) we built optimal by the E/T criterion schedules using an exact algorithm. The optimal schedules and the latest start times of the task execution obtained by Algorithm A2 coincided with the results obtained by exact algorithms.

Below we give examples of E/T 3 problem solving by the developed Algorithm A2, tables containing an optimal schedule and the latest start times of the task execution obtained by an exact algorithm, and graphs showing the relationships between the functional value (5.3) and the start times of the task execution.

5.4.4 Examples of E/T 3 Problem Solving by Algorithm A2

Example 5.2 Initial data for Example 5.2 is in Table 5.4.

We obtain an optimal schedule for this problem (Table 5.5) using the PSC-algorithm for the TT criterion with $s = 0$.

In this schedule: $N_R = 8$, $N_T = 4$, $E_{\min} = 7$. Set $s = 7$, solve the problem again by the TT criterion and obtain an optimal schedule (Table 5.6).

In this schedule: $N_R = 7$, $N_T = 5$, $E_{\min} = 1$. Set $s = 7 + 1 = 8$ (Table 5.7).

Table 5.4 Initial data for Example 5.2

j	1	2	3	4	5	6	7	8	9	10	11	12
l_j	50	79	80	83	88	96	100	102	121	130	147	150
d_j	500	534	700	604	987	951	900	668	528	605	537	999

Table 5.5 A schedule optimal by the TT criterion at zero start time

j	l_j	d_j	C_j	T_j	E_j
1	50	500	50		450
2	79	534	129		405
4	83	604	212		392
9	121	528	333		195
11	147	537	480		57
10	130	605	610	5	
3	80	700	690		10
8	102	668	792	124	
7	100	900	892		8
5	88	987	980		7
6	96	951	1076	125	
12	150	999	1226	227	

The functional value $f = 481$

Table 5.6 A schedule optimal by the TT criterion at $s = 7$

j	l_j	d_j	C_j	T_j	E_j
1	50	500	57		443
2	79	534	136		398
4	83	604	219		385
9	121	528	340		188
11	147	537	487		50
10	130	605	617	12	
3	80	700	697		3
8	102	668	799	131	
7	100	900	899		1
5	88	987	987	0	
6	96	951	1083	132	
12	150	999	1233	234	

The functional value $f = 509$

Table 5.7 A schedule optimal by the TT criterion at $s = 8$

j	l_j	d_j	C_j	T_j	E_j
1	50	500	58	0	442
2	79	534	137	0	397
4	83	604	220	0	384
9	121	528	341	0	187
11	147	537	488	0	49
10	130	605	618	13	0
3	80	700	698	0	2
8	102	668	800	132	0
7	100	900	900	0	0
5	88	987	988	1	0
6	96	951	1084	133	0
12	150	999	1234	235	0

The functional value $f = 514$

$N_R = N_T = 6$ in this schedule. Further increase in start time will lead to an increase in N_T. We select $s = 8$ and obtain a schedule optimal by the E/T criterion using Algorithm A1 at $s = 8$ (Table 5.8).

We obtained several optimal schedules for Example 5.2 (Table 5.9).

Example 5.3 Initial data for Example 5.3 is in Table 5.10.

We obtain an optimal schedule for this problem (Table 5.11) using the PSC-algorithm for the TT criterion with $s = 0$.

We have $N_R = 7$, $N_T = 3$, $E_{\min} = 42$. We set $s = 42$ (Table 5.12).

Now $N_R = 6$, $N_T = 4$, $E_{\min} = 15$. We set $s = 42 + 15 = 57$ (Table 5.13).

Table 5.8 A schedule optimal by the E/T criterion for Example 5.2

j	l_j	d_j	C_j	$\lvert C_j - d_j \rvert$
11	147	537	155	382
10	130	605	285	320
9	121	528	406	122
1	50	500	456	44
2	79	534	535	1
4	83	604	618	14
3	80	700	698	2
8	102	668	800	132
7	100	900	900	0
5	88	987	988	1
6	96	951	1084	133
12	150	999	1234	235

The functional value $f = 1386$

Table 5.9 Schedules optimal by the E/T criterion for Example 5.2 for different start times

s	σ_{opt}	f_{opt}	s	σ_{opt}	f_{opt}
0	11, 10, 9, 1, 2, 4, 3, 8, 7, 6, 5, 12	1307	10	11, 10, 9, 1, 2, 4, 3, 8, 7, 5, 6, 12	1289
1	The same as above	1305	20	The same as above	1329
2	The same as above	1303	30	The same as above	1369
3	11, 10, 9, 1, 2, 4, 3, 8, 7, 5, 6, 12	1301	40	The same as above	1409
4	The same as above	1297	50	The same as above	1449
5	The same as above	1293	60	11, 9, 2, 1, 10, 4, 3, 8, 6, 5, 7, 12	1492
6	The same as above	1289	70	The same as above	1506
7	The same as above	1285	80	11, 9, 2, 1, 4, 8, 3, 10, 6, 5, 7, 12	1527
8	The same as above	1285	90	The same as above	1535
9	The same as above	1287	100	11, 9, 2, 1, 4, 8, 3, 7, 6, 5, 10, 12	1541

Table 5.10 Initial data for Example 5.3

j	1	2	3	4	5	6	7	8	9	10
l_j	50	79	83	88	96	101	102	121	130	147
d_j	499	534	604	987	951	900	668	528	605	537

Table 5.11 A schedule optimal by the TT criterion at zero start time

j	l_j	d_j	C_j	T_j	E_j
1	50	499	50		449
2	79	534	129		405
3	83	604	212		392
8	121	528	333		195
10	147	537	480		57
9	130	605	610	5	
7	102	668	712	44	
6	101	900	813		87
5	96	951	909		42
4	88	987	997	10	

The functional value $f = 59$

Table 5.12 A schedule optimal by the TT criterion at $s = 42$

j	l_j	d_j	C_j	T_j	E_j
1	50	499	92		407
2	79	534	171		363
3	83	604	254		350
8	121	528	375		153
10	147	537	522		15
9	130	605	652	47	
7	102	668	754	86	
6	101	900	855		45
5	96	951	951	0	
4	88	987	1039	52	

The functional value $f = 185$

Table 5.13 A schedule optimal by the TT criterion at $s = 57$

j	l_j	d_j	C_j	T_j	E_j
1	50	499	107		392
2	79	534	186		348
3	83	604	269		335
8	121	528	390		138
10	147	537	537	0	
9	130	605	667	62	
7	102	668	769	101	
6	101	900	870		30
5	96	951	966	15	
4	88	987	1054	67	

The functional value $f = 245$

Table 5.14 A schedule optimal by the TT criterion at $s = 87$

j	l_j	d_j	C_j	T_j	E_j
1	50	499	137		362
2	79	534	216		318
8	121	528	337		191
10	147	537	484		53
3	83	604	567		37
7	102	668	669	1	
9	130	605	799	194	
6	101	900	900	0	
4	88	987	988	1	
5	96	951	1084	133	

The functional value $f = 329$

Here $N_R = N_T = 5$, $E_{\min} = 30$. We set $s = 57 + 30 = 87$ (Table 5.14).

$N_R = N_T = 5$ in this schedule. Further increase in start time will lead to an increase in N_T. We select $s = 87$ and obtain a schedule optimal by the E/T criterion using Algorithm A1 at $s = 87$ (Table 5.15).

$N_R = N_T = 5$ in this schedule. Task 1 has the minimum earliness $E_{\min} = 15$. This allows to increase the start time of the tasks execution by 15 (up to 103) without a change in the functional value. We obtained several optimal schedules for Example 5.3 (Table 5.16).

Example 5.4 Initial data for Example 5.4 is in Table 5.17.

We obtain an optimal schedule for this problem (Table 5.18) using the PSC-algorithm for the TT criterion with $s = 0$.

We have $N_R = 6$, $N_T = 3$, $E_{\min} = 23$. Set $s = 23$ (Table 5.19).

Table 5.15 A schedule optimal by the E/T criterion for Example 5.3

j	l_j	d_j	C_j	$\lvert C_j - d_j \rvert$
10	147	537	234	303
8	121	528	355	173
2	79	534	434	100
1	50	499	484	15
3	83	604	567	37
7	102	668	669	1
9	130	605	799	194
6	101	900	900	0
4	88	987	988	1
5	96	951	1084	133

The functional value $f = 957$

Table 5.16 Schedules optimal by the E/T criterion for Example 5.3 for different start times

s	σ_{opt}	f_{opt}	s	σ_{opt}	f_{opt}
0	10, 9, 8, 1, 2, 3, 7, 6, 5, 4	1095	102	10, 8, 2, 1, 3, 7, 9, 6, 4, 5	957
70	10, 8, 2, 1, 9, 3, 7, 6, 5, 4	982	103	The same as above	959
80	The same as above	966	104	The same as above	961
85	10, 8, 2, 1, 9, 3, 7, 6, 4, 5	960	110	The same as above	973
86	10, 8, 2, 1, 3, 7, 9, 6, 4, 5	958	120	10, 8, 1, 2, 3, 7, 9, 6, 4, 5	986
87	The same as above	957	130	The same as above	998
88	The same as above	957	140	10, 8, 1, 2, 3, 7, 9, 5, 4, 6	1020
90	The same as above	957	150	The same as above	1054
100	The same as above	957			

Table 5.17 Initial data for Example 5.4

j	1	2	3	4	5	6	7	8	9
l_j	65	81	92	96	116	118	132	133	150
d_j	180	334	303	724	500	400	650	499	691

Table 5.18 A schedule optimal by the TT criterion at zero start time

j	l_j	d_j	C_j	T_j	E_j
1	65	180	65		115
2	81	334	146		188
3	92	303	238		65
6	118	400	356		44
5	116	500	472		28
8	133	499	605	106	
4	96	724	701		23
7	132	650	833	183	
9	150	691	983	292	

The functional value $f = 581$

Table 5.19 A schedule optimal by the TT criterion at $s = 23$

j	l_j	d_j	C_j	T_j	E_j
1	65	180	88		92
2	81	334	169		165
3	92	303	261		42
6	118	400	379		21
5	116	500	495		5
8	133	499	628	129	
4	96	724	724	0	
7	132	650	856	206	
9	150	691	1006	315	

The functional value $f = 650$

Table 5.20 A schedule optimal by the E/T criterion for Example 5.4

j	l_j	d_j	C_j	$\lvert C_j - d_j \rvert$
3	92	303	115	188
1	65	180	180	0
2	81	334	261	73
6	118	400	379	21
5	116	500	495	5
8	133	499	628	129
4	96	724	724	0
7	132	650	856	206
9	150	691	1006	315

The functional value $f = 937$

$N_R = 5$, $N_T = 4$ in this schedule. We obtain a schedule optimal by the E/T criterion using Algorithm A1 at $s = 23$ (Table 5.20).

We obtained several optimal schedules for Example 5.4 (Table 5.21).

The corresponding correlation graphic is at Fig. 5.1.

Analysis of Statistical Studies We have carried out numerous studies of problems with up to 20 tasks. We obtained optimal schedules and the latest start times of the tasks execution using Algorithm A2. They coincided with the results obtained by exact algorithms.

We can conditionally split a curve on the graphs of correlation between the functional value (5.3) and the start time (e.g., Fig. 5.1) into sections of three types:

Table 5.21 Schedules optimal by the E/T criterion for Example 5.4 for different start times

s	σ_{opt}	f_{opt}	s	σ_{opt}	f_{opt}
0	6, 1, 3, 2, 8, 5, 7, 4, 9	938	13	3, 1, 2, 6, 8, 5, 4, 7, 9	954
1	The same as above	941	14	The same as above	953
2	The same as above	944	15	The same as above	952
3	The same as above	947	20	3, 1, 2, 6, 5, 8, 4, 7, 9	946
4	The same as above	950	22	The same as above	940
5	The same as above	953	23	The same as above	937
6	The same as above	954	24	The same as above	938
7	The same as above	955	25	The same as above	939
8	The same as above	956	30	The same as above	948
9	The same as above	957	35	2, 1, 3, 6, 5, 8, 4, 7, 9	963
10	3, 1, 2, 6, 8, 5, 4, 7, 9	957	40	1, 3, 2, 6, 5, 8, 4, 7, 9	971
11	The same as above	956	45	1, 3, 2, 6, 5, 7, 4, 8, 9	978
12	The same as above	955			

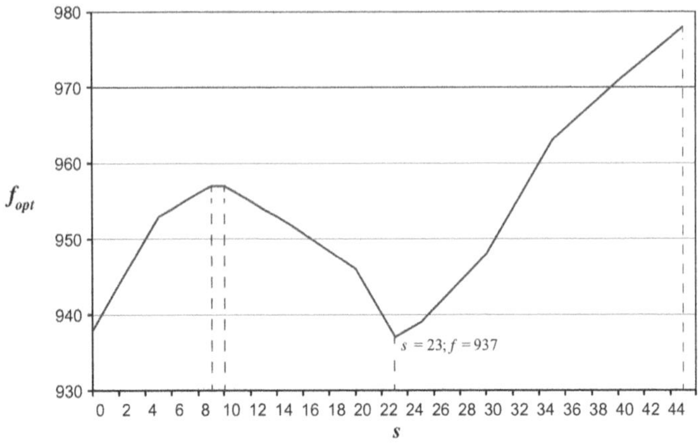

Fig. 5.1 Correlation between the functional value and the start times for Example 5.4

1. The functional value decreases with the increase in the start time. This occurs due to a decrease in the earliness of early tasks, since their number is greater than the number of tardy tasks (Statement 5.46, section [10, 23] in Fig. 5.1).
2. As the start time of the tasks increases, the functional value may increase in some cases, since the earliness of early tasks decreases and long tasks may not use the reserves to decrease their tardiness (section [0, 9] in Fig. 5.1).
3. After obtaining the latest start time of the tasks execution at which the optimal functional value (5.3) is reached, we observe a monotonous growth of the total earliness and tardiness (section [23, 45] in Fig. 5.1).

5.4.5 The Approximation Algorithm for E/T 2 Problem Solving (Algorithm A3: Determination of the Maximum Start Time of the Tasks Execution in a Given Time Segment)

1. Build an optimal by the TT criterion schedule σ_0 for start time $s_0 = t_1$.
2. Use Algorithm A2 to find the set M of optimal start times for which the minimum functional value (5.3) is achieved (see Example 5.3). If $M \cap [t_1, t_k] \neq \varnothing$, then the set of actual start times of the tasks execution is $M_a = M \cap [t_1, t_k]$, go to step 3. Otherwise, if $\forall s \in M$ we have $s < t_1$, then the actual start time of the tasks execution is $s_a = t_1$, go to step 3. Otherwise, if $\forall s \in M$ we have $s > t_k$, then execute *Modified Algorithm A2* (see below) to build a schedule optimal by the E/T criterion, the algorithm terminates.
3. Use Algorithm A1 to build a schedule optimal by the E/T criterion for the start time s_a or any of start times $s \in M_a$. Check heuristic SSOs for Algorithm A1.

Modified Algorithm A2 has the following three steps.
Step 1 has the three sub-steps:

1.1. Determine values N_R, N_T, and $E_{\min}$ for the schedule σ_0, set $i = 0$.
1.2. If $N_R > N_T$ (or $N_R - N_T > 1$ if the number of tasks is odd), then go to step 1.3, else go to step 3.
1.3. $s_{i+1} = \min(s_i + E_{\min}, t_k)$; $i = i + 1$. Go to step 2.

Step 2 Use the PSC-algorithm for the TT problem to decrease the total tardiness in the schedule σ_0. Go to step 1.2.
Step 3 For each $s_i \in [t_1, t_k]$ (including start times t_1 and t_k) build an optimal by the E/T criterion schedules σ_i using Algorithm A1. Check heuristic SSOs for Algorithm A1. Select the start time s_i^* at which we achieve the minimum functional value. If $N_R = N_T$ (or $N_R - N_T = 1$ if the number of tasks is odd) in the corresponding schedule σ_i^*, then find a task with the minimum earliness $E^*_{\min}$. If $E^*_{\min} = k > 0$, then all schedules with start times $s \in \left[s_i^*, \min\left(s_i^* + k, t_k\right)\right]$ will have the same functional value (Statement 5.47). The end of the algorithm.

Justification of Algorithm A3 A decrease in the functional value occurs due to two factors: use of reserves of early tasks by longer tasks (Algorithm A1) and an increase in the start time of the tasks execution which decreases the total earliness.

Statement 5.49 After achieving an optimal start time of the tasks execution, the functional value monotonically increases with an increase in the start time.

Proof In this case, both the factors get worse: the reserves (earliness) of early tasks are used less efficiently and the number of tardy tasks becomes greater than that of early tasks (Statement 5.46). ∎

If $M \cap [t_1, t_k]$ is not empty, then this intersection is the set of actual start times of the tasks execution M_a. If the set M is before t_1 (Example 5.5 below), then the actual start time of the tasks execution is t_1 (Statement 5.49, point 3 of Analysis of Statistical Studies). If the set M is after t_k (Example 5.6 below), then we execute Modified Algorithm A2 because increase as well as decrease in the functional value is possible depending on the start time of the tasks execution (point 2 of Analysis of Statistical Studies). This occurs because there are large reserves (earliness) in such schedules that are decreased with an increase in the start time. After that, some long tasks may no further use the reserves.

5.4.6 *Examples of E/T 2 Problem Solving*

Consider Example 5.4 for E/T 3 problem to solve E/T 2 problem. Initial data and the solving process is in Tables 5.17–5.21 and at Fig. 5.1.

Example 5.5 Given the time interval $[t_1, t_k] = [28, 40]$ (highlighted in gray at Fig. 5.2). Using Algorithm A2, we obtain the optimal start time of the tasks

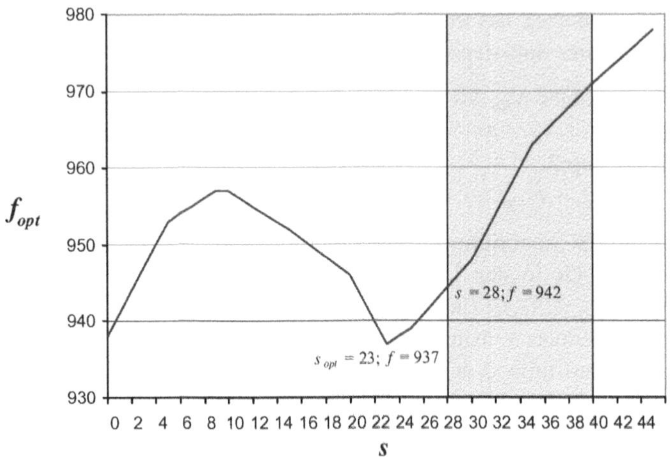

Fig. 5.2 Correlation between the functional value and the start times for Example 5.5 (E/T 2 problem)

execution $s_{opt} = 23 < 28$. Therefore, the actual start time of the tasks execution is $s_a = 28$ because the minimal functional value (5.3) is reached with this start time.

Example 5.6 Given the time interval $[t_1, t_k] = [0,\ 12]$ (highlighted in gray at Fig. 5.3). Using Algorithm A2, we obtain the optimal start time of the tasks execution $s_{opt} = 23 > 12$. Therefore, we execute Modified Algorithm A2. We check the start times 0 and 12, select $s_a = 0$ because the minimal functional value (5.3) is reached with this start time.

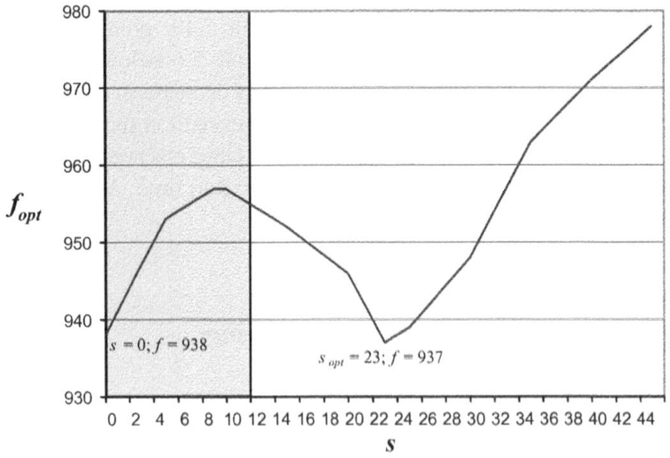

Fig. 5.3 Correlation between the functional value and the start times for Example 5.6 (E/T 2 problem)

References

1. Zgurovsky, M.Z., Pavlov, A.A.: Prinyatie Resheniy v Setevyh Sistemah s Ogranichennymi Resursami (Принятие решений в сетевых системах с ограниченными ресурсами; Decision Making in Network Systems with Limited Resources). Naukova dumka, Kyiv (2010) (in Russian)
2. Pavlov, A.A., Misura, E.B.: PDS-algoritm minimizacii summarnogo operejeniya i zapazdyvaniya pri vypolnenii rabot odnim priborom, esli moment zapuska pribora fiksirovannyi (ПДС-алгоритм минимизации суммарного опережения и запаздывания при выполнении работ одним прибором, если момент запуска прибора фиксированный; PDC-algorithm for total earliness and tardiness minimization on single machine for the case if the machine start time is fixed). Paper Presented at the 21st International Conference on Automatic Control Automatics-2014, National Technical University of Ukraine, Kyiv, 23–27 Sept 2014 (in Russian)
3. Pavlov, A.A., Misura, E.B.: PDS-algoritmy resheniya zadach sostavleniya raspisaniy po kriteriyu operejeniya/zapazdyvaniya na odnom pribore (ПДС-алгоритмы решения задач составления расписаний по критерию опережения/запаздывания на одном приборе; PDC-algorithms to solve single machine earliness/tardiness scheduling problems). Visnyk NTUU KPI Inform. Oper. Comput. Sci. **60**, 4–19 (2014) (in Russian)
4. Pavlov, A.A., Misura, E.B., Kostik, D.Y.: Minimizaciya summarnogo zapazdyvaniya pri nalichii zadanii s otricatel'nymi znacheniyami direktivnyh srokov (Минимизация суммарного запаздывания при наличии заданий с отрицательными значениями директивных сроков; Minimizing total tasks tardiness in the presence of negative deadline values). Visnyk NTUU KPI Inform. Oper. Comput. Sci. **53**, 3–5 (2011) (in Russian)
5. Pavlov, A.A., Misura, E.B., Khalus, E.A.: Skladannya rozkladu vikonannya zavdan' na odnomu priladi z metoyu minimizacii sumarnogo vyperedjennya ta znahodjennya maksimal'nogo pizn'ogo momentu pochatku vikonannya zavdan' v dopustimomu rozkladi (Складання розкладу виконання завдань на одному приладі з метою мінімізації сумарного випередження та знаходження максимального пізнього моменту початку виконання завдань в допустимому розкладі; Single machine scheduling to minimize the total earliness and find the latest start time of tasks in a feasible schedule). Paper Presented at the 21st International Conference on Automatic Control Automatics-2014, National Technical University of Ukraine, Kyiv, 23–27 Sept 2014 (in Ukrainian)
6. Feldmann, M., Biskup, D.: Single-machine scheduling for minimizing earliness and tardiness penalties by meta-heuristic approaches. Comput. Ind. Eng. **44**(2), 307–323 (2003). https://doi.org/10.1016/s0360-8352(02)00181-x
7. Valente, J.M.S., Alves, R.A.F.S.: Improved heuristics for the early/tardy scheduling problem with no idle time. Comput. Oper. Res. **32**(3), 557–569 (2005). https://doi.org/10.1016/j.cor.2003.08.003
8. Wodecki, M.: A block approach to earliness-tardiness scheduling problems. Int. J. Adv. Manuf. Technol. **40**(7–8), 797–807 (2008). https://doi.org/10.1007/s00170-008-1395-7
9. Kanet, J.J.: Minimizing the average deviation of job completion times about a common due date. Naval Res. Logistics Q. **28**(4), 643–651 (1981). https://doi.org/10.1002/nav.3800280411
10. Lawler, E.L.: A fully polynomial approximation scheme for the total tardiness problem. Oper. Res. Lett. **1**(6), 207–208 (1982). https://doi.org/10.1016/0167-6377(82)90022-0
11. T'kindt, V., Billaut, J.-C.: Multicriteria Scheduling: Theory, Models and Algorithms. Springer, Berlin (2002). https://doi.org/10.1007/978-3-662-04986-0
12. Baker, K.R., Scudder, G.D.: Sequencing with earliness and tardiness penalties: a review. Oper. Res. **38**(1), 22–36 (1990). https://doi.org/10.1287/opre.38.1.22

13. Jin, S., Mason, S.J.: Minimizing earliness and tardiness costs on a single machine with uncommon job due dates. Preprint. Bell Engineering Center, University of Arkansas, Fayetteville (2004)
14. Mason, S.J., Jin, S., Jampani, J.: A moving block heuristic for minimizing earliness and tardiness. J. Manuf. Syst. **24**(4), 328–338 (2005). https://doi.org/10.1016/s0278-6125(05)80017-2
15. Garey, M.R., Tarjan, R.E., Wilfong, G.T.: One-processor scheduling with symmetric earliness and tardiness penalties. Math. Oper. Res. **13**(2), 330–348 (1988). https://doi.org/10.1287/moor.13.2.330
16. Yano, C.A., Kim, Y.-D.: Algorithms for a class of single-machine weighted tardiness and earliness problems. Eur. J. Oper. Res. **52**(2), 167–178 (1991). https://doi.org/10.1016/0377-2217(91)90078-a
17. Davis, J.S., Kanet, J.J.: Single-machine scheduling with early and tardy completion costs. Naval Res. Logistics **40**(1), 85–101 (1993). https://doi.org/10.1002/1520-6750(199302)40:1<85:aid-nav3220400106>3.0.co;2-c
18. Ow, P.S., Morton, T.E.: The single machine early/tardy problem. Manag. Sci. **35**(2), 177–191 (1989). https://doi.org/10.1287/mnsc.35.2.177
19. Szwarc, W., Mukhopadhyay, S.K.: Optimal timing scheduling in earliness-tardiness single machine sequencing. Naval Res. Logistics **42**(7), 1109–1114 (1995). https://doi.org/10.1002/1520-6750(199510)42:7<1109:aid-nav3220420709>3.0.co;2-5
20. Sridharan, V., Zhou, Z.: A decision theory based scheduling procedure for single-machine weighted earliness and tardiness problem. Eur. J. Oper. Res. **94**(2), 292–301 (1996). https://doi.org/10.1016/0377-2217(96)00133-6
21. Wan, G., Yen, B.P.C.: Tabu search for single-machine scheduling with distinct due windows and weighted earliness/tardiness penalties. Eur. J. Oper. Res. **142**(2), 271–281 (2002). https://doi.org/10.1016/s0377-2217(01)00302-2
22. Szwarc, W.: Adjacent orderings in single-machine scheduling with earliness and tardiness penalties. Naval Res. Logistics **40**(2), 229–243 (1993). https://doi.org/10.1002/1520-6750(199303)40:2<229:aid-nav3220400207>3.0.co;2-r
23. Lee, C.Y., Choi, J.Y.: A generic algorithm for job sequencing problem with distinct due dates and general early-tardy penalty weights. Comput. Oper. Res. **22**(8), 857–869 (1995). https://doi.org/10.1016/0305-0548(94)00073-h
24. Bank, J., Werner, F.: Heuristic algorithm for unrelated parallel machine scheduling with a common due date, release dates, and linear earliness and tardiness penalties. Math. Comput. Model. **33**(4–5), 363–383 (2001). https://doi.org/10.1016/s0895-7177(00)00250-8
25. Gordon, V., Proth, J.-M., Chu, C.: A survey of the state-of-art of common due date assignment and scheduling research. Eur. J. Oper. Res. **139**(1), 1–25 (2002). https://doi.org/10.1016/s0377-2217(01)00181-3
26. Valente, J.M.S., Alves, R.A.F.S.: Filtered and recovering beam search algorithms for the early/tardy scheduling problem with no idle time. Comput. Ind. Eng. **48**(2), 363–375 (2005). https://doi.org/10.1016/j.cie.2005.01.020
27. Tsai, T.-I.: A genetic algorithm for solving the single machine earliness/tardiness problem with distinct due dates and ready times. Int. J. Adv. Manuf. Technol. **31**(9–10), 994–1000 (2007). https://doi.org/10.1007/s00170-005-0261-0
28. Hoogeveen, J.A., Van de Velde, S.L.: A branch-and-bound algorithm for single-machine earliness–tardiness scheduling with idle time. INFORMS J. Comput. **8**(4), 402–412 (1996). https://doi.org/10.1287/ijoc.8.4.402
29. Tanaka, S., Fujikuma, S., Araki, M.: An exact algorithm for single-machine scheduling without machine idle time. J. Sched. **12**(6), 575–593 (2009). https://doi.org/10.1007/s10951-008-0093-5
30. Du, J., Leung, J.Y.-T.: Minimizing total tardiness on one processor is NP-hard. Math. Oper. Res. **15**(3), 483–495 (1990). https://doi.org/10.1287/moor.15.3.483

31. Tanaev, V.S., Shkurba, V.V.: Vvedenie v Teoriju Raspisaniy (Введение в теорию расписаний; Introduction to Scheduling Theory). Nauka, Moscow (1975) (in Russian)
32. Lawler, E.L.: A "pseudopolynomial" algorithm for sequencing jobs to minimize total tardiness. Ann. Discr. Math. **1**, 331–342 (1977). https://doi.org/10.1016/S0167-5060(08)70742-8
33. Potts, C.N., Van Wassenhove, L.N.: A decomposition algorithm for the single machine total tardiness problem. Oper. Res. Lett. **1**, 177–181 (1982). https://doi.org/10.1016/0167-6377(82)90035-9

Chapter 6
The Total Tardiness of Tasks Minimization on Identical Parallel Machines with Arbitrary Fixed Times of Their Start and a Common Due Date

Abstract We solve a NP-hard problem of constructing a schedule for identical parallel machines that minimizes the total tardiness of tasks in relation to a common due date in case when the start times of machines are fixed at arbitrary time points less than the due date. We present an efficient PSC-algorithm of its solving which is a generalization of our previously developed results: for the problem with equal start times of machines we have derived two sufficient signs of optimality of a feasible solution and constructed two PSC-algorithms. Each of the algorithms checks one of these signs. In this chapter we propose a generalized PSC-algorithm for equal start times of machines that combines the best properties of both PSC-algorithms. We have obtained a modification of the generalized PSC-algorithm for the case of arbitrary start times of machines, its complexity is determined by $O(n^2m)$ function. The first polynomial component of the PSC-algorithm coincides with its second polynomial component. We obtain an efficient estimate of the deviation from an optimal solution for an approximation algorithm of the problem solving. We also present the statistical studies of the PSC-algorithm that showed its high efficiency (efficient exact solutions have been obtained for problems with tens of thousands of variables which is unique for NP-hard combinatorial optimization problems).

6.1 The Problem with the Common for All Machines (Fixed) Start Time of the Tasks Execution

This chapter contains our new original results based on our earlier works [1–3].

The Problems Statement Intractable combinatorial optimization problem "Minimizing the total tardiness of independent tasks with equal weights and a common due date on parallel machines" (TTP) is formulated as follows.

Given a set of n tasks J and m identical parallel machines. We know a processing time l_j for each task $j \in J$. All tasks have a common due date d. We assume that all

M. Z. Zgurovsky and A. A. Pavlov, *Combinatorial Optimization Problems in Planning and Decision Making*, Studies in Systems, Decision and Control 173,
https://doi.org/10.1007/978-3-319-98977-8_6

tasks of the set J and the machines are available after the same time point, we can start processing of any task at any time point and it will continue without interruptions until the task is completed. We need to build such a schedule σ of the tasks $j \in J$ processing on m machines that minimizes the functional

$$f(\sigma) = \sum_{j \in J} \max\big(0,\ C_j(\sigma) - d\big)$$

where $C_j(\sigma)$ is the completion time of task j in the schedule σ.

The problem belongs to the class of NP-hard problems [4]. It can be solved in polynomial time at $m = 2$.

The TTP problem is widely used in the development of production planning systems, project management, construction management and other areas. In particular, we use the algorithm for this problem in the mathematics of the four-level model of planning (including operative planning) and decision making (Chap. 9).

The following results are presented in [5].

Suppose that the tasks of a set $J = \{1,\ 2,\ \ldots,\ n\}$ are arranged in non-decreasing order of processing times l_j. And let $J_k \in J$ be the set of tasks which are processed by machine k. $J_k = \{\Psi_k, R_k\}$. $j_i \in \Psi_k$ if $C_i - l_i < d$. Otherwise, $j_i \in R_k$.

Theorem 6.1 (Tanaev and Shkurba [5]) *There is an optimal schedule in which*

1. $\Psi = \{1,\ 2,\ \ldots,\ |\Psi|\}$ *where* $\Psi = \bigcup_{k=1,\ldots,m} \Psi_k$;
2. *If* $|\Psi| < n$, *then* $\sum_{j \in \Psi_k} l_j \ge d$ *and* R_k *contains those and only those elements* $R = \{\,|\Psi| + 1,\ \ldots,\ n\,\}$ *that differ from* $|\Psi| + k$ *by an amount that is a multiple of* m, $k = 1,\ \ldots,\ m$.

Let $\Theta(\mu)$ be a class of schedules corresponding to the conditions of Theorem 6.1 and additional conditions: $|\Psi| = \mu$ where μ is a natural number. An optimal schedule θ^* belongs to at least one of the classes $\Theta(\mu)$ at some $\mu = \mu^*$.

Theorem 6.2 (Tanaev and Shkurba [5]) *A schedule* $\theta \in \Theta(\mu)$ *is optimal in* $\Theta(\mu)$ *if the value of* $Z(\mu) = \sum_{i=1}^{c(\mu)} \max(0, Z_i)$ *in this schedule reaches its minimum. Here,* $Z_i = \sum_{j \in \Psi_i} l_j - d > 0$, $i = 1 \ldots m$; $c(\mu)$ *is the remainder of* $(n - \mu)$ *divided by* m *if* $n - \mu > 0$ *or* $c(\mu) = m$ *otherwise.*

We develop the results proposed in [5] and give a PSC-algorithm to solve the formulated problem. The algorithm has the following properties: on the basis of the theoretical properties of the TTP problem, there are logical-analytical conditions—the sufficient signs of optimality (SSOs) of feasible[1] solutions. Fulfillment of a SSO during execution of a given polynomial time computational scheme yields an optimal solution.

The algorithm for the TTP problem is the PSC-algorithm of the second class.

[1] All schedules obtained during the considered problem solving are feasible.

6.1.1 Main Theoretical Thesis

Let $\overline{a,b}$ denote the interval of integer numbers from a to b, that is, $\overline{a,b} = \mathbb{Z} \cap [a, b] = a,\ a + 1,\ \ldots,\ b$. Renumber the tasks of the set $J = \{1,\ 2,\ \ldots,\ n\}$ in non-decreasing order of l_j and split them into (optional nonempty) subsets $J_1,\ J_2,\ \ldots, J_i,\ \ldots, J_m$ that do not have common elements. Here, J_i is the set of tasks processed by ith machine, $i = \overline{1, m}$.

The search for an optimal schedule can be limited to consideration of schedules in which each machine process its tasks in the increasing order of their numbers, according to the well-known theorem [5]. Indeed, if for some tasks j_k, j_{k+1} in an optimal sequence $\sigma_i^* = (j_1^*, j_2^*, \ldots, j_n^*)$ we have $l_{j_k} > l_{j_{k+1}}$, then a sequence σ_i that differs from σ_i^* in transposition of elements j_k^* and j_{k+1}^* corresponds to a smaller functional value. This contradicts the assumption of optimality of σ_i^*.

Let us split the set of tasks J_i into subsets $P_i(\sigma)$, $S_i(\sigma)$, $Q_i(\sigma)$ as follows:

$P_i(\sigma)$ is the set of non-tardy tasks in the schedule of machine i;
$S_i(\sigma)$ is the set of tardy tasks in the schedule of machine i for which

$$C_j - l_j < d,\ C_j > d,\ \forall j \in S_i(\sigma)$$

where $C_j - l_j$ is the start time of a task j;

$Q_i(\sigma)$ is the set of tardy tasks in the schedule of machine i for which

$$C_j - l_j \geq d,\ \forall j \in Q_i(\sigma);$$

$$P = \bigcup_{i=\overline{1,m}} P_i;\ S = \bigcup_{i=\overline{1,m}} S_i;\ Q = \bigcup_{i=\overline{1,m}} Q_i;$$

$R_i(\sigma)$ is the time reserve of machine i in a schedule σ:

$$R_i(\sigma) = d - \sum_{j \in P_i(\sigma)} l_j;$$

$\Delta_i(\sigma)$ is the tardiness of task $j \in S_i(\sigma)$ in regard to the due date:

$$\Delta_i(\sigma) = \sum_{j \in P_i(\sigma) \bigcup S_i(\sigma)} l_j - d.$$

Using for convenience the notation we adopted, let us formulate a theorem that determines the class of schedules containing an optimal solution on the set of all possible schedules constructed for a fixed set of tasks J.

Theorem 6.3 *There is an optimal schedule that satisfies the conditions:*

1. $P \cup S = \{1, 2, \ldots, |P \bigcup S|\}$;
2. *If* $|P \bigcup S| < n$, *then* $\sum_{j \in P_i \bigcup S_i} l_j \ge d$ *and* Q_i *contains those and only those elements which differ from* $|P \cup S| + i$ *by an amount that is a multiple of m,* $i = \overline{1, m}$.

Let Ψ_{PS} denote a class of schedules that correspond to the conditions of Theorem 6.3. And let $\Psi_{PS}(\mu)$ be a class of schedules corresponding to the condition: $|P(\sigma) \cup S(\sigma)| = \mu$ where μ is a natural number. We can write that $\Psi_{PS}(\mu) \subset \Psi_{PS}$. An optimal schedule σ^* belongs to at least one of the classes $\Psi_{PS}(\mu)$ at some $\mu = \mu^*$. The number of different non-empty classes $\Psi_{PS}(\mu)$ does not exceed m [5].

Let $\bar{\mu}$ ($\underline{\mu}$) be the greatest (smallest) value of μ at which $\Psi_{PS} \ne$. Let $c(\sigma)$ denote the remainder of$n - |P(\sigma) \cup S(\sigma)|$ divided by m if $n - |P(\sigma) \cup S(\sigma)| > 0$ or $c(\sigma) = m$ otherwise. Theorem 6.4 gives the sufficient conditions under which a schedule $\sigma \in \Psi_{PS}(\mu)$ corresponds to the minimum total tardiness among all schedules of the class $\Psi_{PS}(\mu)$, $\underline{\mu} \le \mu \le \bar{\mu}$ [5].

Theorem 6.4 (Tanaev and Shkurba [5]) *A schedule* $\sigma \in \Psi_{PS}(\mu)$ *is optimal in* $\Psi_{PS}(\mu)$ *if the value of* $z(\sigma) = \sum_{i=1}^{c(\sigma)} \Delta_i(\sigma)$ *reaches its minimum in it.*

Let σ^* denote an optimal schedule for the set of tasks J. The following theorem allows to evaluate the deviation of a schedule from the optimal one.

Theorem 6.5 (Tanaev and Shkurba [5]) *If a schedule* $\sigma \in \Psi_{PS}$, *then* $f(\sigma) - f(\sigma^*) \le z(\sigma)$.

Corollary 6.1 If $\sigma \in \Psi_{PS}$ and $z(\sigma) = 0$, then the schedule σ is optimal.

Theorem 6.4 formulates the conditions of local optimality, i.e., the optimality in the class $\Psi_{PS}(\mu)$, $\underline{\mu} \le \mu \le \bar{\mu}$. Therefore, for arbitrary schedules $\sigma' \in \Psi_{PS}(\mu')$ and $\sigma'' \in \Psi_{PS}(\mu'')$ where $\mu' \ne \mu''$, $z(\sigma') \ne 0$, and $z(\sigma'') \ne 0$, fulfillment of the inequality $f(\sigma') < f(\sigma'')$ does not follow from fulfillment of $z(\sigma') < z(\sigma'')$.

Example 6.1 Given $m = 4$, $d = 9$, processing times l_j are in Table 6.1.

Compare the two following schedules that show the processing times of tasks.

Schedule σ': $J_1(\sigma') = 3, 10, 15$; $J_2(\sigma') = 3, 10, 15$; $J_3(\sigma') = 3, 10$; $J_4(\sigma') = 3, 15$.

Schedule σ'': $J_1(\sigma'') = 10, 15$; $J_2(\sigma'') = 10, 15$; $J_3(\sigma'') = 10, 15$; $J_4(\sigma'') = 3, 3, 3, 3$; $z(\sigma') = 2 < z(\sigma'') = 39$; $f(\sigma') = 41 > f(\sigma'') = 39$.

We distinguish from the class Ψ_{PS} a class of schedules $\Psi_P \subset \Psi_{PS}$ satisfying the following additional conditions:

1. $P = \{1, 2, \ldots, |P|\}$;
2. $\min_{j \in S(\sigma)} l_j > \max_{i=\overline{1,m}} R_i(\sigma)$;
3. If $l_{j_k} \le l_{j_l}$, then $C_{j_k} - l_{j_k} \le C_{j_l} - l_{j_l}$ $\forall j_k, j_l \in S(\sigma)$.

Table 6.1 Initial data for Example 6.1

j	1	2	3	4	5	6	7	8	9	10
l_j	3	3	3	3	10	10	10	15	15	15

Suppose $|P| < n$. Let $P_{\min}$ denote the minimum number of tasks in the set P for which $\Psi_P \neq \varnothing$. Let $P_{\max}$ be the maximum number of tasks in the set P.

Statement 6.1 The following is true for all possible schedules $\sigma \in \Psi_P$ constructed on the set of tasks J:

$$P_{\max} - P_{\min} < m.$$

Proof For any schedule $\sigma \in \Psi_P$, we have:

$$\sum_{j=1}^{|P(\sigma)|} l_j = d \cdot m - \sum_{i=1}^{m} R_i(\sigma).$$

Therefore, such a schedule σ in which $\sum_{i=1}^{m} R_i(\sigma)$ has the maximum (minimum) possible value among all schedules in class Ψ_P, satisfies $|P(\sigma)| = P_{\min}$ (or $P_{\max}$, respectively). But $R_i(\sigma) < \min_{j \in S(\sigma)} l_j$ is true for each machine i in a schedule $\sigma \in \Psi_P$.

Hence,

$$\sum_{i=1}^{m} R_i(\sigma) < m \min_{j \in S(\sigma)} l_j \leq \sum_{i=1}^{m} l_{|P(\sigma)|+i} \text{ and } P_{\max} - P_{\min} < m.$$

Statement 6.2 Moves of tasks in consequence of directed permutations are only possible between the sets $P(\sigma)$ and $S(\sigma)$ when constructing an optimal schedule.

Statement 6.3 If a task $j \in P(\sigma)$ moves from a machine i_k with a greater number of tardy tasks to a machine i_l with a smaller number of tardy tasks, then Δ_{i_k} decreases by l_j.

Statement 6.4 The maximum difference in the number of tardy tasks in schedules $\sigma \in \Psi_P$ does not exceed one.

Proof Validity of this statement is obvious and directly follows from definition of the class Ψ_P.

Definition 6.1 A schedule with the same number of tardy tasks on all machines is called an *even schedule*.

Theorem 6.6, SSO #1 *An even schedule* $\sigma \in \Psi_P$ *is optimal.*

Proof Let L_i denote the number of tardy tasks on machine i. Consider the following cases.

1. $R_i = 0,\ i \in \overline{1,k};\ R_i > 0,\ i \in \overline{k+1,m};\ L_i = L \forall i \in \overline{1,m}$. The schedule σ is optimal in this case because

$$z(\sigma) = \sum_{i=1}^{c(\sigma)} \Delta_i = \sum_{i=1}^{k} \Delta_i = 0.$$

2. $R_i = 0\ \forall i \in \overline{1,m}$; $L_i = L \forall i \in \overline{1,m}$. The schedule σ is optimal in this case because $c(\sigma) = 0$.
3. $R_i > 0\ \forall i \in \overline{1,m}$; $L_i = L \forall i \in \overline{1,m}$. The schedule σ is optimal in this case because $c(\sigma) = 0$. ∎

We introduce some notation. Let $L_{\max}$ be the maximum number of tardy tasks on all machines; $L_{\min}$ be the minimum number of tardy tasks on all machines. Let numbers $i = \overline{1,k}$ correspond to machines with the number of tardy tasks $L_{\max}$. Denote by P the cardinality of the set $P(\sigma)$: $P = |P(\sigma)|$. Let us also make the following notation: $\Delta_\Sigma(\sigma) = \sum_{i=1}^{k} \Delta_i$, $R_\Sigma(\sigma) = \sum_{i=k+1}^{m} R_i$, $\Omega_\Sigma(\sigma) = \min(R_\Sigma(\sigma), \Delta_\Sigma(\sigma))$.

Theorem 6.7 *If $L_{\max}$ is the same in schedules $\sigma \in \Psi_P$ and $\sigma' \in \Psi_P$ constructed on the given set of tasks J, then from $R_\Sigma(\sigma) \neq 0$ and $R_\Sigma(\sigma') \neq 0$ follows $R_\Sigma(\sigma) - R_\Sigma(\sigma') = \Delta_\Sigma(\sigma) - \Delta_\Sigma(\sigma')$.*

Proof Suppose $md = R^0$ and $R_1 \geq R_2 \geq R_3 \geq \ldots R_m$. Then,

$$\begin{aligned} R^0 &= \sum_{j \in P(\sigma)} l_j + \sum_{j \in S(\sigma)} l_j - \sum_{i=1}^{m} \Delta_i = \sum_{j \in P(\sigma)} l_j - \sum_{i=1}^{k} \Delta_i(\sigma) \\ &+ \sum_{j=p+1}^{p+k} l_j + \left(\sum_{j=p+k+1}^{p+m} l_j - \sum_{i=k+1}^{m} \Delta_i \right) = \sum_{j \in P(\sigma)} l_j \\ &+ \left(\sum_{j=p+1}^{p+k} l_j - \Delta_\Sigma(\sigma) \right) + R_\Sigma(\sigma) = \text{const.} \end{aligned} \tag{6.1}$$

Consider the following cases:

1. $P(\sigma) = P(\sigma')$. Theorem is obviously valid according to (6.1).
2. $P(\sigma) < P(\sigma')$. In this case, tasks $j \in S(\sigma)$, $j = \overline{p+1, p+k'}$, $k' < k$, became non-tardy, that is, moved to the set $P(\sigma')$:

$$R^0 = \left(\sum_{j \in P(\sigma)} l_j + \sum_{j=p+1}^{p+k'} l_j \right) + \left(\sum_{j=p+k'+1}^{p+k} l_j - \Delta_\Sigma(\sigma') \right) + R_\Sigma(\sigma').$$

Theorem is valid according to (6.1).

3. $P(\sigma) > P(\sigma'')$, $\Delta_\Sigma(\sigma) < \Delta_\Sigma(\sigma'')$. In this case, the number of tardy tasks on machines $i = \overline{k+1, k+k''}$, $k'' < m$, increased because some tasks from the set $P(\sigma)$ became tardy:

$$R^0(\sigma) = \sum_{j \in P(\sigma)} l_j - \sum_{j=p-k''+1}^{p} l_j + \left(\sum_{j=p-k''+1}^{p} l_j + \sum_{j=p+1}^{p+k} l_j - \Delta_\Sigma(\sigma') \right) + R_\Sigma(\sigma').$$

Theorem is valid according to (6.1). ∎

Theorem 6.8 *The following is true for any two schedules* $\sigma \in \Psi_P$ *and* $\sigma' \in \Psi_P$:

$$f(\sigma) - f(\sigma') = \Omega_\Sigma(\sigma) - \Omega_\Sigma(\sigma'). \tag{6.2}$$

Proof Consider an arbitrary schedule σ constructed on a given fixed set of tasks J. Suppose $P(\sigma) = \{1, 2, \ldots, p-1, p\}$ in this schedule.

Renumber the machines so that $R_1 \geq R_2 \geq R_3 \geq \ldots R_m$. Suppose that the number of tardy tasks is L for each machine $i \in \overline{1,k}$ and $L-1$ for machines $i \in \overline{k+1,m}$.

The following cases are possible during construction of new schedules $\sigma' \in \Psi_P$, $\sigma'' \in \Psi_P$, $\sigma''' \in \Psi_P$ on the set of tasks J:

1. $L_{\min} = L-1$;
2. Even schedule;
3. $L_{\min} = L-2$.

Consider the value of the total tardiness. In case (1), for schedule $\sigma' \in \Psi_P$:

$$\begin{aligned}
f(\sigma') = {} & \left(l_{p+1} - R_1\right) + \left(l_{p+1} - R_1 + l_{p+1+m}\right) + \left(l_{p+1} - R_1 + l_{p+1+m} + l_{p+1+2m}\right) \\
& + \cdots + \left(l_{p+1} - R_1 + l_{p+1+m} + l_{p+1+2m} + \cdots + l_{p+1+(L-1)m}\right) \\
& + \left(l_{p+2} - R_2\right) + \left(l_{p+2} - R_2 + l_{p+2+m}\right) + \left(l_{p+2} - R_2 + l_{p+2+m} + l_{p+2+2m}\right) \\
& + \cdots + \left(l_{p+2} - R_2 + l_{p+2+m} + l_{p+2+2m} + \cdots + l_{p+2+(L-1)m}\right) \\
& + \cdots + \left(l_{p+k} - R_k\right) + \left(l_{p+k} - R_k + l_{p+k+m}\right) + \left(l_{p+k} - R_k + l_{p+k+m} + l_{p+k+2m}\right) \\
& + \cdots + \left(l_{p+k} - R_k + l_{p+k+m} + l_{p+k+2m} + \cdots + l_{p+k+(L-1)m}\right) \\
& + \left(l_{p+k+1} - R_{k+1}\right) + \left(l_{p+k+1} - R_{k+1} + l_{p+k+1+m}\right) \\
& + \left(l_{p+k+1} - R_{k+1} + l_{p+k+1+m} + l_{p+k+1+2m}\right) \\
& + \cdots + \left(l_{p+k+1} - R_{k+1} + l_{p+k+1+m} + l_{p+k+1+2m} + \cdots + l_{p+k+1+(L-2)m}\right) \\
& + \cdots + \left(l_{p+m} - R_m\right) + \left(l_{p+m} - R_m + l_{p+2m}\right) + \left(l_{p+m} - R_m + l_{p+2m} + l_{p+3m}\right) \\
& + \cdots + \left(l_{p+m} - R_m + l_{p+2m} + l_{p+3m} + \cdots + l_{p+m+(L-2)m}\right) \\
= {} & L \cdot l_{p+1} + L \cdot l_{p+2} + \cdots + L \cdot l_{p+k} \\
& + (L-1) \cdot l_{p+1+m} + (L-1) \cdot l_{p+2+m} + \cdots + (L-1) \cdot l_{p+k+m} \\
& + (L-2) \cdot l_{p+1+2m} + (L-2) \cdot l_{p+2+2m} + \cdots + (L-2) \cdot l_{p+k+2m} + \ldots \\
& + l_{p+1+(L-1)m} + l_{p+2+(L-1)m} + \cdots + l_{p+k+(L-1)m} - L(R_1 + R_2 + \cdots + R_k) \\
& + (L-1) \cdot l_{p+k+1} + (L-1) \cdot l_{p+k+2} + \cdots + (L-1) \cdot l_{p+m} \\
& + (L-2) \cdot l_{p+k+1+m} + (L-2) \cdot l_{p+k+2+m} + \cdots + (L-2) \cdot l_{p+2m} + \ldots \\
& + l_{p+k+1+(L-2)m} + l_{p+k+2+(L-2)m} + \cdots \\
& + l_{p+(L-1)m} - (L-1)(R_{k+1} + R_{k+2} + \cdots + R_m)
\end{aligned}$$

Let us denote: $l_{p+r} - R_r = \Delta_r$, $r = \overline{1,k}$, $\sum_{r=1}^{k} \Delta_r = \Delta_\Sigma(\sigma)$, $\sum_{r=k+1}^{m} R_r = R_\Sigma(\sigma)$.

We can represent the total tardiness $f(\sigma)$ corresponding to any schedule $\sigma \in \Psi_P$ and satisfying the conditions of the case (1) as follows:

$$\begin{aligned} f(\sigma) = & L \cdot \Delta_\Sigma(\sigma) + (L-1) \cdot R_\Sigma(\sigma) + (L-1) \cdot l_{p+1+m} + (L-1) \cdot l_{p+2+m} \\ & + \cdots + (L-1) \cdot l_{p+k+m} + (L-2) \cdot l_{p+1+2m} \\ & + (L-2) \cdot l_{p+2+2m} + \cdots + (L-2) \cdot l_{p+k+m} \\ & + \cdots + l_{p+1+(L-1)m} + l_{p+2+(L-1)m} + \ldots l_{p+k+(L-1)m} \\ & + (L-1) \cdot l_{p+k+1} + (L-1) \cdot l_{p+k+2} + \cdots \\ & + (L-1) \cdot l_{p+m} + (L-2) \cdot l_{p+k+1+m} \\ & + (L-2) \cdot l_{p+k+2+m} + \cdots + (L-2) \cdot l_{p+2m} \\ & + \cdots + l_{p+k+1+(L-2)m} + l_{p+k+2+(L-2)m} + \cdots + l_{p+(L-1)m} \end{aligned}$$

or $f(\sigma) = L \cdot \Delta_\Sigma(\sigma) - (L-1)R_\Sigma(\sigma) + \tilde{F}_\Sigma(\sigma)$ where $\tilde{F}_\Sigma(\sigma) = \text{const}$.

Consider two arbitrary schedules $\sigma \in \Psi_P$ and $\sigma' \in \Psi_P$ that have the characteristics $f(\sigma)$, $R_\Sigma(\sigma)$, $\Delta_\Sigma(\sigma)$, $f(\sigma')$, $R_\Sigma(\sigma')$, $\Delta_\Sigma(\sigma')$, respectively.

Suppose $R_\Sigma(\sigma') = R_\Sigma(\sigma) - \delta_R$. Then, $\Delta_\Sigma(\sigma') = \Delta_\Sigma(\sigma) - \delta_R$;

$$\begin{aligned} f(\sigma) - f(\sigma') &= L\Delta_\Sigma(\sigma) - (L-1)R_\Sigma(\sigma) + \tilde{F}_\Sigma(\sigma) \\ & \quad - L(\Delta_\Sigma(\sigma) - \delta_R) + (L-1)(R_\Sigma(\sigma) - \delta_R) - \tilde{F}_\Sigma(\sigma) \\ &= \delta_R \Omega_\Sigma(\sigma) = \min(R_\Sigma(\sigma), \Delta_\Sigma(\sigma)) \end{aligned}$$

Hence,

$$\Omega_\Sigma(\sigma) - \Omega_\Sigma(\sigma') = \min(R_\Sigma(\sigma), \Delta_\Sigma(\sigma)) - \min(R_\Sigma(\sigma'), \Delta_\Sigma(\sigma')).$$

Suppose $\min(R_\Sigma(\sigma), \Delta_\Sigma(\sigma)) = R_\Sigma(\sigma)$. Then, $\min(R_\Sigma(\sigma'), \Delta_\Sigma(\sigma')) = R_\Sigma(\sigma) - \delta_R$;

$$\Omega_\Sigma(\sigma) - \Omega_\Sigma(\sigma') = R_\Sigma(\sigma) - R_\Sigma(\sigma) + \delta_R = \delta_R.$$

The reasoning is similar for $\min(R_\Sigma(\sigma), \Delta_\Sigma(\sigma)) = \Delta_\Sigma(\sigma)$. Therefore, in case (1) for arbitrary schedules $\sigma \in \Psi_P$ and $\sigma' \in \Psi_P$ we have:

$$f(\sigma) - f(\sigma') = \Omega_\Sigma(\sigma) - \Omega_\Sigma(\sigma').$$

Consider an even schedule [case (2)]. A particular case of an even schedule is the schedule $\sigma'' \in \Psi_P$ obtained from the schedule $\sigma \in \Psi_P$ [case (1)] at $\Delta_\Sigma(\sigma) = 0$. In this case, the number of tardy tasks on all machines is $L-1$; $f(\sigma) - f(\sigma'') = \Omega_\Sigma(\sigma) - \Omega_\Sigma(\sigma'') = \Omega_\Sigma(\sigma)$. Consider a schedule $\sigma'' \in \Psi_P$ with the number of tardy tasks on all machines equal to L. We have obtained this schedule from the initial schedule $\sigma \in \Psi_P$ with the following characteristics: $f(\sigma)$, $R_\Sigma(\sigma)$, $\Delta_\Sigma(\sigma)$, $L_{\max} = L$, $\Delta_\Sigma(\sigma) > R_\Sigma(\sigma)$. Let us find the change in the total tardiness:

$$\begin{aligned} f(\sigma) - f(\sigma'') &= L\Delta_\Sigma(\sigma) - (L-1)R_\Sigma(\sigma) + \tilde{F}_\Sigma(\sigma) \\ & \quad - L\big(\Delta_\Sigma(\sigma) - R_\Sigma(\sigma) - \delta_{\Delta_\Sigma(\sigma)}\big) + (L-1)(R_\Sigma(\sigma) - R_\Sigma(\sigma)) - L\delta_{\Delta_\Sigma(\sigma)} - \tilde{F}_\Sigma(\sigma) \\ &= L\Delta_\Sigma(\sigma) - LR_\Sigma(\sigma) + R_\Sigma(\sigma) + \tilde{F}_\Sigma(\sigma) - L\Delta_\Sigma(\sigma) + LR_\Sigma(\sigma) \\ & \quad + L\delta_{\Delta_\Sigma(\sigma)} - L\delta_{\Delta_\Sigma(\sigma)} - \tilde{F}_\Sigma(\sigma) = R_\Sigma(\sigma) \end{aligned}$$

Since $R_\Sigma(\sigma) < \Delta_\Sigma(\sigma)$, then

$$f(\sigma) - f(\sigma'') = \Omega_\Sigma(\sigma) - \Omega_\Sigma(\sigma'') = \Omega_\Sigma(\sigma).$$

Consider case (3): the minimum number of tardy tasks on machines is $L - 2$. Compare two arbitrary schedules $\sigma \in \Psi_P$, $\sigma''' \in \Psi_P$. The schedule σ''' differs from the schedule σ in that a part of tasks $j \in S_j(\sigma)$ became non-tardy.

Suppose $P(\sigma) = \{1, 2, \ldots, |P|\}$;

$$P(\sigma''') = \{1, 2, \ldots, |P|, |P|+1, \ldots, |P|+k, \ldots, |P|+k+f\};$$

$$\Delta_\Sigma(\sigma) = \sum_{i=1}^{k} \Delta_i;\ R_\Sigma(\sigma) = \sum_{i=k+1}^{m} R_i.$$

Remark 6.1 The following is true for arbitrary schedules $\sigma^1 \in \Psi_P$, $\sigma^2 \in \Psi_P$. The set of tardy tasks assigned to one machine is the same for all schedules, except for the tasks that become non-tardy in consequence of permutations.

Lemma 6.1 *The following is true for schedules $\sigma \in \Psi_P$ in case (3):*

$$R_\Sigma(\sigma) > \Delta_\Sigma(\sigma),\ \Delta_\Sigma(\sigma''') > R_\Sigma(\sigma''')$$

Proof We have

$$\sum_i R_i(\sigma) \geq \sum_{j=p+1}^{p+k} l_j + \sum_{j=p+k+1}^{p+k+f} l_j.$$

But

$$R_\Sigma(\sigma) - \Delta_\Sigma(\sigma) = \sum_i R_i(\sigma) - \sum_{j=p+1}^{p+k} l_j.$$

Therefore, $R_\Sigma(\sigma) > \Delta_\Sigma(\sigma)$.

Suppose $P(\sigma''') = p + k + f;\ f \geq 1$. Now we compare $\Delta_\Sigma(\sigma''')$ and $R_\Sigma(\sigma''')$.

The number of machines with a greater number of tardy tasks in the schedule σ''' is $m - f$. To satisfy $R_\Sigma(\sigma''') \geq \Delta_\Sigma(\sigma''')$, we need the following to be true:

$$\sum_i R_i(\sigma''') \geq \sum_{j=p+k+f+1}^{p+k+f+m-f} l_j.$$

But

$$\sum_i R_i(\sigma''') = \sum_i R_i(\sigma) - \sum_{j=p+1}^{p+k+f} l_j.$$

According to Statement 6.1,

$$\sum_i R_i(\sigma''') \le \sum_{j=p+1}^{p+m-1} l_j.$$

Therefore,

$$\sum_{j=p+1}^{p+m-1} l_j - \sum_{j=p+1}^{p+k+f} l_j \ge \sum_{j=p+k+f+1}^{p+k+f+m-f} l_j;$$

$$\sum_{j=p+1}^{p+m-1} l_j \ge \sum_{j=p+k+f+1}^{p+k+f+m-f} l_j + \sum_{j=p+1}^{p+k+f} l_j;$$

$$\sum_{j=p+1}^{p+m-1} l_j \ge \sum_{j=p+1}^{p+k+f+m-f} l_j.$$

We came to a contradiction. Consequently, $R_\Sigma(\sigma''') < \Delta_\Sigma(\sigma''')$.

Let us compare $f(\sigma)$ and $f(\sigma''')$

$$\begin{aligned}
f(\sigma) - f(\sigma''') = {} & Ll_{p+1} + (L-1)l_{p+m+1} + (L-2)l_{p+2m+1} + \cdots + l_{p+1+(L-1)m} \\
& + Ll_{p+2} + (L-1)l_{p+m+2} + (L-2)l_{p+2m+2} + \cdots + l_{p+2+(L-1)m} \\
& + \cdots + Ll_{p+k} + (L-1)l_{p+m+k} + (L-2)l_{p+2m+k} + \cdots + l_{p+k+(L-1)m} \\
& + (L-1)l_{p+k+1} + (L-2)l_{p+m+k+1} + \cdots + l_{p+k+1+(L-2)m} \\
& + \cdots + (L-1)l_{p+k+f} + (L-2)l_{p+m+k+f} + \cdots + l_{p+k+f+(L-2)m} \\
& + \cdots + (L-1)l_{p+k+f+1} + (L-2)l_{p+m+k+f+1} + \cdots + l_{p+k+f+1+(L-2)m} \\
& + \cdots + (L-1)l_{p+m} + (L-2)l_{p+2m} + \cdots + l_{p+(L-1)m} \\
& + \cdots - L(R_1 + R_2 + \cdots + R_k) - (L-1)(R_{k+1} + R_{k+2} + \cdots + R_m) \\
& - (L-1)l_{p+k+f+1} - (L-2)l_{p+m+k+f+1} - \ldots - l_{p+k+f+1+(L-2)m} \\
& - \cdots - (L-1)l_{p+m} - (L-2)l_{p+2m} - \ldots - l_{p+(L-1)m} \\
& - \cdots - (L-1)l_{p+m+1} - (L-2)l_{p+2m+1} - \ldots - l_{p+1+(L-1)m} \\
& - \cdots - (L-1)l_{p+m+k} - (L-2)l_{p+2m+k} - \ldots - l_{p+k+(L-1)m} \\
& - (L-2)l_{p+m+k+1} - \ldots - l_{p+k+1+(L-2)m} - \ldots \\
& - (L-2)l_{p+m+k+f} - \ldots - l_{p+k+f+(L-2)m} \\
& + (L-1)\left(R_1''' + R_2''' + \cdots + R_{m-f}'''\right) + (L-2)\left(R_{m-f+1}''' + R_{m-f+2}''' + \cdots + R_m'''\right) \\
= {} & Ll_{p+1} + Ll_{p+2} + \cdots + Ll_{p+k} + (L-1)l_{p+k+1} + \cdots + (L-1)l_{p+k+f} \\
& - L(R_1 + R_2 + \cdots + R_k) - (L-1)(R_{k+1} + R_{k+2} + \cdots + R_m) \\
& + (L-1)\left(R_1''' + R_2''' + \cdots + R_{m-f}'''\right) + (L-2)\left(R_{m-f+1}''' + R_{m-f+2}''' + \cdots + R_m'''\right) \\
= {} & Ll_{p+1} + Ll_{p+2} + \cdots + Ll_{p+k} + (L-1)l_{p+k+1} + \cdots + (L-1)l_{p+k+f} \\
& - (L-1)(R_1 + R_2 + \cdots + R_m) - (R_1 + R_2 + \cdots + R_k) \\
& + (L-1)(R_1''' + R_2''' + \cdots + R_m''') - (L-2)\left(R_{m-f+1}''' + R_{m-f+2}''' + \cdots + R_m'''\right)
\end{aligned}$$

$$
\begin{aligned}
&= Ll_{p+1} + Ll_{p+2} + \cdots + Ll_{p+k} + (L-1)l_{p+k+1} + \cdots + (L-1)l_{p+k+f} \\
&\quad - \left(R'''_{m-f+1} + R'''_{m-f+2} + \cdots + R'''_{m}\right) - (R_1 + R_2 + \cdots + R_k) \\
&= l_{p+1} + l_{p+2} + \cdots + l_{p+k} - (R_1 + R_2 + \cdots + R_k) \\
&\quad - \left(R'''_{m-f+1} + R'''_{m-f+2} + \cdots + R'''_{m}\right) \\
&= \Delta_\Sigma(\sigma) - R_\Sigma(\sigma''')
\end{aligned}
$$

According to Lemma 6.1,

$$R_\Sigma(\sigma) > \Delta_\Sigma(\sigma), \quad R_\Sigma(\sigma''') < \Delta_\Sigma(\sigma''').$$

Therefore,

$$f(\sigma) - f(\sigma''') = \Delta_\Sigma(\sigma) - R_\Sigma(\sigma''') = \Omega_\Sigma(\sigma) - \Omega_\Sigma(\sigma''').$$

This completes the proof of the theorem. ∎

Theorem 6.9, SSO #2 *If* $\Omega_\Sigma(\sigma) = \min(R_\Sigma(\sigma), \Delta_\Sigma(\sigma)) = 0$ *in a schedule* $\sigma \in \Psi_P$, *then the schedule* σ *is optimal.*

Proof Consider some schedule $\sigma \in \Psi_P$ in which

$$
\begin{gathered}
R_1(\sigma) \geq R_2(\sigma) \geq \cdots \geq R_m(\sigma); \\
\Delta_\Sigma(\sigma) = \sum_{i=1}^{k} \Delta_i; \; R_\Sigma(\sigma) = \sum_{i=k+1}^{m} R_i.
\end{gathered}
$$

Suppose $\Omega_\Sigma(\sigma) = \Delta_\Sigma(\sigma) = 0$. In this case,

$$z(\sigma) = \sum_{i=1}^{(\sigma)} \Delta_i(\sigma) = \sum_{i=1}^{k} \Delta_i = 0.$$

According to Corollary 6.1, the schedule σ is optimal.

Otherwise, suppose $\Omega_\Sigma(\sigma) = R_\Sigma(\sigma) = 0$. Consider a schedule σ' in which $R_\Sigma(\sigma') > 0$. According to Theorem 6.7,

$$\Delta_\Sigma(\sigma') - \Delta_\Sigma(\sigma) = R_\Sigma(\sigma') - R_\Sigma(\sigma)$$

and, according to Eq. (6.2), $f(\sigma') > f(\sigma)$. Hence, the schedule σ is optimal.

The value of $\Omega_\Sigma(\sigma) = \min(R_\Sigma(\sigma), \Delta_\Sigma(\sigma))$ is the main characteristic of a schedule $\sigma \in \Psi_P$ where $\Delta_\Sigma(\sigma)$ shows how much we theoretically able to decrease the functional value in order to obtain an optimal schedule. The total reserve $R_\Sigma(\sigma)$ shows the amount of existing reserves to obtain an optimal schedule.

Statement 6.5 If $z(\sigma) = 0$ in a schedule $\sigma \in \Psi_P$, then $\Omega_\Sigma(\sigma) = 0$.

Proof Consider a schedule $\sigma \in \Psi_P$ in which

$$\Delta_1(\sigma) \leq \Delta_2(\sigma) \leq \cdots \leq \Delta_m(\sigma).$$

In this schedule,

$$z(\sigma) = \sum_{i=1}^{c(\sigma)} \Delta_i(\sigma) = 0$$

in the following cases:

1. The schedule is even (because $c(\sigma) = 0$ in it);
2. $R_i(\sigma) = 0\,\forall i = \overline{1,\ c(\sigma)}$;
3. $\Delta_i(\sigma) = 0\,\forall i = \overline{1,\ c(\sigma)}$.

But if any of these conditions is true, we have
$\Omega_\Sigma(\sigma) = \min(R_\Sigma(\sigma), \Delta_\Sigma(\sigma)) = 0$.
This completes the proof. ∎

Let σ^* denote an optimal schedule for the set of tasks J.

Theorem 6.10 *The following expression is true for any schedule* $\sigma \in \Psi_P$:

$$f(\sigma) - f(\sigma^*) \leq \Omega_\Sigma(\sigma).$$

Proof Let us call a schedule σ_T^* with $z(\sigma_T^*) = 0$ a *theoretically optimal* schedule. Suppose that there is a theoretically optimal schedule $\sigma_{T_1}^* \in \Psi_P$ for the set of tasks J. In this case, according to Statement 6.5, $\Omega_\Sigma\left(\sigma_{T_1}^*\right) = 0$. Therefore,

$$f(\sigma) - f\left(\sigma_{T_1}^*\right) = \Omega_\Sigma(\sigma).$$

Suppose that there is a theoretically optimal schedule $\sigma_{T_2}^* \in \Psi_{PS} \backslash \Psi_P$ for the set of tasks J. In this case, $f\left(\sigma_{T_2}^*\right) = f\left(\sigma_{T_1}^*\right)$. Therefore, we also have

$$f(\sigma) - f\left(\sigma_{T_2}^*\right) = \Omega_\Sigma(\sigma).$$

Suppose now that there is a theoretically optimal schedule σ^* with $z(\sigma^*) \neq 0$ for the considered set of tasks J. According to definition of $z(\sigma)$, we have $z(\sigma) = \Delta_\Sigma(\sigma)$ in any schedule σ with $z(\sigma) > 0$. Therefore,

$$z(\sigma^*) = \Delta_\Sigma(\sigma); \quad f(\sigma) - f(\sigma^*) \leq \Omega_\Sigma(\sigma)$$

In the general case, according to Theorem 6.5, $f(\sigma) - f(\sigma^*) \leq z(\sigma)$ and, consequently, $f(\sigma) - f(\sigma^*) \leq \Omega_\Sigma(\sigma)$. ∎

Thus, the value of $\Omega_\Sigma(\sigma)$ is the estimate of deviation of the functional value from the total tardiness in an optimal schedule $\sigma \in \Psi_P$.

Statement 6.6 $\Omega_\Sigma(\sigma) \le z(\sigma)$ in schedules $\sigma \in \Psi_P$.

Proof As we have shown above, if $z(\sigma) = 0$, then $\Omega_\Sigma(\sigma) = 0$, and if $z(\sigma) > 0$, then $z(\sigma) = \Delta_\Sigma(\sigma)$. Therefore, $\Omega_\Sigma(\sigma) \le z(\sigma)$ for any schedule $\sigma \in \Psi_P$. ∎

Let $I_R(\sigma)$ be the set of machine numbers with the smaller number of tardy tasks in a schedule σ, $I_\Delta(\sigma)$ be the set of machine numbers with the greater number of tardy tasks in a schedule σ.

Let us introduce the new class $\Psi(\sigma_P) \subset \Psi_{PS}$ that consists from arbitrary schedules σ obtained as a result of sequential directed permutations that we perform in an arbitrary order decreasing $\Delta_\Sigma(\sigma)$ and, consequently, $R_\Sigma(\sigma)$. We do these permutations sequentially in the current schedule σ_k, starting from some schedule $\sigma \in \Psi_P$, moving non-tardy tasks between the machines $I_\Delta(\sigma_k)$ and $I_R(\sigma_k)$. Doing this, we do not change the order of tasks processing on other machines than those mentioned above. We obtain a schedule σ_{k+1}. The permutation may change the number of tardy tasks only on one machine with the number $i_1 \in I_\Delta(\sigma_k)$ and on one machine with the number $i_2 \in I_R(\sigma_k)$. Permutation is forbidden if the number of tasks on machine i_2 is greater than the number of tasks on machine i_1 in the schedule σ_{k+1}.

Let us prove the following theorem.

Theorem 6.11 *The following estimate of deviation of the functional value from the optimum is valid for any schedule* $\sigma \in \Psi(\sigma_P)$:

$$f(\sigma) - f(\sigma^*) \le \Omega_\Sigma(\sigma).$$

Proof At the first step, we perform the permutations defined above in a schedule $\sigma_1 \in \Psi_P$ for any two machines from this schedule σ_1.

Validity of Theorem 6.11 is based on Theorems 6.7 and 6.8. To prove the theorem in the general case, consider any two schedules $\sigma' \in \Psi(\sigma_P)$ and $\sigma'' \in \Psi(\sigma_P)$.

Constructing the schedule $\sigma'' \in \Psi(\sigma_P)$ by means of directed permutations in the schedule $\sigma' \in \Psi(\sigma_P)$, we obtain a schedule of one of two types:

1. $L_{\max}(\sigma'') = L_{\max}(\sigma')$;
2. An even schedule.

Consequently, it is not difficult to show that Theorem 6.7 proved for schedules $\sigma \in \Psi_P$ also holds for schedules $\sigma \in \Psi(\sigma_P)$.

Let us write the change in the total tardiness value for schedules $\sigma' \in \Psi(\sigma_P)$ and $\sigma'' \in \Psi(\sigma_P)$:

$$\begin{aligned} f(\sigma') - f(\sigma'') &= L\Delta_\Sigma(\sigma') - (L-1)R_\Sigma(\sigma') + \tilde{F}_\Sigma(\sigma') \\ &\quad - L(\Delta_\Sigma(\sigma') - \delta_\Sigma(\sigma')) + (L-1)(R_\Sigma(\sigma') - \delta_\Sigma(\sigma')) - \tilde{F}_\Sigma(\sigma') = \delta_\Sigma(\sigma') \end{aligned}$$

Therefore, $f(\sigma') - f(\sigma'') = \Omega_\Sigma(\sigma') - \Omega_\Sigma(\sigma'')$ *and* $f(\sigma) - f(\sigma^*) \le \Omega_\Sigma(\sigma)$. ∎

Corollary 6.2 Theorems 6.9, 6.10, and Statement 6.6 hold for schedules $\sigma \in \Psi(\sigma_P)$.

Hence, the change in the functional value is determined by values of $\Delta_\Sigma(\sigma)$ and $R_\Sigma(\sigma)$ in schedules $\sigma \in \Psi(\sigma_P)$ as well as in schedules $\sigma \in \Psi_P$.

Theorem 6.12 *If* $\Omega_\Sigma(\sigma) = \min(R_\Sigma(\sigma), \Delta_\Sigma(\sigma))$ *reaches its minimum in a schedule* $\sigma \in \Psi(\sigma_P)$, *then the schedule* σ *is optimal.*

Proof It follows from the proof of Theorem 6.11 that if $\Omega_\Sigma(\sigma') < \Omega_\Sigma(\sigma'')$, then $f(\sigma') < f(\sigma'')$ by the same amount. ∎

Corollary 6.2 Theorems 6.9, 6.10 and Statement 6.6 hold for schedules $\sigma \in \Psi(\sigma_P)$.

Corollary 6.3 Theorem 6.12 also holds for the class of schedules Ψ_P.

Corollary 6.4 SSO #2 also follows from Theorem 6.12.

Corollary 6.5 According to Theorem 6.12, the logic of an approximation algorithm construction is based on a step-by-step decreasing of $\Omega_\Sigma(\sigma)$ where σ is the current schedule. The upper bound for the deviation of the obtained solution from an optimum is determined by Theorems 6.10 and 6.11. The actual estimate of deviation is less than the upper bound because the value of $\Omega_\Sigma(\sigma^*)$ is unknown.

In the following statements with obvious proof we formulate the properties of tasks $j \in S \cup Q$ in schedules $\sigma \in \Psi(\sigma_P)$.

Statement 6.7 Suppose $S'_k(\sigma) \cup Q'_k(\sigma)$ and $S''_l(\sigma) \cup Q''_l(\sigma)$ are the sets of tardy tasks on machines k and l, respectively. Perform the permutation of the sets of tardy tasks between the machines, so that $S''_k(\sigma) \cup Q''_k(\sigma)$ and $S'_l(\sigma) \cup Q'_l(\sigma)$ are satisfied. The deviation of the functional value from the optimum in the obtained schedules is determined by $\Omega_\Sigma(\sigma) = \min(R_\Sigma(\sigma), \Delta_\Sigma(\sigma))$.

Suppose that the tasks $j \in S \cup Q$ on machines k, l, r in the schedule σ are numbered this way: $j_k, j_{k+m}, j_{k+2m}, \ldots, j_l, j_{l+m}, j_{l+2m}, \ldots, j_r, j_{r+m}, j_{r+2m}, \ldots$.

Definition 6.2 The tasks j_k, j_l, j_r, or $j_{k+m}, j_{l+m}, j_{r+m}, \ldots$, or $j_{k+2m}, j_{l+2m}, j_{r+2m}, \ldots$ are called the *tardy tasks of the same level.*

Statement 6.8 We can interchange the tardy tasks of the same level on the machines with the same number of tardy tasks. The functional value does not change with such permutations.

Statement 6.9 We can always move from a schedule $\sigma \in \Psi(\sigma_P)$ to a schedule $\sigma \in \Psi_{PS}$ with the same functional value permuting the tardy tasks of the same level.

6.1.2 *Scheme of the Problem Solving*

We have constructed the computational scheme for finding an optimal schedule on the basis of the theorems and statements proved above. We can distinguish five stages in this scheme.

Stage I Construction of initial schedule $\sigma_0 \in \Psi_P$. If $\Omega_\Sigma(\sigma_0) = 0$, then the schedule is optimal, terminate. Otherwise, if $R_\Sigma(\sigma_0) \geq \Delta_\Sigma(\sigma_0)$, go to Stage II. Else, if $R_\Sigma(\sigma_0) < \Delta_\Sigma(\sigma_0)$, go to Stage III.

Stage II Construction of an even schedule $\sigma \in \Psi(\sigma_P)$ with the number of tardy tasks on each machine equal to $L(\sigma) = L(\sigma_0) - 1$ where $L(\sigma_0)$ is the maximum number of tardy tasks on a machine in the schedule σ_0. If such schedule is found, then it is optimal, terminate. Otherwise, go to Stage IV.

Stage III Construction of an even schedule $\sigma \in \Psi(\sigma_P)$ with the number of tardy tasks on each machine equal to $L(\sigma) = L(\sigma_0)$.

Stage IV Construction of a schedule $\sigma \in \Psi(\sigma_P)$ for which $\Omega_\Sigma(\sigma) < \Omega_\Sigma(\sigma_0)$. If $\Omega_\Sigma(\sigma) = 0$, then the schedule is optimal, terminate. Otherwise, go to Stage V.

Stage V Construction of a schedule $\sigma \in \Psi_{PS}$ for which $\Omega_\Sigma(\sigma) < \Omega_\Sigma(\sigma_0)$. The end of the algorithm.

Let $W(\sigma)$ denote an estimate of deviation from the optimum for the schedule σ.

To implement the five stages, we use the following types of permutations in the algorithm below.

Permutation $1P - 0P - \Delta$. We move a task j from machine $h \in I_\Delta(\sigma)$ to machine $r \in I_R(\sigma)$. The task j is such that $l_j \geq \Delta_h(\sigma)$, $l_j \leq R_r(\sigma)$.
Permutation $1P - 1P - \Delta$. We interchange a task j' from machine $h \in I_\Delta(\sigma)$ and a task j'' from machine $r \in I_R(\sigma)$. The tasks j', j'' are such that $l'_j - l''_j \geq \Delta_h(\sigma)$, $l'_j - l''_j \leq R_r(\sigma)$.

We use the permutations given above during construction of an even schedule $\sigma \in \Psi(\sigma_P)$ at Stage II. After each of the permutations we obtain the schedule σ' for which

$$\begin{aligned}
I_\Delta(\sigma') &= I_\Delta(\sigma) \backslash \{h\}; \\
|P(\sigma')| &= |P(\sigma)| + 1; \\
\Delta_\Sigma(\sigma') &= \Delta_\Sigma(\sigma) - \Delta_h(\sigma); \\
\Delta_\Sigma(\sigma') &= \Delta_\Sigma(\sigma) - \Delta_h(\sigma).
\end{aligned}$$

We use the following two permutations at Stage III.

Permutation $1P - 0P - R\Delta$. We move a task j from machine $h \in I_\Delta(\sigma)$ to machine $r \in I_R(\sigma)$. The task j is such that $l_j < \Delta_h(\sigma)$, $l_j > R_r(\sigma)$.
Permutation $1P - 1P - R\Delta$. We interchange a task j' from machine $h \in I_\Delta(\sigma)$ and a task j'' from machine $r \in I_R(\sigma)$. The tasks j', j'' are such that $l'_j - l''_j < \Delta_h(\sigma)$, $l'_j - l''_j > R_r(\sigma)$.

After each of the permutations we obtain the schedule σ' for which

$$\begin{aligned} I_\Delta(\sigma') &= I_\Delta(\sigma) \cup \{r\}; \\ |P(\sigma')| &= |P(\sigma)| - 1; \\ \Delta_\Sigma(\sigma') &= \Delta_\Sigma(\sigma) - R_r(\sigma); \\ f(\sigma') &= f(\sigma) - R_r(\sigma). \end{aligned}$$

We use the following two permutations at Stage IV.

Permutation $1P - 0P - R$. We move a task j from machine $h \in I_\Delta(\sigma)$ to machine $r \in I_R(\sigma)$. The task j is such that $l_j \le \Delta_h(\sigma)$, $l_j \le R_r(\sigma)$.

After this permutation we obtain the schedule σ' for which $I_\Delta(\sigma') = I_\Delta(\sigma) \backslash \{h\}$ if $l_j = \Delta_h(\sigma)$ or $I_\Delta(\sigma') = I_\Delta(\sigma)$ if $l_j < \Delta_h(\sigma)$; $R_\Sigma(\sigma') = R_\Sigma(\sigma) - l_j$; $f(\sigma') = f(\sigma) - l_j$.

Permutation $1P - 1P - R$. We interchange a task j' from machine $h \in I_\Delta(\sigma)$ and a task j'' from machine $r \in I_R(\sigma)$. The tasks j', j'' are such that $l'_j - l''_j \le \Delta_h(\sigma)$, $l'_j - l''_j \le R_r(\sigma)$.

After this permutation we obtain the schedule σ' for which $I_\Delta(\sigma') = I_\Delta(\sigma) \backslash \{h\}$ if $l'_j - l''_j = \Delta_h(\sigma)$ or $I_\Delta(\sigma') = I_\Delta(\sigma)$ if $l'_j - l''_j < \Delta_h(\sigma)$;

$$\begin{aligned} R_\Sigma(\sigma') &= R_\Sigma(\sigma) - \left(l'_j - l''_j\right); \\ f(\sigma') &= f(\sigma) - \left(l'_j - l''_j\right). \end{aligned}$$

We use the following three permutations at Stage V to obtain the schedule $\sigma \in \Psi_{PS}$. Tasks of the two sets $P(\sigma)$ and $S(\sigma)$ take part in these permutations with the aim to decrease the value of $W(\sigma)$.

Permutation $1P - 1S - W1$. We interchange a task $j' \in P_r(\sigma)$, $r \in I_R(\sigma)$, and a task $j'' \in S_h(\sigma)$, $h \in I_\Delta(\sigma)$. The tasks j', j'' are such that $l''_j \le l'_j + R_r(\sigma)$ and $l'_j < R_h(\sigma)$.

After this permutation we obtain the schedule σ' for which

$$\begin{aligned} I_\Delta(\sigma') &= I_\Delta(\sigma) \backslash \{h\}; \\ f(\sigma') &= f(\sigma) - \left(l''_j - R_h(\sigma)\right); \\ W(\sigma') &= W(\sigma) - \left(l''_j - R_h(\sigma)\right) \end{aligned}$$

Permutation $1P - 1S - W2$. *We interchange a task* $j' \in P_r(\sigma)$, $r \in I_R(\sigma)$, *and a task* $j'' \in S_h(\sigma)$, $h \in I_\Delta(\sigma)$. *The tasks* j', j'' *are such that* $l''_j \le l'_j + R_r(\sigma)$ *and* $l'_j > R_h(\sigma)$.

After this permutation we obtain the schedule σ' for which

$$I_\Delta(\sigma') = I_\Delta(\sigma);$$
$$W(\sigma') = W(\sigma) - \left(l''_j - R_h(\sigma)\right);$$
$$W(\sigma') = W(\sigma) - \left(l''_j - l'_j\right).$$

Permutation $1P - 1S - W3$. We interchange a task $j' \in P_r(\sigma)$, $r \in I_R(\sigma)$, and a task $j'' \in S_h(\sigma)$, $h \in I_\Delta(\sigma)$. The tasks j', j'' are such that $l''_j > l'_j + R_r(\sigma)$ and $l'_j < R_h(\sigma)$.

After this permutation we obtain the schedule σ' for which

$$I_\Delta(\sigma') = I_\Delta(\sigma);$$
$$f(\sigma') = f(\sigma) - \left(l'_j + R_r(\sigma) - R_h(\sigma)\right);$$
$$W(\sigma') = W(\sigma) - \left(l'_j + R_r(\sigma) - R_h(\sigma)\right).$$

These types of permutations may be used during the algorithm execution for individual tasks as well as for task groups $j \in J_i$, $i = \overline{1,m}$. In the latter case, we should take into account the total processing time of the tasks $j \in J_i$, $i = \overline{1,m}$ when checking the conditions for permutations performing. And in this case we execute an approximation subalgorithm consisting in the search for various arrangements of tasks on the machines.

The sufficient signs of optimality of a feasible solution in the proposed algorithm for the TTP problem solving are the result of the problem properties research. They are valid and efficient both in case of approximation algorithms and in case of fulfillment of the first polynomial component of the algorithm. This differs the algorithm for the TTP problem solving from that for the TT problem.

6.1.3 Description of PSC-Algorithms for the Problem Solving

We have constructed PSC-algorithms A1 and A2 for the problem solving in accordance with the computational scheme given above. Algorithm A1 is more accurate (the deviation of the solution from the optimum is less than that for Algorithm A2 when the approximation subalgorithm is executed). But it is more time consuming than Algorithm A2. We have built the latter one on the basis of efficiency research of Algorithm A1. We have determined the most frequently performed permutations during the execution of the first polynomial component of Algorithm A1. Also, we have developed a more efficient algorithm for an initial schedule construction. This allowed us to reduce the complexity of the algorithm significantly.

The PSC-algorithms for the TTP problem solving include the first polynomial component and the second polynomial component (the approximation algorithm). We have built them solely on directed permutations. The first polynomial component of the algorithm is the deterministic procedure of sequential execution of directed permutations. The total number of permutations is limited by the polynomial of the number of tasks and the number of machines. After the problem solving, we obtain either a strictly optimal solution by the first polynomial component of the algorithm (if any SSO was satisfied during the computation) or an approximate one with an upper bound for the deviation from the optimum.

Algorithm A1$_0$. Construction of Initial Schedule $\sigma_0 \in \Psi_P$ It has six steps.

1. Renumber the tasks of the set J in non-decreasing order of processing times.
2. Initiate the release times of machines: $C_i = 0$, $i = \overline{1, m}$.
3. Set $j = 1$, $i = 1$.
4. If $l_j + C_i > d$, then go to step 6. Otherwise, assign task j to machine i, set $C_i = C_i + l_j$.
5. Continue with the next task: $j = j + 1$. If $j > n$, then the algorithm terminates. Otherwise, go to step 4.
6. Continue with the next machine: $i = i + 1$. If $i \le m$, then go to step 4. Otherwise, set $Q(\sigma_0) = \{j, j+1, \ldots, n\}$ and execute Algorithm AQ. The end of algorithm A1$_0$.

Algorithm AQ has six steps.

1. Determine the reserves of machines: $R_i(\sigma_0) = d - \sum_{j \in P_i(\sigma_0)} l_j$, $i = \overline{1, m}$.
2. Renumber the machines so that $R_1(\sigma_0) \ge R_2(\sigma_0) \ge \ldots R_m(\sigma_0)$.
3. Set $i = 1$.
4. Assign task j to machine i.
5. Set $j = j + 1$. If $j > n$, then Algorithm AQ terminates, we have obtained the schedule σ_0. Otherwise, go to step 6.
6. Set either $i = i + 1$ if $i < m$ or $i = 1$ if $i = m$. Go to step 4.

Algorithm A1 has five stages.

Stage I Construction of initial schedule $\sigma_0 \in \Psi_P$.

1. Construct an initial schedule $\sigma_0 \in \Psi_P$ using Algorithm A1$_0$.
2. Determine $\Omega_\Sigma(\sigma_0)$. If $\Omega_\Sigma(\sigma_0) = 0$, then the schedule σ_0 is optimal, terminate. Otherwise, go to step 3.
3. If $R_\Sigma(\sigma_0) \ge \Delta_\Sigma(\sigma_0)$, then go to step 4. Otherwise, set $\sigma = \sigma_0$, go to step 9.

Stage II Construction of an even schedule $\sigma \in \Psi(\sigma_P)$ with the number of tardy tasks on each machine equal to $L(\sigma) = L(\sigma_0) - 1$.

4. Set $\sigma = \sigma_0$, $h = 1$, $f = F(\sigma)$ where $F(\sigma)$ is the number of machines with the greater number of tardy tasks.
5. If a permutation $1P - 0P - \Delta$ is possible for machine h, then perform it, we obtain a schedule σ', go to step 7. Otherwise, go to step 6.
6. If a permutation $1P - 1P - \Delta$ is possible for machine h, then perform it and go to step 7. Otherwise (if the permutation is not possible), go to step 7.

7. Set $h = h + 1$. If $h \leq f$, then go to step 5, else go to step 8.
8. If $\Omega_\Sigma(\sigma_0) = 0$, then the first polynomial component of the algorithm is fulfilled, the schedule σ is optimal, terminate. Otherwise, go to step 14.

Stage III Construction of an even schedule $\sigma \in \Psi(\sigma_P)$ with the number of tardy tasks on each machine equal to $L(\sigma) = L(\sigma_0)$.

9. Set $r = F(\sigma) + 1$.
10. If a permutation $1P - 0P - R\Delta$ is possible for machine r, then perform it, we obtain a schedule σ', go to step 12. Otherwise, go to step 11.
11. If a permutation $1P - 1P - R\Delta$ is possible for machine r, then perform it, we obtain a schedule σ'. Go to step 12.
12. Set $r = r + 1$. If $r \leq m$, then go to step 10, else go to step 13.
13. If $\Omega_\Sigma(\sigma_0) = 0$, then the first polynomial component of the algorithm is fulfilled, the schedule σ is optimal, terminate. Otherwise, go to step 14.

Stage IV Construction of a schedule $\sigma \in \Psi(\sigma_P)$ for which $\Omega_\Sigma(\sigma) < \Omega_\Sigma(\sigma_0)$.

14. Set $r = F(\sigma) + 1$.
15. If a permutation $1P - 0P - R$ is possible for machine r, then perform it, we obtain a schedule σ', go to step 17. Otherwise, go to step 16.
16. If a permutation $1P - 1P - R$ is possible for machine r, then perform it, we obtain a schedule σ'. Go to step 17.
17. Set $r = r + 1$. If $r \leq m$, then go to step 15, else go to step 18.
18. If $\Omega_\Sigma(\sigma_0) = 0$, then the first polynomial component of the algorithm is fulfilled, the schedule σ is optimal, terminate. Otherwise, go to step 19.

Stage V Construction of a schedule $\sigma \in \Psi_{PS}$ for which $\Omega_\Sigma(\sigma) < \Omega_\Sigma(\sigma_0)$.

19. Redistribute tasks of the set $J \backslash P(\sigma')$ to machines without a change in the machines numbering. Set $\sigma = \sigma'$.
20. If a permutation $1P - 1S - W1$ is possible for machine h, then perform it, we obtain a schedule σ'. Go to step 21.
21. If a permutation $1P - 1S - W2$ is possible for machine h, then perform it, we obtain a schedule σ'. Go to step 22.
22. If a permutation $1P - 1S - W3$ is possible for machine h, then perform it, we obtain a schedule σ'. Go to step 23.
23. Set $h = h + 1$. If $h \leq f$, then go to step 20. Otherwise, determine the functional value $f(\sigma')$ and the upper bound of the deviation of the solution from the optimum $W(\sigma')$. The end of Algorithm A1.

Let us determine the complexity of Algorithm A1. At Stage I, we execute Algorithm $\mathrm{A1}_0$, its complexity is determined by the function $O(mn \log n)$. The computational complexity of Stages III–V does not exceed the complexity of Stage II. Consider Stage II. $1P - 1P - \Delta$ is the very time consuming permutation at this stage. Checking the possibility to perform this permutation is as complex as $O(n^2 m)$. Consequently, the complexity of Algorithm A1 is bounded with a polynomial function $O(n^2 m)$.

Algorithm A2$_0$. Construction of Initial Schedule $\sigma_0 \in \Psi_P$ It has six steps.

1. Renumber the tasks of the set J in non-decreasing order of processing times.
2. Initiate the release times of machines: $C_i = 0$, $i = \overline{1, m}$.
3. Set $j = 1$, $i = 1$.
4. Assign task j to a machine i with the minimum release time C_i.
5. Set $C_i = C_i + l_j$.
6. Continue with the next task: $j = j + 1$. If $j > n$, then Algorithm A2$_0$ terminates. Otherwise, go to step 4.

Algorithm A2 has three stages.

Stage I Construction of initial schedule $\sigma_0 \in \Psi_P$.

1. Construct an initial schedule $\sigma_0 \in \Psi_P$ using Algorithm A2$_0$.
2. Determine $\Omega_\Sigma(\sigma_0)$. If $\Omega_\Sigma(\sigma_0) = 0$, then the schedule σ_0 is optimal, terminate. Otherwise, go to step 3.
3. If $R_\Sigma(\sigma_0) \geq \Delta_\Sigma(\sigma_0)$, then go to step 4. Otherwise, set $\sigma = \sigma_0$, go to step 9.

Stage II Construction of an even schedule $\sigma \in \Psi(\sigma_P)$ with the number of tardy tasks on each machine equal to $L(\sigma) = L(\sigma_0) - 1$.

4. Set $\sigma = \sigma_0$, $h = 1$, $f = F(\sigma)$ where $F(\sigma)$ is the number of machines with the greater number of tardy tasks.
5. If a permutation $1P - 0P - \Delta$ is possible for machine h, then perform it, we obtain a schedule σ', go to step 6. Otherwise, go to step 7.
6. Set $h = h + 1$. If $h \leq f$, then go to step 5, else go to step 8.
7. If a permutation $1P - 0P - R$ is possible for machine h, then perform it. Go to step 6.
8. If $\Omega_\Sigma(\sigma) = 0$, then the first polynomial component of the algorithm is fulfilled, the schedule σ is optimal, determine the functional value $f(\sigma)$. Otherwise, determine the value of deviation of the solution from the optimum $W(\sigma)$. The end of Algorithm A2.

Stage III Construction of an even schedule $\sigma \in \Psi(\sigma_P)$ with the number of tardy tasks on each machine equal to $L(\sigma) = L(\sigma_0)$.

9. Set $r = F(\sigma) + 1$.
10. If a permutation $1P - 0P - R\Delta$ is possible for machine r, then perform it, we obtain a schedule σ', go to step 11. Otherwise, go to step 12.
11. Set $r = r + 1$. If $r \leq m$, then go to step 10, else go to step 13.
12. If a permutation $1P - 0P - R$ is possible for machine r, then perform it, we obtain a schedule σ'. Go to step 11.
13. If $\Omega_\Sigma(\sigma) = 0$, then the first polynomial component of the algorithm is fulfilled, the schedule σ is optimal, determine the functional value $f(\sigma)$. Otherwise, determine the value of deviation of the solution from the optimum $W(\sigma)$. The end of Algorithm A2.

The complexity of Algorithm A2 is determined by the polynomial $O(mn \log n)$.

6.2 The Problem with Arbitrary Fixed Start Times of Machines Which Are Less Than the Common Due Date

6.2.1 *The Problem Formulation*

Given a set of n tasks J and m identical parallel machines. We know a processing time l_j for each task $j \in J$. All tasks have a common due date d. We assume that machine i, $i = \overline{1, m}$, can start to process any task of the set J after the time point s_i. The start times of the machines s_i satisfy $s_i < d$, $i = \overline{1, m}$ and may be different. The machines idle times are forbidden. We need to build such a schedule σ of the tasks $j \in J$ processing on m machines that minimizes the functional

$$f(\sigma) = \sum_{j \in J} \max\left(0,\ C_j(\sigma) - d\right)$$

where $C_j(\sigma)$ is the completion time of task j in the schedule σ.

This problem (we will call it TTPL) is NP-hard [4].

6.2.2 *Studying the Properties of the TTPL Problem and PSC-Algorithm for Its Solving*

Let us show that Statement 6.1 and Theorem 6.7 are valid for the TTPL problem.

Statement 6.10 (Pavlov, Misura, and Lisetsky [3]) The following is true for all possible schedules $\sigma \in \Psi_P$ constructed on the set of tasks J:

$$P_{\max} - P_{\min} < m.$$

Proof For any schedule $\sigma \in \Psi_P$, we have:

$$\sum_{j=1}^{|P(\sigma)|} l_j = d \cdot m - \sum_{i=1}^{m} s_i - \sum_{i=1}^{m} R_i(\sigma).$$

Therefore, such a schedule σ in which $\sum_{i=1}^{m} R_i(\sigma)$ has the maximum (minimum) possible value among all schedules in class Ψ_P, satisfies $|P(\sigma)| = P_{\min}$ (or $P_{\max}$, respectively). But $R_i(\sigma) < \min_{j \in S(\sigma)} l_j$ is true for each machine i in a schedule $\sigma \in \Psi_P$.

Hence,

$$\sum_{i=1}^{m} R_i(\sigma) < m \min_{j\in S(\sigma)} l_j \le \sum_{i=1}^{m} l_{|P(\sigma)|+i} \text{ and } P_{\max} - P_{\min} < m. \qquad \blacksquare$$

Theorem 6.13 (Pavlov, Misura, and Lisetsky [3]) *If the maximum number of tardy tasks* $L_{\max}$ *is the same in schedules* $\sigma \in \Psi_P$ *and* $\sigma' \in \Psi_P$ *constructed on the given set of tasks J, then from* $R_\Sigma(\sigma) \ne 0$ *and* $R_\Sigma(\sigma') \ne 0$ *follows* $R_\Sigma(\sigma) - R_\Sigma(\sigma') = \Delta_\Sigma(\sigma) - \Delta_\Sigma(\sigma')$.

Proof Suppose that $md - \sum_{i=1}^{m} s_i = R^0$ and $R_1 \ge R_2 \ge R_3 \ge \ldots R_m$. Then,

$$\begin{aligned}
R^0 &= \sum_{j\in P(\sigma)} l_j + \sum_{j\in S(\sigma)} l_j - \sum_{i=1}^{m} \Delta_i = \sum_{j\in P(\sigma)} l_j \\
&\quad - \sum_{i=1}^{k} \Delta_i(\sigma) + \sum_{j=p+1}^{p+k} l_j + \left(\sum_{j=p+k+1}^{p+m} l_j - \sum_{i=k+1}^{m} \Delta_i \right) \\
&= \sum_{i=1}^{k} \Delta_i(\sigma) + \sum_{j=p+1}^{p+k} l_j + \left(\sum_{j=p+k+1}^{p+m} l_j - \sum_{i=k+1}^{m} \Delta_i \right) = \text{const}
\end{aligned} \tag{6.3}$$

Consider the following cases:

1. $P(\sigma) = P(\sigma')$. Theorem is obviously valid according to (6.3).
2. $P(\sigma) < P(\sigma')$. In this case, tasks $j \in S(\sigma)$, $j = \overline{p+1, p+k'}$, $k' < k$, became non-tardy, that is, moved to the set $P(\sigma')$:

$$R^0 = \left(\sum_{j\in P(\sigma)} l_j + \sum_{j=p+1}^{p+k'} l_j \right) + \left(\sum_{j=p+k'+1}^{p+k} l_j - \Delta_\Sigma(\sigma') \right) + R_\Sigma(\sigma').$$

   Theorem is valid according to (6.3).
3. $P(\sigma) > P(\sigma'')$, $\Delta_\Sigma(\sigma) < \Delta_\Sigma(\sigma'')$. In this case, the number of tardy tasks on machines $i = \overline{k+1, k+k''}$, $k'' < m$, increased because some tasks from the set $P(\sigma)$ became tardy:

$$R^0(\sigma) = \sum_{j\in P(\sigma)} l_j - \sum_{j=p-k'\prime+1}^{p} l_j + \left(\sum_{j=p-k'\prime+1}^{p} l_j + \sum_{j=p+1}^{p+k} l_j - \Delta_\Sigma(\sigma') \right) + R_\Sigma(\sigma').$$

   Theorem is valid according to (6.3).

PSC-algorithm for the problem solving includes the preliminary stage (Algorithm A0 given below) and the optimization stage. Any of the algorithms listed in Sect. 6.1.3 (A1 or A2) may serve as the optimization stage.

We specify for this problem a polynomial restriction on the amount of computations dependent on the problem size. The value of $\Omega_\Sigma(\sigma) = \min(R_\Sigma(\sigma), \Delta_\Sigma(\sigma))$ is an estimate of the deviation of the functional value from the optimum. If it is zero, then the first polynomial component of the algorithm is fulfilled, and the obtained schedule is optimal.

Since we have different start times of machines in the TTPL problem, in contrast to the TTP problem, let us give a new algorithm for construction of initial schedule $\sigma^{ord} \in \Psi_P$.

Algorithm A0. Construction of Initial Schedule $\sigma^{ord} \in \Psi_P$ It has six steps.

1. Renumber the tasks of the set J in non-decreasing order of processing times.
2. Renumber the machines in non-decreasing order of start times s_i.
3. Initiate the release times of machines: $C_i = s_i \; \forall i = \overline{1, m}$.
4. Select an unassigned task j with the minimum processing time l_j. Assign the task j to a machine i with the minimum release time C_i.
5. Determine the new release time of the machine i: $C_i = C_i + l_j$.
6. If all tasks were assigned, the algorithm terminates. Otherwise, go to step 4.

The first polynomial component of the PSC-algorithm coincides with its approximation algorithm. As a result of the problem solving, we obtain either a strictly optimal solution by the first polynomial component of the algorithm (if at least one SSO, that is, the conditions of Theorem 6.6 or 6.9, was satisfied) or an approximate solution with an upper bound for the deviation of the functional value from the optimum.

The complexity of the first polynomial component of the PSC-algorithm is determined by the complexity of the algorithm (A1 or A2) used in the problem solving.

6.3 Illustrative Example

Example 6.2 to Algorithm A1 The set of tasks J is in Table 6.2. $m = 4$, $d = 14$.

Table 6.3 contains an initial schedule σ_0. $f(\sigma_0) = 74$.

We perform the permutation $1P - 1P - \Delta$ for $h = 4$, $r = 1$, $j' = 4$, $j'' = 1$. We get an even schedule σ^* (Table 6.4), $f(\sigma^*) = 73$. So, the schedule σ^* is optimal.

Table 6.2 Initial data for Example 6.2

j	l_j	j	l_j	j	l_j	j	l_j	j	l_j	j	l_j
1	1	5	2	9	3	13	5	17	6	21	7
2	1	6	2	10	4	14	5	18	6	22	8
3	1	7	3	11	4	15	5	19	7	23	8
4	2	8	3	12	4	16	6	20	7	24	8

Table 6.3 Initial schedule for Example 6.2

i	j	l_j	d	$C_j(\sigma_0)$	$T_j(\sigma_0)$	i	j	l_j	d	$C_j(\sigma_0)$	$T_j(\sigma_0)$
1	1	1	14	1	0	3	3	1	14	1	0
1	5	2	14	3	0	3	7	3	14	4	0
1	9	3	14	6	0	3	11	4	14	8	0
1	13	5	14	11	0	3	15	5	14	13	0
1	17	6	14	17	3	3	19	7	14	20	6
1	21	7	14	24	10	3	23	8	14	28	14
2	2	1	14	1	0	4	4	2	14	2	0
2	6	2	14	3	0	4	8	3	14	5	0
2	10	4	14	7	0	4	12	4	14	9	0
2	14	5	14	12	0	4	16	6	14	15	1
2	18	6	14	18	4	4	20	7	14	22	8
2	22	8	14	26	12	4	24	8	14	30	16

i machine's number, j task's number, $T_j(\sigma_0)$ tardiness of task j in the schedule σ_0

Table 6.4 Optimal schedule for Example 6.2

i	j	l_j	d	$C_j(\sigma_0)$	$T_j(\sigma_0)$	i	j	l_j	d	$C_j(\sigma_0)$	$T_j(\sigma_0)$
1	4	2	14	2	0	3	3	1	14	1	0
1	5	2	14	4	0	3	7	3	14	4	0
1	9	3	14	7	0	3	11	4	14	8	0
1	13	5	14	12	0	3	15	5	14	13	0
1	17	6	14	18	4	3	19	7	14	20	6
1	21	7	14	25	11	3	23	8	14	28	14
2	2	1	14	1	0	4	1	1	14	1	0
2	6	2	14	3	0	4	8	3	14	4	0
2	10	4	14	7	0	4	12	4	14	8	0
2	14	5	14	12	0	4	16	6	14	14	0
2	18	6	14	18	4	4	20	7	14	21	7
2	22	8	14	26	12	4	24	8	14	29	15

6.4 Statistical Research of PSC-Algorithms for the TTP Problem

The complexity of PSC-algorithm for the TTPL problem is determined by the complexity of PSC-algorithm for the TTP problem solving. Therefore, to evaluate the efficiency of PSC-algorithm for the TTPL problem, we present the statistical data we obtained for the TTP problem.

To study the algorithms efficiency, we used a modeling system written in C# in Microsoft Visual Studio 2010 environment. We randomly generated the tasks sets with uniform distribution of parameters. Processing times of the tasks were chosen from interval $\overline{1,200}$. the due date was calculated as $0.7L/m$ where L is the sum of processing times of all tasks. We carried out 2000 tests for each pair (n, m).

We tested the TTP problem instances with a dimension of up to $n = 40{,}000$ tasks and up to $m = 30$ machines. We studied Algorithms A1 and A2 presented in Sect. 6.1.3. Their complexities are $O(n^2m)$ and $O(mn\log n)$, respectively. We did the research on a PC with Pentium Core 2 Duo processor (3.0 GHz clock speed) and 2 GB of random access memory.

Average (for 100 runs) solving time has appeared in the range:

- from 1.87 ms at $n = 3000$ and $m = 5$ to 2361.71 ms at $n = 40{,}000$ and $m = 15$ for Algorithm A1;
- from 2.18 ms at $n = 3000$ and $m = 30$ to 34.7 ms at $n = 40{,}000$ и $m = 20$ for Algorithm A2.

The average decrease in the functional value after one permutation for 2000 runs at $n = 3000$ was 22.153 for Algorithm A1 and 8.214 for Algorithm A2.

$1P - 0P - \Delta$ and $1P - 0P - R$ appeared to be the most efficient permutations. Their peculiarity is that they decrease the value of a solution optimality estimate $\Omega_\Sigma(\sigma)$ reducing the total reserve $R_\Sigma(\sigma)$, i.e., with the maximum use of the existing reserves of machines.

The average frequency of obtaining an optimal solution was 89.2% for Algorithm Al and 74.1% for Algorithm A2. The statistical probability of fulfilling of the first polynomial component of the algorithms increases with an increase in the number of tasks. The average deviation from an optimal solution was 0.000283 for Algorithm A1 and 0.00016 for Algorithm A2. After additional types of permutations introducing, Algorithm A2 is more efficient than Algorithm A1. In this case, we achieve an optimal functional value by the first polynomial component of the algorithm for 92% of instances.

References

1. Zgurovsky, M.Z., Pavlov, A.A.: Prinyatie Resheniy v Setevyh Sistemah s Ogranichennymi Resursami (Принятие решений в сетевых системах с ограниченными ресурсами; Decision Making in Network Systems with Limited Resources). Naukova dumka, Kyiv (2010) (in Russian)
2. Pavlov, A.A., Misura, E.B.: Minimizaciya summarnogo zapazdyvaniya pri vypolnenii nezavisimyh zadanii s obschim direktivnym srokom identichnymi parallelnymi priborami, momenty zapuska kotoryh proizvolny (Минимизация суммарного запаздывания при выполнении независимых заданий с общим директивным сроком идентичными параллельными приборами, моменты запуска которых произвольны; The total tardiness minimization when processing independent tasks with a common due date on the identical parallel machines with an arbitrary starting times) // Visnyk NTUU KPI Inform. Oper. Comput. Sci. **59**, 28–34 (2013) (in Russian)
3. Pavlov, A.A., Misura, E.B., Lisetsky T.N.: Sostavlenie raspisaniya vypolneniya nezavisimyh zadanii identichnymi parallelnymi priborami, momenty zapuska kotoryh menshe obschego

direktivnogo sroka (Составление расписания выполнения независимых заданий идентичными параллельными приборами, моменты запуска которых меньше общего директивного срока; Scheduling independent tasks on the identical parallel machines which starting times are smaller than a common due date). Visnyk NTUU KPI Inform. Oper. Comput. Sci. **58**, 24–28 (2013) (in Russian)

4. Garey, M.R., Johnson, D.S.: Computers and Intractability: A Guide to the Theory of NP-Completeness. W.H. Freeman and Co, San Francisco (1979) https://doi.org/10.1137/1024022
5. Tanaev, V.S., Shkurba, V.V.: Vvedenie v Teoriju Raspisaniy (Введение в теорию расписаний; Introduction to Scheduling Theory). Nauka, Moscow (1975) (in Russian)

Chapter 7
The Total Weighted Completion Time of Tasks Minimization with Precedence Relations on a Single Machine

Abstract We consider the problem of constructing a schedule for a single machine that minimizes the total weighted completion time of tasks when the restrictions on their processing order are given by an arbitrary oriented acyclic graph. The problem is *NP*-hard in the strong sense. Efficient polynomial algorithms for its solving are known only for cases when the oriented acyclic graph is a tree or a series-parallel graph. We give a new efficient PSC-algorithm of its solving. It is based on our earlier theoretical and practical results and solves the problem with precedence relations specified by an oriented acyclic graph of the general form. The first polynomial component of the PSC-algorithm contains sixteen sufficient signs of optimality. One of them will be statistically significantly satisfied at each iteration of the algorithm when solving randomly generated problem instances. In case when the sufficient signs of optimality fail, the PSC-algorithm is an efficient approximation algorithm. If the sufficient signs of optimality are satisfied at each iteration then the algorithm becomes exact. We present the empirical properties of the PSC-algorithm on the basis of statistical studies.

7.1 The Problem Formulation

This chapter contains our new original results based on our earlier works [1–6].

We can formulate the intractable combinatorial optimization problem "The single machine total weighted completion time of tasks minimization with precedence relations given by an oriented acyclic graph" (TWCT) as follows.

The Problem Statement A partially ordered set of tasks $J = (j_1, j_2, \ldots, j_n)$ is processed on a single machine starting at the start time $s = 0$. We know a processing time $l_j > 0$ and a weight ω_j for each task $j \in J$. The machine processes the tasks without interruptions and not more than one at a time. We specify the partial order $\prec$ between the tasks by an oriented acyclic graph [3].

M. Z. Zgurovsky and A. A. Pavlov, *Combinatorial Optimization Problems in Planning and Decision Making*, Studies in Systems, Decision and Control 173,
https://doi.org/10.1007/978-3-319-98977-8_7

We need to find a sequence of tasks in which their total weighted completion time is minimal:

$$F = \sum_{k=1}^{n} \omega_{j_{[k]}} C_{j_{[k]}} \to \min \tag{7.1}$$

where $C_{j_{[k]}}$ is the completion time of the task in position k in a feasible[1] schedule:

$$C_{j_{[k]}} = \sum_{s=1}^{k} l_{j_{[s]}}.$$

The problem is NP-hard in the strong sense and will remain such even if all processing times or all weights are equal to one. It is solvable in polynomial time if the order $\prec$ is a "forest" or a series-parallel graph. If we introduce individual release times for tasks instead of the precedence relations, then the obtained problem will be NP-hard in the strong sense, even if all weights are equal to one.

7.2 The Main Theoretical Thesises

Consider the functional (7.1). Let us use the following notation. Let i or j_i be the number of task corresponding to the indexation specified by the functional; $j_{[g]}$ be the number of a task in position g in a feasible schedule; $\overline{a,b}$ be the interval of integer numbers from a to b, that is, $\overline{a,b} = \mathbb{Z} \cap [a,b] = a,\ a+1,\ \ldots,\ b$. Let $\Re$ denote the set of all permutations $\pi_r = (j_1, j_2, \ldots, j_r)$, $r = \overline{0,n}$, of elements of the set $N = \{1, 2, \ldots, n\}$ where r is the length of a permutation π_r, $\{\pi_0\} = \varnothing$. If $\Re \subseteq \hat{\Re}$, then we will denote by $Q[\Re]$ such a set of all those permutations $\pi^{(q)} \in \hat{\Re}$, $\pi^{(q)} \neq \pi_0$ for which there are no such permutations $\pi \in \Re$ and $\pi^{(1)}, \pi^{(2)} \in \hat{\Re}$ that $\pi = \left(\pi^{(1)}, \pi^{(q)}, \pi^{(2)}\right)$.

Statement 7.1 (Pavlov et al. [3]) Suppose that a functional $F(\pi) = \sum_{k=1}^{n} \omega_{j_{[k]}} C_{j_{[k]}}$ is given on a set $\Re' \subseteq \hat{\Re}$. Let $\Re \subseteq \Re'$. Then, for any permutations

$$\pi' = \left(\pi^{(1)}, \pi^{(a)}, \pi^{(b)}, \pi^{(2)}\right) \quad \text{and} \quad \pi'' = \left(\pi^{(1)}, \pi^{(b)}, \pi^{(a)}, \pi^{(2)}\right) \tag{7.2}$$

belonging to $\Re$ there is a functional $f(\pi)$ defined on a set $Q[\pi]$ with the following property: $F(\pi') < F(\pi'')$ follows from $f\left(\pi^{(a)}\right) > f\left(\pi^{(b)}\right)$ and $F(\pi') = F(\pi'')$ follows from $f\left(\pi^{(a)}\right) = f\left(\pi^{(b)}\right)$.

[1] A schedule is feasible if it does not violate the precedence relations.

Proof Let us introduce the functional

$$\Phi(\pi, R) = \sum_{k=1}^{r} \omega_{j_{[k]}} \left(R + \sum_{s=1}^{k} l_{j_{[s]}} \right) \tag{7.3}$$

where R is a real number. It is obvious that $F(\pi) = \Phi(\pi, 0)$, $\pi^{(a)}$, and $\pi^{(b)}$ are the subsequences of tasks which do not relate among themselves by the precedence relation. Then,

$$F\left(\pi^{(1)}, \pi^{(a)}, \pi^{(b)}, \pi^{(2)}\right) = \Phi\left(\pi^{(1)}, 0\right) + \Phi\left(\pi^{(a)}, t\left(\pi^{(1)}\right)\right) + \Phi\left(\pi^{(b)}, t\left(\pi^{(1)}, \pi^{(a)}\right)\right) + \Phi\left(\pi^{(2)}, t\left(\pi^{(1)}, \pi^{(a)}, \pi^{(b)}\right)\right);$$

$$F\left(\pi^{(1)}, \pi^{(b)}, \pi^{(a)}, \pi^{(2)}\right) = \Phi\left(\pi^{(1)}, 0\right) + \Phi\left(\pi^{(b)}, t\left(\pi^{(1)}\right)\right) + \Phi\left(\pi^{(a)}, t\left(\pi^{(1)}, \pi^{(b)}\right)\right) + \Phi\left(\pi^{(2)}, t\left(\pi^{(1)}, \pi^{(a)}, \pi^{(b)}\right)\right).$$

Since

$$t\left(\pi^{(1)}, \pi^{(a)}, \pi^{(b)}\right) = t\left(\pi^{(1)}, \pi^{(b)}, \pi^{(a)}\right),$$

then it is necessary and sufficient that

$$\Phi\left(\pi^{(a)}, t\left(\pi^{(1)}\right)\right) + \Phi\left(\pi^{(b)}, t\left(\pi^{(1)}, \pi^{(a)}\right)\right) \le \Phi\left(\pi^{(b)}, t\left(\pi^{(1)}\right)\right) + \Phi\left(\pi^{(a)}, t\left(\pi^{(1)}, \pi^{(b)}\right)\right) \tag{7.4}$$

be true to satisfy the inequality

$$F\left(\pi^{(1)}, \pi^{(a)}, \pi^{(b)}, \pi^{(2)}\right) \le F\left(\pi^{(1)}, \pi^{(b)}, \pi^{(a)}, \pi^{(2)}\right).$$

In other words,

$$\sum_{j \in \{\pi^{(b)}\}} \omega_j \sum_{j \in \{\pi^{(a)}\}} l_j \le \sum_{j \in \{\pi^{(a)}\}} \omega_j \sum_{j \in \{\pi^{(b)}\}} l_j. \tag{7.5}$$

Denote by $p(\pi) = \sum_{j \in \{\pi\}} \omega_j \Big/ \sum_{j \in \{\pi\}} l_j$ the *priority* of a permutation π. It is obvious that the equality (7.5) is true for $l_j \ge 0$ and the values of ω_j defined by arbitrary real numbers. Then, we can rewrite the Eq. (7.5) in the form:

$$p\left(\pi^{(b)}\right) \leq p\left(\pi^{(a)}\right). \tag{7.6}$$

This completes the proof. ■

Definition 7.1 Suppose that a partial order is given with an oriented acyclic graph on the set of tasks J. If $j_k \prec j_l$ and there does not exist a task j_z such that $j_k \prec j_z$, $j_z \prec j_l$, then we will call j_k an *immediate predecessor* of j_l and denote this fact as $j_k \prec\!\prec j_l$.

Definition 7.2 A sequence of tasks $\sigma = \left(j_{[1]}, j_{[2]}, \ldots, j_{[g]}\right)$ is called *feasible* if for any $j_{[k]} \prec j_{[l]}$ processing of the task $j_{[k]}$ completes before processing of the task $j_{[l]}$, according to this sequence.

Definition 7.3 A *priority ordered (p-ordered)* schedule is such a feasible sequence of tasks that satisfies the following properties:

1. (a) A subset $S_1 \subset J$ that has the maximum possible priority is processed first, i.e.,

$$p(S_1) > p(U), \quad \forall U \subset J \backslash S_1. \tag{7.7}$$

   (b) The subset S_1 contains minimal possible number of tasks subject to (7.7).
   (c) A subset $S_2 \subset J \backslash S_1$ satisfying the properties (a) and (b) is processed second, etc.

2. Let $\alpha_i(S_i)$ denote a sequence of tasks of a set S_i. Then, a set of tasks of any beginning subsequence of $\alpha_i(S_i)$ satisfies the property 1.

Theorem 7.1 *An optimal schedule is p-ordered.*

Proof of this theorem follows from successive application of Lemma 3 [7]. The proof of the lemma must be reformulated taking into account the definition of the priority of a task set G in the form $p(G) = \Omega(G)/L(G)$. ■

Theorem 7.2 *A p-ordered schedule is optimal for a partial order specified by a series-parallel graph if all permutations of the form* (7.2) *were completed.*

Proof of the theorem follows from the algorithm of a p-ordered schedule construction for a series-parallel graph [8]. Indeed, different p-ordered schedules may differ only in permutations of the form (7.2) subject to $p\left(\pi^{(a)}\right) = p\left(\pi^{(b)}\right)$. ■

Definition 7.4 A subsequence of tasks $\alpha\left(j_{[g]}\right) = \left(j_{[r]}, j_{[r+1]}, \ldots, j_{[g]}\right)$ in a sequence σ is called a *chain of tasks* if

1. $j_{[r]} \prec\!\prec j_{[r+1]} \prec\!\prec \cdots \prec\!\prec j_{[g]}$;
2. $p\left(j_{[r]}, \ldots, j_{[g-k-1]}\right) \leq p\left(j_{[g-k]}, \ldots, j_{[g-1]}, j_{[g]}\right)$ for each allowable natural k where $p\left(\alpha\left(j_{[g]}\right)\right) = \Omega\alpha\left(j_{[g]}\right) / L\alpha\left(j_{[g]}\right)$ is the priority of the chain $\alpha\left(j_{[g]}\right)$; $\Omega\alpha\left(j_{[g]}\right) = \sum_{j_{[l]} \in \alpha\left(j_{[g]}\right)} \omega_{j_{[l]}}$ is the weight of the chain $\alpha\left(j_{[g]}\right)$; $L\alpha\left(j_{[g]}\right) = \sum_{j_{[l]} \in \alpha\left(j_{[g]}\right)} l_{j_{[l]}}$ is its processing time.

Definition 7.5 A subsequence of tasks $\alpha(j_{[g]}) = (j_{[r]}, j_{[r+1]}, \ldots, j_{[g-1]}, j_{[g]})$ in a sequence σ is called an *elementary construction of tasks* if each task $j_{[l]} \in \alpha(j_{[g]})$ satisfies: $j_{[l]} \prec j_{[g]}$, $j_{[r-1]} \bar{\prec} j_{[g]}$, $l = \overline{r, g-1}$. Here, $j_{[g]}$ is a task that occupies the latest position g in the elementary construction of tasks $\alpha(j_{[g]})$; l and r are the numbers of positions occupied by tasks $j_{[l]}$, $j_{[r]}$ in the sequence σ. Parameters of an elementary construction of tasks $\alpha(j_{[g]})$ are defined through the parameters of the tasks $j_{[l]} \in \alpha(j_{[g]})$ this way: $\Omega\alpha(j_{[g]}) = \sum_{j_{[l]} \in \alpha(j_{[g]})} \omega_{j_{[l]}}$ is the weight of the elementary construction of tasks $\alpha(j_{[g]})$; $L\alpha(j_{[g]}) = \sum_{j_{[l]} \in \alpha(j_{[g]})} l_{j_{[l]}}$ is its processing time.

In the particular case, the elementary construction may have a form of a single task or a chain of tasks.

We use structures based on task chains (Definition 7.4) to present the theoretical part of this chapter. And for the convenience of describing the algorithm implementation, it is efficient to use elementary constructions (Definition 7.5). Such a representation of the various task structures does not contradict each other. They are based on unified principles, and one structure includes the other.

Definition 7.6 Two chains (or two elementary constructions) of tasks $\alpha(j_{[g]})$, $\beta(j_{[f]})$ are called *independent* (from each other) if any pair of tasks $j_{[k]} \in \alpha(j_{[g]})$ and $j_{[l]} \in \beta(j_{[f]})$ does not exist in the precedence relations.

Definition 7.7 A *construction* K is a feasible sequence of tasks with the structure shown at Fig. 7.1. Here, $\beta, \alpha_1, \ldots, \alpha_k$ are chains or elementary constructions; $f \prec \alpha_i$, $\forall i = \overline{1, k}$; $p(\alpha_1) \geq p(\alpha_2) \geq \cdots \geq p(\alpha_k)$.

Definition 7.8 A *complex construction* $\tilde{K}$ is such a construction in which:

- either a construction K^* (of the type K) takes place instead of any chain α_t and the task f immediately precedes the first task of the construction K^*;
- or, similar, the first condition may be satisfied in the construction $\tilde{K}$ for any chain or an elementary construction $\tilde{\alpha}_j$ in $\tilde{K}$, and so on.

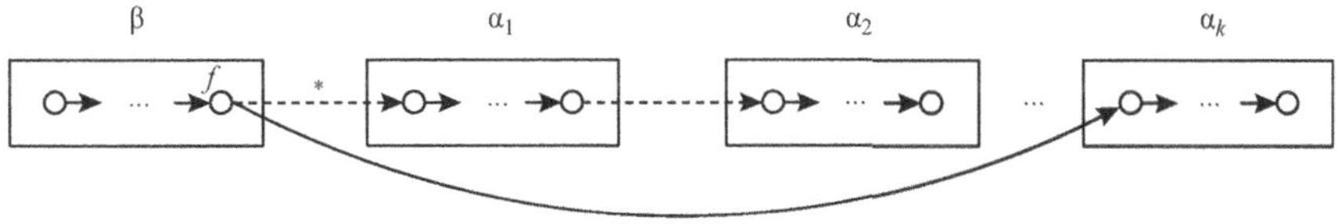

* Each chain α_j must relate with the task f either directly (continuous line) or through another chain(s) (dashed lines).

Fig. 7.1 A construction of tasks

K or $\tilde{K}$ should satisfy this: in any partitioning of tasks' sequence of the construction into two disjoint parts, the priority of the beginning part is less than or equal to the priority of the final part (an analog to the chain property).

Definition 7.9 An *interval of move* in a sequence σ for a separate task $j_{[k]}$ or a subsequence of tasks is the maximum time interval satisfying the condition: if we move the task $j_{[k]}$ or a subsequence of tasks by this interval into an earlier position, then it does not violate the feasibility of the sequence σ.

For example, the interval of move of a task $j^*_{[g]}$ in a sequence $\sigma = \left(j^1_{[1]}, j^2_{[2]}, \ldots, j^p_{[p]}, j^{p+1}_{[p+1]}, j^{p+2}_{[p+2]}, \ldots, j^{g-1}_{[g-1]}, j^*_{[g]}\right)$ is defined by the value of $I_{j^*_{[g]}} = \sum_{k=p+1}^{g-1} l_{j_{[k]}}$ where $p+1$ is the minimum position of a task $j_{[p+1]}$ for which $j_{[k]} \bar{\prec} j^*_{[g]}$ $\forall k = \overline{p+1, g-1}$, $j_{[p]} \prec j^*_{[g]}$. After the move, the sequence of tasks σ transforms into a sequence $\sigma_1 = \left(j^1_{[1]}, j^2_{[2]}, \ldots, j^p_{[p]}, j^*_{[p+1]}, j^{p+1}_{[p+2]}, \ldots, j^{g-2}_{[g-1]}, j^{g-1}_{[g]}\right)$.

The interval of move for a chain of tasks $\alpha(j_{[g]}) = (j_{[r]}, j_{[r+1]}, \ldots, j_{[g-1]}, j_{[g]})$ is $I_\alpha = \sum_{k=p+1}^{r-1} l_{j_{[k]}}$ where $j_{[k]} \bar{\prec} \alpha(j_{[g]})$ $\forall k = \overline{p+1, r-1}$, $j_{[p]} \prec \alpha(j_{[g]})$.

Definition 7.10 An *interval of efficient move* for a task $j_{[g]}$ is defined as the interval $I^{eff}_{j_{[g]}} = \sum_{k=\rho}^{g-1} l_{j_{[k]}}$ where ρ is the position within the interval of move where the difference $\omega_{j_{[g]}} \sum_{k=\rho}^{g-1} l_{j_{[k]}} - l_{j_{[g]}} \sum_{k=\rho}^{g-1} \omega_{j_{[k]}} > 0$ reaches its maximum. The interval of efficient move for a chain or an elementary constructions of tasks $\alpha(j_{[g]})$ is $I^{eff}_{\alpha(j_{[g]})} = \sum_{k=q}^{r-1} l_{j_{[k]}}$ where q is the position where the difference $\Omega_{\alpha(j_{[g]})} \sum_{k=q}^{r-1} l_{j_{[k]}} - L_{\alpha(j_{[g]})} \sum_{k=q}^{r-1} \omega_{j_{[k]}} > 0$ reaches its maximum. Here, $\Omega_{\alpha(j_{[g]})} \sum_{k=q}^{r-1} l_{j_{[k]}}$ is a decrease in the functional value for the chain of tasks as a result of the move; $L_{\alpha(j_{[g]})} \sum_{k=q}^{r-1} \omega_{j_{[k]}}$ is an increase in the functional value for tasks $j_{[k]}$, $k = \overline{r-1, q}$, moved into later positions. The interval of efficient move for a construction K is $I^{eff}_{K(j_{[g]})} = \sum_{k=i}^{s-1} l_{j_{[k]}}$ where i is the position where the difference $\Omega K \sum_{k=i}^{s-1} l_{j_{[k]}} - LK \sum_{k=i}^{s-1} \omega_{j_{[k]}} > 0$ reaches its maximum; s is the position of the first task of the construction K before the efficient move. The interval of efficient move for a complex construction $\tilde{K}$ is defined similarly.

Statement 7.2 Suppose that we have a subsequence α and a set of subsequences $G(\alpha_1, \ldots, \alpha_k)$ where $\alpha_1, \ldots, \alpha_k$ are subsequences of tasks not related with each other by the precedence relation graph. Then, a combination of maximum priority subsequences (MPSSes) that includes α is defined by the following rule:

1. Construct a sequence of tasks G_1:

$$G_1 = \left(\alpha_{(1)}, \alpha_{(2)}, \ldots, \alpha_{(k_1)}\right) \tag{7.8}$$

where $p\left(\alpha_{(1)}\right) \geq p\left(\alpha_{(2)}\right) \geq \cdots \geq p\left(\alpha_{(k_1)}\right), p\left(\alpha_j\right) \geq p(\alpha), j = \overline{1, k_1}$, and $p\left(\alpha_j\right) \leq p(\alpha)$ $\forall \alpha_j \in G \backslash G_1$.

2. Determine the combination of subsequences $\alpha, \alpha_1, \ldots, \alpha_p$ that has the maximum total priority $p_{\max}$ where

$$p_{\max} = p\left(\alpha + \sum_{j=1}^{p} \alpha_j\right) \geq p\left(\alpha + \sum_{j=1}^{p-t} \alpha_j\right), \quad t = \overline{1, p-1},$$

$$p\left(\alpha + \sum_{j=1}^{p} \alpha_j\right) > p\left(\alpha + \sum_{j=1}^{p+1} \alpha_j\right).$$

Proof

(a) Suppose $p(\alpha_1) > p(\alpha_2)$, then

$$p(\alpha_2) < p(\alpha_1 + \alpha_2) < p(\alpha_1), \tag{7.9}$$

the inequality (7.9) is obvious.

(b) It follows from inequality (7.9) that the procedure (7.8) is valid.
(c) Consider a set $G_2 \subset G_1$ consisting of all $\alpha_j, p\left(\alpha_j\right) > p_{\max}$. Suppose that

$$p\left(\alpha + \sum_{\alpha_j \in G_2} \alpha_j\right) < p_{\max}. \tag{7.10}$$

Then, there are no such a set of subsequences of tasks $G_3 \subset G_2$ that

$$p\left(\alpha + \sum_{\alpha_j \in G_3} \alpha_j\right) \geq p_{\max}. \tag{7.11}$$

We will prove the inequality (7.11) by contradiction. Suppose that a set G_3 exists. Then, the inequality (7.10) is not satisfied because Eq. (7.9) is true. Suppose

$$p\left(\alpha + \sum_{\alpha_j \in G_2} \alpha_j\right) > p_{\max}. \tag{7.12}$$

If G_2 is a subset of $\alpha, \alpha_1, \ldots, \alpha_p$, then by definition of $p_{\max}$ we arrive at a contradiction.

Suppose that $\alpha, \alpha_1, \ldots, \alpha_p$ is a subset of G_2. But $p(\alpha + G_2) < p_{\max}$ by definition. Therefore, there is no set G_2 that satisfies the inequality (7.12). ∎

7.3 PSC-Algorithm for TWCT Problem Solving with the Precedence Relations Specified by an Arbitrary Oriented Acyclic Graph

We have developed a PSC-algorithm of the second class for the problem solving. Let us give its substantiation with sufficient signs of optimality (SSOs) of the obtained schedules.

7.3.1 *Polynomially-Solvable Subclasses of TWCT Problem on a Set of Maximum Priority*

Let us determine the conditions for the algorithm to construct an optimal schedule in polynomial time (*p*-conditions) [2, 5, 9]. We impose the conditions on the precedence constraints and the priority relation in the maximum priority set.

Suppose that we have constructed a *p*-ordered schedule $\sigma = \{M_1, M_2, \ldots, M_t, g_{[k]}\}$ at iteration $k-1$. Here, M_i, $i = \overline{1, t-1}$, is the set which has the highest priority from all task sets $M_i, \ldots, M_t$; $g_{[k]}$ is the task included in the schedule at iteration k. We create a chain $\alpha(g_{[k]})$ and perform its efficient move [6]. An elementary construction of tasks can also serve as a chain. As a result, we obtain the schedule $\sigma^1 = \{M_1, M_2, \ldots, M_i, \alpha(g_{[k]}), M_{i+1}, \ldots, M_t\}$ where $f \bar{\prec} \alpha(g_{[k]})$: $\forall f \in M_s$, $s = \overline{i+1, t}$.

If $p(\alpha(g_{[k]})) \le p(M_i)$, then we have built an optimal schedule [6].

Suppose that $M_i \prec \alpha(g_{[k]})$ and $p(\alpha(g_{[k]})) > p(M_i)$. M_i is the maximum priority set and, consequently, it has the structure of a general construction [6]. Consider the priority relation and precedence constraints of tasks in M_i. Without loss of generality, we assume that the last task of the chain β precedes the first task of arbitrary form constructions, and $p(K_1) \ge \cdots \ge p(K_t) > p(\beta)$ (Fig. 7.2).

Case 1. Suppose that $p(K_i) \ge p(\alpha(g_{[k]}))$ and $p(K_{i+1}) < p(\alpha(g_{[k]}))$, and, moreover, the tasks of the constructions K_s, $s = \overline{i+1, t}$, and the chain $\alpha(g_{[k]})$ do not have

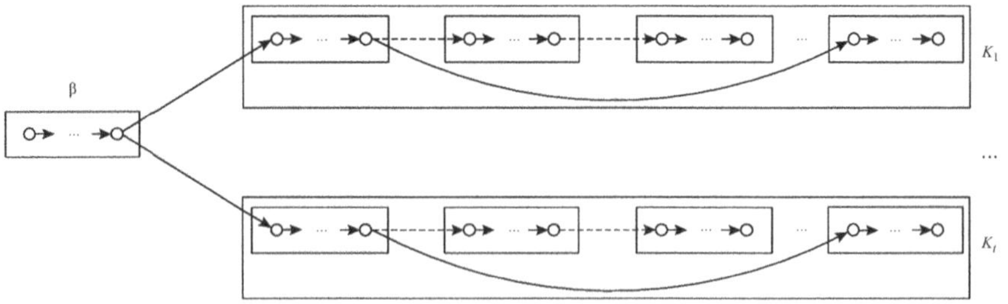

Fig. 7.2 Precedence constraints of tasks in the maximum priority set M_i

common arcs on the graph of partial ordering. Then, an optimal schedule on the set $M_i \cup \alpha(g_{[k]})$ has the form

$$\beta, K_1, \ldots, K_i, \alpha(g_{[k]}), K_{i+1}, \ldots, K_t. \tag{7.13}$$

Moreover, the following precedence relations are possible in M_i which violate the precedence conditions of the tasks in the construction: the arcs defining the precedence relations from constructions K_s, $s = \overline{1, i}$, to the chain $\alpha(g_{[k]})$ may be arbitrary.

Case 2. $p(K_{i+1}) < p(\alpha(g_{[k]}))$, $K_s \precsim \alpha(g_{[k]})$, $s = \overline{i+1, t}$; $p(K_i) < p(\alpha(g_{[k]}))$, $K_i \prec \alpha(g_{[k]})$. Consider the structure of a construction K_i and the priority relation of tasks in it:

1. Check if the chain $\alpha(g_{[k]})$ as a result of its efficient move took a position in K_i according to its priority, and the tasks immediately preceding it belong to the sets of a higher priority. In this case, we build a new construction K_i' which includes the chain $\alpha(g_{[k]})$ and is the maximum priority set on the set $K_i \cup \alpha(g_{[k]})$. If $K_i \cup \alpha(g_{[k]})$ is one priority set, then K_i' has the form

$$K_i' = \beta_i, K_i^1, \ldots, K_i^s, \alpha(g_{[k]}), K_i^{s+1}, \ldots, K_i^z$$

where $p(K_i^s) \geq p(\alpha(g_{[k]}))$, $p(K_i^{s+1}) < p(\alpha(g_{[k]}))$. Otherwise,

$$K_i' = \beta_i, K_i^1, \ldots, K_i^s, \alpha(g_{[k]}), K_i^{s+1}, \ldots, K_i^f$$

where $p(K_i') \geq p(K_i^{f+1})$ and constructions K_i^l, $l = \overline{f+1, z}$ form non-adjacent priority ordered sets. If $p(K_i') \leq p(K_{i-1})$, then we have constructed an optimal schedule in the form

$$\beta, K_1, \ldots, K_{i-1}, K_i', K_i^{f+1}, \ldots, K_i^z, K_{i+1}, \ldots K_t. \tag{7.14}$$

Otherwise, we perform the efficient move for the construction K_i'. We arrange tasks of the constructions or chains $K_{s+1}, \ldots, K_{i-1}$ shifted after the move of K_i', and the tasks of the constructions K_i^l, $l = \overline{f+1, z}$. Thus, if the graph specifying the precedence relation on the set of tasks $K_{s+1}, \ldots, K_{i-1}, K_i^l$, $l = \overline{f+1, z}$, satisfies p-conditions, then an optimal schedule has the form

$$\beta, K_1, \ldots, K_s, K_i', \overline{K_{i,s}}, K_{i+1}, \ldots K_t \tag{7.15}$$

where $p(K_s) \geq p(K_i')$, $p(K_{s+1}) < p(K_i')$, $K_l \precsim K_i'$, $l = \overline{s+1, i-1}$, and $\overline{K_{i,s}}$ are the p-ordered sets of tasks $K_{s+1}, \ldots, K_{i-1}, K_i^l$, $l = \overline{f+1, z}$.

Remark 7.1 If $p(K_s) < p(K_i^l)$, $K_s \prec K_i^l$, then we build priority ordered schedule on the set of tasks $K_i^l \cup K_s$, etc.

The schedule (7.14) or (7.15) is also optimal for the following cases:

- for arbitrary precedence relations between the tasks of constructions K_f, $f = \overline{1,s}$ and the chain $\alpha(g_{[k]})$;
- if the first task of the chain $\alpha(g_{[k]})$ is preceded only by the tasks of the construction K_i^l which belong to chains and construction of higher priority, i.e., $\forall \gamma \prec\prec \alpha(g_{[k]})$: $p(K_i^s) \geq p(\alpha(g_{[k]}))$, $\gamma \in K_i^s$;
- suppose that the priority-ordered maximum priority set M_i has the form $\beta_i^1, \ldots, \beta_i^j, K_i^1, \ldots, K_i^t$. Here, the chains $\beta_i^1, \ldots, \beta_i^j$ are not related pairwise and are priority ordered, and their last tasks immediately precede the first tasks of the constructions $K_1, \ldots, K_t$ and the chain $\alpha(g_{[k]})$. Then, the precedence of other tasks of the chains $\beta_i^1, \ldots, \beta_i^j$ to the chain $\alpha(g_{[k]})$ can be arbitrary.

Justification The proof of the schedules' optimality follows from the fact that we can eliminate from the graph all arcs which define the precedence relations and are not the arcs of a construction. This schedule is also optimal for the obtained subgraph which has the form of a complex construction and, therefore, satisfies the p-conditions. ■

2. The schedule (7.14) or (7.15) obtained in item 1 is also optimal:

   - If preceding p-conditions for chains β_i and $\alpha(g_{[k]})$ are violated, but the condition of immediate precedence of the last task of β_i is satisfied [9].
   - If the condition of immediate precedence of the last task of $\beta_i = (\beta_i^1, \ldots, \beta_i^j)$ to the chain $\alpha(g_{[k]})$ is violated, but

$$p(\beta_i^1, \ldots, \beta_i^j) \geq p(\alpha(g_{[k]})),\ \beta_i^q \bar{\prec} \alpha(g_{[k]}),\ q = \overline{l,j},\ \beta_i^{l-1} \prec\prec \alpha(g_{[k]}). \qquad (7.16)$$

This condition is also true if K_i has the form $\beta_i^1, \ldots, \beta_i^j, K_i^1, \ldots, K_i^t$. In this case, the condition (7.16) must be satisfied for each chain β_i which violates the precedence condition of the last task.

Justification The schedule $\beta_i, \alpha(g_{[k]})$ is optimal for a chain or a construction $\beta_i \cup \alpha(g_{[k]})$. And the schedule $\beta_i^1, \ldots, \beta_i^j, \alpha(g_{[k]})$ is optimal for a construction $\beta_i^1, \ldots, \beta_i^j \cup \alpha(g_{[k]})$. Since the priority of each of the constructions $K_i^1, \ldots, K_i^s$ is greater than or equal to the priority of $\alpha(g_{[k]})$, the schedule (7.14) is optimal. ■

Remark 7.2 If the subgraph of $(\beta_i, \alpha(g_{[k]}))$ or $(\beta_i^1, \ldots, \beta_i^j, \alpha(g_{[k]}))$ does not satisfy p-conditions, then we eliminate all arcs except $\beta_i^{l-1} \prec\prec \alpha(g_{[k]})$. The schedule $\sigma' = \beta_i, \alpha(g_{[k]})$ or $\sigma' = \beta_i^1, \ldots, \beta_i^j, \alpha(g_{[k]})$ will also be optimal for this subgraph.

- If the construction K_i has the form $\beta_i, K_i^1, \ldots, K_i^t$, then the position of $\alpha(g_{[k]})$ in the optimal schedule on the set $K_i \cup \alpha(g_{[k]})$ determines the priority value of the sets $\alpha(g_{[k]})$ and $(\beta_i^l, \ldots, \beta_i^j, K_i^1, \ldots, K_i^s)$ where $p(K_i^s) \geq p(\alpha(g_{[k]}))$ and $p(K_i^{s+1}) < p(\alpha(g_{[k]}))$. If $p(\beta_i^l, \ldots, \beta_i^j, K_i^1, \ldots, K_i^s) < p(\alpha(g_{[k]}))$, then the schedule

$$\beta_i^1, \ldots, \beta_i^{l-1},\ \alpha(g_{[k]}),\ \beta_i^l, \ldots, \beta_i^j,\ K_i^1, \ldots, K_i^s \tag{7.17}$$

is optimal.

3. Suppose that the constraints that define precedence relations of the chains β_i or $\beta_i^l, \ldots, \beta_i^j$ to $\alpha(g_{[k]})$ are arbitrary. But tasks of a construction K_i belonging to the chains and constructions of higher priority (for which p-conditions regarding the chains $\beta_i^l, \ldots, \beta_i^j$ are not violated) precede $\alpha(g_{[k]})$. Then, the schedule (7.14) or (7.15) obtained in item 1 is also optimal.

Justification The schedule (7.14) or (7.15) is optimal for subgraph $(K_i, \alpha(g_{[k]}))$. We obtain the subgraph after eliminating constraints that define precedence relations of the tasks of chains $\beta_i^l, \ldots, \beta_i^j$ and all tasks of constructions $K_i^1, \ldots, K_i^s$ to $\alpha(g_{[k]})$ which are not the arcs of a construction. ■

We use a similar scheme to obtain an optimal schedule if the construction K_i has a multi-level structure. If we meet the precedence conditions and the priority relation defined in item 1 with respect to some subconstruction K_i, then we proceed to the consideration of the precedence relations and the priority values of this subconstruction, similarly to item 1.

Remark 7.3 It is impossible to determine, outside the logic of the algorithm, the constraints on the precedence relations and a problem instance parameters satisfying of which in the general case yields an optimal schedule in polynomial time.

Remark 7.4 We have written the conditions of items 1–3 only for tasks of $\alpha(g_{[k]})$ since the sequences of all other tasks are p-ordered at the previous iterations.

Definition 7.11 We will call a subsequence β' *locking* if it satisfies the conditions $\beta' \prec\!\prec K_i,\ i = \overline{1,q}$, on the set $\{\beta', K_1, \ldots, K_q\}$. We call the subsequences $K_1, \ldots, K_q$ *successors of the first level.*

If the chains $\beta_i^1, \ldots, \beta_i^j$ are non-adjacent priority-ordered locking chains on the set M_i, then, in accordance with the structure of M_i [3], we will obtain the following subsequence after the efficient move:

$$\beta_i^1, \ldots, \beta_i^j,\ K_i^1, \ldots, K_i^s,\ \alpha(g_{[k]}),\ K_i^{s+1}, \ldots, K_i^t \tag{7.18}$$

where $p(K_i^1) \geq \cdots \geq p(K_i^s) \geq p(\alpha(g_{[k]})) \geq p(K_i^{s+1}) \geq \cdots \geq p(K_i^t)$. The schedule (7.18) is optimal for above items 1–3.

We remove in the constraints on the precedence relations in items 2–3 the requirement that the last task of chains $\{\beta_i^1, \ldots, \beta_i^j\}$ must immediately precede all first tasks of constructions and chains of the first level. Let us remove the requirement that the locking chains immediately precede the first task of $\alpha(g_{[k]})$. Then, we consider the constraints on the priority relations of constructions and chains in M_i which allow to obtain an optimal schedule in polynomial time.

Suppose that $\alpha(g_{[k]}) = (g_{[s]}, \ldots, g_{[k]})$ and $\beta_i^t \prec\prec g_{[f+1]}$, $t = \overline{1,j}$. Let $\alpha_{end}(g_{[k]}) = (g_{[f+1]}, \ldots, g_{[k]})$ and $\alpha_{begin}(g_{[f]}) = (g_{[s]}, \ldots, g_{[f]})$. Then, the following statement holds.

Statement 7.3 Suppose that $\alpha_{begin}(g_{[f]})$ and the tasks of $\beta_i^1, \ldots, \beta_i^j, K_i^1, \ldots, K_i^t$ have no common arcs in the partial ordering graph. Then, if $p(\beta_i^1, \ldots, \beta_i^j, K_i^1, \ldots, K_i^t) \geq p(\alpha_{begin}(g_{[f]}))$, $p(K_i^s) \geq p(\alpha_{end}(g_{[k]}))$ where $p(\beta_i^l) < p(\alpha_{begin}(g_{[f]}))$ and $p(\beta_i^{l-1}) \geq p(\alpha_{begin}(g_{[f]}))$, then the schedule (7.18) is optimal on the set of tasks $M_i \cup \alpha(g_{[k]})$. If $p(\beta_i^l, \ldots, \beta_i^j, K_i^1, \ldots, K_i^s) < p(\alpha_{begin}(g_{[f]}))$, $p(K_i^s) \geq p(\alpha_{end}(g_{[k]}))$, then the schedule

$$\beta_i^1, \ldots, \beta_i^{l-1},\ \alpha_{begin}(g_{[f]}),\ \beta_i^l, \ldots, \beta_i^j, K_i^1, \ldots, K_i^s,\ \alpha_{end}(g_{[k]}), \ldots, K_i^t. \tag{7.19}$$

is optimal.

Proof Inequality $p(\alpha_{begin}(g_{[f]})) \leq p(\alpha(g_{[k]}))$ corresponds to the properties of a chain, by its definition. Consider two possible situations that determine the positions of the chain $\alpha_{begin}(g_{[f]})$ in an optimal schedule:

1. Suppose that the chain $\alpha_{begin}(g_{[f]})$, in accordance with its priority, occupies a position between the tasks of the constructions $K_i^1, \ldots, K_i^s$ in an optimal schedule, i.e., it is the successor of the first level. This condition is equivalent to introducing the following additional precedence relations to the first task of $\alpha_{begin}(g_{[f]})$: $\beta_i^1 \prec\prec \alpha_{begin}(g_{[f]})$, …, $\beta_i^j \prec\prec \alpha_{begin}(g_{[f]})$. The obtained precedence constraints define the construction for which the schedule (7.18) is optimal.
2. Suppose that $\alpha_{begin}(g_{[f]})$, in accordance with its priority, occupies a position between the chains $\beta_i^1, \ldots, \beta_i^j$ in an optimal schedule, i.e., it is a locking chain. All the first tasks of the subsequences $K_i^1, \ldots, K_i^s$, $\alpha_{end}(g_{[k]})$,…, K_i^t are preceded by the last tasks of the chains $\beta_i^1, \ldots, \beta_i^j$. Thus, only the chain $\alpha_{begin}(g_{[f]})$ can be locking. Tasks of $\beta_i^1, \ldots, \beta_i^j$ are priority-ordered non-adjacent chains and, therefore, the sequence $\beta_i^1, \ldots, \beta_i^{l-1}, \alpha_{begin}(g_{[f]}), \beta_i^l, \ldots, \beta_i^j$ is optimal on the set $\beta_i^1, \ldots, \beta_i^j, \alpha_{begin}(g_{[f]})$. This position of the chain $\alpha_{begin}(g_{[f]})$ is equivalent to introducing additional restrictions $\alpha_{begin}(g_{[f]}) \prec\prec K_i^1$, …, $\alpha_{begin}(g_{[f]}) \prec\prec K_i^t$. The resulting maximum priority set satisfies the requirements of the construction for which the schedule (7.19) is optimal. The schedules (7.18) and (7.19) are different in a position of adjacent subsequences of tasks (Statement 7.1). Therefore, the optimal schedule is determined by the priority values of the subsequences $\beta_i^l, \ldots, \beta_i^j, K_i^1, \ldots, K_i^t$ and $\alpha_{begin}(g_{[f]})$. ∎

Corollary 7.1 If $p(\beta_i^l,\ldots,\beta_i^j,K_i^1,\ldots,K_i^t)<p(\alpha_{begin}(g_{[f]}))$ where $p(\beta_i^l)<p(\alpha_{begin}(g_{[f]}))$, $p(\beta_i^{l-1})\geq p(\alpha_{begin}(g_{[f]}))$, then a schedule is optimal if $\alpha_{begin}(g_{[f]})$ is locking in it.

Proof The schedules (7.18) and (7.19) are different in a position of the adjacent subsequences of tasks (Statement 7.1 defines the permutation). Therefore, the functional value of the schedule (7.19) is smaller. If we remove the constraint $p(K_i^s)\geq p(\alpha_{end}(g_{[k]}))$, then the schedule

$$\beta_i^1,\ldots,\beta_i^{l-1},\ \alpha_{begin}(g_{[f]}),\ \beta_i^l,\ldots,\beta_i^j,K_i^1,\ldots,K_i^r,\ \alpha_{end}(g_{[k]}),\ K_i^{r+1},\ldots,K_i^t,$$

is optimal for situation 2 of Statement 7.3. Here, $p(K_i^r)\geq p(\alpha_{end}(g_{[k]})), p(K_i^{r+1})<p(\alpha_{end}(g_{[k]}))$ if the tasks of $\alpha_{end}(g_{[k]})$ and K_i^{r+1}, …, K_i^s are not adjacent. But this schedule also differs from (7.19) in the positions of adjacent task sequences. Therefore, it has a smaller functional value than that of the schedule (7.19). In the presence of precedence constraints between the tasks of K_i^r and $\alpha_{end}(g_{[k]})$ where $p(K_i^r)<p(\alpha_{end}(g_{[k]}))$, the procedure for obtaining an optimal schedule for the situation 2 of Statement 7.3 has a different structure [3]. But the functional value of the obtained schedule, since it is optimal, is less than that of the schedule (7.19). Thus, an optimal schedule obtained for the case where $\alpha_{begin}(g_{[f]})$ is the locking chain is also optimal on the set of tasks $M_i\cup\alpha(g_{[k]})$. ∎

Corollary 7.2 If $p(\beta_i^l,\ldots,\beta_i^j,K_i^1,\ldots,K_i^s)\geq p(\alpha_{begin}(g_{[f]}))$ and $p(K_i^s)\geq p(\alpha_{end}(g_{[k]}))$ where $p(\beta_i^l)<p(\alpha_{begin}(g_{[f]}))$ and $p(\beta_i^{l-1})\geq p(\alpha_{begin}(g_{[f]}))$, $\alpha_{begin}(g_{[f]})=(g_{[s]},\ldots,g_{[f]})$ and $\exists g_{[f+1]}\in\alpha(g_{[k]})$: $\beta_i^t \prec\!\prec g_{[f+1]}$, $t=\overline{1,j}$, then the schedule (7.18) is optimal on the set of tasks $M_i\cup\alpha(g_{[k]})$. The precedence relations from tasks β_i^t, $t=\overline{1,j}$ to tasks of $\alpha_{begin}(g_{[f]})$ can be arbitrary.

Proof Remove in the partial ordering graph all the arcs defining the precedence relations from tasks β_i^t, $t=\overline{1,j}$, to tasks of $\alpha_{begin}(g_{[f]})$. We obtain a subgraph of the initial problem that satisfies the conditions of Statement 7.3. Consequently, the schedule (7.18) is optimal on the considered set $M_i\cup\alpha(g_{[k]})$. ∎

Remark 7.5 If $\exists v\in K_i^\rho$, $\rho\geq s$: $v\prec\!\prec g_{[h]}$, $g_{[h]}\in\alpha_{begin}(g_{[f]})$, then we verify the conditions of Statement 7.3 for the chain $\alpha_{begin}(g_{[h-1]})=(g_{[s]},\ldots,g_{[h-1]})$. Note that if $\exists v\in K_i^\rho$, $\rho\geq s$, such that $v\prec\!\prec g_{[s]}$, then the situation 3 takes place and, therefore, the schedule (7.18) is optimal on the set of tasks under consideration.

Consider the generalization of this statement. Suppose that M_i is in the form

$$M_i=\beta_1,\ldots,\beta_i,K_1,\alpha_1,K_2\ldots,K_j,\alpha_j,K_{j+1},\ldots,K_p,\alpha_p \tag{7.20}$$

where the sequence $K_1,\alpha_1,K_2\ldots,K_j,\alpha_j,K_{j+1},\ldots,K_p,\alpha_p$ consists of priority-ordered structures and chains; $\beta_j\prec\!\prec K_s$, $j=\overline{1,i}$; $\beta_j\prec\alpha_s$, $s=\overline{1,p}$. The chains and constructions order is determined by the priority relations. Here, a set K_j may be a combination of non-adjacent priority-ordered structures and chains. To simplify the proof, we assume that the violation of the preceding relations to the first task on the set M_i is only for chains α_i. Partition the chain α_i: $\alpha_i = \left(\alpha_i^{begin}, \alpha_i^{end}\right)$ where α_i^{begin} is the beginning part of the chain α_i not adjacent to any task of M_i, $\beta_t \prec\prec \alpha_i^{end}$, $t = \overline{1, i}$.

Statement 7.4 Suppose that the tasks of M_i satisfy the following priority relations: $p\left(\alpha_1^{begin}\right) \geq p\left(\alpha_2^{begin}\right) \geq \cdots \geq p\left(\alpha_j^{begin}\right)$ and $p(K_1) \geq p(\alpha_1^{end})$, …, $p(K_j) \geq p\left(\alpha_j^{end}\right)$. Then if $p(\beta_f, \ldots, \beta_i, K_1) \geq p\left(\alpha_1^{begin}\right)$ where $p(\beta_f) < p\left(\alpha_1^{begin}\right)$ and $p(\beta_{f-1}) \geq p\left(\alpha_1^{begin}\right)$, then the schedule (7.20) is optimal on the set of tasks M_i.

Proof The proof contains two parts.

1. Only sequences $\left\{\alpha_j^{begin}\right\}$ may occupy positions adjacent to chains $\{\beta_f\}$, and thus, be locking chains in an optimal schedule. Let us prove that no arbitrary sequence α_j^{begin} can occupy positions immediately preceding all constructions of the set M_i. Preceding to all constructions of M_i is equivalent to fulfillment of the condition that α_j^{begin} precedes to all first tasks of constructions and chains of the first level. Therefore, α_j^{begin} would satisfy the requirement to be a locking chain in the construction. The condition of the immediate precedence of all tasks of $\{\beta_f\}$ to the chains α_l^{begin}, $l = \overline{1, p}$, $l \neq j$ is equivalent to setting the constraints of successors of the first level for α_l^{begin}. Thus, we have defined a construction on the set M_i. Therefore, it is true for an arbitrary α_j^{begin} that

$$p(\beta_f, \ldots, \beta_i, K_1, \alpha_1, K_2, \ldots, K_j) > p(\beta_f, \ldots, \beta_i, K_1) \geq p\left(\alpha_1^{begin}\right).$$

Denote $\tau\left(\alpha_j^{begin}\right) = (\beta_g, \ldots, \beta_i, K_1, \alpha_1, K_2, \ldots, K_j)$ where

$$p\left(\alpha_j^{begin}\right) \geq p(\beta_g) \quad \text{and} \quad p\left(\alpha_j^{begin}\right) < p(\beta_{g-1}).$$

Let $p\left(\alpha_j^{begin}\right) < p(\beta_f)$. Then, because of a priority property (the sum of two priorities is less than the greater of them), the following relationships are true:

$$p(\beta_f, \ldots, \beta_i) \leq p(\beta_f, \ldots, \beta_i, K_1, \alpha_1, K_2, \ldots, K_j),$$

this implies

$$p(\beta_f, \ldots, \beta_i, K_1, \alpha_1, K_2, \ldots, K_j) < p\left(\tau\left(\alpha_j^{begin}\right)\right)$$

because $p\left(\alpha_1^{begin}\right) \geq p(\beta_f, \ldots, \beta_{g-1})$. Therefore,

$$p\left(\tau\left(\alpha_j^{begin}\right)\right) > p(\beta_f, \ldots, \beta_i, K_1, \alpha_1, K_2, \ldots, K_j) \geq p\left(\alpha_1^{begin}\right) \geq p\left(\alpha_j^{begin}\right)$$

since $p\left(\alpha_1^{begin}\right) \geq p\left(\alpha_j^{begin}\right)$. Thus, we have

$$p\left(\tau\left(\alpha_j^{begin}\right)\right) \geq p\left(\alpha_j^{begin}\right). \tag{7.21}$$

Consequently, no arbitrary chain α_j^{begin} can precede the subsequence K_1 in an optimal schedule.

2. Consider a schedule σ' optimal under the condition that some subset of chains $\left\{\alpha_{i_j}^{begin}\right\}$ in it is locking and, therefore, precedes all the constructions of M_i in order. Also, the rest of tasks of α_j^{begin} should occupy positions in the schedule according to their priorities among the constructions $\{K_j\}$. The imposed restrictions correspond to the requirements of the construction. Therefore, due to the priority relations in the set M_i, this schedule has the form

$$\begin{gathered}\beta_1, \ldots, \beta_{j_1}, \alpha_{i_1}^{begin}, \beta_{j_1+1}, \ldots, \beta_{j_s}, \alpha_{i_s}^{begin}, \beta_{j_s+1}, \ldots, \beta_i, K_1, \alpha_1, K_2, \ldots, K_{i_1}, \alpha_{i_1}^{end}, \\ K_{i_1+1}, \alpha_{i_1+1}, \ldots, K_{i_s}, \alpha_{i_s}^{end}, K_{i_s+1}, \ldots, K_p, \alpha_p\end{gathered} \tag{7.22}$$

where $p(\beta_{j_1}) \geq p\left(\alpha_{i_1}^{begin}\right) > p(\beta_{j_1+1})$, …, $p(\beta_{j_s}) \geq p\left(\alpha_{i_s}^{begin}\right) > p(\beta_{j_s+1})$.

Let us compare the functional values of the schedule (7.22) and the optimal schedule, provided that $\left\{\alpha_{i_1}^{begin}, \ldots, \alpha_{i_{s-1}}^{begin}\right\}$ are locking chains in the set M_i and $\alpha_{i_s}^{begin}$ takes the position of a successor of the first level. These requirements to M_i also satisfy the properties of a construction. The schedule has the form

$$\begin{gathered}\beta_1, \ldots, \beta_{j_1}, \alpha_{i_1}^{íá÷}, \beta_{j_1+1}, \ldots, \beta_{j_{s-1}}, \alpha_{i_{s-1}}^{íá÷}, \beta_{j_{s-1}+1}, \ldots, \beta_{j_s}, \beta_{j_s+1}, \ldots, \beta_i, K_1, \\ \alpha_1, K_2, \ldots, K_{i_1}, \alpha_{i_1}^{êîí}, K_{i_1+1}, \alpha_{i_1+1}, \ldots, K_{i_{s-1}}, \alpha_{i_{s-1}}^{êîí}, K_{i_{s-1}+1}, \ldots, K_{i_s}, \alpha_{i_s}, \\ K_{i_s+1}, \ldots, K_p, \alpha_p\end{gathered} \tag{7.23}$$

The schedules (7.22) and (7.23) differ from each other in positions of adjacent task sequences $\beta_{j_s+1}, \ldots, \beta_i, K_1, \ldots, K_{i_s}$ and $\alpha_{i_s}^{begin}$. In view of the relation (7.21) and the fact that $p(\alpha_i^{end}) > p(\alpha_i)$, we have $p(\beta_{j_s+1}, \ldots, K_{i_s}) > p\left(\alpha_{i_s}^{begin}\right)$. Consequently, the functional value of the schedule (7.23) is less than that of the schedule (7.22). Continuing in a similar way to compare the schedule (7.23) and a schedule in which the locking chains are $\alpha_{i_1}^{begin}, \ldots, \alpha_{i_{s-2}}^{begin}$, we derive that the latter one has a lower functional value. In the last step of the comparison we find that the schedule (7.20) has the smallest functional value. Since the choice of chains $\left\{\alpha_{i_j}^{begin}\right\}$ is arbitrary, the schedule (7.20) is optimal. ∎

Remark 7.6 In Statement 7.4, the violation of precedence conditions to the first task of a successor of the first level does not depend on its type. Therefore, we can replace the chains α_i with arbitrarily nested constructions.

Suppose that the precedence conditions and the priority relations of Statement 7.4 are valid on the set of tasks M_i. Consider the additional precedence constraints under which the schedule (7.20) is also optimal:

- under arbitrary precedence of tasks of the chains β_f, $f = \overline{1, i}$, to the chains α_j^{begin}, $j = \overline{1, p}$;
- under arbitrary precedence of tasks of the chains α_j^{begin} to the chains α_l^{begin}, $l > j$;
- under arbitrary precedence of tasks of the constructions K_j to tasks of the chains α_l^{begin}, $l > j$.

The proof of optimality follows from the fact that the task set of the initial set's subgraph without the above eliminated constraints satisfies the conditions of Statement 7.4.

Thus, we have shown that we can obtain an optimal schedule in polynomial time for maximum priority sets which are not generally combinations of constructions if we satisfy the following restrictions:

1. If the last task of the chain β precedes all the first tasks of chains and constructions of the first level. Arbitrary precedence relations are possible from tasks of β to arbitrary constructions and chains in M_i [9].
2. If condition 1 is violated but $\alpha(g_{[k]})$ satisfies the requirement of condition (7.16). This condition defines the precedence and priority relations for the chains β and $\alpha(g_{[k]})$. It is valid for an arbitrary construction or chain in M_i.
3. Arbitrary precedence relations are possible between priority-ordered chains and constructions, from a higher priority construction to a smaller priority one. The precedence of the chain β to the chain $\alpha(g_{[k]})$ can be arbitrary in this case.

We have derived the restrictions on the structures of the maximum priority sets of TWCT problem instances satisfying of which allows to obtain an optimal schedule in a polynomial time. Statements 7.3 and 7.4 define polynomially solvable subclasses of the original problem. The conditions of the statements are defined for one priority subsets. Checking these conditions during the PSC-algorithm execution allows us to significantly expand the class of polynomially solvable TWCT problems.

7.3.2 The Structure of PSC-Algorithm

The PSC-algorithm for the TWCT problem solving consists of two units [10]:

- an initial feasible schedule construction unit (an algorithm see [1]);
- optimization unit.

The optimization unit implements an optimal schedule construction for the entire set of tasks. We use a modified algorithm of p-ordered schedule construction for a series-parallel graph. We check fifteen SSOs developed on the basis of the conditions derived in Sect. 7.3.1. Also we check another SSO #16 during the problem solving. If the SSO #16 is satisfied, then we exclude the exponential search of constructions and build them according to the rules for a series-parallel graph.

We have developed an iterative algorithm for the optimization unit. The problem solving at each current iteration consists in analyzing the priorities and performing an efficient move of a task chain (or elementary, simple, or complex construction). The efficient move defines the position of its insertion into the p-ordered sequence obtained at the previous iteration. At each iteration k we construct a p-ordered schedule in the following way. We add the next task from the initial feasible sequence to the optimal schedule obtained at the iteration $k-1$. Then we build a chain (or an elementary construction) of tasks on the basis of the added task, perform its efficient move and check the fifteen SSOs (Sect. 7.3.3.1) for the current subsequence. If it satisfies at least one of the SSOs, then the constructed subsequence of tasks is optimal and obtained by a polynomial subalgorithm, we proceed to the next iteration. Otherwise, we build an elementary construction of tasks, according to certain rules. We insert it into the earlier positions in the p-ordered schedule, in accordance with its priority, without violating the entire sequence's feasibility. In the next steps of the algorithm, we build more complex structures of interrelated tasks (simple or nested constructions) on the basis of the elementary construction. By the same rules, we insert them into the p-ordered schedule of the current iteration. Then, we perform optimization of tasks shifted to later positions in consequence of the constructions' insertion. Thus, we build a p-ordered schedule for the entire set of tasks. According to the definition, the schedule is an ordered sequence of maximum priority subsets.

At the end of each iteration, we check the SSO #16. If it is satisfied, then the current subsequence of tasks is optimal and obtained by a polynomial subalgorithm. Otherwise, we continue to solve the problem by this iterative algorithm as an approximation algorithm, without checking the SSOs.

If at least one of the sixteen SSOs is satisfied at each iteration of the algorithm, then the schedule for the entire set of tasks is also optimal and obtained by the first polynomial component of the PSC-algorithm. Otherwise, the schedule for the entire set of tasks is approximate and obtained by the second polynomial component of the algorithm. Thus, the first polynomial component of the PSC-algorithm coincides with its second polynomial component.

In contrast to the PSC-algorithm given in [1], we have heuristically implemented the procedure for the task set partitioning into the maximum priority subsets and the ordering procedure within each subset. This turned out to be statistically more efficient since the decomposition procedure from [1] requires large computational efforts. If any SSO is satisfied, then this algorithm becomes exact.

We have constructed the optimization unit of the PSC-algorithm so that the functional value for the entire set of tasks decreases (or remains unchanged) after each iteration. This allows us to stop the solving when reaching a time limit and to estimate the deviation of the obtained solution from the searched optimal one. The latter is possible since we have estimates obtained for the solved instances of this problem, and the estimates' values depend on the current solution [4].

Let us note one more property of the developed algorithm. The structure of the algorithm allows us to "rough up" the proposed solutions, i.e., gain in time due to loss of accuracy of the results obtained. We do it excluding from the algorithms the procedures that influence the functional value only insignificantly, by statistics. This property allows us to build efficient approximate solutions.

7.3.3 Theoretical Foundations of PSC-Algorithm

7.3.3.1 Sufficient Signs of Optimality of a Feasible Subsequence of Tasks

We now formulate SSOs of a feasible subsequence on a maximum priority set based on Sect. 7.3.1.

Suppose that the conditions of the first polynomial component of the algorithm fulfilled till the iteration $k - 1$. We have the following schedule in the beginning of iteration k: $\sigma = \{M_1, M_2, \ldots, M_t, g_{[k]}\}$. Here, M_i, $i = \overline{1, t-1}$, is the set which has the highest priority from all task sets $M_i, \ldots, M_t$; $g_{[k]}$ is the task included in the schedule on iteration k. We create a chain $\alpha(g_{[k]})$ and perform its efficient move [6]. An elementary construction of tasks can also serve as a chain. Suppose that we obtain a schedule

$$\sigma^1 = \{M_1, M_2, \ldots, M_i, \alpha(g_{[k]}), M_{i+1}, \ldots, M_t\} \quad (Structure\ \ S_1)$$

SSO #1[2] If Structure S_1 and the priority relations in it satisfy the cases considered in Algorithms 1–3 of the polynomial algorithm Π1 [1] which allows to

[2]All fifteen SSOs are the signs of optimality for feasible subsequences obtained at the current iterations. We give more detailed justification for the implementation of insertion procedures for separate tasks or their constructions in the PSC-algorithm description (Sect. 7.3.4). We base the justification on an analysis of the priority and precedence relations. The presented logic of justifying the rules for constructing a p-ordered schedule in the PSC-algorithm causes the necessity of a more detailed investigation of these relations.

reduce the structure of the task set to a series-parallel graph, then the schedule obtained at the current iteration is optimal.

Suppose that $f \bar{\prec} \alpha(g_{[k]})$: $\forall f \in M_s$, $s = \overline{i+1, t}$ (*Structure* S_2).

SSO #2 If the structure of the task set corresponds to Structure S_2, and $p(\alpha(g_{[k]})) \le p(M_i)$, then the current schedule is optimal [6].

Suppose that $M_i \prec \alpha(g_{[k]})$ and $p(\alpha(g_{[k]})) > p(M_i)$ (*Condition 1*). M_i is the maximum priority set and, consequently, it has the structure of a general construction [6]. Consider the priority relation and precedence constraints of tasks in M_i. Without loss of generality, we assume that the last task of the chain β precedes the first task of arbitrary form constructions, and $p(K_1) \ge \cdots \ge p(K_t) > p(\beta)$ (*Structure* S_3, see Fig. 7.2).

Case 1. Suppose that $p(K_i) \ge p(\alpha(g_{[k]}))$ and $p(K_{i+1}) < p(\alpha(g_{[k]}))$, and, moreover, the tasks of the constructions K_s, $s = \overline{i+1, t}$, and the chain $\alpha(g_{[k]})$ do not have common arcs on the graph of partial ordering (*Condition 2*).

Moreover, the following precedence relations are possible M_i which violate the precedence conditions of the tasks in the construction: the arcs defining the precedence relations from constructions K_s, $s = \overline{1, i}$, to the chain $\alpha(g_{[k]})$, may be arbitrary (*Structure* S_4).

SSO #3 If M_i has either the form of Structure S_3 or S_4 and Conditions 1 and 2 are met, then we have built an optimal schedule on the set $M_i \cup \alpha(g_{[k]})$ in the form

$$\beta, K_1, \ldots, K_i, \alpha(g_{[k]}), K_{i+1}, \ldots, K_t. \tag{7.24}$$

Case 2. $p(K_{i+1}) < p(\alpha(g_{[k]}))$, $K_s \bar{\prec} \alpha(g_{[k]})$, $s = \overline{i+1, t}$; $p(K_i) < p(\alpha(g_{[k]}))$, $K_i \prec \alpha(g_{[k]})$ (*Condition 3*). Consider the structure of a construction K_i and the priority relation of tasks in it:

1. Check if the chain $\alpha(g_{[k]})$ as a result of its efficient move took a position in K_i according to its priority, and the tasks immediately preceding it belong to higher priority sets. In this case, we build a new construction K_i' that includes the chain $\alpha(g_{[k]})$ and is the maximum priority set on the set $K_i \cup \alpha(g_{[k]})$ (*Action 1*). If $K_i \cup \alpha(g_{[k]})$ is one priority set, then K_i' has the form

$$K_i' = \beta_i, K_i^1, \ldots, K_i^s, \alpha(g_{[k]}), K_i^{s+1}, \ldots, K_i^z$$

where $p(K_i^s) \ge p(\alpha(g_{[k]}))$, $p(K_i^{s+1}) < p(\alpha(g_{[k]}))$. Otherwise,

$$K_i' = \beta_i, K_i^1, \ldots, K_i^s, \alpha(g_{[k]}), K_i^{s+1}, \ldots, K_i^f$$

where $p(K_i') \ge p(K_i^{f+1})$, and constructions K_i^l, $l = \overline{f+1, z}$ form non-adjacent priority ordered sets. If $p(K_i') \le p(K_{i-1})$ (*Condition 5*), then we have constructed an optimal schedule in the form

$$\beta, K_1, \ldots, K_{i-1}, K_i', K_i^{f+1}, \ldots, K_i^z, K_{i+1}, \ldots K_t. \tag{7.25}$$

Otherwise, we perform the efficient move for the construction K_i' and arrange tasks of the constructions or chains $K_{s+1}, \ldots, K_{i-1}$ shifted after the move of K_i', and the tasks of the constructions K_i^l, $l = \overline{f+1, z}$ (*Action 2*). Thus, if the graph of the precedence relation on the set of tasks $K_{s+1}, \ldots, K_{i-1}, K_i^l$, $l = \overline{f+1, z}$, satisfies *p*-conditions (*Condition 6*), then an optimal schedule has the form

$$\beta, K_1, \ldots, K_s, K_i', \overline{K_{i,s}}, K_{i+1}, \ldots K_t \tag{7.26}$$

where $p(K_s) \geq p(K_i')$, $p(K_{s+1}) < p(K_i')$, $K_l \not\prec K_i'$, $l = \overline{s+1, i-1}$, and $\overline{K_{i,s}}$ are the *p*-ordered sets of tasks $K_{s+1}, \ldots, K_{i-1}, K_i^l$, $l = \overline{f+1, z}$.

SSO #4 If M_i has either the form of Structure S_3 or S_4 and Conditions 3 and 4 are met, then we perform Action 1. If Condition 5 is met, then we have built an optimal schedule in the form (7.25). Otherwise, if Condition 6 is met after performing Action 2, then we have built an optimal schedule in the form (7.26).

Remark 7.7 If $p(K_s) < p(K_i')$, $K_s \prec K_i'$, then we build priority ordered schedule on the set of tasks $K_i' \cup K_s$, etc.

The schedule (7.25) or (7.26) is also optimal for the following cases:

- for arbitrary precedence relations between the tasks of constructions K_f, $f = \overline{1, s}$ and the chain $\alpha(g_{[k]})$ (*Structure S_5*);
- if the first task of the chain $\alpha(g_{[k]})$ is preceded only by the tasks of the construction K_i' which belong to chains and construction of higher priority, i.e., $\forall \gamma \prec\prec \alpha(g_{[k]})$: $p(K_i^s) \geq p(\alpha(g_{[k]}))$, $\gamma \in K_i^s$ (*Structure S_6*);
- suppose that the priority-ordered maximum priority set M_i has the form $\beta_i^1, \ldots, \beta_i^j, K_i^1, \ldots, K_i^t$. Here, the chains $\beta_i^1, \ldots, \beta_i^j$ are not related pairwise and are priority ordered, and their last tasks immediately precede the first tasks of the constructions $K_1, \ldots, K_t$ and the chain $\alpha(g_{[k]})$. Then, the precedence of other tasks of the chains $\beta_i^1, \ldots, \beta_i^j$ to the chain $\alpha(g_{[k]})$ can be arbitrary (*Structure S_7*).

SSO #5 If M_i has the form of any of Structures S_5, S_6, S_7 and Conditions 3 and 4 are met, then we perform Action 1. If Condition 5 is met, then we have built an optimal schedule in the form (7.25). Otherwise, if Condition 6 is met after performing Action 2, then we have built an optimal schedule in the form (7.26).

2. The schedule (7.25) or (7.26) obtained in item 1 is also optimal:

   - If preceding *p*-conditions for chains β_i and $\alpha(g_{[k]})$ are violated, but the condition of immediate precedence of the last task of β_i is satisfied (*Structure S_8*);
   - If the condition of immediate precedence of the last task of $\beta_i = (\beta_i^1, \ldots, \beta_i^j)$ to the chain $\alpha(g_{[k]})$ is violated, but

$$p\left(\beta_i^1, \ldots, \beta_i^j\right) \ge p\left(\alpha\left(g_{[k]}\right)\right),\ \beta_i^q \bar{\prec} \alpha\left(g_{[k]}\right),\ q = \overline{l,j},$$

$$\beta_i^{l-1} \prec\prec \alpha\left(g_{[k]}\right)\ \ (\textit{Structure } S_9) \tag{7.27}$$

This condition is also true for the case when K_i has the form $\beta_i^1, \ldots, \beta_i^j$, $K_i^1, \ldots, K_i^t$. In this case, the condition (7.27) must be satisfied for each chain β_i which violates the precedence condition of the last task (*Structure* S_{10}).

SSO #6 If M_i has the form of any of Structures S_8, S_9, S_{10} and Conditions 3 and 4 are met, then we perform Action 1. If Condition 5 is met, then we have built an optimal schedule in the form (7.25). Otherwise, if Condition 6 is met after performing Action 2, then we have built an optimal schedule in the form (7.26).

- If the construction K_i has the form $\beta_i, K_i^1, \ldots, K_i^t$ (*Structure* S_{11}), then the position of $\alpha\left(g_{[k]}\right)$ in the optimal schedule on the set $K_i \cup \alpha\left(g_{[k]}\right)$ determines the priority value of the sets $\alpha\left(g_{[k]}\right)$ and $\left(\beta_i^l, \ldots, \beta_i^j, K_i^1, \ldots, K_i^s\right)$ where $p\left(K_i^s\right) \ge p\left(\alpha\left(g_{[k]}\right)\right)$ and $p\left(K_i^{s+1}\right) < p\left(\alpha\left(g_{[k]}\right)\right)$. If $p\left(\beta_i^l, \ldots, \beta_i^j, K_i^1, \ldots, K_i^s\right) < p\left(\alpha\left(g_{[k]}\right)\right)$ (*Condition 7*), then an optimal schedule has the form

$$\beta_i^1, \ldots, \beta_i^{l-1},\ \alpha\left(g_{[k]}\right),\ \beta_i^l, \ldots, \beta_i^j,\ K_i^1, \ldots, K_i^s. \tag{7.28}$$

SSO #7 If M_i has the form of any of Structures S_3, …, S_{10}; K_i has the form of Structure S_{11}; and Condition 7 is met, then we have built an optimal schedule in the form (7.28).

3. Suppose that the constraints that define precedence relations of the chains β_i or $\beta_i^l, \ldots, \beta_i^j$ to $\alpha\left(g_{[k]}\right)$ are arbitrary. But the tasks of a construction K_i belonging to the chains and constructions of higher priority (for which p-conditions regarding the chains $\beta_i^l, \ldots, \beta_i^j$ are not violated) precede $\alpha\left(g_{[k]}\right)$ (*Structure* S_{12}). Then, the schedule (7.25) or (7.26) obtained in item 1 is also optimal.

SSO #8 If M_i has the form of Structure S_{12} and Conditions 3 and 4 are met, then we perform Action 1. If Condition 5 is met, then we have built an optimal schedule in the form (7.25). Otherwise, if Condition 6 is met after performing Action 2, then we have built an optimal schedule in the form (7.26).

If the chains $\beta_i^1, \ldots, \beta_i^j$ are non-adjacent priority-ordered locking chains on the set M_i, then, in accordance with the structure of M_i [3], we will obtain the following subsequence after the efficient move:

$$\beta_i^1, \ldots, \beta_i^j,\ K_i^1, \ldots, K_i^s,\ \alpha\left(g_{[k]}\right),\ K_i^{s+1}, \ldots, K_i^t \tag{7.29}$$

where $p\left(K_i^1\right) \ge \cdots \ge p\left(K_i^s\right) \ge p\left(\alpha\left(g_{[k]}\right)\right) \ge p\left(K_i^{s+1}\right) \ge \cdots \ge p\left(K_i^t\right)$ (*Condition 8*). The schedule (7.29) is optimal for above items 1–3.

We remove in the constraints on the precedence relations in items 2–3 the requirement that the last task of chains $\{\beta_i^1, \ldots, \beta_i^j\}$ must immediately precede all first tasks of constructions and chains of the first level.

Let us denote as Structures $S'_3, \ldots, S'_{10}, S'_{12}$ the structures of the set M_i corresponding to Structures $S_3, \ldots, S_{10}, S_{12}$ described in items 1–3 with taking into account the above restrictions.

SSO #9 If M_i has the form of any of Structures $S_3, \ldots, S_{10}, S_{12}, S'_3, \ldots, S'_{10}, S'_{12}$ and Condition 8 is met, then we have built an optimal schedule in the form (7.29).

Let us remove the requirement that the locking chains immediately precede the first task of $\alpha(g_{[k]})$. Then, we consider the constraints on the priority relations of constructions and chains in M_i which allow to obtain an optimal schedule in polynomial time.

Suppose that $\alpha(g_{[k]}) = (g_{[s]}, \ldots, g_{[k]})$ and $\beta_i^t \prec\!\prec g_{[f+1]}$, $t = \overline{1,j}$ (*Condition 9*). Let $\alpha_{end}(g_{[k]}) = (g_{[f+1]}, \ldots, g_{[k]})$ and $\alpha_{begin}(g_{[f]}) = (g_{[s]}, \ldots, g_{[f]})$.

According to Statement 7.3, suppose that $\alpha_{begin}(g_{[f]})$ and the tasks of $\beta_i^1, \ldots, \beta_i^j$, $K_i^1, \ldots, K_i^t$ have no common arcs in the partial ordering graph (*Condition 10*). Then, if $p(\beta_i^1, \ldots, \beta_i^j, K_i^1, \ldots, K_i^t) \ge p(\alpha_{begin}(g_{[f]}))$, $p(K_i^s) \ge p(\alpha_{end}(g_{[k]}))$ where $p(\beta_i^l) < p(\alpha_{begin}(g_{[f]}))$ and $p(\beta_i^{l-1}) \ge p(\alpha_{begin}(g_{[f]}))$ (*Condition 11*), then the schedule (7.29) is optimal on the set of tasks $M_i \cup \alpha(g_{[k]})$. If $p(\beta_i^l, \ldots, \beta_i^j$, $K_i^1, \ldots, K_i^s) < p(\alpha_{begin}(g_{[f]}))$, $p(K_i^s) \ge p(\alpha_{end}(g_{[k]}))$ (*Condition 12*), then an optimal schedule has the form

$$\beta_i^1, \ldots, \beta_i^{l-1}, \ \alpha_{begin}(g_{[f]}), \ \beta_i^l, \ldots, \beta_i^j, K_i^1, \ldots, K_i^s, \ \alpha_{end}(g_{[k]}), \ldots, K_i^t. \tag{7.30}$$

SSO #10 If M_i has the form of any of Structures $S_3, \ldots, S_{10}, S_{12}, S'_3, \ldots, S'_{10}, S'_{12}$ and Conditions 9–11 are met, then we have built an optimal schedule in the form (7.29). If Conditions 9, 10, and 12 are met on these structures, then we have built an optimal schedule in the form (7.30).

If $p(\beta_i^l, \ldots, \beta_i^j, K_i^1, \ldots, K_i^t) < p(\alpha_{begin}(g_{[f]}))$ where $p(\beta_i^l) < p(\alpha_{begin}(g_{[f]}))$, $p(\beta_i^{l-1}) \ge p(\alpha_{begin}(g_{[f]}))$ (*Condition 13*), then the schedule (7.29) is optimal.

SSO #11 If M_i has the form of any of Structures $S_3, \ldots, S_{10}, S_{12}, S'_3, \ldots, S'_{10}, S'_{12}$ and Conditions 9–11 and 13 are met, then we have built an optimal schedule in the form (7.29).

If $p(\beta_i^l, \ldots, \beta_i^j, K_i^1, \ldots, K_i^s) \ge p(\alpha_{begin}(g_{[f]}))$ and $p(K_i^s) \ge p(\alpha_{end}(g_{[k]}))$ (*Condition 14*) where $p(\beta_i^l) < p(\alpha_{begin}(g_{[f]}))$ and $p(\beta_i^{l-1}) \ge p(\alpha_{begin}(g_{[f]}))$ (*Condition 15*), $\alpha_{begin}(g_{[f]}) = (g_{[s]}, \ldots, g_{[f]})$ and $\exists g_{[f+1]} \in \alpha(g_{[k]})$: $\beta_i^t \prec\!\prec g_{[f+1]}$, $t = \overline{1,j}$ (*Condition 16*), then the schedule (7.29) is optimal on the set of tasks $M_i \cup \alpha(g_{[k]})$. The precedence relations from tasks β_i^t, $t = \overline{1,j}$ to $\alpha_{begin}(g_{[f]})$ can be arbitrary.

SSO #12 If M_i has the form of any of Structures $S_3, \ldots, S_{10}, S_{12}, S'_3, \ldots, S'_{10}, S'_{12}$, taking into account an arbitrary preceding from tasks β_i^t, $t = \overline{1,j}$ to $\alpha_{begin}(g_{[f]})$, and

Conditions 9–11 and 14–16 are met, then we have built an optimal schedule in the form (7.29).

If $\exists v \in K_i^{\rho}$, $\rho \geq s$: $v \prec\prec g_{[h]}$, $g_{[h]} \in \alpha_{begin}\left(g_{[f]}\right)$ (*Condition 17*), then we verify the conditions of Statement 7.3 for the chain $\alpha_{begin}\left(g_{[h-1]}\right) = \left(g_{[s]}, \ldots, g_{[h-1]}\right)$. Note that if $\exists v \in K_i^{\rho}$, $\rho \geq s$, such that $v \prec\prec g_{[s]}$, then the situation 3 takes place and, therefore, the schedule (7.29) is optimal on the set of tasks under consideration.

SSO #13 If M_i has the form of any of Structures $S_3, \ldots, S_{10}, S_{12}, S'_3, \ldots, S'_{10}, S'_{12}$ and Condition 17 is met, then we check Conditions 9–12 for the chain $\alpha_{begin}\left(g_{[h-1]}\right) = \left(g_{[s]}, \ldots, g_{[h-1]}\right)$. If Conditions 9–11 are met, then we have built an optimal schedule in the form (7.29). If Conditions 9, 10, and 12 are met, then we have built an optimal schedule in the form (7.30).

Let M_i has the form (7.20) where the sequence $K_1, \alpha_1, K_2 \ldots, K_j, \alpha_j, K_{j+1}, \ldots, K_p, \alpha_p$ consists of priority-ordered structures and chains; $\beta_j \prec\prec K_s$, $j = \overline{1, i}$; $\beta_j \prec \alpha_s$, $s = \overline{1, p}$. The chains and constructions order is determined by the priority relations. Here, a set K_j may be a combination of non-adjacent priority-ordered structures and chains. We assume that the violation of the preceding relations of the first task on the set M_i is only for chains α_i. Denote this structure of M_i as *Structure* S_{13}. Partition the chain α_i: $\alpha_i = \left(\alpha_i^{begin}, \alpha_i^{end}\right)$ where α_i^{begin} is the beginning part of the chain α_i not adjacent to any task of M_i, $\beta_t \prec\prec \alpha_i^{end}$, $t = \overline{1, i}$.

According to Statement 7.4, if the tasks of M_i satisfy the following priority relations: $p\left(\alpha_1^{begin}\right) \geq p\left(\alpha_2^{begin}\right) \geq \cdots \geq p\left(\alpha_j^{begin}\right)$ and $p(K_1) \geq p\left(\alpha_1^{end}\right)$, …, $p(K_j) \geq p\left(\alpha_j^{end}\right)$ (*Condition 18*), then if $p(\beta_f, \ldots, \beta_i, K_1) \geq p\left(\alpha_1^{begin}\right)$ where $p(\beta_f) < p\left(\alpha_1^{begin}\right)$ and $p(\beta_{f-1}) \geq p\left(\alpha_1^{begin}\right)$ (*Condition 19*), then the schedule (7.20) is optimal on the set of tasks M_i.

SSO #14 If M_i has the form of Structure S_{13} and Conditions 18 and 19 are met, then we have built an optimal schedule in the form (7.20).

Remark 7.8 In Statement 7.4, the violation of the precedence conditions to the first task of the successor of the first level does not depend on its type. Therefore, we can replace the chains α_i in Structure S_{13} with arbitrarily nested constructions (denote the obtained structure as *Structure* S_{14}).

Suppose that the precedence conditions and the priority relations of Statement 7.4 are valid on the set of tasks M_i, that is, the structure of M_i corresponds to either the form of Structure S_{13} or S_{14}. Consider the additional precedence constraints under which the schedule (7.20) is also optimal:

- under arbitrary precedence of tasks of the chains β_f, $f = \overline{1, i}$, to the chains α_j^{begin}, $j = \overline{1, p}$ (*Structure* S_{15});
- under arbitrary precedence of tasks of the chains α_j^{begin} to the chains α_l^{begin}, $l > j$ (*Structure* S_{16});
- under arbitrary precedence of tasks of the constructions K_j to tasks of the chains α_l^{begin}, $l > j$ (*Structure* S_{17}).

SSO #15 If M_i has the form of any of Structures S_{14}, …, S_{17} and Conditions 18 and 19 are met, then we have built an optimal schedule in the form (7.20).

Algorithm for Checking Sufficient Signs of Optimality of the Current Feasible Subsequence has the following 16 steps.

1. Create a chain (or an elementary construction) of tasks $\alpha(g_{[k]})$ and perform its efficient move. If the obtained Structure S_1 and the priority relations in it satisfy the cases considered in Algorithms 1–3 of the polynomial algorithm П1 [1], then the current schedule is optimal, terminate the checking algorithm. Otherwise, go to step 2.
2. If the structure of the obtained task set corresponds to Structure S_2 and $p(\alpha(g_{[k]})) \le p(M_i)$, then the current schedule is optimal, terminate the checking algorithm. Otherwise, go to step 3.
3. If M_i has either the form of Structure S_3 or S_4 and Conditions 1 and 2 are met, then we have built an optimal schedule on the set $M_i \cup \alpha(g_{[k]})$ in the form (7.24), terminate the checking algorithm. Otherwise, go to step 4.
4. If M_i has either the form of Structure S_3 or S_4 and Conditions 3 and 4 are met, then we perform Action 1. If Condition 5 is met, then we have built a schedule in the form (7.25), go to step 5. Otherwise, if Condition 6 is met after performing Action 2, then we have built a schedule in the form (7.26), go to step 5. Otherwise, go to step 6.
5. If $p(K_s) < p(K_i')$, $K_s \prec K_i'$, then return to step 4 considering there the construction K_i' as a chain $\alpha(g_{[k]})$ and the construction K_s as the construction K_i and build in a similar way a priority-ordered schedule on the set of tasks $K_i' \cup K_s$. Otherwise, the current schedule of the form (7.25) or (7.26) is optimal, terminate the checking algorithm.
6. If M_i has the form of any of Structures S_5, S_6, S_7 and Conditions 3 and 4 are met, then we perform Action 1. If Condition 5 is met, then we have built an optimal schedule in the form (7.25), terminate the checking algorithm. Otherwise, if Condition 6 is met after performing Action 2, then we have built an optimal schedule in the form (7.26), terminate the checking algorithm. Otherwise, go to step 7.
7. If M_i has the form of any of Structures S_8, S_9, S_{10} and Conditions 3 and 4 are met, then we perform Action 1. If Condition 5 is met, then we have built an optimal schedule in the form (7.25), terminate the checking algorithm. Otherwise, if Condition 6 is met after performing Action 2, then we have built an optimal schedule in the form (7.26), terminate the checking algorithm. Otherwise, go to step 8.
8. If M_i has the form of any of Structures S_3, …, S_{10}; K_i has the form of Structure S_{11}; and Condition 7 is met, then we have built an optimal schedule in the form (7.28), terminate the checking algorithm. Otherwise, go to step 9.
9. If M_i has the form of Structure S_{12} and Conditions 3 and 4 are met, then we perform Action 1. If Condition 5 is met, then we have built an optimal schedule in the form (7.25), terminate the checking algorithm. Otherwise, if Condition 6

is met after performing Action 2, then we have built an optimal schedule in the form (7.26), terminate the checking algorithm. Otherwise, go to step 10.

10. If M_i has the form of any of Structures S_3, …, S_{10}, S_{12}, S'_3, …, S'_{10}, S'_{12} and Condition 8 is met, then we have built an optimal schedule in the form (7.29), terminate the checking algorithm. Otherwise, go to step 11.
11. If M_i has the form of any of Structures S_3, …, S_{10}, S_{12}, S'_3, …, S'_{10}, S'_{12} and Conditions 9–11 are met, then we have built an optimal schedule in the form (7.29), terminate the checking algorithm. If Conditions 9, 10, and 12 are met on these structures, then we have built an optimal schedule in the form (7.30), terminate the checking algorithm. Otherwise, go to step 12.
12. If M_i has the form of any of Structures S_3, …, S_{10}, S_{12}, S'_3, …, S'_{10}, S'_{12} and Conditions 9–11 and 13 are met, then we have built an optimal schedule in the form (7.29), terminate the checking algorithm. Otherwise, go to step 13.
13. If M_i has the form of any of Structures S_3, …, S_{10}, S_{12}, S'_3, …, S'_{10}, S'_{12}, taking into account an arbitrary preceding from tasks β_i^t, $t = \overline{1,j}$ to $\alpha_{begin}\left(g_{[f]}\right)$, and Conditions 9–11 and 14–16 are met, then we have built an optimal schedule in the form (7.29), terminate the checking algorithm. Otherwise, go to step 14.
14. If M_i has the form of any of Structures S_3, …, S_{10}, S_{12}, S'_3, …, S'_{10}, S'_{12} and Condition 17 is met, then check Conditions 9–12 for the chain $\alpha_{begin}\left(g_{[h-1]}\right) = \left(g_{[s]}, \ldots, g_{[h-1]}\right)$. If Conditions 9–11 are met, then we have built an optimal schedule in the form (7.29), terminate the checking algorithm. If Conditions 9, 10, and 12 are met, then we have built an optimal schedule in the form (7.30), terminate the checking algorithm. Otherwise, go to step 15.
15. If M_i has the form of Structure S_{13} and Conditions 18 and 19 are met, then we have built an optimal schedule in the form (7.20), terminate the checking algorithm. Otherwise, go to step 16.
16. If M_i has the form of any of Structures S_{14}, …, S_{17} and Conditions 18 and 19 are met, then we have built an optimal schedule in the form (7.20), terminate the checking algorithm. Otherwise, none of the fifteen above SSOs was satisfied, terminate the checking algorithm.

7.3.3.2 The Main Statements to Substantiate the Rules of a *p*-Ordered Schedule Construction When the Precedence Relations Are Specified by an Arbitrary Oriented Acyclic Graph

The algorithm is based on the following statements [4] generalizing the positions given in [1] and presented in a form convenient to describe the PSC-algorithm.

Statement 7.5 Suppose that in a feasible sequence σ the following condition is true for independent elementary constructions of tasks $\alpha\left(j_{[g]}\right) = \left(j_{[r]}, j_{[r+1]}, \ldots, j_{[g-1]}, j_{[g]}\right)$ and $\beta\left(j_{[f]}\right) = \left(j_{[s]}, j_{[s+1]}, \ldots, j_{[f-1]}, j_{[f]}\right)$ where $f < r$:

$$p\left(\alpha\left(j_{[g]}\right)\right) \le p\left(\beta\left(j_{[f]}\right)\right). \tag{7.31}$$

If $\forall \upsilon < s,\ \forall k > g$ we have:

$$\text{if } j_{[\upsilon]} \prec \beta\left(j_{[f]}\right), \text{ then } p\left(j_{[\upsilon]}\right) \ge p\left(\beta\left(j_{[f]}\right)\right) \tag{7.32}$$

$$\text{if } \alpha\left(j_{[g]}\right) \prec j_{[k]}, \text{ then } p\left(j_{[k]}\right) \le p\left(\alpha\left(j_{[g]}\right)\right) \tag{7.33}$$

then the elementary construction of tasks $\beta\left(j_{[f]}\right)$ must be processed before the elementary construction of tasks $\alpha\left(j_{[g]}\right)$ in a p-ordered schedule.

Proof If we meet the conditions (7.31)–(7.33), then it is impossible, without violation of the feasibility of the considered sequence, to form such two independent elementary constructions α_1 and β_1 where α_1 includes $\alpha\left(j_{[g]}\right)$, β_1 includes $\beta\left(j_{[f]}\right)$, and $p(\alpha_1) > p(\beta_1)$. ∎

Suppose that in a feasible schedule $\sigma(J_1) = \left(j_{[1]}, j_{[2]} \ldots, j_{[g]}\right)$ the first $g-1$ tasks $j_{[1]}, j_{[2]}, \ldots, j_{[g-1]}$ are p-ordered tasks for the task set $\left\{j_{[1]}, j_{[2]}, \ldots, j_{[g-1]}\right\}$. Let σ^1 denote this initial subsequence. The tasks $j_{[r]}, j_{[r+1]}, \ldots, j_{[g-1]}, j_{[g]}$ form an elementary construction $\alpha\left(j_{[g]}\right)$. There is an elementary construction $\beta\left(j_{[f]}\right)$,

$$\beta\left(j_{[f]}\right) = \left(j_{[s]}, j_{[s+1]}, \ldots, j_{[f-1]}, j_{[f]}\right), \quad f < r, \quad \beta\left(j_{[f]}\right) \prec \alpha\left(j_{[g]}\right).$$

For all other tasks $j \in \sigma^1$ we have:

$$j \not\prec \alpha\left(j_{[g]}\right) \tag{7.34}$$

Statement 7.6 If

$$p\left(\beta\left(j_{[f]}\right)\right) \ge p\left(\alpha\left(j_{[g]}\right)\right), \tag{7.35}$$

then any task $j_{[k]} \in \alpha\left(j_{[g]}\right)$ takes the position $t > f$ in a p-ordered schedule.

Proof t is less than f only if we can organize a construction of tasks including elementary constructions $\alpha\left(j_{[g]}\right)$, $\beta\left(j_{[f]}\right)$, and tasks in positions between them with a greater priority than that of the elementary construction $\beta\left(j_{[f]}\right)$. This is impossible due to p-ordering of σ^1 and condition (7.35). ∎

Statement 7.7 If $p\left(\beta\left(j_{[f]}\right)\right) < p\left(\alpha\left(j_{[g]}\right)\right)$ and $\forall j_{[q]} \in \sigma^1,\ f < q < r,$ such that $\beta\left(j_{[f]}\right) \prec j_{[q]}$, we have

$$p\left(\beta\left(j_{[f]}\right)\right) > p\left(j_{[q]}\right), \tag{7.36}$$

then any task $j_{[k]} \in \alpha(j_{[g]})$ will take a position $t \le f + n_1$ in a p-ordered schedule where n_1 is the number of tasks in $\alpha(j_{[g]})$.

Proof According to Statement 7.5 and the condition $j_{[l]} \prec \alpha(j_{[g]})$ formulated above, the tasks $j_{[q]}$ will be processed after the elementary construction $\alpha(j_{[g]})$ in a p-ordered schedule. The latter construction after its efficient move becomes an elementary construction $\alpha(j_{[f+n_1]})$.

Due to the fact that $p(\alpha(j_{[f+n_1]})) > p(\beta(j_{[f]}))$ and $\beta(j_{[f]}) \prec \alpha(j_{[f+n_1]})$, we organize an elementary construction $\gamma(j_{[f+n_1]}) = \beta(j_{[f]}) \cup \alpha(j_{[f+n_1]})$ which will take a position $t \le f + n_1$ in a p-ordered schedule. ∎

Statement 7.8 Suppose that the sequence including $\sigma^1 \cup j_{[g]}$ is p-ordered. If $p(\alpha(j_{[g]})) > p(\beta(j_{[f]}))$, then the positions between the elementary constructions of tasks $\beta(j_{[f]})$ and $\alpha(j_{[g]})$ in a p-ordered schedule can only occupy the elementary constructions of tasks $z(j_{[q]})$ for which

$$p(z(j_{[q]})) > p(\alpha(j_{[g]})), \quad \beta(j_{[f]}) \prec z(j_{[q]}). \tag{7.37}$$

Proof Let us analyze the elementary constructions of tasks $z(j_{[q]})$ occupying positions between elementary constructions $\beta(j_{[f]})$ and $\alpha(j_{[g]})$. Suppose that $\beta(j_{[f]}) \overline{\prec} z(j_{[q]})$. If $p(z(j_{[q]})) > p(\beta(j_{[f]}))$, then, due to p-ordering of the sequence σ^1, these elementary constructions occupy earlier positions than the elementary construction $\beta(j_{[f]})$. If $p(z(j_{[q]})) < p(\alpha(j_{[g]}))$, then, according to Statement 7.5 and the conditions (7.34), (7.36), tasks of $z(j_{[q]})$ will appear after the elementary constructions of tasks $\alpha(j_{[g]})$ in a p-ordered schedule of the sequence σ^1.

Suppose that $\beta(j_{[f]}) \prec z(j_{[q]})$. If $p(z(j_{[q]})) \ge p(\alpha(j_{[g]}))$, then the tasks of $z(j_{[q]})$ occupy positions between the elementary constructions $\beta(j_{[f]})$ and $\alpha(j_{[g]})$. If $p(z(j_{[q]})) < p(\alpha(j_{[g]}))$, then the tasks of $z(j_{[q]})$ occupy the positions following the elementary construction $\alpha(j_{[g]})$. ∎

The following statement shows the main property of a series-parallel graph which we use in the algorithm.

Statement 7.9 (Pavlov et al. [4]) Suppose that the sequence of tasks $\sigma = (j_{[1]}, j_{[2]}, \ldots, j_{[g-1]}, j_{[g]})$ is feasible. If the precedence relations are specified by a series-parallel graph, then the following is true. If $j_{[k]} \prec (j_{[x]}, j_{[y]}, j_{[z]})$ where $j_{[x]}, j_{[y]}, j_{[z]}$ are independent tasks, then there is no such task $j_{[l]}$, $l = \overline{1, k-1}$, in the sequence σ which does not relate to the task $j_{[k]}$ and satisfies only one of the conditions: either $j_{[l]} \prec j_{[x]}$, or $j_{[l]} \prec j_{[y]}$, or $j_{[l]} \prec j_{[z]}$.

7.3.3.3 Rules of a *p*-Ordered Schedule Construction for the Case of a Series-Parallel Graph

The basis of the PSC-algorithm is the modified Algorithm A (presented below) of constructing a *p*-ordered schedule for a series-parallel graph. In contrast to the algorithm given in [1] and based on task chains, we base Algorithm A on elementary constructions, all tasks of which do not have immediate predecessors. This makes their interval of move always greater than zero, which ensures a higher efficiency of the algorithm.

We now formulate the rules for constructing a *p*-ordered schedule on a set of tasks $\{j_{[1]}, j_{[2]}, \ldots, j_{[g]}\}$ based on the above statements and used in Algorithm A.

Consider the initial sequence of tasks $\sigma^1 = (j_{[1]}, j_{[2]}, \ldots, j_{[g-2]}, j_{[g-1]})$ and the oriented subgraph which was built from the initial graph and determines the precedence relations on the task set of this sequence.

Suppose that this sequence is *p*-ordered for the graph under consideration and the specified functional. Let us add to this sequence one task $j_{[g]}$ feasible by precedence relations. Accordingly, we expand the initial oriented subgraph adding to it the arcs corresponding to the precedence relations of the tasks $j_{[1]}, j_{[2]}, \ldots, j_{[g]}$. Denote the obtained sequence as $\sigma^2 = (j_{[1]}, j_{[2]}, \ldots, j_{[g]})$.

Consider the following cases.

1. Suppose that the tasks $j_{[r]}, j_{[r+1]}, \ldots, j_{[g-1]}, j_{[g]}$ form an elementary construction $\alpha(j_{[g]})$. And suppose that $j \overline{\prec} \alpha(j_{[g]})$ for all tasks $j \in \sigma^2 \backslash \alpha(j_{[g]})$. Then, to construct a *p*-ordered schedule from the considered sequence, it is sufficient to insert the elementary construction of tasks $\alpha(j_{[g]})$ into position l determined by the interval of efficient move. Let $\alpha(j_{[l+n_1-1]})$ denote the elementary construction $\alpha(j_{[g]})$ after its insertion. Here, n_1 is the number of tasks in the elementary construction, $l+n_1-1$ is the number of the position occupied by the last task of the elementary construction.

Proof The decrease in the functional value for tasks $j \in \alpha(j_{[l+n_1-1]})$ in consequence of the efficient move is

$$\Omega\alpha(j_{[l+n_1-1]}) \sum_{k=r-1}^{l} l_{j_{[k]}}.$$

Accordingly, the increase in the functional value for tasks that have taken up later positions is equal to

$$L\alpha(j_{[l+n_1-1]}) \sum_{k=r-1}^{l} \omega_{j_{[k]}}.$$

According to the definition of the interval of efficient move, the difference

$$\Omega\alpha\left(j_{[l+n_1-1]}\right)\sum_{k=r-1}^{l} l_{j_{[k]}} - L\alpha\left(j_{[l+n_1-1]}\right)\sum_{k=r-1}^{l} \omega_{j_{[k]}}$$

has a maximum value.

Consequently, tasks (or elementary constructions) within the interval $\overline{1, l-1}$ have a higher priority than that of the elementary construction $\alpha\left(j_{[l+n_1-1]}\right)$. Correspondingly, the elementary construction $\alpha\left(j_{[l+n_1-1]}\right)$ is followed by tasks (in general, elementary constructions) of the lower priority. Let us show that the obtained sequence is p-ordered.

Let σ_1^1, σ_2^1, …, σ_k^1 be the MPSSes that constitute the p-ordered schedule σ^1. Then, the elementary construction $\alpha\left(j_{[l+n_1-1]}\right)$, as a result of its efficient move, immediately follows one of the subsequences σ_j^1 and cannot be inserted into either of them. Indeed, any subsequence σ_j^1 cannot be represented in the form $\sigma_{j_1}^1 \cup \sigma_{j_2}^1$ of such subsequences that $p\left(\sigma_{j_1}^1\right) > p\left(\sigma_{j_2}^1\right)$. Otherwise, in accordance with the definition of p-ordered sequence, the subsequence $\sigma_{j_1}^1$ would be in place of the subsequence σ_j^1. Consequently, $p\left(\sigma_{j_2}^1\right) \geq p\left(\sigma_{j_1}^1\right)$, and we cannot replace $\sigma_{j_2}^1$ with the elementary construction $\alpha\left(j_{[l+n_1-1]}\right)$ as a result of its efficient move. The elementary construction $\alpha\left(j_{[l+n_1-1]}\right)$ is a subsequence σ_{j+1}^1, since it has the highest priority among all the subsequences specified on the set of tasks $J \backslash J_1$ ready to be processed after the subsequence σ_j^1. Here, $J = \left\{j_{[1]}, j_{[2]}, \ldots, j_{[g]}\right\}$, J_1 is the set of tasks in the subsequences σ_1^1, …, σ_j^1.

Due to p-ordering of the schedule σ^1, tasks that appear after the elementary construction $\alpha\left(j_{[l+n_1-1]}\right)$ as a result of its efficient move are also p-ordered on the set of tasks $J \backslash \{J_1 \cup J_2\}$. Here, J_2 is the set of tasks in the elementary construction $\alpha\left(j_{[l+n_1-1]}\right)$. Consequently, the considered sequence obtained as a result of the efficient move of the elementary construction $\alpha\left(j_{[g]}\right)$ is p-ordered. ∎

Statement 7.10 The sequence σ^2 is p-ordered if the conditions formulated in the case 1 are satisfied and $I\alpha\left(j_{[g]}\right) = 0$.

2. Consider the next case. Suppose that we have performed the efficient move for the elementary construction of tasks $\alpha\left(j_{[g]}\right)$. We have obtained a sequence σ^3 which contains an elementary construction

$$\beta\left(j_{[f]}\right) = \left(j_{[s]}, j_{[s+1]}, \ldots, j_{[f]}\right), \quad \beta\left(j_{[f]}\right) \prec \alpha\left(j_{[g]}\right).$$

For the remaining tasks $j \in \sigma^3 \backslash \left\{\beta\left(j_{[f]}\right) \cup \alpha\left(j_{[g]}\right)\right\}$, we meet the condition $j \bar{\prec} \alpha\left(j_{[g]}\right)$. Suppose that $p\left(\beta\left(j_{[f]}\right)\right) \geq p\left(\alpha\left(j_{[g]}\right)\right)$. In this case, in order to construct the

p-ordered schedule of the sequence under consideration, according to Statement 7.6, it is sufficient to insert the elementary construction $\alpha(j_{[g]})$ into the position determined by its interval of efficient move. Hence, the sequence σ^3 is a p-ordered sequence due to p-ordering of σ^1. Suppose that $p(\beta(j_{[f]})) < p(\alpha(j_{[g]}))$, and the following is true for all tasks $z(j_{[q]}), f < q < r$, in the sequence σ^3:

$$p(\beta(j_{[f]})) \geq p(z(j_{[q]})). \tag{7.38}$$

Since the sequence $\sigma^2 \backslash j_{[g]}$ is p-ordered, and $p(\beta(j_{[f]})) < p(\alpha(j_{[g]}))$, then the elementary construction $\alpha(j_{[g]})$ after its efficient move takes in the sequence σ^3 positions immediately following the elementary construction $\beta(j_{[f]})$, i.e., $f + n_1$.

Let J_1^{eff} denote a set of tasks moved into later positions as a result of this procedure. The following relation is true:

$$p(\beta(j_{[f]})) \geq \sum_{j \in J_1^{eff}} \omega_j \Big/ \sum_{j \in J_1^{eff}} l_j.$$

According to Statement 7.7, we organize an elementary construction of tasks $\gamma(j_{[f+n_1]})$ in the sequence σ^3. Here, n_1 is the number of tasks in the elementary construction $\alpha(j_{[g]})$ for which we also perform the efficient move. If the sequence σ^3 is within the interval $\overline{1, f + n_1}$, we arrive at the case 1. Suppose that we have inserted the elementary construction of tasks $\gamma(j_{[f+n_1]})$ into the positions $\overline{k, t}$ as a result of its efficient move. Here, t is the position occupied by the last task of the above elementary construction. Let $\gamma(j_{[t]})$ denote the elementary construction, J_2^{eff} be a set of tasks moved to later positions as a result of this procedure, and σ^4 be a sequence obtained.

The following is true:

$$p(\gamma(j_{[t]})) > \sum_{j \in J_2^{eff}} \omega_j \Big/ \sum_{j \in J_2^{eff}} l_j > \sum_{j \in J_1^{eff}} \omega_j \Big/ \sum_{j \in J_1^{eff}} l_j. \tag{7.39}$$

Due to the facts that:

- according to the proof of p-ordering for the case 1, the subsequences of interrelated tasks in the sequence σ^4 in positions before the elementary construction $\gamma(j_{[t]})$ have a higher priority;
- the relation (7.39) is true;
- due to p-ordering of the sequence σ^1, it is impossible to organize among the tasks $j \in J_2^{eff}$, $j \in J_1^{eff}$ new elementary constructions which have a higher priority than the elementary construction $\gamma(j_{[t]})$,

hence, the sequence σ^4 is p-ordered by definition.

It is clear that if the conditions of case 2 are satisfied and $I^{eff}_{\gamma(j_{[t]})} = 0$, the sequence σ^3 is p-ordered.

3. Consider the case when there exist elementary constructions of tasks $z(j_{[q]})$, $\beta(j_{[f]}) \prec z(j_{[q]})$, in the sequence σ^2 for which $p(\beta(j_{[f]})) < p(z(j_{[q]}))$. These elementary constructions in the sequence σ^2 are ordered according to their priorities, since the sequence σ^1 is p-ordered.

Consider the sequence σ^3 obtained as a result of the efficient move of the elementary construction $\alpha(j_{[g]})$.

3.1. Suppose that the elementary construction $\alpha(j_{[g]})$ took positions immediately following the elementary construction $\beta(j_{[f]})$ i.e., $f + n_1$ where n_1 is the number of tasks in $\alpha(j_{[g]})$. The following is true:

$$p(\alpha(j_{[f+n_1]})) > \sum_{j \in J_1^{eff}} \omega_j \Big/ \sum_{j \in J_1^{eff}} l_j. \tag{7.40}$$

According to Statement 7.7, we organize a new elementary construction of tasks $\gamma(j_{[f+n_1]}) = \beta(j_{[f]}) \cup \alpha(j_{[f+n_1]})$. Since

$$p(\gamma(j_{[f+n_1]})) > p(\beta(j_{[f]})), \tag{7.41}$$

we perform the efficient move for the elementary construction $\gamma(j_{[f+n_1]})$. J_2^{eff} is a set of tasks moved into later positions as a result of this procedure.

Suppose that the elementary construction of tasks $\gamma(j_{[f+n_1]})$ has become an elementary construction $\gamma(j_{[f+n_1-n_2]})$ as a result of its efficient move. Here, n_2 is the number of tasks $j \in J_2^{eff}$. The following relations are true for the obtained sequence σ^4:

$$\sum_{j \in J_1^{eff}} \omega_j \Big/ \sum_{j \in J_1^{eff}} l_j < p(\alpha(j_{[g]}));$$

$$\sum_{j \in J_2^{eff}} \omega_j \Big/ \sum_{j \in J_2^{eff}} l_j < p(\gamma(j_{[f+n_1-n_2]})) < p(\alpha(j_{[g]})).$$

Since, in the general case, $\exists j_k \in J_1^{eff}$ and $\exists j_l \in J_2^{eff}$ such that $p(j_k) > p(j_l)$, we should reorder tasks within the interval $Q_1 = \overline{f + n_1 - n_2 + 1, g}$ according to their priorities. Let us denote the resulting sequence as σ^5.

Since the relations (7.40), (7.41) are true, we can increase the priority of the elementary construction $\gamma(j_{[f+n_1-n_2]})$ by including in it elementary constructions

$z(j_{[q]}) \in Q_1$. These tasks immediately follow the elementary construction $\gamma(j_{[f+n_1-n_2]})$ in the sequence σ^5, So, the last of the tasks takes the position $f+n_1-n_2+n_3$ where n_3 is the number of tasks $j_{[q]} \in Q_1$.

We organize a construction of tasks K_1:

$$p(K_1) = \frac{\Omega\gamma(j_{[f+n_1-n_2]}) + \sum_{j\in Q_2} \omega_j}{L\gamma(j_{[f+n_1-n_2]}) + \sum_{j\in Q_2} l_j}$$

where Q_2 is a subset of tasks $j \in Q_1$ on which we reach

$$\max \frac{\sum_{i=1}^{n_3} \omega_{j_{[f+n_1-n_2+i]}} + \Omega\gamma(j_{[f+n_1-n_2]})}{\sum_{i=1}^{n_3} l_{j_{[f+n_1-n_2+i]}} + L\gamma(j_{[f+n_1-n_2]})}.$$

Since $p(K_1) > p(\gamma(j_{[f+n_1-n_2]}))$, then we perform the efficient move, in turn, for the construction of tasks K_1. Denote the resulting sequence as σ^6.

Suppose that the construction K_1 took positions $\overline{p,\ t}$ after this move. Here, t is the position of the last task $j_{[t]} \in K_1$.

Let J_3^{eff} denote the set of tasks moved into earlier positions as a result of the move of the construction K_1. The following relations hold:

$$\sum_{j\in J_3^{eff}} \omega_j \Big/ \sum_{j\in J_3^{eff}} l_j < p(K_1); \quad \sum_{j\in J_3^{eff}} \omega_j \Big/ \sum_{j\in J_3^{eff}} l_j > p(\gamma(j_{[f+n_1-n_2]}))$$

In the sequence σ^6, the construction K_1 is followed by tasks $j_{[q]} \in Q_1$, $p(j_{[q]}) > p(\gamma(j_{[f+n_1-n_2]}))$, not included in the construction. Therefore, we should insert the tasks $j \in J_3^{eff}$ into positions corresponding to their priorities in the sequence σ^6. Denote the resulting sequence as σ^7. All tasks j within the interval $\overline{1,\ t}$ in this sequence are p-ordered, in accordance with p-ordering of σ^1. The proof of this statement is similar to the proof of the above case 1 if we consider the construction of tasks K_1 as an elementary construction $\alpha(j_{[g]})$.

Elementary constructions of tasks within the interval $\overline{t+1,\ g}$ have lower priority than that of the construction K_1. Otherwise, they would be included in it by construction. These elementary constructions are also p-ordered. Hence, the sequence σ^7 is p-ordered. Obviously, if the conditions of sub-item 3.1 fulfill and $I^{eff}_{\gamma(j_{[f+n_1-n_2]})} = 0$, $I^{eff}_{K_1} = 0$, then the sequence is p-ordered.

3.2. Suppose that the elementary construction of tasks $\alpha(j_{[g]})$ has become an elementary construction of tasks $\alpha(j_{[\upsilon]})$ after its efficient move within the sequence σ^3. υ is the position occupied by the last task $j_{[\upsilon]} \in \alpha(j_{[\upsilon]})$. Consider a case when there are other tasks between the elementary constructions $\alpha(j_{[\upsilon]})$ and $\beta(j_{[f]})$. According to Statement 7.8, these are the tasks or elementary constructions of tasks $z(j_{[q]})$ which satisfy: $p(z(j_{[q]})) \geq p(\alpha(j_{[\upsilon]}))$; $\beta(j_{[f]}) \prec z(j_{[q]})$. Let Q_3 denote the set of such tasks. We organize the construction of tasks K_2 including the elementary constructions of tasks $\beta(j_{[f]})$, $\alpha(j_{[\upsilon]})$, and tasks $j \in Q_3$:

$$p(K_2) = \frac{\Omega\beta(j_{[f]}) + \sum\limits_{j \in Q_3} \omega_j + \Omega\alpha(j_{[\upsilon]})}{L\beta(j_{[f]}) + \sum\limits_{j \in Q_3} l_j + L\alpha(j_{[\upsilon]})}.$$

If we consider the special case of the construction of tasks K_2, the elementary construction of tasks $\gamma(j_{[f+n_1]})$, as the construction of tasks K_2, we arrive at the case considered in sub-item 3.1.

Therefore, in order to construct a p-ordered schedule, it is sufficient to perform for the construction of tasks K_2 the procedures performed for the elementary construction $\gamma(j_{[f+n_1]})$ described in sub-item 3.1.

Obviously, if the conditions of sub-item 3.2 are satisfied and $I^{eff}_{K_2} = 0$, the sequence under consideration is p-ordered.

Thus, sub-item 3.2 generalizes all cases of constructing a p-ordered schedule for the sequence σ^3 defined in case 2.

4. Suppose that we have performed the efficient move for the elementary construction of tasks $\alpha(j_{[g]})$. Consider the general case when there is an elementary construction $\beta(j_{[f]}) = (j_{[s]}, j_{[s+1]}, \ldots, j_{[f-1]}, j_{[f]})$ in the obtained sequence, $\beta(j_{[f]}) \prec \alpha(j_{[g]})$, and the elementary constructions of tasks $\delta(j_{[l]}) \in \sigma^3 \backslash \{\beta(j_{[f]}) \cup \alpha(j_{[g]})\}$ satisfy: $\delta(j_{[l]}) \prec j_{[g]}$, $\forall l, s$.

Suppose that we have created, by the rules stated in sub-item 3.1, the construction of tasks K_1 of the maximum priority, including the elementary constructions of tasks $\alpha(j_{[g]})$, $\beta(j_{[f]})$, and elementary constructions of tasks $z(j_{[q]}) \in Q_1$ which satisfy $p(z(j_{[q]})) > p(\beta(j_{[f]}))$.

Suppose that we have performed the efficient move for the construction of tasks K_1. Tasks that occupy positions following the construction K_1 are p-ordered. Considering the construction of tasks K_1 as an elementary construction $\alpha(j_{[\upsilon]})$, we arrive at the case described in sub-item 3.2. Thus, by performing the procedures described in sub-item 3.2, we construct a p-ordered sequence of tasks of the sequence σ^2 in the general case. Obviously, if no permutation was done in case 4, then the sequence σ^3 is p-ordered.

7.3.3.4 Algorithm a for the Problem Solving When the Precedence Relations Are Specified by a Series-Parallel Graph

We now formulate an algorithm for constructing a p-ordered schedule for the case when the strict order of precedence relations is specified by a series-parallel graph. The algorithm is based on the above rules for constructing a p-ordered schedule on a feasible sequence σ defined above.

Algorithm A contains three units: ordering unit (Algorithm A1), optimization unit (Algorithm A2), and the unit of reordering on intervals of efficient move (Algorithm A1.1).

Algorithm A1 (Ordering Unit) has the following nine steps.

1. Arrange all tasks in non-increasing order of priorities ω_j/l_j. Denote the obtained sequence as σ_{ord}.
2. Consider the sequence σ_{ord} from the beginning.
3. Select the first task $j_{[1]}$ of the sequence σ_{ord} and move it into the first position in the new sequence σ_{ord}^{new}.
4. Select from the set of tasks $J\backslash j_{[1]}$ the tasks $j_{[r]}, r = \overline{2, n}$, which immediately precede $j_{[1]}$.
5. Insert the tasks $j_{[r]}$ before the task $j_{[1]}$ in σ_{ord}^{new} according to their priorities, without violating the feasibility of the sequence σ_{ord}^{new}.
6. For each moved task, in turn, find immediate predecessors and in a similar way (see steps 4, 5) insert them into the new sequence.
7. Repeat steps 4–6 for the remaining tasks of the sequence σ_{ord} until all the tasks necessary for completing the task $j_{[1]}$ are included.
8. Proceed to the next task $j_{[2]}$ of the sequence σ_{ord}. If this task is not in the sequence σ_{ord}^{new}, include it in this sequence after the task $j_{[1]}$. Go to step 4 and in a similar way insert before the task $j_{[2]}$ its predecessors excluding the tasks that exist at the earlier positions in the sequence σ_{ord}^{new}.
9. If all the tasks of the sequence σ_{ord} are considered, terminate. We have constructed the sequence σ_{ord}^{new}.

The sequence σ_{ord}^{new} is feasible by construction. If all the tasks of this sequence were in a non-decreasing order of their priorities, then the sequence would be optimal. In the general case, lower priority tasks may appear in the sequence σ_{ord}^{new} at earlier positions than those of higher priority tasks. Therefore, we perform the optimization unit to build an optimal schedule.

Algorithm A2 (Optimization Unit) At this stage, we insert separate tasks or elementary constructions of tasks into earlier positions. Such insertion procedures, being carried out according to certain rules, lead to a decrease in the functional value. This allows us to construct an optimal sequence by the TWCT criterion. The insertion procedures are as follows.

1. Suppose that we have already performed $g - 1$ iteration. Consider iteration g and the next task $j_{[g]}$ which is in position g in the sequence σ_{ord}^{new}. Consider the tasks occupying the previous positions within interval $\overline{1, g-1}$ and build an elementary construction of dependent tasks $\alpha(j_{[g]}) = (j_{[r]}, j_{[r+1]}, \ldots, j_{[g-1]}, j_{[g]})$. In the particular case, the elementary construction may consist of only one task $j_{[g]}$.
2. Determine the parameters of the elementary construction of tasks $\alpha(j_{[g]})$: $L\alpha(j_{[g]}) = \sum_{j \in \alpha(j_{[g]})} l_j$; $\Omega\alpha(j_{[g]}) = \sum_{j \in \alpha(j_{[g]})} \omega_j$.
3. Determine the interval of move for the elementary construction of tasks $\alpha(j_{[g]})$. For this purpose, consider the tasks within interval $\overline{1, r-1}$ searching for the first task preceding the elementary construction $\alpha(j_{[g]})$. Let it be a task $j_{[p]}$, then the interval of move $I\alpha(j_{[g]})$ for the elementary construction of tasks $\alpha(j_{[g]})$ is

$$I\alpha(j_{[g]}) = \sum_{k=p+1}^{r-1} l_{j_{[k]}}.$$

4. Determine the interval of efficient move for the elementary construction of tasks $\alpha(j_{[g]})$. Check the following inequality for each position l within the interval of move:

$$p(\alpha(j_{[g]})) > \sum_{k=l}^{r-1} \omega_{j_{[k]}} \Big/ \sum_{k=l}^{r-1} l_{j_{[k]}}.$$

   Determine the difference

$$\Omega\alpha(j_{[g]}) \sum_{k=l}^{r-1} l_{j_{[k]}} - L\alpha(j_{[g]}) \sum_{k=l}^{r-1} \omega_{j_{[k]}} > 0. \tag{7.42}$$

   According to Definition 7.10, we need to find the position l within the interval of move in which the difference (7.42) reaches its maximum value. Suppose that this difference is maximal in a position $f + 1$. Go to step 5.

5. Insert the elementary construction $\alpha(j_{[g]})$ into the position $f + 1$ determined as the interval of efficient move so that the first task $j_{[r]}$ of the construction $\alpha(j_{[g]}) = (j_{[r]}, j_{[r+1]}, \ldots, j_{[g-1]}, j_{[g]})$ occupies the position $f + 1$. The following cases are possible:

   - an elementary construction $\beta(j_{[f]}) = (j_{[s]}, j_{[s+1]}, \ldots, j_{[f]})$, $\beta(j_{[f]}) \prec \alpha(j_{[g]})$, occupies the positions $\overline{s, f}$;
   - a task $j_{[f]} \not\prec \alpha(j_{[g]})$ occupies the position f.

In the first case, go to step 6. Otherwise, go to step 16 considering there an elementary construction $\alpha(j_{[g]})$ as the construction K_1.

6. Check the condition:

$$p(\beta(j_{[f]}))<p(\alpha(j_{[g]})). \tag{7.43}$$

If Eq. (7.43) is true, go to step 7. Otherwise, go to step 16 considering there an elementary construction $\alpha(j_{[g]})$ as the construction K_1.

7. Build an elementary construction of tasks $\gamma(j_{[f+n_1]})$ including elementary constructions $\beta(j_{[f]})$ and $\alpha(j_{[f+n_1]})$ where n_1 is the number of tasks in the elementary construction $\alpha(j_{[g]})$. Determine the priority of the elementary construction of tasks $\gamma(j_{[f+n_1]})$:

$$p\left(\gamma\left(j_{[f+n_1]}\right)\right)=\frac{\Omega\gamma\left(j_{[f+n_1]}\right)}{L\gamma\left(j_{[f+n_1]}\right)}$$

where $L\gamma(j_{[f+n_1]})=L\alpha(j_{[f+n_1]})+L\beta(j_{[f]})$; $\Omega\gamma(j_{[f+n_1]})=\Omega\alpha(j_{[f+n_1]})+\Omega\beta(j_{[f]})$.

8. Determine the interval of efficient move for the elementary construction $\gamma(j_{[f+n_1]})$ by the rules described in steps 3 and 4.
9. Insert the elementary construction $\gamma(j_{[f+n_1]})$ into positions determined by the interval of efficient move. As a result of this move, the elementary construction $\gamma(j_{[f+n_1]})$ becomes an elementary construction $\gamma(j_{[f+n_1-n_2]})$ where n_2 is the number of tasks $j\in I^{eff}_{\gamma(j_{[f+n_1]})}$.
10. Reorder all the tasks that follow the elementary construction $\gamma(j_{[f+n_1-n_2]})$, within the interval $\overline{f+n_1-n_2+1,\ g}$, according to Algorithm A1.1 given below.
11. Build the construction of tasks K_1 by the following rules. For each position k within the interval $\overline{f+n_1-n_2+1,\ g}$ find the value of the expression

$$\frac{\sum\limits_{l=f+n_1-n_2+1}^{k}\omega_{j_{[l]}}+\Omega\gamma\left(j_{[f+n_1-n_2]}\right)}{\sum\limits_{l=f+n_1-n_2+1}^{k}l_{j_{[l]}}+L\gamma\left(j_{[f+n_1-n_2]}\right)}-p\left(\gamma\left(j_{[f+n_1-n_2]}\right)\right) \tag{7.44}$$

and determine its maximum positive value. Suppose that the maximum of Eq. (7.44) is reached in a position t. Go to step 12. If you could not find a position with a positive value of Eq. (7.44), go to step 16.

12. Build the construction of tasks K_1. It includes the elementary construction $\gamma\left(j_{[f+n_1-n_2]}\right)$ and tasks within the interval $\overline{f+n_1-n_2+1,\ t}$. Determine the priority of the construction K_1.
13. Determine the interval of efficient move for the construction K_1 by the rules described in steps 3 and 4.
14. Insert the construction K_1 into a position determined by the interval of efficient move.
15. Reorder all the tasks that follow the construction K_1 up to the task $j_{[g]}$ in accordance with Algorithm A1.1. Proceed to step 5 considering there the construction K_1 as an elementary construction $\alpha\left(j_{[g]}\right)$.
16. Organize a new construction of tasks K_2 as follows.
    - sequentially consider the tasks that occupy the previous positions in regard to the construction K_1 starting from the position immediately preceding the construction K_1;
    - search for a task $j_{[z]} \prec K_1$.

    The following cases are possible:

    - a task $j_{[z]} \prec K_1$ is not found. Go to step 23;
    - the task $j_{[z]}$ is found and occupies the position z. Go to step 17.
17. Organize an elementary construction of tasks $\beta\left(j_{[z]}\right)$ by the above rules.
18. Determine the parameters of an elementary construction of tasks $\beta\left(j_{[z]}\right)$:

$$\Omega\beta\left(j_{[z]}\right) = \sum_{j\in\beta\left(j_{[z]}\right)} \omega_j; \quad L\beta\left(j_{[z]}\right) = \sum_{j\in\beta\left(j_{[z]}\right)} l_j.$$

19. Check the condition:

$$p\left(\beta\left(j_{[z]}\right)\right) < p(K_1). \tag{7.45}$$

    If the condition is false, go to step 16 and search for the next task $j_{[z]} \prec K_1$ satisfying Eq. (7.45). If Eq. (7.45) is true, go to step 20. If such a task is not found, go to step 22.

20. Build a construction K_2 including the construction K_1, the elementary construction $\beta\left(j_{[z]}\right)$, and the tasks in the intermediate positions between K_1 and $\beta\left(j_{[z]}\right)$.
21. Determine the interval of efficient move for the construction K_2 by the rules described in steps 3 and 4. Move the construction K_2 into positions determined by the interval of efficient move. Reorder all the tasks that follow the construction K_2 up to the position g in accordance with Algorithm A1.1. Go to step 5 considering there the construction K_2 as an elementary construction $\alpha\left(j_{[g]}\right)$.

22. If $g = n$ (the entire sequence σ_{ord}^{new} is considered), the algorithm terminates. Otherwise, $g = g + 1$, go to step 1, new iteration $g + 1$ begins.

Algorithm A1.1 (Unit of Reordering on Intervals of Efficient Move) Suppose that the set of tasks J_1^{eff} was moved into later positions as a result of the efficient move of the elementary construction of tasks $\alpha(j_{[f+n_1]})$. As a result of constructing the elementary construction $\gamma(j_{[f+n_1]}) = \beta(j_{[f]}) \cup \alpha(j_{[f+n_1]})$ and performing the efficient move for it, the set of tasks J_2^{eff} within the interval of its efficient move has also appeared at later positions in comparison with the initially occupied ones. There are such tasks $j_k \in J_1^{eff}$ for which $p(j_k) > p(\beta(j_{[f]}))$, $\beta(j_{[f]}) \prec j_k$. Let us denote these tasks as j_k^*.

Since, in the general case, $\exists j_k^* \in J_1^{eff}$ and $\exists j_l \in J_2^{eff}$ such that $p(j_k^*) > p(j_l)$, the tasks j_k^* can take earlier positions during a p-ordered schedule construction. The purpose of Algorithm A1.1 is to arrange the tasks $j \in J_1^{eff}$ and $j \in J_2^{eff}$ in accordance with their priorities. Rules for Algorithm A1.1 development are based on above Statement 7.9. It follows from this statement that $\forall j_k^* \in J_1^{eff}$ and $\forall j_r \in J_2^{eff}$ we have $j_r \bar{\prec} j_k^*$. Therefore, it is sufficient to insert the tasks j_k^* into the subsequence of tasks $j \in J_2^{eff}$ in accordance with their priorities during a p-ordered schedule construction on the sets of tasks J_1^{eff} and J_2^{eff}. This is implemented by the efficient move procedure.

Algorithm A1.1 Sequentially consider tasks $j \in J_1^{eff}$ from the beginning. Suppose that the current task we consider is $j_{[k]}^* \in J_1^{eff}$.

1. Consider tasks $j \in J_1^{eff}$ occupying earlier positions and organize an elementary construction of dependent tasks $\alpha\left(j_{[k]}^*\right)$.
2. Determine the parameters of the elementary construction of tasks $\alpha\left(j_{[k]}^*\right)$:

$$L\alpha\left(j_{[k]}^*\right) = \sum_{j \in \alpha\left(j_{[k]}^*\right)} l_j; \quad \Omega\alpha\left(j_{[k]}^*\right) = \sum_{j \in \alpha\left(j_{[k]}^*\right)} \omega_j.$$

3. Determine the interval of move for the elementary construction of tasks $\alpha\left(j_{[k]}^*\right)$.
4. Determine the interval of efficient move for the elementary construction of tasks $\alpha\left(j_{[k]}^*\right)$.
5. Move the elementary construction $\alpha\left(j_{[k]}^*\right)$ into a position determined by the interval of efficient move.
6. Proceed to the next task $j_{[k+1]}^* \in J_1^{eff}$, go to step 1. If we have considered all the tasks $j \in J_1^{eff}$, the algorithm terminates.

Justification of Algorithm A We can formulate the following statements [4] based on the above algorithm for the case when the strict order of precedence relations is specified by a series-parallel graph.

Statement 7.11 Suppose that we build any feasible sequence σ^2 by adding a feasible by precedence relations task $j_{[g]}$ to a p-ordered sequence σ^1. Algorithm A transforms such sequence σ^2 into a p-ordered sequence.

Proof of this statement is obvious. ∎

Theorem 7.3 *All p-ordered schedules built on the task set J with a precedence relation specified by an oriented acyclic graph can differ in structure, but they have the same functional value.*

Proof Suppose that we can organize k different subsequences $\sigma_1, \ldots, \sigma_k$ on the set of tasks J. The subsequences should correspond the requirements to a p-ordered schedule, be ready for processing at the zero time point, and have the highest priority of all possible feasible subsequences ready for processing at the zero time point. Consider the following three cases.

1. All the subsequences $\sigma_1, \ldots, \sigma_k$ have no common tasks. Suppose that we first assign a subsequence σ_i, $i = \overline{1, k}$, when constructing a p-ordered schedule. Then, according to the definition of a p-ordered schedule and due to the independence of the subsequences $\sigma_1, \ldots, \sigma_k$, we will next assign a subsequence $\sigma_j \neq \sigma_i$, etc., until all the sequences $\sigma_1, \ldots, \sigma_k$ are assigned one after another in a random order. As a result, all the initial p-ordered schedules contain all tasks from the set J_1 which is the set of all tasks in the subsequences $\sigma_1, \ldots, \sigma_k$. So, the total processing time of all tasks is the same with the same value of the common component of the functional.
2. The subsequences $\sigma_1, \ldots, \sigma_k$ are not independent. In this case, common tasks of two or more subsequences $\sigma_1, \ldots, \sigma_k$ are the beginning of each subsequence. All remaining tasks of each subsequence should have the maximum total priority and be independent among themselves (tasks may be dependent within each subsequence). Then, the beginning subsequence of an arbitrary p-ordered schedule consists of tasks belonging to the set J_1 with a total priority equal to the maximum priority and, therefore, the total processing time of all tasks is the same with the same functional value.
3. The most general case of the structure of beginning MPSSes. The set of non-coinciding beginning MPSSes contains all feasible combinations of beginning subsequences. Any beginning subsequence of a p-ordered schedule contains tasks from the set J_1. That is, it contains all the different tasks of the beginning subsequences with a total priority equal to the maximum priority, the total processing time of the tasks, and the same value of the corresponding component of the functional.

The further procedure for constructing a p-ordered schedule consists in constructing a p-ordered schedule on the set of tasks $J \backslash J_1$. The tasks are related by the corresponding oriented acyclic graph obtained from the initial one by eliminating vertices of the set J_1 and arcs incident to them. Consequently, the problem of constructing a new p-ordered schedule for the remaining tasks reduces to the cases considered above. ∎

Theorem 7.4 Algorithm A, *working with any series-parallel graph and starting with an arbitrary feasible schedule for the set of tasks J, yields a p-ordered schedule.*

Proof By construction, Algorithm A consists of uniform iterations. On each iteration, we add to the p-ordered schedule of the previous iteration a next task feasible by precedence relations. According to Statement 7.11, a result of the iteration is a p-ordered schedule of the subsequence under consideration.

Consequently, a result of Algorithm A execution for the entire sequence σ is a p-ordered schedule for the considered set of tasks J. ∎

The following statement is obvious.

Statement 7.12 If the initial feasible sequence σ is not transformed by the above Algorithm A, that is, the algorithm performs no permutations, then the sequence is p-ordered.

Statement 7.13 An optimal schedule constructed by algorithm A is p-ordered.

Proof Consider an optimal schedule as a feasible schedule. This schedule cannot be changed at any iteration of the above algorithm, since any change in the feasible schedule improves the functional value. When we described the algorithm, we have shown that the absence of permutations indicates the p-ordering of the schedule part being considered. Hence, an optimal schedule is p-ordered. ∎

Corollary 7.3 Algorithm A is optimal for any series-parallel graph.

Theorem 7.5 (Pavlov et al. [4]) *Algorithm A is a polynomial algorithm. Its complexity is determined by the function* $O(n^2)$.

7.3.3.5 Theoretical Foundations of Generalization of Algorithm A for the Case When the Precedence Relations Are Specified by an Arbitrary Oriented Acyclic Graph. Formalization of SSO #16 of a Feasible Subsequence

Suppose that the precedence relations on the task set J are specified by an *arbitrary* oriented acyclic graph. As we said above, such a problem is NP-hard in the strong sense, which implies the exponential complexity of obtaining an exact solution. Let us investigate the possibility of applying Algorithm A to solve the problem when the partial order is specified by an arbitrary oriented acyclic graph. In this case, Statement 7.9 is not valid. Therefore, to obtain an exact solution of the problem in the general form, we need to modify Algorithm A taking into account the following factors:

- we should perform p-ordering of tasks within the intervals of efficient move by Algorithm A (instead of Algorithm A1.1);
- when building constructions, we should include into them, in addition to the tasks determined by the rules of Algorithm A, the tasks relating to them by the immediate precedence relations and occupying the earlier positions in the sequence σ.

In this regard, an enumeration of a large number of different constructions is possible when building maximum priority constructions. This can lead to an exponential complexity during searching for an exact solution of the problem.

Let us formulate the conditions necessary to avoid the enumeration when building maximum priority constructions.

Consider the sequence

$$\sigma^m = \left(j_{[1]}, \ldots, j_{[k]}, j_{[k+1]}, \ldots, j_{[k+l]}, \beta\left(j_{[f]}\right), \alpha\left(j_{[f+y]}\right), \rho\left(j_{[q_1]}\right), \rho\left(j_{[q_2]}\right), \ldots, \rho\left(j_{[q_r]}\right)\right).$$

Suppose that the sequence σ^m satisfies the following requirements:

- the beginning part of the sequence σ'_m of tasks within the interval $\overline{1, f+y}$ is p-ordered;
- for all $\rho\left(j_{[q_k]}\right)$, $k = \overline{1, r}$, we have: $\alpha\left(j_{[f+y]}\right) \overline{\prec} \rho\left(j_{[q_k]}\right)$, $\beta\left(j_{[f]}\right) \prec \rho\left(j_{[q_k]}\right)$; the elementary constructions $\rho\left(j_{[q_k]}\right)$ are independent among themselves;
- $j_{[s]} \overline{\prec} \gamma\left(j_{[f+y]}\right)$, $s = \overline{1, k+l}$; $\gamma\left(j_{[f+y]}\right) = \beta\left(j_{[f]}\right) \cup \alpha\left(j_{[f+y]}\right)$.

Let J_s denote the set of these tasks $j_{[s]}$, $s = \overline{1, k+l}$. We organize the construction of tasks K_1^1 including the elementary construction $\gamma\left(j_{[f+y]}\right)$ and one of the elementary constructions $\rho\left(j_{[q_k]}\right)$, $k = \overline{1, r}$. For definiteness, let it be the elementary construction $\rho\left(j_{[q_1]}\right)$. By construction, the construction K_1^1 must include the tasks or elementary constructions of tasks $j_s \in J_s$ for which:

- $j_s \prec \rho\left(j_{[q_1]}\right)$;
- there are no other tasks between the tasks j_s and the elementary construction $\gamma\left(j_{[f+y]}\right)$ in the sequence σ^m.

Let $J_{s_1} \subset J_s$ denote the set of these tasks j_s.

Determine the parameters of the construction K_1^1 as follows:

$$\Omega K_1^1 = \Omega\gamma\left(j_{[f+y]}\right) + \Omega\rho\left(j_{[q_1]}\right) + \sum_{j \in J_{S_1}} \omega_j; \quad LK_1^1 = L\gamma\left(j_{[f+y]}\right) + L\rho\left(j_{[q_1]}\right) + \sum_{j \in J_{S_1}} l_j.$$

In order to satisfy $p\left(K_1^1\right) > p\left(\gamma\left(j_{[f+y]}\right)\right)$, we must meet the following conditions:

$$p\left(\rho\left(j_{[q_1]}\right)\right) > p\left(\gamma\left(j_{[f+y]}\right)\right), \tag{7.46}$$

$$p\left(\rho\left(j_{[q_1]}\right)\right) > \sum_{j \in J_{S_1}} \omega_j \Big/ \sum_{j \in J_{S_1}} l_j, \tag{7.47}$$

$$\sum_{j \in J_{S_1}} \omega_j \Big/ \sum_{j \in J_{S_1}} l_j < \frac{\Omega\gamma\left(j_{[f+y]}\right) + \Omega\rho\left(j_{[q_1]}\right)}{L\gamma\left(j_{[f+y]}\right) + L\rho\left(j_{[q_1]}\right)}. \tag{7.48}$$

If any of the conditions (7.46)–(7.48) is not satisfied, then

$$p\left(K_1^1\right) < \sum_{j \in J_{S_1}} \omega_j \Big/ \sum_{j \in J_{S_1}} l_j. \tag{7.49}$$

The construction K_1^1 that satisfies the condition (7.49) does not meet the requirements of a p-ordered schedule. Therefore, we can organize the construction K_1^1 only if we meet all the conditions (7.46)–(7.48) when building a p-ordered schedule on the specified set of tasks.

The conditions (7.46)–(7.48) hold if we choose any elementary construction $\rho\left(j_{[q_k]}\right)$ instead of the elementary construction $\rho\left(j_{[q_1]}\right)$.

Let us formulate the conditions necessary to organize a construction containing simultaneously v elementary constructions $\rho\left(j_{[q_1]}\right), \ldots, \rho\left(j_{[q_v]}\right)$.

We organize a construction K_1^2 including the elementary constructions $\gamma\left(j_{[f+y]}\right)$, $\rho\left(j_{[q_1]}\right), \ldots, \rho\left(j_{[q_v]}\right)$, and the tasks $j_s \in J_s$ that satisfy at least one of the two following conditions:

- $j_s \prec \rho\left(j_{[q_1]}\right), \ldots, j_s \prec \rho\left(j_{[q_v]}\right)$;
- there are no other tasks between the tasks j_s and the elementary construction $\gamma\left(j_{[f+y]}\right)$.

In order for the construction K_1^2 to contain at the same time $v = \overline{1, r}$ elementary constructions $\rho\left(j_{[q_1]}\right), \ldots, \rho\left(j_{[q_v]}\right)$, the following conditions must be satisfied:

$$p\left(\rho\left(j_{[q_1]}\right)\right) > p\left(\gamma\left(j_{[f+y]}\right)\right), \ldots, p\left(\rho\left(j_{[q_v]}\right)\right) > p\left(\gamma\left(j_{[f+y]}\right)\right);$$

$$\sum_{j \in J_{S_v}} \omega_j \Big/ \sum_{j \in J_{S_v}} l_j < \frac{\Omega\gamma\left(j_{[f+y]}\right) + \Omega\rho\left(j_{[q_1]}\right) + \cdots + \Omega\rho\left(j_{[q_v]}\right)}{L\gamma\left(j_{[f+y]}\right) + L\rho\left(j_{[q_1]}\right) + \cdots + L\rho\left(j_{[q_v]}\right)}. \tag{7.50}$$

The above set requirements for constructions building severely restrict the search of various constructions. This allows us to formulate the conditions under which the modified Algorithm A achieves an optimal solution in a polynomial time. The following statements allows us to distinguish those cases when the exhaustive search of various constructions is excluded during their building.

Consider a sequence σ^r which is p-ordered within the interval $\overline{1, f+y}$.

Statement 7.14 Let K_1 be a construction built according to the rules described in Algorithm A. The construction K_1 includes the elementary constructions $\gamma\left(j_{[f+y]}\right)$, $\rho\left(j_{[q_k]}\right)$, $k = \overline{1, r}$; $\gamma\left(j_{[f+y]}\right) \prec \rho\left(j_{[q_k]}\right)$. Suppose that a MPSS σ_k occupies positions $\overline{p-l, p-1}$ in the sequence σ where p is the position occupied by the first task of the elementary construction $\gamma\left(j_{[f+y]}\right)$. Let J_k denote the set of tasks of the sequence σ_k. If

$$p(K_1) \le \sum_{j \in J_k} \omega_j \Big/ \sum_{j \in J_k} l_j, \tag{7.51}$$

then it is impossible to build a new construction on this set of tasks with a priority higher than that of the subsequence σ_k.

Proof The validity of this statement follows from p-ordering of the sequence under consideration within the interval $\overline{1, f+y}$. ∎

Statement 7.15 Suppose that during organizing the construction K_2 described in Algorithm A the elementary construction of tasks $\beta\left(j_{[z]}\right)$ is a part of a MPSS σ_r^1. Let J_r^1 denote the set of tasks of this subsequence. If

$$\sum_{j \in J_r^1} \omega_j \Big/ \sum_{j \in J_r^1} l_j > p(K_1), \tag{7.52}$$

then we cannot organize a construction K_2 corresponding to the requirements of p-ordering.

Proof The validity of this statement is based on p-ordering of the sequence under consideration within the interval $\overline{1, t-1}$ by construction. Here, t is the position occupied by the first task of the construction K_1. ∎

Statement 7.16 Suppose that we have additionally included tasks $j \in J_{s_1}$, $j \in J_{s_2}$, …, $j \in J_{s_r}$, $j \in J_s$ when building a construction which includes elementary constructions $\gamma\left(j_{[f+y]}\right)$, $\rho\left(j_{[q_1]}\right)$, $\rho\left(j_{[q_2]}\right)$, …, $\rho\left(j_{[q_r]}\right)$. If

$$\forall j \in J_s,\ j \prec \gamma\left(j_{[f+y]}\right);\ \forall j \in J_{s_1},\ j \prec \rho\left(j_{[q_1]}\right); \ldots; \forall j \in J_{s_r},\ j \prec \rho\left(j_{[q_r]}\right)$$

$$J_s = J_{s_1} = J_{s_2} = \ldots = J_{s_r} \text{ or } J_s \cup J_{s_1} \cup J_{s_2} \cup \ldots \cup J_{s_r} = \varnothing, \tag{7.53}$$

then we should build the construction K_1 according to the rules for constructions building described in Algorithm A.

Proof The statement is valid because we meet the conditions of Statement 7.6 in the subsequence under consideration. ∎

Statement 7.17 The modified Algorithm A allows to build a p-ordered schedule for any oriented acyclic graph if we meet the following conditions during the problem solving:

1. When building the construction K_1:
   - we do not meet the conditions (7.46)–(7.48), (7.50) $\forall p\left(j_{[q_r]}\right)$, $k = \overline{1, r}$;
   - we meet the conditions (7.46)–(7.48), (7.50) and also at least one of the conditions (7.51) or (7.53).
2. When building the construction K_2: either we do not have tasks $j_{[z]} \prec K_1$, or $p\left(\beta\left(j_{[z]}\right)\right) \geq p(K_1)$. If we do not meet both these conditions, then we meet the condition (7.52).

Proof The validity of this statement follows from the above presented justification of the rules for constructing a p-ordered schedule and Statements 7.14–7.16. If we do not meet the conditions of Statement 7.17, then we should develop heuristic procedures to solve the problem in the general form. Such heuristic procedures should limit the number of constructions we build, depending on the desired accuracy of the resulting solution. ∎

SSO #16 If each time the constructions K_1 and K_2 met the corresponding Conditions 1 and 2 of Statement 7.17 during an iteration of optimization, then the subsequence considered at this iteration is optimal.

7.3.4 Description of PSC-Algorithm for TWCT Problem Solving When the Precedence Relations Are Specified by an Arbitrary Oriented Acyclic Graph

We have developed a modification of the optimization unit A2 for a graph in the general form on the basis of the above statements. The PSC-algorithm for the problem solving is a modified Algorithm A. It includes the ordering unit A1 given above and the modified optimization unit A3.

Algorithm A3 is based on Algorithm A2 for a series-parallel graph with the following modifications allowing to obtain an exact solution of the problem in the general form:

- we perform p-ordering of tasks within the intervals of efficient move by Algorithm A3 (instead of Algorithm A1.1);
- when building constructions, we include into them, in addition to the tasks determined by the rules of Algorithm A2, the tasks relating to them by the immediate precedence relations and occupying the earlier positions in the current sequence.

Algorithm A3 checks on the current subsequence the SSOs given in Sect. 7.3.3.1. We check these conditions as follows. Suppose that we consider an iteration k. We add the next task $j_{[g]}$ from the feasible sequence to the schedule σ^{k-1} obtained at the iteration $k-1$. Then, we build a chain (or an elementary construction) of tasks $\alpha(j_{[g]})$ and insert it in accordance with certain rules (Sect. 7.3.3.1) into the sequence σ^{k-1} without violating its feasibility and taking into account the tasks' priorities. We analyze the obtained structure. If it corresponds to at least one SSO, then the obtained subsequence of tasks is optimal, we proceed to the next task and the next iteration. Thus, we begin checking the SSOs on the current subsequence in Algorithm A3 when we add the next task $j_{[g]}$ to the optimal sequence obtained at the previous iteration. If at least one of the fifteen SSOs is satisfied, then the current subsequence of tasks is optimal, we go to the next iteration. Otherwise, we perform the current iteration of the Algorithm A3. At the end of the iteration we check the SSO #16. If it is satisfied, then the current subsequence of tasks is optimal. Otherwise, we continue the problem solving by the iterative Algorithm A3 as an approximate algorithm, without checking the SSOs.

Algorithm A3 (Modified Optimization Unit) has the following 27 steps.

1. Suppose that we have already performed $g-1$ iteration, consider iteration g. The subsequence within the interval $\overline{1, g-1}$ is p-ordered. Add to it the next task $j_{[g]}$ which is in position g in the sequence σ^{new}_{ord}. So, a subsequence in the form $\sigma = \{M_1, M_2, \ldots, M_t, j_{[g]}\}$ is within the interval $\overline{1, g}$ where $t \geq 0$.
2. Starting with iteration 2, perform the algorithm for checking the fifteen SSOs described in Sect. 7.3.3.1 for the subsequence σ considering there the task $j_{[g]}$ as the task $g_{[k]}$. If at least one of the fifteen SSOs is satisfied, then an optimal subsequence within the interval $\overline{1, g}$ is obtained, proceed to step 26 (begin the next iteration). Otherwise, return to the initial subsequence σ and go to step 3.
3. Consider the tasks occupying the previous positions within interval $\overline{1, g-1}$ and build an elementary construction of dependent tasks $\alpha(j_{[g]}) = (j_{[r]}, j_{[r+1]}, \ldots, j_{[g-1]}, j_{[g]})$. In the particular case, the elementary construction may consist of only one task $j_{[g]}$. Determine the parameters of the elementary construction of tasks $\alpha(j_{[g]})$: $L\alpha(j_{[g]}) = \sum_{j \in \alpha(j_{[g]})} l_j$; $\Omega\alpha(j_{[g]}) = \sum_{j \in \alpha(j_{[g]})} \omega_j$.
4. Determine the interval of move for the elementary construction of tasks $\alpha(j_{[g]})$. For this purpose, consider the interval $\overline{1, r-1}$ searching for the first task preceding the elementary construction $\alpha(j_{[g]})$. Let it be a task $j_{[p]}$, then the interval of move $I\alpha(j_{[g]})$ for the elementary construction of tasks $\alpha(j_{[g]})$ is

$$I\alpha(j_{[g]}) = \sum_{k=p+1}^{r-1} l_{j_{[k]}}.$$

5. Determine the interval of efficient move for the elementary construction of tasks $\alpha(j_{[g]})$. We check the following inequality for each position l within the interval of move:

$$p(\alpha(j_{[g]})) > \sum_{k=l}^{r-1} \omega_{j_{[k]}} \Big/ \sum_{k=l}^{r-1} l_{j_{[k]}}.$$

Determine the difference

$$\Omega\alpha(j_{[g]}) \sum_{k=l}^{r-1} l_{j_{[k]}} - L\alpha(j_{[g]}) \sum_{k=l}^{r-1} \omega_{j_{[k]}} > 0. \tag{7.54}$$

According to Definition 7.10, we need to find the position l within the interval of move in which the difference (7.54) reaches its maximum value. Suppose that this difference is maximal in a position $f + 1$. Go to step 6.

6. Insert the elementary construction $\alpha(j_{[g]})$ into the position $f + 1$ determined as the interval of efficient move so that the first task $j_{[r]}$ of the construction $\alpha(j_{[g]}) = (j_{[r]}, j_{[r+1]}, \ldots, j_{[g-1]}, j_{[g]})$ occupies the position $f + 1$. Check the following cases:

   - an elementary construction $\beta(j_{[f]}) = (j_{[s]}, j_{[s+1]}, \ldots, j_{[f]})$, $\beta(j_{[f]}) \prec \alpha(j_{[g]})$, occupies the positions $\overline{s,f}$;
   - a task $j_{[f]} \bar{\prec} \alpha(j_{[g]})$ occupies the position f.

   In the first case, go to step 7. Otherwise, go to step 18 considering there an elementary construction $\alpha(j_{[g]})$ as the construction K_1.

7. Check the condition:

$$p(\beta(j_{[f]})) < p(\alpha(j_{[g]})). \tag{7.55}$$

   If Eq. (7.55) is true, go to step 8. Otherwise, go to step 18 considering there an elementary construction $\alpha(j_{[g]})$ as the construction K_1.

8. Build an elementary construction of tasks $\gamma(j_{[f+n_1]})$ including the elementary constructions $\beta(j_{[f]})$ and $\alpha(j_{[f+n_1]})$ where n_1 is the number of tasks in the elementary construction $\alpha(j_{[g]})$. Determine the priority of the elementary construction of tasks $\gamma(j_{[f+n_1]})$:

$$p\left(\gamma\left(j_{[f+n_1]}\right)\right) = \frac{\Omega\gamma\left(j_{[f+n_1]}\right)}{L\gamma\left(j_{[f+n_1]}\right)}$$

   where $L\gamma(j_{[f+n_1]}) = L\alpha(j_{[f+n_1]}) + L\beta(j_{[f]})$; $\Omega\gamma(j_{[f+n_1]}) = \Omega\alpha(j_{[f+n_1]}) + \Omega\beta(j_{[f]})$.

9. Determine the interval of efficient move for the elementary construction $\gamma(j_{[f+n_1]})$ by the rules described in steps 4 and 5.
10. Insert the elementary construction $\gamma(j_{[f+n_1]})$ into positions determined by the interval of efficient move. As a result of this move, the elementary construction $\gamma(j_{[f+n_1]})$ becomes an elementary construction $\gamma(j_{[f+n_1-n_2]})$ where n_2 is the number of tasks $j \in I^{eff}_{\gamma(j_{[f+n_1]})}$.
11. Reorder all the tasks that follow the elementary construction $\gamma(j_{[f+n_1-n_2]})$ within the interval $\overline{f+n_1-n_2+1,\ g}$, according to Algorithm A3.
12. Build the construction of tasks K_1 by the following rules. For each position k within the interval $\overline{f+n_1-n_2+1,\ g}$ find the value of the expression

$$\frac{\sum_{l=f+n_1-n_2+1}^{k}\omega_{j_{[l]}}+\Omega\gamma\left(j_{[f+n_1-n_2]}\right)}{\sum_{l=f+n_1-n_2+1}^{k}l_{j_{[l]}}+L\gamma\left(j_{[f+n_1-n_2]}\right)}-p\left(\gamma\left(j_{[f+n_1-n_2]}\right)\right) \tag{7.56}$$

and determine its maximum positive value. Suppose that the maximum of Eq. (7.56) is reached in a position t. Go to step 13. If we could not find a position with a positive value of Eq. (7.56), go to step 18.

13. Build the construction of tasks K_1. It includes:

- the elementary construction $\gamma(j_{[f+n_1-n_2]})$;
- such independent elementary constructions of tasks $\rho(j_{[q_k]})$ within the interval $\overline{f+n_1-n_2+1,\ t}$ that $\alpha(j_{[f+n_1-n_2]}) \overline{\prec} \rho(j_{[q_k]})$, $\beta(j_{[f-n_2]}) \prec \rho(j_{[q_k]})$, $\forall k=\overline{1,r}$;
- such tasks $j_{[s]} \overline{\prec} \gamma(j_{[f+n_1-n_2]})$, $s=\overline{1,f-n_2-n_3}$ where n_3 is the number of tasks in the elementary construction $\beta(j_{[f-n_2]})$, that each $j_{[s]}$ precedes at least one $\rho(j_{[q_k]})$ and there are no other tasks between each task $j_{[s]}$ and the chain $\gamma(j_{[f+n_1-n_2]})$ in the current sequence.

14. Check Condition 1 of Statement 7.17 for the construction K_1 (a sufficient sign of optimality for the construction K_1 during a p-ordered schedule construction). If Condition 1 of Statement 7.17 is satisfied, go to step 15, otherwise go to step 25 (the resulting solution will be approximate).
15. Determine the priority of the construction K_1. Determine the interval of efficient move for the construction K_1 by the rules described in steps 4 and 5.
16. Insert the construction K_1 into a position determined by the interval of efficient move.
17. Reorder all the tasks that follow the construction K_1 up to the task $j_{[g]}$ in accordance with Algorithm A3. Proceed to step 6 considering there the construction K_1 as an elementary construction $\alpha(j_{[g]})$.

18. Organize a new construction of tasks K_2 as follows.
    - sequentially consider the tasks that occupy the previous positions in regard to the construction K_1 starting from the position immediately preceding the construction K_1;
    - search for a task $j_{[z]} \prec K_1$.

    The following cases are possible:

    - a task $j_{[z]} \prec K_1$ is not found. Go to step 25;
    - the task $j_{[z]}$ is found and occupies the position z. Go to step 19.

19. Organize an elementary construction of tasks $\beta\left(j_{[z]}\right)$ by the above rules.
20. Determine the parameters of an elementary construction of tasks $\beta\left(j_{[z]}\right)$:

$$\Omega\beta\left(j_{[z]}\right) = \sum_{j \in \beta\left(j_{[z]}\right)} \omega_j; \quad L\beta\left(j_{[z]}\right) = \sum_{j \in \beta\left(j_{[z]}\right)} l_j.$$

21. Check the condition:

$$p\left(\beta\left(j_{[z]}\right)\right) < p(K_1). \tag{7.57}$$

    If the condition is false, go to step 18 and search for the next task $j_{[z]} \prec K_1$ satisfying Eq. (7.57). If Eq. (7.57) is true, go to step 22. If such a task is not found, go to step 25.

22. Build a construction K_2 including the construction K_1, the elementary construction $\beta\left(j_{[z]}\right)$, and the tasks in the intermediate positions between K_1 and $\beta\left(j_{[z]}\right)$.
23. Check Condition 2 of Statement 7.17 for the construction K_2 (a sufficient sign of optimality for the construction K_2 during a p-ordered schedule construction). If Condition 2 of Statement 7.17 is satisfied, go to step 24, otherwise go to step 25 (the resulting solution will be approximate).
24. Determine the interval of efficient move for the construction K_2 by the rules described in steps 4 and 5. Move the construction K_2 into positions determined by the interval of efficient move. Reorder all the tasks that follow the construction K_2 up to the position g in accordance with Algorithm A3. Proceed to step 6 considering there the construction K_2 as an elementary construction $\alpha\left(j_{[g]}\right)$.
25. If SSO #16 is satisfied, then the subsequence considered at the iteration g is optimal, go to step 26. Otherwise, the first polynomial component of the PSC-algorithm is not fulfilled, the obtained solution is approximate, go to step 27 continuing Algorithm A3 execution without checking the SSOs (without performing steps 2, 14, 23, 25, and 26).

26. Partition the current sequence into MPSSes using Algorithm AMPSS given below. Go to step 27.
27. If $g = n$ (the entire sequence σ_{ord}^{new} is considered), the algorithm terminates. Otherwise, $g = g + 1$, go to step 1, new iteration $g + 1$ begins.

Thus, the logic of the algorithm during building an optimal schedule at a current iteration consists in the following. We add the next task from the feasible sequence or groups of tasks immediately related with it and perform their efficient move. We insert them into the optimal schedule obtained at the previous iteration, into positions corresponding to their priorities. This necessitates *p*-ordering of the schedule at the previous iteration.

Algorithm for Partitioning a Sequence Into the Maximum Priority Subsequences AMPSS We use this algorithm to partition the sequence of tasks obtained at the current iteration into MPSSes without changing their order. If at least one of the SSOs is satisfied, then the current sequence is optimal and feasible. Then, Algorithm AMPSS constructs a *p*-ordered sequence. The first positions in it occupies a subsequence of tasks of the highest possible priority on the entire sequence of tasks. Then a subsequence follows that has the maximum possible priority on the rest of the sequence, etc., until the entire current sequence is considered. The partition obtained by Algorithm AMPSS is a *p*-ordered sequence by construction, since the initial sequence is optimal and the constructed subsequences correspond to the requirements of *p*-ordering.

Algorithm AMPSS has the following five steps.

1. Consider the sequence from the beginning. Set $l = 1$, $N = 1$, the first subsequence includes the first task: $M_1 = \{j_{[1]}\}$.
2. Proceed to the next task of the sequence: set $l = l + 1$. Start to form the next MPSS: set $N = N + 1$, $M_N = \{j_{[l]}\}$.
3. Check the possibility of combining the subsequence M_N with the previous subsequence as follows. If the priority of the subsequence M_N (sum of the weights of its tasks divided to the sum of their processing times) is greater than the priority of the subsequence M_{N-1}, go to step 4, else go to step 5.
4. Set $M_{N-1} = M_{N-1} \cup M_N$, recalculate the priority of the subsequence M_{N-1}. Set $N = N - 1$. If $N = 1$, go to step 5, otherwise go to step 3 to check the possibility of combining the new subsequence M_N with the previous subsequence.
5. If all tasks are considered ($l = n$), the algorithm terminates. Otherwise, go to step 2.

The Sufficient Sign of Optimality of a Feasible Solution Obtained by the PSC-algorithm for the TWCT Problem Solving Suppose that at least one of the sixteen SSOs of a feasible subsequences given above was satisfied during each optimization iteration. Then, the sequence obtained by Algorithm A3 on the entire set of tasks is optimal, the solution is obtained by the first polynomial component of the PSC-algorithm. Otherwise, the first polynomial component of the PSC-algorithm is not fulfilled, the solution is approximate and obtained by the second polynomial component of the algorithm.

7.3.5 Example of the Problem Solving

Consider building a sequence of tasks $j \in J$ on a single machine optimal by the TWCT criterion. We specify the precedence relationships in Fig. 7.3. The processing times l_j and weights ω_j of tasks are in Table 7.1.

We arrange the tasks in non-increasing order of priorities ω_j/l_j and obtain the sequence $\sigma_{ord} = \{6, 13, 17, 9, 12, 2, 5, 8, 10, 14, 4, 3, 15, 18, 1, 19, 20, 11, 16, 7\}$ (Table 7.2).

Then we build the sequence σ_{ord}^{new} by Algorithm A1: $\sigma_{ord}^{new} = \{8, 9, 5, 6, 12, 14, 13, 17, 1, 2, 10, 3, 4, 15, 18, 19, 20, 11, 16, 7\}$ (Table 7.3).

Execute Algorithm A3. Tasks 8, 9, 5, and 6 remain in their positions by construction of the sequence σ_{ord}^{new}. The priority of the MPSS {8, 9, 5, 6} is 780/145 ≈ 5.38.

Iteration 1. Task 12 has priority 5 < 5.38. Its interval of move is zero, we cannot insert it into an earlier position. Task 14 has an even lower priority than task 12, so it remains in its position. Task chain {12, 13} is preceded by task 14 which we should move together with it. Tasks {12, 14, 13} form an elementary construction $\alpha(13)$, its priority is 665/120 ≈ 5.54. According to the interval of efficient move, we insert the elementary construction $\alpha(13)$ into the position 1. The subsequence {12, 14, 13, 8, 9, 5, 6} (see Table 7.4) satisfies SSO #2, so it is *p*-ordered.

Iteration 2. Task 17 has priority 5.4. Chain $\alpha(17)$ in this case consists of a single task. According to the interval of efficient move, we insert it after task 13 and obtain the following subsequence (Table 7.5).

The current subsequence satisfies SSO #2, so it is also optimal.

Iteration 3. Tasks 1 and 2 remain in their positions. Task 10 has a higher priority than the chain {1, 2}. According to the interval of efficient move, we insert task 10 before task 1. We obtain the subsequence shown in Table 7.6.

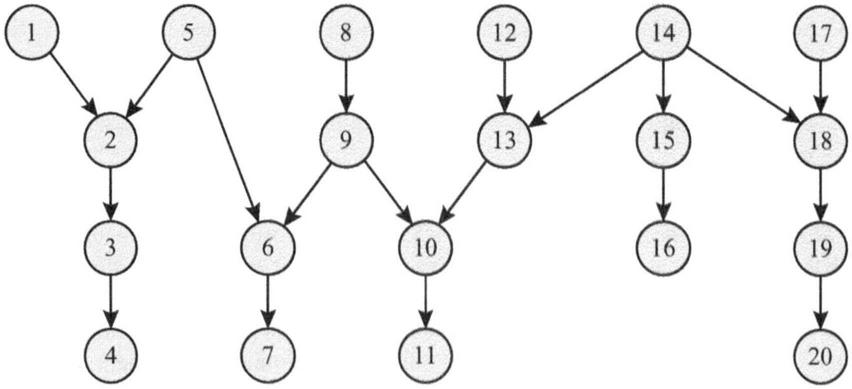

Fig. 7.3 Precedence relationships graph for the problem instance

Table 7.1 Initial data for the example

j	l_j	ω_j	$p(j)$	j	l_j	ω_j	$p(j)$	j	l_j	ω_j	$p(j)$	j	l_j	ω_j	$p(j)$
1	30	20	0.67	6	25	300	12.0	11	100	15	0.15	16	100	15	0.15
2	25	75	3.00	7	70	7	0.10	12	70	350	5.00	17	5	27	5.40
3	15	15	1.00	8	40	100	2.50	13	40	300	7.50	18	15	15	1.00
4	60	80	1.33	9	70	350	5.00	14	10	15	1.50	19	50	10	0.20
5	10	30	3.00	10	20	40	2.00	15	10	10	1.00	20	100	20	0.20

Table 7.2 The priority ordered sequence σ_{ord}

j	l_j	ω_j	$p(j)$	j	l_j	ω_j	$p(j)$	j	l_j	ω_j	$p(j)$	j	l_j	ω_j	$p(j)$
6	25	300	12.0	2	25	75	3.00	4	60	80	1.33	19	50	10	0.20
13	40	300	7.50	5	10	30	3.00	3	15	15	1.00	20	100	20	0.20
17	5	27	5.40	8	40	100	2.50	15	10	10	1.00	11	100	15	0.15
9	70	350	5.00	10	20	40	2.00	18	15	15	1.00	16	100	15	0.15
12	70	350	5.00	14	10	15	1.50	1	30	20	0.67	7	70	7	0.10

Table 7.3 The sequence σ_{ord}^{new}

j	l_j	ω_j	$p(j)$	j	l_j	ω_j	$p(j)$	j	l_j	ω_j	$p(j)$	j	l_j	ω_j	$p(j)$
8	40	100	2.50	14	10	15	1.50	10	20	40	2.00	19	20	40	2.00
9	70	350	5.00	13	40	300	7.50	3	15	15	1.00	20	15	15	1.00
5	10	30	3.00	17	5	27	5.40	4	60	80	1.33	11	60	80	1.33
6	25	300	12.0	1	30	20	0.67	15	10	10	1.00	16	10	10	1.00
12	70	350	5.00	2	25	75	3.00	18	15	15	1.00	7	15	15	1.00

Table 7.4 Iteration 1

j	l_j	ω_j	$p(j)$	MPSS	p(MPSS)
12	70	350	5.00	1	5.54
14	10	15	1.50	1	
13	40	300	7.50	1	
8	40	100	2.50	2	5.38
9	70	350	5.00	2	
5	10	30	3.00	2	
6	25	300	12.0	2	

Table 7.5 Iteration 2

j	l_j	ω_j	$p(j)$	MPSS	p(MPSS)
12	70	350	5.00	1	5.54
14	10	15	1.50	1	
13	40	300	7.50	1	
17	5	27	5.40	2	5.40
8	40	100	2.50	3	5.38
9	70	350	5.00	3	
5	10	30	3.00	3	
6	25	300	12.0	3	

Table 7.6 Iteration 3

j	l_j	ω_j	$p(j)$	MPSS	p(MPSS)
12	70	350	5.00	1	5.54
14	10	15	1.50	1	
13	40	300	7.50	1	
17	5	27	5.40	2	5.40
8	40	100	2.50	3	5.38
9	70	350	5.00	3	
5	10	30	3.00	3	
6	25	300	12.0	3	
10	20	40	2.00	4	2.00
1	30	20	0.67	5	1.73
2	25	75	3.00	5	

Table 7.7 Optimal sequence

j	l_j	ω_j	C_j	MPSS	p(MPSS)	$F(j)$	j	l_j	ω_j	C_j	MPSS	p(MPSS)	$F(j)$
12	70	350	70	1	5.54	24,500	2	25	75	345	5		25,875
14	10	15	80	1		1200	3	15	15	360	6	1.27	5400
13	40	300	120	1		36,000	4	60	80	420	6		33,600
17	5	27	125	2	5.40	3375	15	10	10	430	7	1.00	4300
8	40	100	165	3	5.38	16,500	18	15	15	445	8	0.27	6675
9	70	350	235	3		82,250	19	50	10	495	8		4950
5	10	30	245	3		7350	20	100	20	595	8		11,900
6	25	300	270	3		81,000	11	100	15	695	9	0.15	10,425
10	20	40	290	4	2.00	11,600	16	100	15	795	10	0.15	11,925
1	30	20	320	5	1.73	6400	7	70	7	865	11	0.10	6055

This current subsequence satisfies SSO #2, so it is optimal. All the next tasks of the sequence σ_{ord}^{new} cannot be moved into earlier positions, according to their priorities. The current sequence is optimal (Table 7.7). The functional value is 391,280.

7.4 Statistical Research of the Algorithm for the Problem Solving

Statistical studies have shown high efficiency of the developed PSC-algorithm. We performed the researches on a PC with Intel Pentium CPU with 3.6 GHz clock frequency and 2 GB of RAM. We have made 7700 tests for the TWCT problem, in 70% of cases the resulting schedule was optimal.

The percentage of an optimal solution obtaining by the first polynomial component of the PSC-algorithm decreases with increasing the problem size (the number of tasks n). It also increases with increasing the completeness C of the precedence relation graph (the number of arcs in percentage of that number in the complete graph $n(n-1)/2$). This characteristic of the algorithm was almost independent of whether the task priorities differ significantly. At $C = 10\%$, we have got the percentage of the first polynomial component fulfillment from 62% (at $n = 50$) to 41% (at $n = 500$). At $C = 70\%$, the results were from 100% (at $n = 50$) to 95% (at $n = 500$), respectively.

Thus, obtaining an optimal schedule depends significantly on the value of the graph completeness C. For a fixed number of tasks, the frequency of the first polynomial component fulfillment is quite high at values of C in the ranges 0–25% and 40–100%. For $C = 70\%$ it reaches a value of more than 95%.

Fulfillment of the first polynomial component insignificantly depends on the specified parameters (weights and processing times). If the priorities of the tasks differ significantly, there is a slightly higher probability of fulfilling the first polynomial component than in the opposite case.

Analysis of the statistical results shows that the first polynomial component of the PSC-algorithm for the TWCT problem is statistically significant for randomly generated problem instances.

With an increase in the graph's completeness C and the number n of vertices in it, the time required to execute the algorithm also increases. But even at $n = 500$, depending on the value of C, the calculation time varies only from about 11 (at $C = 10\%$) to 19 s (at $C = 70\%$). Thus, the performance of the algorithm is good enough to use it in automated control systems in real time mode.

References

1. Zgurovsky, M.Z., Pavlov, A.A.: Prinyatie Resheniy v Setevyh Sistemah s Ogranichennymi Resursami (Принятие решений в сетевых системах с ограниченными ресурсами; Decision Making in Network Systems with Limited Resources). Naukova dumka, Kyiv (2010) (in Russian)
2. Aksionova, L.A.: Novye polinomial'nye podklassy trudnoreshaemoi zadachi "Minimizaciya summarnogo vzveshennogo momenta" dlia mnozhestva odnogo prioriteta (Новые полиномиальные подклассы труднорешаемой задачи «Минимизация суммарного взвешенного момента» для множества одного приоритета; New polynomial subclasses of the intractable problem "Minimizing the total weighted completion time" for one priority set). Upravl. Sist. i Mash. **182**, 21–28 (2002) (in Russian)
3. Pavlov, A.A. (ed.): Konstruktivnye Polinomialnye Algoritmy Resheniya Individualnyh Zadach iz Klassa NP (Конструктивные полиномиальные алгоритмы решения индивидуальных задач из класса NP; Constructive Polynomial Algorithms for Solving Individual Problems from the Class NP). Tehnika, Kyiv (1993) (in Russian)
4. Pavlov, A.A. (ed.): Osnovy Sistemnogo Analiza i Proektirovaniya ASU (Основы системного анализа и проектирования АСУ; Fundamentals of System Analysis and Design of Automated Control Systems). Vyshcha Shkola, Kyiv (1991) (in Russian)

5. Pavlov, A.A., Aksionova, L.A.: Novye usloviya polinomial'noi sostavlyayushchei PDS-algoritma zadachi "Minimizaciya summarnogo vzveshennogo momenta" (Новые условия полиномиальной составляющей ПДС-алгоритма задачи «Минимизация суммарного взвешенного момента»; New conditions for the polynomial component of PSC-algorithm for the problem "Minimization of the total weighted completion time"). Probl. Programmir. **2001**(1), 69–75 (2001) (in Russian)
6. Pavlov, A.A., Pavlova, L.A.: Osnovy metodologii proektirovaniya PDS-algoritmov dlya trudnoreshaemyh kombinatornyh zadach (Основы методологии проектирования ПДС-алгоритмов для труднорешаемых комбинаторных задач; Fundamentals of PDC-algorithms design methodology for intractable combinatorial problems). Probl. Inform. i Upravl. **4**, 135–141 (1995) (in Russian)
7. Sidney J.B.: Decomposition algorithm for single-machine sequencing with precedence relation and deferral costs. Oper. Res. **23**, 283–298 (1975)
8. Pavlov, A.A. (ed.): Sistemy avtomatizirovannogo planirovaniya i dispetchirovaniya gruppovyh proizvodstvennyh processov (Системы автоматизированного планирования и диспетчирования групповых производственных процессов; Automated planning and dispatching of group production processes). Tekhnika, Kyiv (1990)
9. Pavlov, A.A., Pavlova, L.A.: About one subclass of polynomially solvable problems from class "Sequencing jobs to minimize total weighted completion time subject to precedence constraints". Vestn. Mezhdunar. Solomon. Univer. **1999**(1), 109–116 (1999)
10. Zgurovsky, M.Z., Pavlov, O.A., Misura, E.B. PDS-algoritmy i trudnoreshaemye zadachi kombinatornoi optimizacii (ПДС-алгоритмы и труднорешаемые задачи комбинаторной оптимизации; PDC-algorithms and intractable combinatorial optimization problems). Syst. Res. and Inform. Technol. **2009**(4), 14–31 (2009) (in Russian)

Part II
Hierarchical Planning and Decision Making in Network Systems with Limited Resources

Chapter 8
The Four-Level Model of Planning and Decision Making

Abstract We provide an overview of known models, methods and software for scheduling and operational planning for objects with a network representation of technological processes and limited resources. The problem is an operational plan construction to produce a potential order portfolio which is the best in terms of criteria defined by the customer. We make the conclusion that an immediate solution of this problem (multi-stage network scheduling problem) is inefficient. The result of analysis is the four-level model of planning (including operational) and decision making, in which we formalize formal procedures both for obtaining an operational schedule and for its operative adjustment. The four-level model includes the combinatorial optimization problems presented in Chaps. 2–7 as well as the Decision Making Unit, a subsystem that performs decision making functions in case if various events appear during planning. In the Decision Making Unit we use our modified Analytic Hierarchy Process which is based on the research of empirical pairwise comparisons matrices with the help of combinatorial optimization models with weighted components of the additive functional.

8.1 General Description of Systems with Network Representation of Technological Processes and Limited Resources

This chapter contains our new original results based on our earlier work [1].

Advanced representation and solution techniques in production planning and scheduling received significant attention during the past decades, both from the part of research communities and the industry [2]. This interest comes quite natural, regarding that these methods hold out a promise of increased productivity, better service level, higher flexibility, together with lower production costs. It is presumed that the above objectives can be reached by supporting the management to make smarter decisions on various levels of the planning hierarchy.

Despite the attractive prospects, only a few of the recent research results has migrated into everyday practice. Although advances in operations research and

M. Z. Zgurovsky and A. A. Pavlov, *Combinatorial Optimization Problems in Planning and Decision Making*, Studies in Systems, Decision and Control 173,
https://doi.org/10.1007/978-3-319-98977-8_8

artificial intelligence led to the development of novel modeling and solution techniques, industrial applications often require more: on the part of the researchers, richer models and more efficient algorithms.

The most well-studied area of planning and management of complex systems is production planning and management. Its methods and the obtained results are also important for other fields of the production economy.

Production systems with network representation of technological processes and limited resources are widespread: their share makes up to 80% of all types of production all over the world. Such systems, in particular, include:

- discrete type production;
- make-to-order manufacturing;
- job shop;
- batch-type production;
- building industry;
- project scheduling and management systems.

Productions with network representation of technological processes and limited resources are complex types of production. Their complexity approaches that of flexible production systems [3–5]. For most of them, the common characteristics are [4, 6]:

- large and constantly updated product range, their sophistication, large number of operations to be processed;
- network representation of the technological process, i.e., presence of precedence relations between the operations;
- the number of part titles and the labor intensity of their production change in sufficiently short time intervals;
- wide variety of the operational equipment;
- the quantity of products released are not the same among the planning periods;
- significant difference in the technological routes for different products;
- wide unification of parts and assembly units with different applicability in products structure;
- resources required to perform operations are limited, their availability changes in time;
- subject-specific specialization of assembly shops that have common products to release;
- the need in operational control over production of spare parts, cooperation products, new equipment;
- large number of consumers, and only a part of them are permanent; the urgency of orders with possible individual requirements; tight deadlines for the new products development;
- large percentage of the production technical preparation stage in the general production cycle, this necessitates planning and coordination of start and release dates throughout the entire technological production chain;

- existence of long production cycles requires the management of production reserves structure when moving to another planning period.

All these specific characteristics significantly complicate the process of planning and management in modern conditions. In addition, e.g., post-Soviet enterprises in almost all areas of industry have faced the following trends in recent decades [6]:

- dramatic increase in the production mix and the number of products modifications;
- increase in the number of consumers and suppliers of raw materials and components;
- increase in the number of orders simultaneously fulfilled at a manufacture;
- saturation of market demand and, as a result, a decrease in the rate of return and balancing at the level of minimum profitability;
- the level of capacity utilization approaches to the limit values;
- the need in sharp increase of the products' quality, at the same time reducing their cost to maintain market competition, primarily from foreign producers.

These tendencies lead to a management crisis and do not allow any longer to efficiently plan and manage a production using information technologies that implement simplified, that is, inadequate models of planning and management. The demand of the market to make "only what is needed when it is needed and in the amount needed" [7, p. 15] is the reason to develop more efficient tools for automated planning and production management.

Efficient management in socio-economic systems, including industrial enterprises, requires in these conditions to use modern management concepts and response rapidly to a volatile market situation. This in turn is impossible without precise and comprehensive information about the state of production, financial activity, and resources of the enterprise, without established business processes and competent administrative management.

The main way to overcome the crisis are modern information technologies for planning and management. The following basic requirements are imposed on such planning and management systems.

8.2 Requirements for the Creation of Software for Scheduling and Operational Planning in Systems with Network Representation of Technological Processes and Limited Resources

1. Implementation of a *progressive production model* that allows to reduce the amount of work in progress, to shorten the production cycles, setup times and the production preparation stage; to simplify planning, routing, and scheduling;

to increase productivity, efficiency of supply and total flexibility. Examples of such progressive production organization are group technology, cell production, flexible production systems, and others.

2. *Planning hierarchy*. The system must solve the planning problem as a whole, i.e., to have in itself the means of preliminary (strategic), coordinated (tactical) and exact (operational) planning. This implies the *hierarchy of the planning model* and the implementation of *decisions coordination*. Decisions made at a lower level of hierarchy must be coordinated with the higher levels decisions through a feedback. In the hierarchical approach to production planning and control the detailed monolithic formulation is replaced by a sequence of models that are consistent with a hierarchy of decisions that have to be made [8].
3. *Aggregation and disaggregation* as an implementation of hierarchical planning to ensure the ability of working with a huge sized production data. Discrete organizational and production systems, such as small-series production, are complex management systems that are difficult to automate due to the enormous dimensions of the problems to be solved. Design data arrays in productions often reach a dimension of hundreds of thousands of parts and operations interrelated technologically. It is impossible to solve any optimization problem on graphs of such dimension. Therefore, the system must include procedures of aggregation and disaggregation. Aggregate (strategic and tactical) decisions are made first and impose constraints within which more detailed (operational) decisions are made. In turn, the detailed decisions provide the feedback to evaluate the quality of the aggregate decisions. Decisions at the higher levels of the hierarchy are invariably based on aggregate models. The success of the hierarchical approach depends, to a large extent, on the consistency between the aggregation and disaggregation procedures, and on the interaction between the models at the different levels [8]. The requirements for the procedures of aggregation and disaggregation are as follows:

   - they must take into account the main timing constraints (for example, due dates and precedence relations) and the resource constraints;
   - the aggregation procedure must be feasible: disaggregation must give the initially aggregated information;
   - aggregation must reduce the planning problem complexity so that obtaining a solution close to the optimal in a reasonable time becomes possible;
   - aggregation must provide an opportunity to implement the global optimum search strategy;
   - aggregation and disaggregation may be performed automatically but preferably in interactive mode with experts: this would ensure a more efficient implementation of the global optimum search strategy.

4. Availability of *multiple criteria* for schedules optimization. In severe market competition conditions, the system must be capable to plan according to various criteria of optimality, with a choice of criteria according to current market needs.

Important are economic and production criteria related to profit maximization, costs minimization, earliness and tardiness minimization in relation to due dates, orders fulfillment just in time, and others. Various restrictions, as the precedence relations or resources availability, must be taken into account. The developed algorithms must allow to maximize the profit of enterprises in the market competition conditions. The algorithms must ensure the fullest utilization of equipment, energy resources saving through the efficient equipment usage, and the maximum shortening in the production cycle of products.
The main scheduling criteria in market conditions are the following:

- orders fulfillment just in time;
- total weighted tardiness minimization in relation to the due dates of products;
- minimization of the total weighted completion time of products;
- minimization of the total costs of earliness and tardiness in relation to the due dates of products;
- minimization of makespan (the completion time of the last task).

5. *Modularity* of the algorithmic ware, i.e., finding common algorithmic blocks of basic algorithms for various problems and combining them for efficient system design of algorithms on the basis of a clear constructive analysis.
6. *Universality* of the algorithmic ware, i.e., the ability to easily reconfigure from one criterion to another, to include additional constraints, and to adapt to various management systems planning. Implementation of complex systems functioning planning due to creating a system of new highly efficient interrelated algorithms based on common logic of problems solving for various optimality criteria. This will allow to solve planning problems efficiently as a whole.
7. *Efficient exact and approximate methods* must be used to solve the optimization problems of planning. The optimization models are intractable problems of combinatorial optimization of various types. The system's mathematics must adopt state-of-the-art theoretical achievements for solving the intractable combinatorial planning problems under consideration.
8. *Relevance to a real production* in its complexity and diversity. The planning model must reflect the resource constraints, the actual equipment loading, the relationships between operations in the technological process, the large number of various production linkages. The model must take into account the products design complexity, different quantity of products released in different planning periods, different start times of products manufacturing, and other production specificities.

Below we give brief reviews of models for complex systems planning in modern conditions and of known scheduling methods.

8.3 Overview of Known Models, Methods and Software for Scheduling and Operational Planning

Let us consider the most general formulation of the production planning problem. Suppose there is a set of packages of interrelated jobs, the package is called a product. For each product the technology of its production is given. For some products the due dates may be specified. For each job we know its processing time. A set of limited resources (workplaces and equipment units) is used for the jobs processing. We need to build an operational plan for the jobs packages processing with a breakdown to resources. The plan must take into account one of the specified optimality criteria and have a possibility of operative adjusting in the case of its partial violating.

This problem belongs to the class of production planning and scheduling problems with optimization by economically significant criteria. The world scientists work on such problems solving since the 1950s researching approaches to planning as well as methods of optimization.

Let us consider some key problems solved during development and implementation of production planning models [9]. Any planning problem starts with a specification of customer demand that is to be met by the production plan. In most contexts, future demand is at best only partially known, and often is not known at all. To the extent that any forecast is inevitably inaccurate, one must decide how to account for or react to this demand uncertainty.

A key choice is what *planning decisions* to include in the model. By definition, production-planning models include decisions on production and inventory quantities. But in addition, there might be resource acquisition and allocation decision.

The identification of the *relevant costs* is also an important issue. For production planning, one typically needs to determine the variable production costs, including setup-related costs, inventory holding costs, and any relevant resource acquisition costs. There might also be costs associated with imperfect customer service, such as when demand is backordered.

Important is also the selection of the *time period* and *planning horizon.*

Production planning is usually done at an *aggregate level*, for both products and resources. Distinct but similar products are combined into aggregate product families that can be planned together so as to reduce planning complexity. Similarly production resources, such as distinct machines or labor pools, are aggregated into an aggregate machine or labor resource. Care is required when specifying these aggregates to assure that the resulting aggregate plan can be reasonably disaggregated into feasible production schedules.

Volumetric techniques and methods of scheduling are usually used to solve the production planning problems. A linear programming model for production planning optimization given in [9] is an example of a volumetric planning method. For any realistic problem size, we can solve that basic linear model by any good linear-programming algorithm, such as the simplex method. Later we can extend the model introducing additional constraints, e.g., for lost sales, backorders, dependent demand items etc.

Volumetric techniques are used in case when the product manufacturing cycle is much shorter than the planning period. However, these methods do not take into account the technology of products manufacturing. The relationships obtained by volumetric methods between the necessary and available resources does not reflect the actual picture. The reason is that the need for a resource of one or another type does not arise during the entire planning period but only when we start to execute the corresponding job during the product manufacturing. Thus, planning by volumetric techniques leads to uneven loading of equipment [10].

The advantage of *scheduling* models in front of volumetric planning models is that they always have the resource constraints as the main constraints. This ensures that there is no overload of an equipment. Scheduling models are based on a representation of the products manufacturing process in the form of a liaison graph that shows the processing sequence of technological operations.

Algorithms of multi-network planning with limited resources are grounded on the results of scheduling theory (ST). Classical statements of scheduling problems, their complexity, mathematical methods for their solving, and the complexity evaluation for these methods are studied within the ST framework.

Scheduling is the main instrument for synchronization, coordination and support of the work process. Coordinated work of all divisions of the enterprise must only follow a single work plan that objectively takes into account the available resources and the achieved organization level. The main purpose of scheduling is to solve a set of problems related to the jobs distribution in time between available resources, coordination of all subdivisions, tasks assignment to the given planning period, and operational management [11].

A schedule is a set of start times for all operations determined by means of the available resources analysis. The schedule must satisfy all the requirements that follow from specified production constraints. A scheduling problem is, as a rule, the optimization problem: to select from all feasible schedules (that meet the imposed restrictions) the one with the extreme value of a given optimality criterion.

Enterprises form the production plan on the basis of an order portfolio. The production planning problem consists in a convenient set of orders selection for including it into the production plan. “Convenience” of orders means that this set should provide the desired level of some technical and economic indicators, and at the same time, some function (criterion) must reach its extreme value [11].

The enterprise’s production plan formed at the preliminary planning stage should be detailed in time points sense and communicated to specific production units at the operational planning stage.

The most often objective functions used in the general scheduling problem are to minimize the total production cycle length for all parts batches (the throughput time), the total “not in process” time for all parts, the total idle time of equipment [12]. A very important criterion is also to assure the start and release times of the parts batches to be not later than the planned due times. The planned due times are the constraints in the general problem.

The production planning model must include the following tasks:

- the production plan development for the enterprise at the stage of the order portfolio preparation by a manager taking into account different criteria of profit maximization;
- coordinated distribution of the production plan (jobs assignment) to subdivisions and intervals of the planning period;
- operational plans creation for the enterprise's subdivisions.

As an integrated model of planning and management of the production process, the production planning model must solve the problems of preliminary, coordinated and operational manufacturing planning.

The problem of coordinated production plans development for all subdivisions of the enterprise requires the creation of a distributed model of the plans construction for all management levels. But the multi-level management requires clear linking between decisions made at each level. This is possible only if the active elements of the model and the decision maker interact during the problem solving.

Hence, the analysis shows that the most efficient planning models are hierarchical. Hierarchical planning [13–16] represents a philosophy to address complex problems, is quite general, and has been adapted to a wide variety of systems by suitable choice of aggregation and disaggregation schemes and submodels. Hierarchical planning methods are responsive to management needs at each level of the organization structure [8].

Scheduling is one of the most complicated field of mathematics. Intra-shop scheduling and the production process regulation problems are the most difficult in the automation. A practical need in schedules for workplaces stimulated the development of ST as a special discipline.

ST is the common algorithmic basis of production planning systems. A good review of the algorithms for ST problems solving is given in [17]. Study of ST problems has showed that most of them belong to the class of intractable combinatorial optimization problems. The complexity theory of Cook [18] and Karp [19] arose in the early 1970s and became a tool to evaluate the complexity of the scheduling problems solving. A more complete exposition of this theory was given by Gary and Johnson in [20].

One of the most productive periods in the domains of management science and operations research was the 1950s [17]. During this time many problems were solved by the application of crude but effective heuristics which formed the basis of the development of classical ST. The work of Johnson [21] is believed to be one of the earliest works in ST. An optimal algorithm for a two machine flow-shop was developed in this research. Then other efficient methods, solvable in polynomial time have been proposed [22–24].

The main methods to solve the intractable problems of combinatorial optimization are the following [17, 25]:

- branch and bound method (the main enumerative strategy) [26];
- constructive approximation methods: priority dispatch rules [27–29] including fuzzy logic [30], the insertion algorithm of Werner and Winkler [31] with the beam search [32, 33];
- approximation heuristic search procedures [34–36];
- iterative local search methods [37–39] including, in particular, recession vector method [40] and its modifications [41], global equilibrium search [42], shifting bottleneck procedure [43–48];
- iterative methods of artificial intelligence: genetic algorithms [49, 50], genetic local search [51–55]; constraint satisfaction [46, 56–59] including that with neural networks use [60–65]; ant colony optimization [66–68]; large step optimization [69–75]; tabu search [76–82]; problem space methods and heuristics [83] including greedy algorithms [84]; threshold algorithms: iterative improvements and threshold accepting methods, simulated annealing [37, 48, 51, 85, 86];
- relatively new approaches are logic programming by constraints (constraint programming, Gecode packet) [87–89] where the new results are obtained due to novel algorithms propagating integral constraints and scheduling constraints; analytical combinatorics [90] which are the analytical methods of the properties' description for large size combinatorial structures, new results can be obtained directly applying methods for complex problems including "combinatorial monsters" from multi-machine multi-stage (job shop, open shop) planning problems;
- we propose a new perspective approach: PSC-algorithms (earlier, PDC-algorithms) [1, 12, 91], we give many of them in the first part of this book.

Considering the methods of intractable combinatorial optimization problems solving, researches have returned to the heuristics notion of the 1950s however with some novel and elaborate modifications [17]. Many scientists made efforts to systematize the algorithms development, to create a system of basic algorithms to solve many problem classes, and to estimate the algorithms complexity. However, today the problem is much broader. We need to develop systems of novel highly efficient interrelated algorithms that would allow to solve planning and management problems efficiently as a whole and to ensure the complex systems functioning efficiency. Systems with PSC-algorithms meet such requirements.

The current industrial practice in production scheduling is also dominated by heuristic approaches, such as priority rule-based schedulers [2, 32, 92]. In spite of this, well-known formal methods are available to describe what makes a schedule feasible, and also to optimize the schedule according to various criteria. The most promising branch of these methods, constraint-based scheduling emerged in the early eighties [93, 94]. It offers a rich and straightforward representation to model even the finest details of the scheduling problem. However, the solution of the vast instances of the NP-complete combinatorial optimization problems that often arise in practice challenges currently known algorithms [95].

Existing Systems for manufacturing and organizational processes planning on the market can be divided into three groups corresponding to the basic planning conceptions: *MRP*, *MRP II* and *ERP* which are the development of each other.

The need in accounting at all stages of a product life cycle significantly complicates the product designing and manufacturing problem. However, enterprises manage to solve it with the use of automated management systems. Informational support of the manufacturing stage is carried out by computer aided management (CAM) and computer aided control (CAC) systems. CAM systems usually include implementations of material requirements planning (MRP), manufacturing resource planning (MRP II), supply chain management (SCM) systems, and developed from them enterprise resource planning (ERP) systems.

MRP [96] is one of the most popular logistics[1] concepts in the world, appeared in the 1950s but widely spread only with the development of computer technology in the 1970s. A large number of micro-logistical systems were developed and are operated on its basis. Appearing of a more advanced MRP II concept and ERP software development, reducing its cost caused the MRP software products to be very rare, as a rule, a part of obsolete information systems of enterprises.

An implementation of a system working on MRP methodology is a computer program that allows to optimally regulate the components supply to the manufacturing process and to control the stocks and the process itself. The main task of MRP is to ensure the availability of the needed number of needed components and materials at any time within the planning period. Additional task is the possible reduction of permanent stocks and, consequently, the warehouse unloading [97].

MRP systems are used when working with materials, components, semi-finished products and their parts. The demand for those parts depends on that for specific finished products. That is, the demand for raw material resources depends heavily on consumers' demand for final products. MRP system can also work with a wide range of material resources.

The usual practice of using MRP in business is connected with the planning and control of ordering and supplying (purchasing) procedures for a wide range of material resources in mechanical engineering. MRP systems were developed to use at industrial enterprises. It's use is logical and expedient if an enterprise has the discrete type of manufacturing (assembly-to-order, make-to-order, make-to-stock, repeating type etc.). In this case, there is a bill of materials and the article for the releasing products. If an enterprise is a process industry or a continuous-batch processing, then the use of MRP functions is justified in case of a long production cycle [98].

[1]Logistics is the discipline dedicated to the organization of a rational process of goods and services promoting, from raw material suppliers to consumers. It also studies the products and services circulation sphere, stocks and provisions management, and the creation of goods circulation infrastructure. A wider definition of logistics treats it as a teaching about planning, management, and control of material, informational and financial resources movement in different systems.

MRP systems are rarely used to plan material requirements in service, transportation, trade, and other non-manufacturing organizations. Potentially, though, the ideas of MRP systems can be applied with some assumptions in such organizations if their activities require material planning in a relatively long time period.

MRP systems are based on material planning for optimal production organization and include both MRP and capacity resources planning (CRP) functions. Their goal is to create optimal conditions for the production plan execution. The main idea of MRP systems is that any unit of materials or components needed for the production must be available at the right time and in the right amount.

Main goals of MRP are:

- meeting the demand for materials, components and products for production planning and delivery to consumers;
- supporting low stock levels;
- planning of purchasing and production operations, delivery scheduling.

MRP system allows to determine how many products we need to produce and in what time period. Then, the system determines the time and necessary quantities of material resources to meet the needs of the production schedule.

The main advantage of MRP systems is that they form a sequence of manufacturing operations with materials and components. The sequence ensures in time production of units (semi-finished products) for the main production schedule execution and for the finished products release.

The main disadvantages of MRP systems are:

- significant amount of calculations and preliminary data processing;
- increase in logistics costs for order processing and transportation as the firm strives to further reduce inventories of work in progress or to switch to small orders with a high frequency of their acceptation;
- insensitivity to short-term changes in demand;
- a large number of failures due to the large system size and its complexity.

Just in time (JIT) and *Kanban* planning concepts developed in Japan [99], as well as MRP developed in the USA, are focused on the production in time and on the products sales. JIT is the management based on the highest organization of defect-free production, synchronization of production processes including supply operations, and subcontracts fulfillment. It is used mainly at enterprises with a mass character of production [100]. MRP provides a close relationship between long-term and operational planning. However, this system covers basically only inter-shop level of the enterprise's operational management. MRP systems focus on the material flow aspect of production, and assume that products can be manufactured with fixed lead times [2]. Hence, they completely disregard the actual load on production capacities. No wonder that in an age characterized by market fluctuations, the plans generated this way can be barely unfolded to executable detailed schedules. Recently, several approaches have been suggested to couple the capacity and material flow oriented aspects of production planning [101–103]. A common characteristic of these models is that they apply a high-level description of the production activities and their complex interdependencies, which—in practice—has to be encoded manually, by a human expert. The high-level formalism does not always reflect the context of the underlying processes, and it cannot guarantee the feasibility of the production plans. Furthermore, the results depend largely on the proficiency and the mindfulness of the human modeler. We need to find a novel, aggregate formulation of the production planning problem which ensures that the generated plans can be refined into feasible detailed schedules. The representation of the planning problem should be generated automatically, from data readily available in de facto standard production databases.

MRP II systems are the further development of MRP systems and are focused on the efficient planning of all the resources of a manufacturing enterprise. One can typically distinguish the following components of MRP II systems [98, 104]:

- business development planning (business plan drafting and updating);
- master production schedule (MPS) formation;
- material requirements planning (MRP);
- material requirements plan execution;
- capacity resources planning (CRP);
- capacity resources plan execution;
- sales and purchases planning;
- feedback subsystem.

MRP II specifies the principles of detailed production planning for the enterprise including order accounting, capacity utilization planning, demand planning for all manufacturing resources (raw materials, components, equipment, personnel), production costs planning, production process simulation and accounting, products release planning, operative adjustment of the plan [105].

MRP II provides both operational and financial production planning and a wider coverage of company resources than MRP. In contrast to MRP systems, an MRP II system plans in material terms as well as in monetary terms [105, 106].

MRP II systems assume the financial component (business planning) involvement into the information integration. They include special tools for a financial plan formation and budget estimates preparation, forecasting and management of cash flows. The possibility of the production plan execution is determined on their basis in terms of existing and anticipated financial resources [107].

ERP systems, being the further development of integrated information management systems of the enterprise, usually add to the above functionality the distribution resources planning (DRP) and the technological maintenance and repairs resources planning.

DRP systems provide an optimal solution (that is, planning, accounting and management) for the tasks of material and technical resources and finished products transportation [108].

In addition, a special project management subsystem for managing the long-term projects execution is usual for MRP II and ERP systems. The subsystem assumes full-featured planning of material and labor resources and capacities, production scheduling, process management and current projects invoicing [98].

The main difference of ERP systems from MRP II systems developing in their direction is the support for [109]:

- different types of productions;
- resource planning in various activity areas (not only goods production);
- resource planning and management for a complex multi-profile enterprise, a corporation.

Most ERP systems (such as Ultima [110]) are focused on small and medium-sized businesses. "One hundred percent" ERP systems are rare: just a few products [109]. They indeed are high-level systems in terms of functionality and, accordingly, of the intellect brought from numerous developers. ERP systems allow you to plan and manage your business in various areas of activity in a separate enterprise or a multi-profile corporation. They should not be confused with systems performing some functions of multi-dimensional control in some organization.

MRP II and ERP systems implementation, as a rule, involves changes and additions to the activities, revision and correction of management principles. Even partial implementation of a corporate information system can take up to several years accompanied by a demolition of established standards and procedures.

De facto standard in production automation software market is German ERP system SAP R/3. It is a large software and hardware complex for the complete automation of the entire production process and for paperless management technology implementation. The system contains, in particular, the modules for: financial accounting; controlling; material flow management; technical repairs and maintenance of equipment; sales, shipping, billing; projects system; fixed assets management, planning, and control; personnel management [111]. Thus, R/3 is a complete automation system that includes bookkeeping, control and planning of material requirements throughout the enterprise, unified accounting and reporting, management of material flow, stock and warehousing, equipment maintenance and repair, sales and investment activities, and the personnel management.

The disadvantages of the system are the implementation complexity and its high cost. The latter makes the system an unattainable luxury for many small and medium-sized enterprises. According to the well-known American analytical firm Gartner Group, the system is designed for large companies with an annual turnover of $200 million [109].

The implementation of Dutch ERP system Baan costs about 600–800 thousand US dollars for an average enterprise with a staff of 100–150 people (without hardware costs) [109].

As of 2010, according to the analysis of Panorama Consulting [112, 113], ERP systems were divided into three groups in decreasing order of their market share:

1. SAP R/3: 24%; Oracle Applications (absorbed other ERP products PeopleSoft Enterprise and JD Edwards EnterpriseOne): 18%; Microsoft Project: 11%;

2. Products of Epicor, Sage, Infor, IFS, QAD, Lawson, Ross: 11% at all;
3. Products of ABAS, Activant Solutions, Baan, Bowen and Groves, Compiere, Exact, Netsuite, Visibility, Blue Cherry, HansaWorld, Intuitive, Syspro and other suppliers: a total of 36% of the market.

The market share distribution depends on the customers' size. In the ERP segment for companies with revenues of more than $1 billion, SAP had 47%, Oracle had 32%, and Microsoft had 4%. While in the segment for companies with revenues of up to $25 million, SAP had 22%, Oracle had 23%, Microsoft had 16%.

The situation in the regional markets may differ from the world market. So, IDC indicated the following distribution of supplier shares in the Russian market in 2010: SAP had 50.5%; Russian product 1C: Enterprise had 26% [114]; Oracle had 8.2%, Microsoft had 7.4%; Galaxy had 2,4% with a total market volume of 650 million US dollars [115]. In the Ukrainian market: SAP had 43.4%, Information Technologies (IT-Enterprise) [116] had 15.7%, 1C had 13.9%, Oracle had 11.7%, Microsoft had 6.1% with $46.64 million volume [117]. In Brazil, about half of the market belonged to domestic Totvs while SAP had about 30% [118].

The most developed ERP systems perform various business functions related to production planning, purchases, sales, analysis of marketing prospects, financial and personnel management, warehousing, fixed assets accounting, etc. MRP II systems are mainly focused on business functions directly related to production. SCM and MRP II modules can be implemented as subsystems of ERP [119].

Unfortunately, ERP systems architecture oriented to transactional work on the basis of a database management system is not designed for large volumes of calculations with the initial data. That is why separate modules (requiring intensive calculations) of the operational functions unit of ERP systems are often placed in specialized software products and appear as dedicated classes of applications for production management: APS and MES [120–122].

Advanced planning and scheduling (APS) is the concept of production planning. Its main feature is the possibility to build an operational schedule for all equipment of the enterprise.

Here are the main features that APS systems should have [120–122]:

- scheduling to within a second;
- forward planning (attempt to execute all production and purchase orders as soon as possible);
- backward planning (attempt to complete production orders by a certain date, execute purchase orders by the time they are used in production);
- multilevel aggregation;
- a lot of stocks replenishment algorithms;
- inventories tracking;
- control over execution and notifications in a graphical format and as reports.

In relation to ERP class systems, an APS system is an add-on that extends and replaces their functionality in part of scheduling. At the same time, APS uses the information contained in the transaction part of ERP (about actual customers orders, components balances in warehouses, etc.). At the end of the planning process, the APS system passes the relevant results, such as production, purchase and transportation orders, forecasts, etc., to the ERP system [122].

Manufacturing execution system (MES) [123] is a specialized software application designed to solve the problems of synchronization, coordination, analysis, and optimization of some production. MES systems belong to the class of control systems of the job shop level. But they can also be used for integrated production management in the enterprise as a whole. MES systems can serve as a source of information for ERP systems on the status of the production orders in process and equipment launched [122].

APS and MES systems also exist as products separate from ERP systems. In this case, other departments, automated systems of the enterprise or experts input the initial data and analyze the output reports [123].

A small overview of APS systems can be found in [36]. A study carried out by Werner and Yong [124] focused on the evaluation of commercial scheduling software [125]. It was found that most of them were intended to be tools that facilitate production planning. Many packages are not based on any specific scheduling algorithm. For some of those that are, the exact scheduling procedure is kept confidential by the vendor. Only a general description of the underlying philosophy is made known to the customers. Often, this type of packages is treated like a black box that customers are persuaded to purchase mostly due to good marketing.

Pinedo [25] lists 18 selected scheduling systems applied in practice today. In spite of a large number of developed scheduling methods, only a few practical applications have entered into everyday use in industrial reality. Especially for the local scheduling level several knowledge-based systems have been presented [126], e.g., ISIS, OPIS, SONIA [127–130], all using the experience of human experts and problem-specific knowledge of the application domain. An experience obtained in projects addressing real-world scheduling problems showed that not only the applied scheduling methods but also several other features play a significant role for the acceptance of scheduling systems [128, 131].

ERP systems are necessary tools for enterprises. But it became clear at the end of the last century that their management capabilities in today's rapidly developing business world are not sufficient. The authoritative consulting company Gartner Group announced the completion of ERP systems era in 1999. It proposed the concepts of CSRP and ERP II instead [132].

A production planning system of the 21st century has two focuses: on production efficiency and on customer values creation. This new planning paradigm is *customer synchronized resource planning* (CSRP). CSRP uses proven, integrated ERP functionality and redirects production planning from production to the customer. CSRP provides efficient methods and applications for creating products with increased value for the purchaser.

ERP II refers to *enterprise resource and relationship processing*. Its primary idea is to go beyond the tasks of internal processes optimizing. Traditional for ERP systems areas of the enterprise's activities are financial management, accounting, sales and purchases management, relations with debtors and creditors, personnel management, manufacturing, inventory management. In addition, ERP II class systems allow to manage relationships with customer, supply chains, and to trade via the Internet.

Systems of ERP class and, especially, CSRP and ERP II systems, are expensive and difficult to implement. In particular, CSRP implementation requires:

- to optimize production activities (operations) by building an efficient production infrastructure;
- to integrate the buyer and buyer-focused subdivisions of the organization with the main planning and production subdivisions;
- to implement open technologies to create a technological infrastructure that can support the integration of buyers, suppliers and production management applications.

Some alternative approaches to industrial enterprises computerization were briefly described in [116]: separate large-series "boxed" software purchase; own complex systems development; "western" complex automation systems purchase. A resulting conclusion stated that such approaches cannot be applied for the full computerization of the enterprise, are inefficient in terms of the existing level of technology and do not take into account domestic realities.

Thus, we can note that existing in the market multi-level planning and scheduling tools are presented quite fully. However, the optimization criteria and methods are often hidden from the end user or represented by only one or several low complexity criteria in order to perform simple optimization and obtain mediocre results. To solve serious optimization problems and implement new algorithms, we need to develop mathematics and software of a new level for planning and scheduling systems.

In monograph [1] we presented a hierarchical *three-level planning and management model* for complex systems with a network representation of technological processes and limited resources. It is based on the hierarchical planning principle and a set of interrelated mathematical models and methods. We have replaced in it the general mathematical model of the multi-stage scheduling problem with a sequence of discrete mathematical models compatible with the hierarchy of decisions made at each level of management. We have created the system of new highly efficient interrelated algorithms for planning problems solving in modern conditions. Also, we have given an efficient formal procedure to solve a poorly formalized problem of the best plan choosing from a set of plans constructed according to different optimality criteria. The best of existing heuristic algorithms of multi-stage scheduling problems solving contain a linear or random combination of different domination rules. The rules do not guarantee the quality of the solutions obtained. In contrast, we determine the global optimum search strategy during the planning problem solving. This allows us to obtain near-optimal solutions.

The three-level planning model works with a set of packages of interrelated jobs. We call the package a "task" (which is a product). An oriented acyclic graph defines the processing technology for each task. We use a set of limited resources to process the jobs. The set of resources and executors is divided into separate, sufficiently autonomous units: multi-resources. A multi-resource is a stable group of shared resources, for example, a brigade, a group of equipment of the same type, single-profile subdivision. We need to build a coordinated plan of the jobs packages processing on multi-resources and distribute the production plan (assign the jobs) to resources taking into account the following seven optimality criteria:

1. The enterprise's total profit maximization in the absence of due dates of tasks.
2. The total profit maximization subject to the condition: each task has a due date that must not be violated (just in time planning).
3. The total profit maximization subject to the conditions: some tasks have due dates that must not be violated, other tasks do not have due dates (criteria 1 and 2 combination).
4. The total profit maximization subject to the conditions: each task has a due date; the total weighted tardiness of tasks in regard to their due dates must be minimized.
5. The criterion 4 with additional condition: due dates of some tasks must not be violated (criteria 2 and 4 combination).
6. Each task has a due date and a given absolute value of profit for its processing. The profit does not depend from the completion time of the task if it is not tardy in regard to its due date. Otherwise, the enterprise's profit for this task is equal to zero. The problem is to maximize the total profit of the enterprise.
7. All tasks have due dates. We need to minimize the total cost (penalty for the enterprise) both for earliness and tardiness in regard to the due dates.

The three-level planning and management model for complex systems with network representation of technological processes and limited resources has the aggregated, coordinated and operational planning levels. The model's foundation is the mathematics based on our new constructive theory for intractable scheduling problems solving by the criteria of enterprises' profit maximization [6, 133].

According to the three-level model, we have three stages of the operational plan construction. The first stage is an aggregated model constructing. There we represent the planning system as a single machine. We combine the jobs into aggregated jobs which are the sets of jobs that belong to one task and are executed in the same multi-resource within single pass through the multi-resource.

At the *first level* (preliminary or forecast planning) we solve the problem of the order portfolio optimization. We include to the order portfolio the most convenient for the enterprise orders, i.e., such that their execution will ensure obtaining the maximum profit with restrictions on production resources in the market conditions. We clarify strategic plans and opportunities for their implementation at the stage of forecast planning. We solve the problems of efficient resources allocation and bottlenecks identification in production. As a consequence, we can see the need in additional equipment.

Aggregation (grouping) problem belongs to the most methodologically complex ST problems. We can distinguish four interdependent tasks in the aggregation problem solving: to choose the aggregation method; to determine the aggregation level for jobs and resources; to form the structure of aggregated technological models; to form the aggregated specifications.

We use aggregated jobs (operations), resources, and time intervals as aggregated elements of the planning model. We aggregate jobs with the help of constructional, technological, batch aggregation methods, by "generalized operations" and "generalized route" construction.

The choice of jobs aggregation method depends on the detailing level of products article, on uniformity of batch technological routes, on the technological generality of operations.

We do the resources aggregation by combining technologically homogeneous workplaces, capacities of one site, shop, or of the whole enterprise into separate groups. The choice of the resources aggregation method is defined by the production structure of the enterprise and the subdivision, the production type, the durations of the production cycles.

At the forecast planning stage, we perform a preliminary aggregated production scheduling on the basis of the aggregated model by the profit maximization criterion. Profit maximization as a general optimality criterion for all levels of the planning model plays an important role in directing the production. The time factor is important for the enterprise's profitability. The planning in modern market conditions should be done according to the principle of releasing "only what is needed when it is needed and in the amount needed" [7]. If the production costs will not grow, then the price of products, and hence the profit of the enterprise, will vary depending on the release time of products.

Based on the aggregated information, we construct a graph of the tasks' critical paths. Our final objective is to build the schedule according to a chosen optimality criterion. To do it, we solve the special case of important problem of the total weighted completion time of tasks minimization (TWCT). In this special case, only terminal vertices of the precedence relations graph have a non-zero weight coefficient. We construct the graph of the tasks' critical paths and solve the TWCT problem at the first level of the model for each of the seven objective functions formulated above.

The *second level* of the model is for coordinated planning, that is, to coordinate the enterprise units operation. It consists in the construction of a concerted plan for the tasks processing that takes into account the above optimality criteria. The aggregated jobs priorities obtained at the first level serve as additional information that allows to significantly improve the solution's efficiency.

With the help of the production plan model formed at this level, all subsequent detailed plans and schedules are interrelated. The production plan ensures that all the following production schedules and operational plans created on their basis are feasible.

It is expedient to take the batches (assembly units) tardiness minimization in regard to the planned dates of their start and release as the optimality criterion in the coordinated planning problems. The optimization should take into account the available resources and constraints at all stages of the production process. Such criterion will ensure the products release in time and additional production costs reduction. It corresponds to the main goal of operational production management: to implement the release plan in terms of nomenclature, quantity, quality and timing at the lowest cost. This way we will achieve the best results.

The coordinated plans for the tasks processing constructed on the second level of the model by the given optimality criteria are passed to Decision Making Unit (DMU) [134]. We generate a whole series of possible feasible plans in the DMU that differ in a specific type of criterion, due dates, weight coefficients, manufacturing technology. As a result, we choose one plan to pass it to the third level of the model for implementation.

If we could not obtain a plan satisfying the specified requirements, then we pass the information to the first level for adjustment: we can change the production resources (capacities, materials, work force), the output volumes (the composition of the order portfolio), the due dates, etc.

The procedures of the *third level* (exact or operational planning) work in accordance with the obtained plan for aggregated jobs in multi-resources. They allow to build the jobs schedule with their assignment to resources and optimization by criteria agreed with the overall optimality criterion for the enterprise's activity. We introduce components into the optimality criteria for the second and the third levels of the model that take into account various factors:

- components that increase the level of equipment utilizing;
- components that minimize the deviation of the actual release times of parts from planned dates;
- components that reduce the amount of equipment downtime.

The three-level planning model proposed and substantiated in [1] lays the foundation for solving the corresponding multi-stage scheduling problem. The model is hierarchical (contains features of the planning process aggregation and disaggregation). This property allows us to efficiently use single-stage scheduling problems. We can obtain solutions for three intractable combinatorial optimization problems using efficient exact PSC-algorithms described in [1]. A standard ST problem formulation implies a schedule evaluation by a single criterion. Unlike this, we evaluate each of the obtained plans by a set of criteria that define the plan's quality. Therefore, finding plans optimal by one of the criteria is only a preparatory stage for finding an efficient plan. The problem is to find the best plan from a given set in accordance with a non-formalized overall goal.

We have combined in the single three-level planning model [1] the volumetric scheduling and operational planning at an aggregated level with detailing to the coordinated planning level. We build priority ordered sequences of products processing using the solution of the TWCT problem for five basic criteria and their two combinations (we also can expand the set to 31 optimality criteria). On the basis of these priorities, we build a coordinated plan of jobs processing on the second level of the model using sequential disaggregation. The plan takes into account the resource constraints, the network representation of the process and the optimality criterion.

We pass the resulting coordinated plans for products processing to the DMU where we choose the best plan according to the criterion of minimizing the risk of not obtaining the expected profit. We use in the DMU our original mathematics, a modified Analytic Hierarchy Process method by Saaty [135].

Since the time we wrote the monograph [1], we have managed to solve the fundamentally complex problem of constructive formalization of the network technology model. We have developed a new network model that enabled us to include new single-stage problems of the operational planning level to the multi-stage network scheduling problem (MSNSP). We have created new efficient exact PSC-algorithms for these single-stage problems solving. We describe them in the first part of this book. As a result, it became possible to formalize the third level of the model (the exact planning level) as the MSNSP description and to obtain the operational plan as a solution to this problem, as well as to perform its operative adjustment if necessary.

Thus, in Sect. 8.4 we propose a new four-level model of planning which is the development of the three-level model. To implement it, we have created a new network model, formalized the operational and operative planning levels, and solved the single-stage scheduling problems described in the first part of the book and used at the third level of the model. The new model also formally describes the procedure for solving the problem by 31 optimality criteria.

We have shown above that none of the considered planning *systems* satisfies all the requirements given in Sect. 8.2. So, to implement the original and efficient multi-level planning models described above, we need to develop a new software based on the new methodology for multi-level planning in complex systems.

Based on the market analysis and the possibilities of methods, models, and production planning systems, we see a new APS system creation based on the new planning methodology in market conditions as the most preferable option.

As a result of the theoretical studies, we have created the planning methodology (Chap. 9) and based on it a software system for production planning in complex systems (APS system) that corresponds to all requirements stated in Sect. 8.2 (Sect. 9.9).

8.4 Characteristic Features of the Four-Level Model of Planning and Decision Making

Based on the three-level model of hierarchical planning described in Sect. 8.3, we have developed a four-level model of planning (including operative planning) and decision making (FLM). The FLM was created for organizational and manufacturing systems with a network representation of technological processes. It includes:

- an aggregated model construction;
- a coordinated plan construction for aggregated jobs processing according to 31 criteria of optimality;
- the best plan choosing in the DMU;
- exact (operational) planning;
- operative planning in case of partial failure of the obtained operational plan.

We have formalized the third level of the model as MSNSP on the basis of our original methodology.

In the FLM, experts create a virtual network of a special type with specified properties on the basis of data from a real organizational or manufacturing process of a discrete type. They break the physical equipment into a number of virtual machines. To solve the problem, experts build a network with the minimum necessary amount of physical equipment. A MRP II or ERP system can perform some or all of the expert functions.

Algorithmic ware of the model consists of five blocks. Chapter 9 contains their detailed description. We determine the actions of experts in the DMU (Unit 1) and implement the formal models in Units 2–5:

- Unit 1: Decision Making Unit (it may be subsystem of an ERP system or a connection with an ERP system).
- Unit 2: the first level model construction, preliminary planning.
- Unit 3: the second level model construction, coordinated planning.
- Unit 4: the third level model construction, exact operational planning.
- Unit 5: the fourth level of the model, operative planning.

Decision Making Unit is how we call a subsystem of the enterprise's general ERP system, for simplicity of understanding. The subsystem performs decision making functions in the event of different situations during the planning. Experts can be real decision makers, or we can obtain the information for decision making from other subsystems of the ERP system.

We present an outline of algorithmic ware of the FLM in Fig. 8.1.

Functions of the DMU are:

1. The order portfolio composition (passing it to the planning system);
2. The planning criterion specification: the choice of objective; specification of the coefficients that define the functional; setting the minimum expected profit;
3. Specification of the technology to process the order portfolio. The topology associated with the technology is specified in the form of N separate subnets where N is the number of basic criteria in the overall objective function. Subnets can have common equipment. For a synthetic criterion, therefore, we consider a special case when the initial network is represented as independent subnets with common virtual machines.
4. Achieving an agreement with the customer on all changes to the plan (if a feasible schedule is not obtained).

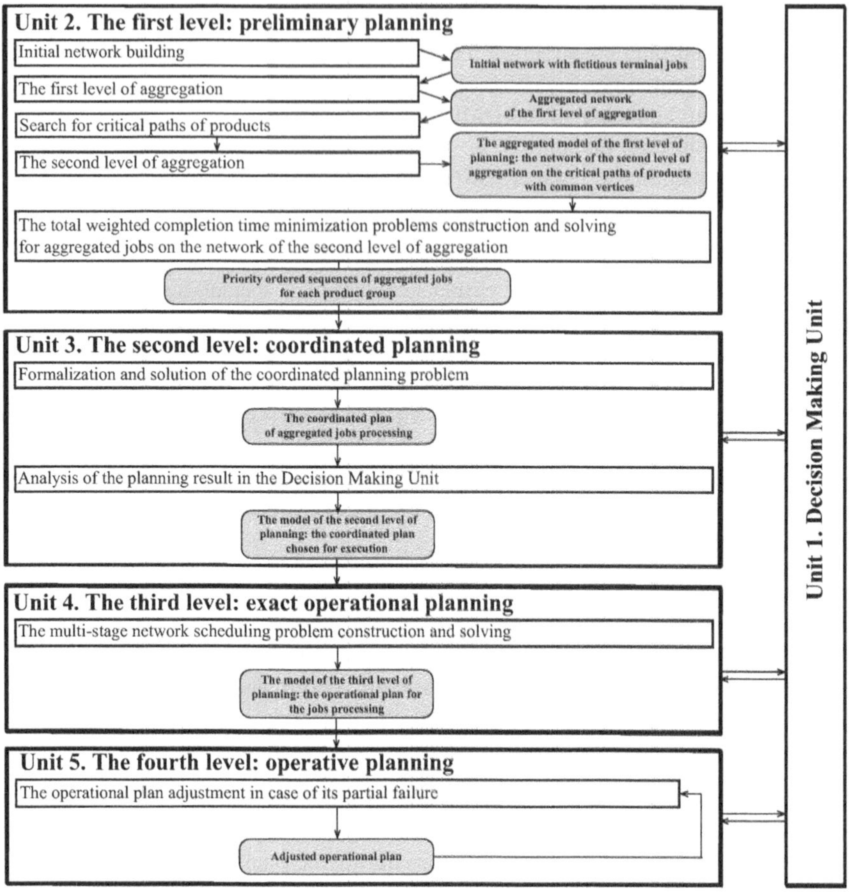

Fig. 8.1 Outline of algorithmic ware of the four-level model of planning and decision making

5. A new order portfolio composition (if the obtained profit is less than the specified minimum expected value).
6. Choosing the best plan according to the criterion of minimizing the risk of not obtaining the expected profit. We choose the plan from the set of alternative plans obtained at the third level of planning.
7. Organization of interaction between Units 2–5.
8. Operative planning execution using models of Unit 5.
9. If necessary, experts set the desired completion times for products during planning by a basic optimality criterion. The times are passed to the model of the third level of planning as the due dates.
10. Determination of the operational plan adjustment time at the fourth level and setting new evaluation criteria in regard to the approved due dates.

We considered the procedure for a well-grounded choice of a production plan from the set of feasible plans in [1]. The procedure was developed for uncertainty conditions and is based on a hierarchy tree construction. We propose the further development of this procedure in Sect. 8.5.

In Unit 2, we construct an aggregated planning model of the first level. Experts in the DMU set the problem parameters: a technological network corresponding to the technology of product manufacturing; one of 31 optimality criteria; minimum expected profit of the enterprise. The technological network is a virtual network model with the specified properties. Experts build it based on data from a real organizational or manufacturing process of a discrete type.

For planning by a basic optimality criterion, the network should have the properties described in Sect. 9.1 with the following restriction. Several virtual machines that correspond to the same real machine may appear in the network only if they are spaced in time by the technology. This makes impossible different jobs processing in the same time interval on the same real machine.

For planning by a synthetic optimality criterion, products should be divided into $N \in [1, 5]$ separate groups. N corresponds to the number of basic components in the synthetic criterion. No product can be a part of two different groups. The topology associated with the technology is specified as N separate subnets corresponding to N separate product groups. Each subnet has its own basic criterion. The properties of each subnet should correspond to Sect. 9.1. Subnets cannot have precedence relations with each other, but they can be interrelated through a common equipment.

After we constructed the initial network, a two-level aggregation follows. The first level of aggregation is the construction of an aggregated representation of the initial network through the aggregation of work and resources. This way we construct the aggregated network of the first level of aggregation. Also, we correct processing times of jobs that were not aggregated with other jobs. The second level of aggregation is the search for products' critical paths and the construction of the critical paths network with common vertices. The rules for combining jobs into the common vertices are defined in the DMU.

Next, we construct and solve optimization problems with aggregated jobs by TWCT criterion. We solve such approximating problem for each basic component of the synthetic optimality criterion included in the specified functional, except for the basic Criterion 1. For the basic Criterion 1, we solve unchanged TWCT problem. We solve all the problems on the network of the second level of aggregation, using the algorithm created on the basis of the PSC-algorithm for the general TWCT problem (Chap. 7). As a result, we obtain priority ordered sequences of the aggregated jobs for each product group.

In Unit 3, we formalize and solve the problem of coordinated planning of the aggregated jobs processing on the network of first level of aggregation. We take into account the specified optimality criterion. To construct the coordinated plans, we complement the priority ordered sequences (TWCT problems solutions) for each group of products with aggregate jobs that are not on the critical paths of products. Thus, we return to the network of the first level of aggregation. Then we assign the aggregated jobs to the aggregated network resources. We build the coordinated plans for aggregated jobs for each possible order of product groups execution.

After the aggregated jobs assignment, experts in the DMU decide what to do with products that violated their due dates. For example, they can decide to shift the start time of their execution, to change the order portfolio, to change the due dates, to increase the amount of resources, etc. After making corresponding changes in the model, we solve the problem again, starting from a certain time moment.

After each distribution, we check the actual combination of jobs in the common vertices. We do it only once: if the composition of at least one common vertex has changed, then we change the aggregated and initial networks, solve the TWCT problem again and redistribute the jobs. Also we perform the technology compatibility check: the jobs combined during the aggregation should be processed without interruptions. If it is not actually so, then we change the aggregated network of the first level of aggregation and the initial network, but without solving the problem again.

After the coordinated planning, we analyze its result in the DMU. We choose from the constructed plans the best one by functional value and the one by the criterion of minimizing the risk of not obtaining the expected profit. We do it on the basis of our original mathematical version of the modified Analytic Hierarchy Process method by T. L. Saaty (Sect. 8.5). If the profit is less than the expected minimum, then experts in the DMU compose a new order portfolio and change the network or adjust the due dates.

In Unit 4, we construct and solve the MSNSP. To do this, we disaggregate the aggregated jobs returning to the initial network model with the changes made during the coordinated planning. We construct the MSNSP by specifying the due dates of the terminal jobs on the modified initial network. The due dates are determined by the completion times of products in the chosen coordinated plan. The optimality criterion is the products execution start time maximization subject to the total earliness minimization and the absence of tardiness in regard to the final due dates. To solve the MSNSP means to determine the due dates for each job and to solve single-stage scheduling problems on the network elements. We use novel solving methods and PSC-algorithms described in the first part of the book. As a solution, we obtain the start and completion times of all jobs in the network, i.e., the operational plan. For a basic optimality criterion, we solve the problem using the method described in Sect. 9.5.1. For a synthetic criterion, we use four alternative methods described in Sect. 9.5.2 and then choose the best plan from the obtained operational plans by the functional value or by other parameters (in the case of equal functional values).

If the solution of the MSNSP is not feasible (the due dates set by the customer are violated), then we recalculate the profit for this schedule by the original functional. If the new calculated profit is not less than the expected minimum, then we need to go to the DMU to coordinate the new due dates with the customer. If the profit is less than expected, then in case of planning according to a basic optimality criterion, we implement the above listed function 9 of the DMU. If the criterion is synthetic, then we change the order portfolio directly in the DMU. If the new profit becomes less than the expected minimum after the implementation of the DMU's function 9, then we also change the order portfolio in the DMU.

Unit 5 implements the operative planning model (the fourth level of the planning model) for the event of a partial non-fulfillment of the operational plan. For a basic criterion of optimality, we solve this problem using the method described in Sect. 9.7.3. We use novel PSC-algorithms we give in the first part of the book to solve the intractable combinatorial problems. For a synthetic criterion, we use a universal method described in Sect. 9.7.2. In a particular case, the method can be used for a basic criterion. We specify a new optimality criterion for each product group: we minimize the tardiness costs or earliness/tardiness costs in regard to the specified due dates. We combine product groups with the same criterion into one subnet, exclude completed jobs from the initial network, and build a new operational plan for uncompleted jobs. Then, we obtain a complete operational plan combining the new plan with the plan for the completed jobs. We analyze the resulting schedule in the DMU. The experts accept the schedule or change the set of jobs assigned to each machine in an expert way. If the profit calculated for the new operational plan is less than the expected minimum, then experts in the DMU adjust the new operational plan so that it can be executed in the accepted due dates. They can change the technology (add an equipment), increase the capacity (productivity, number of shifts), etc.

We consider the algorithmic ware of the FLM in detail in Chap. 9.

8.5 Models and Methods of Decision Making Based on Hierarchical Goal Estimation of Alternatives

Section 8.5 contains our new original results based on our earlier works [1, 136–138].

One of the functions of the DMU (Sect. 8.4) is the choice of a single production plan for execution in the planning period, from a set of feasible production plans. A large number of non-formalized or in principle not measured in advance factors can influence the real profit from the implementation of a feasible production plan. We showed in [1] that it is so in many cases, if the planning period is long enough. We suggested there to choose a feasible (in the projected profit sense) plan corresponding to the minimum risk of not obtaining the expected profit value. We formulated the problem itself within the decision theory framework in the presence of a non-formalized goal. We have modified Analytic Hierarchy Process (AHP) of Saaty [135] and used our modification as the solving method. The classical statement of the problem solved by AHP is as follows.

There is an overall goal and a number of alternatives $A_1 \ldots A_m$. We need to choose the "best" alternative in terms of the overall goal. Suppose we have introduced a unit to measure the overall goal achievement by an alternative. If it were possible to construct a function $f(A_i)$, $i = \overline{1,m}$,[2] that measures the correspondence of an alternative A_i, $i = \overline{1,m}$, to a specified goal in those units, then the problem would have a trivial solution: one chooses the alternative that reaches the maximum $\max_i f(A_i)$.

For most real problems, the problem of direct construction of the function $f(A_i)$ is insoluble because of many factors [139]:

- a poorly formalized problem;
- a combination of quantitative and qualitative factors with the help of which it is formulated;
- a confidence that there are factors specifying the $f(A_i)$ function which the researcher is not aware of or cannot directly measure.

In this case, we use AHP that implements the reconstruction (using linear superposition) of the non-formalized overall goal values on alternatives A_i, $i = \overline{1,m}$.

The principle of identity and decomposition laid down in AHP assumes the problem structuring in the form of a hierarchy or network [139, 140]. Usually, in the most elementary form, a hierarchical structure is built from the top, through intermediate levels to the lowest level. Therefore, we need to decompose the problem and present it in a hierarchical form to solve the multi-criteria choice problem. We have the following decision making problem presented in a hierarchical form: m alternatives $A_1 \ldots A_m$ and s criteria levels E_j^i, $i = \overline{1,s}$, $j = \overline{1,m_i}$.

At subsequent stages of AHP, we compare the hierarchy elements pair-wise in terms of their impact on the overall characteristic (overall goal or the upper level criterion). Using a group of pairwise comparison matrices (PCMs), we form a set of local priorities expressing the relative impact of elements set on an element of the upper level.

We now give one possible description of how a hierarchy tree with weights assigned to each branch reconstructs an unknown expression $f(A_i)$, $i = \overline{1,m}$. Consider the following expressions: $E_j^s(A_i)$ is the value that a criterion E_j^s takes for an alternative A_i, $E_j^s(A_i) = \omega^i_{E_j^s} > 0$, $i = \overline{1,m}$, $j = \overline{1,m_s}$, weights $\omega^i_{E_j^s}$ for a fixed j are measured in the same units and normalized: $\sum_{i=1}^{m} \omega^i_{E_j^s} = 1$; $E_j^{s-1}(A_i) = \sum_{l=1}^{m_s} \omega^{E_l^s}_{E_j^{s-1}} E_l^s(A_i)$ where $E_j^{s-1}(A_i)$ is the value that a criterion E_j^{s-1} takes for an alternative A_i; $\omega^{E_l^s}_{E_j^{s-1}} > 0$ is the contribution (weight) of a criterion E_l^s into a criterion

[2]Here and after, $\overline{a,b}$ denotes the interval of integer numbers from a to b, that is, $\overline{a,b} = \mathbb{Z} \cap [a,b] = a, a+1, \ldots, b$.

E_j^{s-1}, the weights $\omega_{E_j^{s-1}}^{E_l^s}$ for fixed s and j are measured in the same units and normalized: $\sum_{i=1}^{m_s} \omega_{E_j^{s-1}}^{E_l^s} = 1$.

In the general case: $E_t^{s-j}(A_i) = \sum_{l=1}^{m_s-j+1} \omega_{E_t^{s-j}}^{E_l^{s-j+1}} E_l^{s-j+1}(A_i)$ where $E_t^{s-j}(A_i)$ is a value that a criterion E_t^{s-j} takes for an alternative A_i; $\omega_{E_t^{s-j}}^{E_l^{s-j+1}} > 0$ is the contribution (weight) of a criterion E_l^{s-j+1} into a criterion E_t^{s-j}, the weights $\omega_{E_t^{s-j}}^{E_l^{s-j+1}}$ for fixed s, j, and t are measured in the same units and normalized: $\sum_{l=1}^{m_s-j+1} \omega_{E_t^{s-j}}^{E_l^{s-j+1}} = 1$.

Finally,

$$E_1^1(A_i) = \sum_{l=1}^{m_2} \omega_{E_1^1}^{E_l^2} E_l^2(A_i), \quad i = \overline{1,m} \tag{8.1}$$

Equation (8.1) is an approximation of an unknown value of $f(A_i)$, $i = \overline{1,m}$.

We estimate the local weights by the corresponding PCMs.

Classical AHP is grounded on a formal procedure for estimating the weights of alternatives (subcriteria) [135, 141] from the empirical PCM. The procedure reduces to finding positive eigenvalues and the eigenvector of the PCM on the basis of Frobenius theorem [135, 139, 141]. We describe other methods of objects' weights estimation in [1].

Section 8.5.1 contains a brief description, in accordance with [1], of our modified AHP. Our method is based on the use of original combinatorial optimization models for the objects' weights estimation by empirical PCMs. Efficient alternatives weights estimation in AHP and its modifications allows us to formulate the classical problem of a non-formalized overall goal identification in the case when the alternatives are given in the form of real number vectors [1].

In Sect. 8.5.2 we describe two methods for unknown function reconstruction from its known redundant representation (with precision to the coefficients' values). In Sects. 8.5.2.1 and 8.5.2.2, we propose an efficient procedure for univariate and multivariate polynomial regression construction using an active experiment and normalized orthogonal polynomials of G. E. Forsythe (OPFs) [142, 143]. The procedure is based on the results given in [1, 136, 137]. This approach differs from the generally accepted methods for multivariate regression construction [144–148].

In Sect. 8.5.2.3 we propose a strict efficient procedure for an unknown function reconstruction from its redundant representation (within the precision of the coefficients' values) and the results of a passive experiment. The procedure is a generalization of the results given in [1]. The proposed method can be considered as inverse to the well-known classical method of group accounting of arguments [149, 150]. The latter method takes a redundant representation of an unknown function and proposes an original procedure of reasonable exclusion of components that do not belong to the function sought. This method is the most efficient in solving the problem of unknown overall goal reconstruction since it uses results of a passive experiment (we can apply an active experiment in exceptional cases). The results we present in Sects. 8.5.1–8.5.2.3, are of an independent nature and can be used to solve problems that go beyond the scope of the researches presented in this book.

8.5.1 *Modified Analytic Hierarchy Process*

8.5.1.1 Weights Estimation by an Empirical PCM Based on Optimization Models with Weighted Components of an Additive Functional

As we know, the efficiency of AHP [135] directly depends on the precision of local weights estimation by empirical PCMs. In the classical method [135], the weights estimation is based on the Frobenius theorem: if the elements γ_{ij} of an empirical PCM $(\gamma_{ij})_1^n$ define perfect weights $\omega_i > 0$, $i = \overline{1,n}$ (where n is the number of compared objects, $\gamma_{ij} = \omega_i/\omega_j$), then the following equation is true:

$$\begin{pmatrix} \frac{\omega_1}{\omega_1} & \frac{\omega_1}{\omega_2} & \cdots & \frac{\omega_1}{\omega_n} \\ \vdots & \vdots & \vdots & \vdots \\ \frac{\omega_n}{\omega_1} & \frac{\omega_n}{\omega_2} & \cdots & \frac{\omega_n}{\omega_n} \end{pmatrix} \begin{pmatrix} \omega_1 \\ \vdots \\ \omega_n \end{pmatrix} = n \begin{pmatrix} \omega_1 \\ \vdots \\ \omega_n \end{pmatrix}. \tag{8.2}$$

Also, in this case, the pair n, $\bar{\omega}$ is unique (precision of $\bar{\omega}$ is within a positive constant), and we efficiently calculate ω_i from the approximate formula [135]

$$\omega_i = \frac{\sqrt[n]{\prod_{j=1}^{n} \gamma_{ij}}}{\sum_{i=1}^{n} \sqrt[n]{\prod_{j=1}^{n} \gamma_{ij}}}. \tag{8.3}$$

If γ_{ij} do not correspond to the Eq. (8.2), then $\gamma_{\max}$ (the positive eigenvalue to which a positive eigenvector corresponds) of the matrix $(\gamma_{ij})_1^n$ is greater than n [135]. The difference $\gamma_{\max} - n$ characterizes whether the empirical PCM is well-conditioned or ill-conditioned.

It has been found experimentally that we can indeed efficiently estimate the weights by empirical PCMs with $n \leq 9$ using this method. At the planning stage, when we generate virtual alternatives with the goal to choose the best one for implementation, the number of alternatives may be substantially greater than nine. In addition, the matrix $(\gamma_{ij})_1^n$ may contain zeros (experts could not answer how many times the alternative i is better than the alternative j). If it is so, then Eq. (8.2) does not hold even in the perfect case. Finally, it is possible in some cases to obtain good estimates of initial weights even by an ill-conditioned empirical PCM. The problem is to find the criteria and the corresponding mathematical optimization models that guarantee the precision of the estimates obtained. In order to solve these problems, we presented the following results in [1, Chap. 7].

As we indicated in [1], we use a set A of PCM coefficients γ_{ij}, $i,j = \overline{1,n}$, as the initial data to construct and validate criteria and optimization models to estimate the weights ω_i, $i = \overline{1,n}$. The coefficients satisfy the conditions:

$$\left\{\gamma_{ii} \notin A \forall i = \overline{1,n}\right\} \overline{\subset} A;\ \forall \gamma_{ij} \in A\ \forall i \neq j \text{ if } \gamma_{ij} \geq 1;\ \gamma_{ji} \notin A \tag{8.4}$$

where γ_{ij} is the degree of advantage of object i over object j in relation to the specified goal (criterion); n is the number of objects to be compared.

For completely consistent PCMs, we have

$$\frac{\omega_i}{\omega_j} = \gamma_{ij},\quad \omega_i = \gamma_{ij}\omega_j\ \ \forall \gamma_{ij} \in A$$

where ω_i is the weight of object i; γ_{ij} is the empirical degree of advantage of object i over object j in relation to the specified goal (criterion).

Then, we can use the following measures as the consistency degree of ω_i/ω_j and $\gamma_{ij} \in A$:

$$\left|\omega_i - \gamma_{ij}\omega_j\right|; \tag{8.5}$$

$$\frac{1}{\gamma_{ij}}\left|\frac{\omega_i}{\omega_j} - \gamma_{ij}\right|. \tag{8.6}$$

The measure in Eq. (8.6) is more logical. However, its direct use leads to non-linear programming problems, that is, to actually inefficient mathematical models. Nevertheless [1], the measure (8.6) was used indirectly for weights estimation in some linear programming models.

In Chap. 7 of [1], we proposed and investigated 23 combinatorial optimization models to estimate the weights ω_i, $i = \overline{1,n}$, $\forall \gamma_{ij} \in A$. In particular, we use linear programming problems (LPPs):

- models 1 and 2 estimate the weights corresponding to $\min \sum_{(ij)\in|A|} \sum \left|\omega_i - \gamma_{ij}\omega_j\right|$;
- model 3 estimates the weights corresponding to $\min \max_{\forall (ij)\in|A|} \left|\omega_i - \gamma_{ij}\omega_j^*\right|$;
- model 6 guarantees the allowable consistency of given weights ω_{i_l}, $l = \overline{1,k}$ in relation to all objects and minimizes $\max\left|\omega_i - \gamma_{ij}\omega_j\right|$ for the remaining weights;
- model 7 is a sequential solution of LPPs of a special kind, each LPP minimizes the criterion $\sum_{(ij)\in|A|} \sum \frac{1}{\gamma_{ij}} \left|\frac{\omega_i}{\omega_j} - \gamma_{ij}\right|$ with a certain error, the calculation is terminated on the first LPP that has a feasible solution;

- models 8–14 estimate the weights ω_i, $i = \overline{1,n}$, by empirical PCMs with unilateral constraints:
  - (a) $\forall(ij) \in |A|, \frac{\omega_i}{\omega_j} \geq \gamma_{ij}$;
  - (b) $\forall(ij) \in |A|, \frac{\omega_i}{\omega_j} \leq \gamma_{ij}$;
  - (c) mixed case: $\forall\gamma_{ij}$ either (a) or (b) is satisfied;
- models 15–23 estimate the weights ω_i, $i = \overline{1,n}$, by inhomogeneous empirical PCMs, i.e., an arbitrary element γ_{ij} of PCM corresponds to one of four cases:
  - it shows how many times the weight of one object is greater than the weight of another object in relation to the specified goal (criterion);
  - it shows how much the weight of one object is greater than the weight of another object in relation to the specified goal (criterion);
  - it shows that one object is more preferable than the other (one weight is greater than the other);
  - there is no information on paired comparison of the objects.

The ambiguity in the use of optimization models requires to solve the following problem. Let k equally normalized sets of weights ω_i^l, $i = \overline{1,n}$, $l = \overline{1,k}$ be estimated for k different combinatorial optimization problems. We need to formalize the problem of choosing the best of these sets. We have solved this problem in [1] for all types of models 1–23. In particular, for models 1–14, the best set of weights is determined with a procedure that uses the following consistency coefficients:

$$M_1\left(\bar{\omega}^l\right) = \sum_{i=1}^{n} K\left(\omega_i^l\right),\ M_2\left(\bar{\omega}^l\right) = \max_{i,\ i=\overline{1,n}} K\left(\omega_i^l\right),\ l = \overline{1,k}$$

where

$$K\left(\omega_i^l\right) = \frac{1}{n-1} \sum_{\substack{j=1 \\ j \neq i}}^{n} \frac{1}{\gamma_{ij}} \left| \frac{\omega_i^l}{\omega_j^l} - \gamma_{ij} \right|. \tag{8.7}$$

Remark 8.1 If $\gamma_{ij} < 1$ is satisfied for some j, then the corresponding summand in (8.7) is replaced by $\frac{1}{\gamma_{ji}} \left| \frac{\omega_j^l}{\omega_i^l} - \gamma_{ji} \right|$. In other words, the consistency factor $K\left(\omega_i^l\right)$ for ω_i^l is determined by $\gamma_{ij} \geq 1$.

Models 1–6 do not take into account the practically always fulfilled empirical premise: for $\forall\gamma_{ij} \geq 1$ $(i \neq j)$, the more γ_{ij}, the more an expert is mistaken. This factor can be taken into account [1] if we multiply each component of the functional of the corresponding combinatorial optimization problem by a nonnegative coefficient r_{ij}. The more $\gamma_{ij} - 1$, the less the value of r_{ij} where γ_{ij} is taken from the empirical PCM.

In [1, 138], we investigated the following two combinatorial optimization models. Due to the computational experiments [138], these two proved to be more efficient than any of the models 1–7.

Model 24 is the following:

$$\min \sum_{(ij)\in|A|}\sum r_{ij}\cdot y_{ij}, \tag{8.8}$$

$$-y_{ij} \le \omega_i - \gamma_{ij}\omega_j \le y_{ij}, \quad y_{ij} \ge 0, \quad \omega_i \ge a \ge 1, \quad i = \overline{1,n} \tag{8.9}$$
$$(ij)\in|A|$$

where ω_i, $i = \overline{1,n}$, $y_{ij}, \forall(i,j) \in |A|$ are variables of LPP; r_{ij} is a weighted coefficient.

Optimal solution of LPP (8.8), (8.9) also corresponds to such weights ω_i^* $i = \overline{1,n}$, for which we achieve $\min \sum_{(ij)\in|A|}\sum r_{ij}\left|\omega_i - \gamma_{ij}\omega_j\right|$.

Model 25 is the following:

$$\min_{\substack{\Delta_{ij}^1,\,\Delta_{ij}^2\\ \forall(ij)\in|A|}} \Big(\sum_{(ij)\in|A|}\sum r_{ij}\Delta_{ij}^1 - C_{lm}\sum_{(ij)\in|A|}\sum r_{ij}\Delta_{ij}^2\Big), \tag{8.10}$$

$$\begin{aligned}
&\ln\gamma_{ij} + \Delta_{ij}^2 \le W_i - W_j \le \ln\gamma_{ij} + \Delta_{ij}^1, \quad \forall(ij) \in |A|,\\
&0 \le \Delta_{ij}^1 \le \ln(1 + l\cdot\Delta_1(l)), \quad \forall(ij) \in |A|,\\
&0 \ge \Delta_{ij}^2 \ge \ln(1 - m\cdot\Delta_2(m)), \quad \forall(ij) \in |A|,\\
&W_i \ge 0, \quad i = \overline{1,n}
\end{aligned} \tag{8.11}$$

where W_i, Δ_{ij}^1, Δ_{ij}^2 are variables of LPP, $W_i = \ln\omega_i$, $i = \overline{1,n}$; $|A|$ is the set of pairs (ij) each of which is an index for all $\gamma_{ij} \in A$; A is the set of coefficients γ_{ij}, $i,j = \overline{1,n}$, of the PCM that satisfy the conditions (8.4); C_{lm} is a coefficient determined from the expression

$$\ln(1 + l\cdot\Delta_1(l)) = C_{lm}\ln\frac{1}{(1 - m\cdot\Delta_2(m))};$$

γ_{ij} is the degree of advantage of object i over object j in relation to the specified goal (criterion); l and m are natural numbers that successively take values (1, 1), (1, 2), (2, 1), (2, 2) and so on; $\Delta_1(x)$ and $\Delta_2(x)$ are given numerical scalar functions of the natural argument, the functions take negative values; n is the number of objects to be compared.

We terminate the iterations after we obtain the first successful solution of the LPP (8.10), (8.11). After this, we estimate the objects' weights by expression

$$\omega_i^* = e^{W_i^*}, \quad i = \overline{1,n} \tag{8.12}$$

where ω_i^* are the objects' weights; W_i^* is a variable which value we determine when solving the problem (8.10), (8.11).

Remark 8.2 If $\forall r_{ij} \equiv 1$, then Model 24 uses measure (8.5) and Model 25 uses measure (8.6) [1].

To use these optimization models efficiently, we need to solve the following problems.

Problem 1 To determine and justify the consistency degree of a large dimension empirical PCM taking into account its possible incompleteness.

Problem 2 To determine the form and parameters of γ_{ij} perfect values probability distortion laws ($\gamma_{ij} \geq 1, i \neq j$). The perfect values depend on the parameter $\gamma_{ij} - 1$, and we should follow the principle: the larger the number $\gamma_{ij} - 1$, the greater the distortion.

Problem 3 To determine the best optimization model(s) with weighted components of the functional for each type of γ_{ij} perfect values distortion law by the results of statistical modeling. To formulate the rules for calculating weighted coefficients values taking into account that they are calculated from the distorted elements γ_{ij}^* (taken from the empirical PCM) and are the larger, the larger the number $\gamma_{ij}^* - 1$ ($\gamma_{ij}^* \geq 1,\ i \neq j$).

We have solved these problems in [1, 138] the following way.

PCMs' Consistency Degree Determination (Problem 1 Solving) To efficiently estimate alternatives' weights and find criteria by the proposed optimization models, we need to introduce a consistency indicator for large dimension and possibly incomplete PCMs. We proposed in [1] the following solution to this problem.

We use Model 25 ($\forall r_{ij} \equiv 1$) for the analysis. Suppose that we have estimated weights $\hat{\omega}_i$ by the empirical PCM with elements ij equal to $\gamma_{ij}^* \vee 0$ using Model 25 ($\forall r_{ij} \equiv 1$). After this, we estimate M weights by M PCMs. In each of the PCMs we replace $\gamma_{ij}^* > 0$, $\forall (ij) \in |A|$, with arbitrary γ_{ij}^{*l}, $\forall (ij) \in |A|$, $l = \overline{1,M}$, which satisfy the conditions

$$\gamma_{ij}^{*l} - \gamma_{ij}^* = \pm \left| \frac{\hat{\omega}_i}{\hat{\omega}_j} - \gamma_{ij}^* \right|, \quad \forall (ij) \in |A|$$

where M is a sufficiently large natural number.

Remark 8.3 The models coincide with those for which we have estimated $\hat{\omega}_i$, $i = \overline{1,n}$, except for the replacement of γ^*_{ij} with γ^{*l}_{ij}.

Suppose $\hat{\omega}^l_i$, $i = \overline{1,n}$, are the alternatives' weights estimates obtained by l-th optimization model (or l-th set of optimization models), $l = \overline{1,M}$. Weights $\hat{\omega}_i$, $\hat{\omega}^l_i$, $l = \overline{1,M}$, $i = \overline{1,n}$, are equally normalized.

The weights estimates $\hat{\omega}_i$, $i = \overline{1,n}$, are statistically justified if the numbers $\rho(\hat{\omega} - \hat{\omega}^l)$, $l = \overline{1,M}$ are admissible, where $\hat{\omega} = (\hat{\omega}_l, \ldots, \hat{\omega}_n)^{\mathsf{T}}$, $\hat{\omega}^l = (\hat{\omega}^l_1, \ldots, \hat{\omega}^l_n)^{\mathsf{T}}$, $l = \overline{1,M}$, and $\rho(\hat{\omega} - \hat{\omega}^l)$ can be interpreted as the difference degree of vectors $\hat{\omega}$ and $\hat{\omega}^l$ [1].

As indicated in [135], the number 15 is the maximum possible number of alternatives that allows you to correctly solve the practical problem of choosing the best alternative by classical AHP. Saaty proposed the consistency ratio $CR \leq 0.1$ (Eq. (8.14) below) to state that the consistency degree of the PCM is acceptable. So, this ratio can only be used for positive PCMs with dimensions up to 15×15.

In [1] we suggested to use a measure $d(\hat{\omega}, \hat{\omega}^l)$ in the following form as a criterion for a result obtained by the optimization models:

$$d(\hat{\omega}, \hat{\omega}^l) = \sqrt{\sum_{i=1}^{n} \left(\frac{\hat{\omega}_i}{\sqrt{\sum_{j=1}^{n} (\hat{\omega}_j)^2}} - \frac{\hat{\omega}^l_i}{\sqrt{\sum_{j=1}^{n} \left(\hat{\omega}^l_j\right)^2}} \right)^2}, \quad l = \overline{1,M} \tag{8.13}$$

where n is the number of objects being compared; $\hat{\omega}_i$, $i = \overline{1,n}$, are the reference weights of the objects; $\hat{\omega}^l_i$, $i = \overline{1,n}$, are the estimated weights of the objects.

We want to find the relationship between the acceptability indicator of the consistency degree of PCM's elements (the consistency ratio CR) proposed by Saaty (8.14) and the measure $d(\hat{\omega}, \hat{\omega}^l)$ $\forall \hat{\omega}^l$, $l = \overline{1,M}$.

Saaty suggested using the consistency index CI [139] as an indicator for the consistency degree of PCM's elements:

$$CI = \frac{\lambda_{\max} - n}{n - 1}$$

where CI is the consistency index; $\lambda_{\max}$ is the eigenvalue of the PCM; n is the dimension of the PCM.

To state the acceptability of the consistency degree, T. Saaty proposes in [135] to use the consistency ratio CR:

$$CR = \frac{CI}{CIS} \tag{8.14}$$

where CR is the consistency ratio; CI is the consistency index; CIS is the average value of CRs calculated for a large number of randomly generated PCMs in the Fundamental Scale.

We can calculate the value of CIS by the formula [141]:

$$CIS = \frac{1.98 \cdot (n-2)}{n}.$$

Saaty recommended [135] to consider the weights to be reliable if $CR \leq 0.1$.

In [138], we presented the experimental results showing that the condition $CR \leq 0.1$ corresponds to the relation $d(\hat{\omega}, \hat{\omega}^l) \leq 0.12 \; \forall l = \overline{1, M}$. In the experiment, we chose M to be 1000.

Solutions to Problems 2 and 3 with detailed descriptions of computational experiments, conclusions, and recommendations are presented in [138], in particular, for the following distributions:

- normal distribution with expectation $\mu = 0.5$ and variance $\sigma = 0.2$;
- normal distribution with expectation $\mu = 0.25$ and variance $\sigma = 0.2$;
- gamma distribution with parameters $k = 3$, $\theta = 2$;
- beta distribution with parameters $\alpha = 2$, $\beta = 2$;
- beta distribution with parameters $\alpha = 0.5$, $\beta = 0.5$;
- Rayleigh distribution with the parameter $\sigma = 3$;
- exponential distribution with the parameter $\lambda = 1.5$.

In the experiments, we calculated noisy PCM's elements from the relations [1]

$$\gamma_{ij}^* = k_{ij} + \gamma_{ij}, \quad k_{ij} = nc_{\max} \cdot \frac{(\gamma_{ij} - 1)}{8} \cdot rnd$$

where rnd is the random coefficient obtained in accordance with the specified distribution law. Thus, the noisiness of the PCM's element depends on the random coefficient obtained according to a given probability distribution, the expectation of which is the greater, the larger the number $\gamma_{ij} - 1$, $\gamma_{ij} \geq 1$, $i \neq j$. In [138], we have found the bounds for all parameters that specify γ_{ij}^*. The bounds exceeding leads to inefficiency in solving the objects' weights estimation problem.

We have got the following results [138]:

- the choice of the function to calculate r_{ij} does not depend on the parameters of the probability distribution if the noisiness degree of PCM's elements is not greater than 30%;
- for all cases of probabilistic distribution of the noisiness of PCM's elements, we recommend to use one of the following functions to calculate the value of r_{ij}: either $r_{ij} = \frac{a^3}{a^2 + (\gamma_{ij}^* - 1)^2}$, $a = 10$, or $r_{ij} = \frac{1}{\sqrt[3]{(\gamma_{ij}^* - 1)^2}}$. The efficiency of these functions is practically invariant to probability distributions;

- in the case when the noisiness of the PCM's elements corresponds to one of the probabilistic laws being studied, but we do not know to which one, we recommend to use the optimization Model 24. As a result of the experiments [1, 138], Model 25 should be applied only if $\forall r_{ij} \equiv 1$ (Model 7 [1]) for the case when the dependence of the expert information's distortion on the value of $\gamma_{ij} \geq 1$ is unknown. This follows from the fact that in this case the model approximately minimizes the criterion (8.7), and thus, in a particular case, reflects the dependence of the decrease in the corresponding component of the functional on the increase of γ_{ij}.

8.5.1.2 Analytic Hierarchy Process Modification

The use of discrete optimization models allowed us to modify the AHP first for a two-level structure and then generalize the results to an arbitrary hierarchy tree structure [1].

The two-level structure has the form shown in Fig. 8.2 where E_1^1 is the overall goal; $A_1 \ldots A_m$ are the alternatives, and we need to choose the best of them; $E_1^1(A_i) = \omega_i$, $i = \overline{1, m}$.

We estimate the weights ω_i, $i = \overline{1, m}$, by the empirical PCM $\left(\gamma_{ij}\right)_1^m$.

We have developed several modifications of AHP: the first version, the second version implementations 1 and 2, the third version, the fourth version implementations 1–4 [1]. The foundation for all the modifications is the following idea. As we showed by statistical studies [1], optimization models efficiently estimate the weights even by ill-conditioned PCMs (for large values of m). Nevertheless, they may wrongly determine the best alternative. We proposed to estimate the best (not more than sixteen) weights by the initial PCM, to construct for them a new empirical PCM, and to determine the best alternative by the latter PCM.

In Sect. 7.4.2 of [1], we present the modified AHP (the general case). We now give only one of all the results expounded there.

The classical use of the AHP assumes that the values of $\omega_{E_t^{s-j}}^{E_l^{s-j+1}}$ for all j, l, t do not depend on alternatives $A_1 \ldots A_m$. Often this is not the case. Then, the direct use

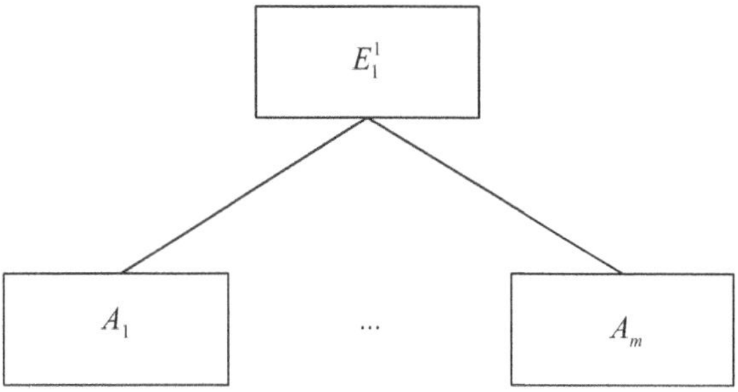

Fig. 8.2 A hierarchy tree with the two-level structure

of classical AHP is reduced to a two-level hierarchy. This can sometimes significantly hinder the efficient solving of the problem. However, the classical scheme is naturally generalized to this case.

Suppose we have estimated the weights $\omega^{i}_{E^s_j}$, $i=\overline{1,m}$, $j=\overline{1,m_s}$, of the last level of the hierarchy tree. Then we construct m sets of weights $\left\{\omega^{E^{s-j+1}_l}_{E^{s-j}_t}\ \forall l,t,j\right\}$ from empirical PCMs under the assumption that each i-th set, $i=\overline{1,m}$, corresponds to the realization of the alternative A_i. That is, we estimate the contribution (weight) of the criterion E^{s-j+1}_l into the criterion E^{s-j}_t assuming that we implement the alternative A_i, i.e., we estimate the weights $\omega^{E^{s-j+1}_l}_{E^{s-j}_t}(A_i)$ by the PCMs. The corresponding normalizations of the weights of each set are the same. Then, for each set of weights $\left\{\omega^{E^{s-j+1}_l}_{E^{s-j}_t}(A_i),\ \forall l,t,j\right\}$, $i=\overline{1,m}$, we estimate the resulting weights of the alternatives $E^1_1(A_i)=\omega_i(A_i)=\sum_{l=1}^{m_2}\omega^{E^2_l}_{E^1_1}(A_i)E^2_l(A_i)$ [formula (8.1)]. The intermediate formula has the form

$$E^{s-j}_t(A_i)=\sum_{l=1}^{m_{s-j+1}}\omega^{E^{s-j+1}_l}_{E^{s-j}_t}(A_i)E^{s-j+1}_l(A_i).$$

We estimate the weights $\omega^{i}_{E^s_j}$, $i=\overline{1,m}$, $j=\overline{1,m_s}$, of the last level of the hierarchy once during $\omega_i(A_i)$ estimation from the i-th set of weights. The maximum weight $\max_i \omega_i(A_i)$ corresponds to the best alternative.

This situation is possible, for example, when the overall goal has the form $f\left(A_i, W^{A_i}_1, \ldots, W^{A_i}_{k_i}\right)$ where $W^{A_i}_1, \ldots, W^{A_i}_{k_i}$ are the factors formalizing the overall goal if we consider the alternative A_i. We assume that the structure of the hierarchy tree does not depend on the alternative, only the weights of subcriteria vary. We give in [1] an example of such a hierarchy tree. A particular case: an alternative defines a subset of subcriteria at each level of the hierarchy, starting with the second level.

8.5.2 *Methods for an Unknown Function Reconstruction from Its Redundant Representation*

8.5.2.1 Univariate Polynomial Regression Construction from a Redundant Representation Using an Active Experiment

General Known Provisions of Univariate Regression Analysis Suppose that the regression model has the form

$$Y(x) = \theta_0 + \theta_1 x + \cdots + \theta_r x^r + E \tag{8.15}$$

where x is a deterministic variable whose value a researcher can set arbitrarily in experiments; θ_i, $i = \overline{0, r}$, are unknown coefficients; E is a random variable with an arbitrary distribution; $ME = 0$ (M is a sign of mean or mathematical expectation); σ_E^2 (variance) is limited.

We have performed n experiments. Their result are two samples of volume n: $\left(x_i, i = \overline{1, n}; Y(x_i) = y_i, i = \overline{1, n}\right)$.

According to (8.15),

$$y_i = \sum_{j=0}^{r} \theta_j x_i^j + \delta_i, \quad i = \overline{1, n} \tag{8.16}$$

where δ_i is an unknown realization of the random variable E in experiment i. We can consider the numbers y_i, δ_i as realizations of random variables Y_i, $i = \overline{1, n}$, and Δ_i, $i = \overline{1, n}$. Here, Δ_i, $i = \overline{1, n}$, are independent random variables that has the same distribution with the random variable E; Y_i, $i = \overline{1, n}$, are independent random variables with the variance σ_E^2; and the relationship between Y_i and Δ_i is

$$Y_i = \sum_{j=0}^{r} \theta_j x_i^j + \Delta_i. \tag{8.17}$$

We estimate unknown coefficients θ_j, $j = \overline{0, r}$, minimizing the expression

$$\min_{\theta_j,\, j = \overline{0, r}} \sum_{i=1}^{n} \left(y_i - \sum_{j=0}^{r} \theta_j x_i^j \right)^2 \tag{8.18}$$

We introduce a matrix notation:

$$A = \begin{pmatrix} 1 x_1 \ldots x_1^r \\ \ldots\ldots\ldots \\ 1 x_n \ldots x_n^r \end{pmatrix}; \quad y = (y_1 \ldots y_n)^\mathsf{T}$$

$$Y = (Y_1 \ldots Y_n)^\mathsf{T}; \; \theta = (\theta_0 \ldots \theta_r)^\mathsf{T};$$

$$\hat{\theta} = \left(\hat{\theta}_0 \ldots \hat{\theta}_r\right)^\mathsf{T}$$

where $\hat{\theta}_j$ are the estimates of θ_j, $j = \overline{0, r}$, according to (8.18).

Then [142, 143], we consider either

$$\hat{\theta} = \left(A^\mathsf{T} A\right)^{-1} A^\mathsf{T} y \tag{8.19}$$

or $\hat{\theta} = \left(A^{\mathsf{T}}A\right)^{-1}A^{\mathsf{T}}Y$ if $\hat{\theta}_j$, $j = \overline{0, r}$, as random variables. The difficulties associated with the matrix $\left(A^{\mathsf{T}}A\right)^{-1}$ inversion disappear if we move from the model (8.15) to the regression model specified by normalized OPFs [142, 143]:

$$\begin{aligned} Y(x) &= \omega_0\theta_0(x) + \omega_1\theta_1(x) + \cdots + \omega_r\theta_r(x) + E; \\ Q_j(x) &= q_{j0} + q_{j1}x + \cdots + q_{jj}x^j; \end{aligned} \tag{8.20}$$

$$\sum_{i=1}^{n} Q_j^2(x_i) = 1, \ \sum_{i=1}^{n} Q_j(x_i)Q_l(x_i) = 0 \ \forall j \neq l, q, l = \overline{0, r}.$$

According to [142, 143], we give the following formulas.

$$\hat{\theta}_j = \hat{\omega}_r q_{rj} + \cdots + \hat{\omega}_j q_{jj}, \quad j = \overline{0, r}$$

where $\hat{\theta}_j$ are the estimates of the regression polynomial coefficients; $\hat{\omega}_j$ are the estimates of weight coefficients obtained by the least squares method; q_{jj} are the coefficients of the normalized OPFs.

The variances of the coefficients' estimates have the form

$$Var\left(\hat{\theta}_j\right) = \sigma^2 \sum_{i=r}^{j} q_{ij}^2. \tag{8.21}$$

We calculate the estimates of weight coefficients by the formula [142, 143]:

$$\hat{\omega}_j = \sum_{i=1}^{n} y_i Q_j(x_i), \quad j = \overline{0, r} \tag{8.22}$$

where $Q_j(x_i)$ is the value of j-th normalized OPF for the argument x_i.

We calculate the coefficients of the normalized OPFs by the recurrent formula (8.23) [142, 143]:

$$\lambda Q_j(x) = xQ_{j-1}(x) - \alpha Q_{j-1}(x) - \beta Q_{j-2}(x) \tag{8.23}$$

where α, β, λ are the coefficients we define as follows:

$$\alpha = \sum_{i=1}^{n} x_i Q_{j-1}^2(x_i), \tag{8.24}$$

$$\beta = \sum_{i=1}^{n} x_i Q_{j-1}(x_i) Q_{j-2}(x_i), \tag{8.25}$$

$$\lambda = \sqrt{\sum\nolimits_{i=1}^{n} \left(x_i Q_{j-1}(x_i) - \alpha Q_{j-1}(x_i) - \beta Q_{j-2}(x_i)\right)^2}. \tag{8.26}$$

We calculate the first two orthogonal polynomials by formulas

$$Q_0(x) = \frac{1}{\sqrt{n}}, \tag{8.27}$$

$$Q_1(x) = -\frac{\bar{x}}{\sqrt{\sum\nolimits_{i=1}^{n} (x_i - \bar{x})^2}} + \frac{x}{\sqrt{\sum\nolimits_{i=1}^{n} (x_i - \bar{x})^2}}. \tag{8.28}$$

The Problem Statement In (8.15), r is a redundant number, and the model (8.15) contains extra terms. We need to reconstruct the true expression for the regression line from the redundant representation (8.15) excluding from (8.15) the terms for which $\theta_j = 0$. If E has normal distribution, then we can solve the problem eliminating the corresponding coefficients $\hat{\omega}_j$ if corresponding to it Fisher criterion falls into the critical region [142, 143]. The problem arises if E has an arbitrary distribution. We need to know the upper bound of the estimate for $\sigma^2 = Var(E)$ to solve the problem. We present in [1] the results of $Var\left(\hat{\theta}_j\right)$ calculation as a function of σ^2 by formula (8.21) at $j = \overline{0,5}$ when the values of x_i, $i = \overline{1,n}$, are distributed with an equal step along the segment (–50, 50). The coefficients q_{jj} were calculated using Eqs. (8.20), (8.23)–(8.28). Analysis of Table 6.1 [1] shows that:

1. The larger j, the smaller $Var\left(\hat{\theta}_j\right)$ for fixed n. Indeed, at $n = 10$ we have

$$Var\left(\hat{\theta}_0\right) = \sigma^2 \times 0.400466, \quad Var\left(\hat{\theta}_1\right) = \sigma^2 \times 0.0024855,$$
$$Var\left(\hat{\theta}_2\right) = \sigma^2 \times 4.26 \times 10^{-6}, \quad \ldots Var\left(\hat{\theta}_5\right) = \sigma^2 \times 1.28 \times 10^{-15},$$

   i.e., increasing j decreases the value of $Var\left(\hat{\theta}_j\right)$ by an order of magnitude. This implies the following result.
2. We can determine the true power of the regression polynomial with minimal number of tests if the unknown non-zero coefficients θ_j are not small numbers in absolute value. In our example at $n = 10$, the variance of the coefficient estimate at x^2 is already equal to $\sigma^2 \times 4.26 \times 10^{-6}$. That is, if the true regression is linear, then the estimates $\hat{\theta}_2$, $\hat{\theta}_3$, $\hat{\theta}_4$, $\hat{\theta}_5$ are actually zero to within the corresponding digits after the decimal point (the three-sigma rule for the normal distribution and the use of the Chebyshev's inequality in the general case).

3. Reliable estimates θ_0, θ_1 require a very large number of experiments n ($Var\left(\hat{\theta}_1\right) = \sigma^2 \times 4.59 \times 10^{-5}$ for $n = 500$). The decision to exclude θ_0 or θ_1 from the Eq. (8.15) using (8.21) for small n is not constructive.
4. We calculate the values of variances $Var\left(\hat{\theta}_j\right)$ before an active experiment if an area $[a,\ b]$ of change of the deterministic argument x and a step Δ were given.

Theoretical and Experimental Substantiation of Recommendations for Conducting an Active Experiment Due to the pt. 4, we can find $Var\left(\hat{\theta}_j\right)$ before performing an active experiment by choosing the change area of the deterministic argument x. Therefore, it becomes necessary to develop valid recommendations for determining the change areas of x in an active experiment and for its most efficient implementation. In [136, 137] we give solutions to the following problems for a univariate polynomial regression reconstruction from its redundant representation.

A Choice of an Active Experiment Area We can distinguish six conditional areas for an experiment:

1. The beginning is greater than zero, a small interval. For example, $[10,\ 20]$.
2. The beginning is greater than zero, a large interval. For example, $[10,\ 1010]$.
3. The beginning is zero, a small interval. For example, $[0,\ 10]$.
4. The beginning is zero, a large interval. For example, $[0,\ 1000]$.
5. A small interval symmetric around zero. For example, $[-5,\ 5]$.
6. A large interval symmetric around zero. For example, $[-500,\ 500]$.

In tables [136], we give the variances for the estimates of regression polynomial coefficients calculated from formulas (8.20), (8.23)–(8.28). It follows from these tables that the wider the interval, the less the variance of the coefficients' estimates. Also, the variances decrease when the experiment interval is symmetric around zero. If negative values are not allowed, then we recommended to start the interval for $\hat{\theta}_j$ at the point zero. The form of the interval significantly affects the variance of the coefficients' estimates, even more than the change in the number of input points.

Because of this, the problem arises to efficiently recalculate before an active experiment the coefficients of the OPFs q_{ij} calculated for x_i, $i = \overline{1, n}$, if we scale the deterministic variable x, i.e. replace it with the deterministic variable $z = kx$ where k is the scaling factor; $z_i = kx_i$, $i = \overline{1, n}$.

In [136], we have obtained the following result: if q^x_{jl} $\forall j = \overline{0, r}$, $\forall l = \overline{0, j}$ are the coefficients of the normalized OPFs calculated with respect to x_i, $i = \overline{1, n}$, and q^z_{jl} $\forall j = \overline{0, r}$, $\forall l = \overline{0, j}$ are the coefficients of the normalized OPFs calculated with respect to z_i, $i = \overline{1, n}$, $z_i = kx_i$, $i = \overline{1, n}$, then

$$\underset{\forall j=\overline{0,r},\ \forall l=\overline{0,j}}{q^z_{jl} = \frac{1}{k^j} q^x_{jl}}, \quad Var\left(\hat{\theta}^z_j\right) = \sigma^2 \sum_{l=r}^{j} \left(\frac{1}{k^j} q^x_{jl}\right)^2 = \left(\frac{1}{k^j}\right)^2 Var\left(\hat{\theta}^x_j\right) \tag{8.29}$$

Equation (8.29) significantly simplifies the procedure for automating the search for the most efficient possible area of change of the deterministic variable x before an active experiment, in terms of minimizing the expressions

$$Var\left(\hat{\theta}_j\right) = \sigma^2 \sum_{l=r}^{j} q_{jl}^2, \quad j = \overline{0, r}. \tag{8.30}$$

Repeating Experiments In practical problems, a researcher has to perform a large number of experiments on a small range of the arguments' values. If the values have the uniform distribution, then some inconvenience arise because we need to perform each subsequent experiment changing the argument's value by some small number Δx. This is hard to implement in practice when studying real physical phenomena. In this case, we consider convenient to submit to the object's input the repeating sequence of the values $x_1, \ldots, x_{r+p}, x_1, \ldots, x_{r+p}, \ldots$ where $p \geq 1$, r is the power of the univariate polynomial defined with a redundant representation. It turns out that in this case we do not need to construct OPFs from the values sequence of the deterministic argument x: $x_1, \ldots, x_{r+p}, x_1, \ldots, x_{r+p}, \ldots$ in which the numerical sequence $x_1, \ldots, x_{r+p}$ is repeated l times.

We have proved in [137] that the coefficients' estimates $\hat{\theta}_j$, $j = \overline{0, r}$, will not change when averaging the results of the experiments if we perform a univariate polynomial regression analysis using normalized OPFs for the input values x_1, x_2, $\ldots$, x_{r+p}. Let us introduce the notation:

$$X = (x_1, x_2, \ldots, x_n), \quad n = r + p,$$
$$X' = (x_{11}, x_{21}, \ldots, x_{l1}, x_{12}, x_{22}, \ldots, x_{l2}, \ldots, x_{1n}, x_{2n}, \ldots, x_{ln}).$$

There are l copies of X in X', i.e., $x_{ki} = x_i \; \forall k = \overline{1, l}$, $\forall i = \overline{1, n}$;

$$Y' = (y_{11}, y_{21}, \ldots, y_{l1}, \ldots, y_{1n}, y_{2n}, \ldots, y_{ln});$$
$$Y = \left(\frac{\sum_{k=1}^{l} y_{k1}}{l}, \frac{\sum_{k=1}^{l} y_{k2}}{l}, \ldots, \frac{\sum_{k=1}^{l} y_{kn}}{l} \right).$$

In other words, instead of two vectors X' and Y', each of dimension $n \times l$, we use vectors X and Y of dimension n for univariate regression analysis where n is the number of different argument values in each series, $i = \overline{1, l}$.

Let $Q_j(x)$, $j = \overline{0, r}$ be the normalized OPFs constructed on X, $Q_j'(x)$, $j = \overline{0, r}$ be the normalized OPFs constructed on a sample X'.

We showed in [137] that

$$Q_j'(x) = \frac{Q_j(x)}{\sqrt{l}}, \; j = \overline{0, r}, \; \left(\underset{k=\overline{1,l}}{x_{ki} = x_i} \right). \tag{8.31}$$

From this it follows [137] that the estimates $\hat{\theta}_j' = \hat{\theta}_j$, $j = \overline{0, r}$ where $\hat{\theta}_j'$ were obtained by X', Y' and $\hat{\theta}_j$ by X, Y. We can formulate the regression problem solved on X, Y data in an equivalent way: $Y(x) = \theta_0 + \theta_1 x + \cdots + \theta_r x^r + E_1$ where x takes the values $x_1, \ldots, x_n$, $n = r + p$, $p \geq 1$; $Var(E_1) = \sigma^2/l$. Therefore, we can minimize $Var\left(\hat{\theta}_j\right)$, $j = \overline{0, r}$, as much as possible before the active experiment varying the parameters. Namely, we can find the most efficient area which we choose the values $x_1, \ldots, x_n$ from and choose the number of repetitions l in the selection of the series $x_1, \ldots, x_n$.

8.5.2.2 Multivariate Polynomial Regression Construction from a Redundant Representation Using an Active Experiment

We presented in [1] a method for a multivariate polynomial regression construction from a redundant representation in the conditions of an active experiment. The method is based on reducing the multivariate regression to a sequence of univariate regressions by fixing all variables except one. The drawback of the method is that univariate regressions have a large number of terms with small powers at x. As we showed above, the coefficients of such polynomials are poorly restored.

In this section we give a modified method for a multivariate polynomial regression construction based on the ideas presented in [1].

Let the multidimensional model be given in the form

$$y(\bar{x}) = \sum_{\forall (i_1 \ldots i_t) \in K} \; \sum_{\forall (j_1 \ldots j_t) \in K(i_1 \ldots i_t)} b_{i_1 \ldots i_t}^{j_1 \ldots j_t} (x_{i_1})^{j_1} \cdot (x_{i_2})^{j_2} \cdot \ldots \cdot (x_{i_t})^{j_t} + E \tag{8.32}$$

where $\bar{x} = (x_1, \ldots, x_n)^{\mathsf{T}}$ is the deterministic vector of input variables; x_i is the i-th component of the vector $\bar{x}$; $b_{i_1 \ldots i_t}^{j_1 \ldots j_t}$ are unknown coefficients; j_l are natural numbers; i_l are natural indices from the set $\{1, \ldots, n\}$; E is a random variable with zero expectation and a limited unknown variance σ_E^2 (as well as in univariate case, we may know the upper bound of σ_E^2).

The model (8.32) is redundant: perhaps some of the coefficients $b_{i_1 \ldots i_t}^{j_1 \ldots j_t}$ are zero.

Suppose that all the values of the arguments change identically: $x_1 = x_2 = \cdots = x_n = x$. Then, under such a scheme, the multivariate regression becomes a univariate regression of the form

$$\theta_0^0 + \theta_1^0 x + \theta_2^0 x^2 + \cdots + \theta_{n_0}^0 x^{n_0}. \tag{8.33}$$

We estimate the coefficients $\theta_0^0, \theta_1^0, \ldots, \theta_{n_0}^0$ using the results of univariate regression analysis.

The model (8.33) allows us to construct a system of $n_0 + 1$ linear equations associating numbers $\hat{\theta}_0^0, \hat{\theta}_1^0, \ldots, \hat{\theta}_{n_0}^0$ with the coefficients $b_{i_1 \ldots i_t}^{j_1 \ldots j_t}$ of the model (8.32).

The left side of the i-th equation includes coefficients at the i-th power of x, the right side contains the estimates $\hat{\theta}_0^0, \hat{\theta}_1^0, \ldots, \hat{\theta}_{n_0}^0$.

Next, we fix one variable $x_1 = x_{1f}$. We equally change the remaining variables: $x_2 = \cdots = x_n = x$. Similarly to the previous step, we have obtained a univariate regression of the form $\theta_0^1 + \theta_1^1 x + \theta_2^1 x^2 + \cdots + \theta_{n_1}^1 x^{n_1}$.

Having obtained the estimates $\hat{\theta}_0^1, \hat{\theta}_1^1, \ldots, \hat{\theta}_{n_1}^1$, we can construct a linear equations system similar to the first case. However, not only the coefficients of the initial model but also the fixed variable x_{1f} will now be on the equations' left side.

Similarly, we successively perform active experiments for $i = \overline{2, n}$ followed by the regression analysis application.

We need to combine the obtained systems and solve them to estimate the original model coefficients.

Let us write the resulting system in the matrix form:

$$Ab = \theta \tag{8.34}$$

where A is the matrix of the system coefficients; b is the vector of unknown coefficients $b_{i_1 \ldots i_t}^{j_1 \ldots j_t}$ of the model (8.32), the dimension of b is r; θ is a vector of coefficients of univariate regressions.

Solving the system (8.34) by the least squares method, we will surely obtain the coefficients' estimates if there are r linearly independent rows in the matrix A (where r is the number of unknown coefficients of the original model excluding the free term). Since the matrix A does not have complex numbers by construction, then we can use the following fact [151] to check the linearly independent rows number:

$$rank\left(A^{\mathrm{T}} A\right) = rank(A) \tag{8.35}$$

If $\det\left(A^{\mathrm{T}} A\right) \neq 0$, then $rank(A) = r$. In this case, the matrix $A^{\mathrm{T}} A$ is nondegenerate, and the solution (8.34) is nondegenerate and unique.

If the system (8.34) has r linearly independent rows, i.e., $rank(A) = r$, then we can state that the coefficients of the original model will be obtained by the least squares method. Otherwise, there is no solution to the problem.

Remark 8.4 We solve the problem of the right-hand sides vector precision as follows. We showed in [1] that the estimates of univariate regressions' coefficients with small powers have a large variance. Thus, choosing linear equations from univariate regressions, it is good to add to the general system not all the equations but only those whose right parts contain θ_i^j near x to at least the second power. This guarantees a high precision of the right-hand sides vector and exact estimation of unknown multivariate regression coefficients. If we cannot collect r linearly independent rows in the system using such an approach, then we can include in it the equations containing θ_i^j near x to the first power at the right side. The system should not contain equations corresponding to the free term in the constructed univariate regressions. The reason is: the estimation of the free term of a univariate regression has a large variance and will worsen the quality of the estimation of the multivariate regression's coefficients. We estimate the free term of the original model using the estimates of the remaining coefficients. It is the arithmetic mean of the deviations from the input data of the regression model values without the free term.

Remark 8.5 In practice, situations arise when the number of linearly independent rows is less than the number of unknown coefficients. In this case, we can find a partial solution by choosing from the obtained system a new nondegenerate subsystem with fewer variables. It is convenient to use a human-machine procedure because a person can see and select such subsystems. Estimation of some coefficients of the initial model significantly simplifies it and allows you to solve a smaller regression problem by any other method.

Remark 8.6 We have proposed a method to reduce the multivariate polynomial regression construction to a sequence of univariate regression constructions and solving a system of linear equations with constant coefficients by two approaches:

1. $x_1 = x_2 = \cdots = x_n = x$;
2. $x_i = const,\ x_1 = \cdots = x_{i-1} = x_{i+1} = \cdots = x_n = x$.

The method is of a universal nature. However, the analysis of a particular redundant representation allows to apply another methods of reducing to univariate polynomial regressions construction. For example, we fix some subset of variables and change the other variables equally. We must always have in mind that the maximum power in a univariate polynomial regression should be greater than one, everything else being equal. At the same time the power must be the lowest possible because the greater the power in a univariate polynomial regression, the more difficult it is to minimize errors in the normalized OPFs construction (if we use sequential implementation of division operations).

Remark 8.7 If the arithmetic roots of the variance of the vector θ's coefficients are sufficiently small in the resulting linear equations system, and the estimates of some coefficients from the redundant representation (8.33) are modulo close to zero, then we can exclude the corresponding terms from the redundant representation (8.32) with a high degree of reliability. In [137], we gave examples and results of statistical experiments that illustrate and confirm the proposed method's efficiency.

Examples We now illustrate the method of substantial simplification of constructing a multivariate polynomial regression given by a redundant representation.

Preliminary Comments Before carrying out an active experiment, we need to determine in advance the area of change of the deterministic argument x using the above results, i.e., formulas (8.21)–(8.28), (8.31). The area should include the values $x > 1$ and guarantee a practically exact estimate of unknown coefficients θ_j, $j \geq 2$, in univariate regressions if we know the upper bound of $Var(E)$. The use of (8.31) allows to substantially simplify this problem solving.

Example 1 The redundant representation of a multivariate regression has the form

$$Y(X) = a_0 + \sum_{i=1}^{4} a_i x_i + a_5 x_1 x_2 + a_6 x_3 x_4^2 + a_7 x_1 x_2 x_3^2 + a_8 x_1 x_2 x_3 x_4 + E.$$

1. Let $x_i = x$, $i = \overline{1,4}$. We obtain:

$$Y(X) = a_0 + x \sum_{i=1}^{4} a_i + a_5 x^2 + a_6 x^3 + (a_7 + a_8) x^4 + E.$$

$$\theta_0 = a_0, \quad \theta_1 = \sum_{i=1}^{4} a_i, \quad \theta_2 = a_5, \quad \theta_3 = a_6, \quad \theta_4 = a_7 + a_8.$$

Having estimated $\hat{\theta}_j$, $j = \overline{0,4}$, we have obtained estimates of a_0, a_5, a_6 and two linear equalities to estimate a_i, $i = \overline{1,4}$, a_7, a_8. If the number of experiments n is large, then the estimates $\hat{\theta}_j$, $j = \overline{1,4}$ are practically reliable, otherwise we should use only estimates $\hat{\theta}_j$, $j = \overline{2,4}$.

2. Let $x_3 = x_{3f}$, $x_1 = x_2 = x_4 = x$. We obtain the univariate regression

$$a_0 + x a_3 x_{3f} + (a_1 + a_2 + a_4) x + \left(a_5 + a_6 x_{3f} + a_7 x_{3f}^2\right) x^2 + a_8 x_{3f} x^3 + E.$$

$$\theta_0 = a_0, \quad \theta_1 = a_1 + a_2 + a_4, \quad \theta_2 = a_5 + a_6 x_{3f} + a_7 x_{3f}^2, \quad \theta_3 = a_8 x_{3f}.$$

We estimate a_7 by the estimate $\hat{\theta}_2$ and a_8 by the estimate $\hat{\theta}_3$. We also obtain the equality $\hat{\theta}_1 = a_1 + a_2 + a_3$. Similarly, we find the required number of equalities to estimate the coefficients a_i, $i = \overline{1,4}$, for the case when n is sufficiently large. In the opposite case, we have reliably estimated the coefficients a_5, a_6, a_7, a_8, eliminated those modulo close to zero and reduced the problem of a nonlinear polynomial regression construction from a redundant representation to the problem of a linear multivariate regression construction from its redundant (linear) representation.

Example 2 (illustration to Remark 8.6). The redundant representation of a multivariate regression has the form

$$Y(X) = a_0 + a_1x_1 + a_2x_2 + a_3x_3 + a_4x_1x_2 + a_5x_1x_3 + a_6x_1^5x_2x_3^2x_4^8 + a_7x_1^3x_2x_3^3x_4^{10} + E.$$

Let $x_1 = x_{1f}$, $x_4 = x_{4f}$, $x_2 = x_3 = x$. The univariate regression has the form

$$a_0 + a_1x_{1f} + (a_2 + a_3 + a_4x_{1f})x + a_5x^2 + a_6x_{1f}^5x_{4f}^8x^3 + a_7x_{1f}^3x_{4f}^{10}x^4 + E.$$

$$\theta_0 = a_0 + a_1x_{1f}, \quad \theta_1 = a_2 + a_3 + a_4x_{1f}, \quad \theta_2 = a_5, \quad \theta_3 = a_6x_{1f}^5x_{4f}^8, \quad \theta_4 = a_7x_{1f}^3x_{4f}^{10}.$$

Hence, $\hat{a}_5 = \hat{\theta}_2$, $\hat{a}_6 = \hat{\theta}_3 \Big/ x_{1f}^5x_{4f}^8$, $\hat{a}_7 = \hat{\theta}_4 \Big/ x_{1f}^3x_{4f}^{10}$.

Setting $x_3 = x_{3f}$, $x_4 = x_{4f}$, $x_1 = x_2 = x$, we obtain a univariate regression in which $\hat{\theta}_2 = \hat{a}_4$ (such an estimate is sufficiently precise for the minimum number of experiments). We have again reduced the problem to a multivariate linear regression construction from a redundant linear representation.

Remark 8.8 We recommend to use repetitive experiments (8.31) to construct an active experiment with a sufficiently large number of tests. This will ensure practically reliable estimation of the coefficients in the linear part of a multivariate polynomial regression (except for the free term).

We should note that we can design an active experiment on the basis of repeated data only when we know a redundant representation for a specific multivariate polynomial regression. This specific representation should allow us to determine maximal powers of univariate polynomial regressions that define the amount of repeated data.

8.5.2.3 Unknown Function Reconstruction from a Redundant Representation and the Results of a Passive Experiment with a Limited Dataset

The methods we present below are a generalization of our previous results [1].

The Problem Statement We consider the problem of reconstructing a numerical scalar function of real arguments that uniquely determines a certain relationship. The source is the analysis of a limited amount of observable data (input–output). In other words, a certain relationship is uniquely defined by a function $f(\bar{x})$ not known to the observer. We have the results of a passive experiment $f(\bar{x}^i) = y_i$, $i = \overline{1, P}$, where P is a small number. We need to reconstruct the relationship, that is, to find the true function $f(\bar{x})$, not its approximation, using the results of the passive experiment.

The problem stated in this form is incorrect. We will solve it in the following special formulation.

We consider the class of functions

$$\sum_{i=1}^{L} a_i \psi_i(\bar{x}) \tag{8.36}$$

where $\psi_i(\bar{x})$ are known basis functions, a_i are unknown coefficients, and $\bar{x} \in R_n$ is a n-dimensional vector. The set $I = \{\psi_i(\bar{x}),\ i = \overline{1,L}\}$ is redundant. $f(\bar{x})$ is an unknown function which is a weighted sum on a subset $I_1 \subset I$. We need to reconstruct $f(\bar{x})$ from the results of precise experiments $\bar{x}^i, f(\bar{x}^i),\ i = \overline{1,P},\ P \geq L$ or $P < L$. We also assume that $P > K$ where K is the number of basis functions in the set I_1. Hence, the desired relationship is represented in the form

$$f(\bar{x}) = \sum_{j=1}^{K} a_{i_j} \psi_{i_j}(\bar{x});\quad P \leq L,\ K \leq P - 1 \text{ or } P > L,\ K \leq L \tag{8.37}$$

where $\psi_{i_j}(\bar{x}) \in I_1$ and are not known in advance.

This problem formulation occurs when the researcher makes various assumptions about a possible analytic representation of an unknown relationship. The proposed methods for the problem solving assume that the representation (8.37) exists and is specified by the researcher.

Criterion for Determining the True Relationship The following criterion may serve to determine the true relationship based on the problem statement.

A set $I_1 = \{\psi_{i_1}(\bar{x}), \ldots, \psi_{i_K}(\bar{x})\}$ that satisfies Eq. (8.37) may be any subset of the set I for which, subject to $K < P$, there exist such coefficients a_{i_j} that

$$\sum_{l=1}^{P} \left[f(\bar{x}^l) - \sum_{j=1}^{K} a_{i_j} \psi_{i_j}(\bar{x}^l) \right] = 0. \tag{8.38}$$

A set I_1 is a subset of the set I satisfying the conditions (8.38) and containing a minimal number of basis functions. We assume that such a representation is unique.

Method 1, Deterministic ($P > L,\ K \leq L$) From the set of basis functions

$$I = \{\psi_i(\bar{x}),\ i = \overline{1,L}\} \tag{8.39}$$

we exclude the basis function $\psi_L(\bar{x})$ and solve the following LPP for the remaining basis functions:

$$\min y; \tag{8.40}$$

$$-y \leq f(\bar{x}^l) - \sum_{j=1}^{L-1} a_j \psi_j(\bar{x}^l) \leq y, \; y \geq 0, \; l = \overline{1, P} \tag{8.41}$$

Here, a_j, $j = \overline{1, L-1}$, y_l, $l = \overline{1, P}$ are LPP (8.40)–(8.41) variables. If the functional value is zero in the optimal solution of the LPP (to within an error of the solution), then we finally exclude the basis function from the set I (8.39). Otherwise, $\psi_L(\bar{x}) \in I_1$ and remains in set I. Let I_1 denote the set I after the first iteration of the algorithm that implements Method 1. We repeat the described procedure for $\psi_{L-2}(\bar{x})$ etc. up to the function $\psi_1(\bar{x})$ inclusive. All the basis functions that have fallen into the set I_1 on the previous iterations participate in the corresponding LPP construction at the current iteration.

Let us formulate the LPP for the current iteration $t \geq 2$. Let I^{t-1} be the set of basis functions obtained from I (8.39) as a result of $t-1$ iterations. We exclude from the set I^{t-1} the basis function $\psi_{L-t+1}(\bar{x})$ and solve the following problem for the basis functions remaining in I^{t-1}:

$$\min y; \tag{8.42}$$

$$-y \leq f(\bar{x}^l) - \sum_{\forall j_m} a_{j_m} \psi_{j_m}(\bar{x}^l) \leq y, \quad y \geq 0, \; l = \overline{1, P} \tag{8.43}$$

where $\{\psi_{j_m}(x)\}$ is the set of basis functions in I^{t-1} after we exclude the function $\psi_{L-t}(\bar{x})$ from it.

If the functional value is zero in the optimal solution of (8.42)–(8.43), then we finally exclude the basis function $\psi_{L-t+1}(\bar{x})$ from I^{t-1}. Else, it belongs to I_1 and is returned to I^{t-1}. We denote as I^t the set I^{t-1} after the t-th iteration.

We obtain the desired function $f(\bar{x})$ as the solution of the LPP constructed on the L-th iteration if $\psi_1(\bar{x}) \in I_1$ (the functional value is not zero). Otherwise, we finally exclude $\psi_1(\bar{x})$ from I^{L-1} that becomes the set of basis functions I_1.

We formulate the following LPP for the constructed set of basis functions I_1:

$$\min y; \tag{8.44}$$

$$-y \leq f(\bar{x}^l) - \sum_{\forall j_k} a_{j_k} \psi_{j_k}(\bar{x}^l) \leq y, \quad y \geq 0, \; l = \overline{1, P} \tag{8.45}$$

where $\{\psi_{j_k}(x)\}$ is the set of basis functions in the constructed set I_1 (the functional value in the optimal solution of LPP (8.44)–(8.45) is necessarily equal to zero, to within an error of the LPP solution).

Thus, we have reconstructed the unknown relationship:

$$f(\bar{x}) = \sum_{\forall j_k} a_{j_k} \psi_{j_k}(\bar{x}), \quad y_l \geq 0, \quad l = \overline{1, P} \tag{8.46}$$

where $\forall a_{j_k}$ are the solution of LPP (8.44)–(8.45); $\{\psi_{j_k}(\bar{x})\} = I_1$.

Method 1a, Deterministic ($P > L_1 + \sum_{i=1}^{L_2} k_i$, $K \leq L_1 + \sum_{i=1}^{L_2} k_i$) We generalize the above problem to the case when we introduce basis functions of two kinds (simple and composite) into the representation (8.40)–(8.41). Let $\bar{\psi}_i(\bar{x})$ be the notation of a simple basis function, $\bar{\bar{\psi}}_i(\bar{x})$ denote a composite basis function which is a linear convolution of auxiliary basis functions $\psi_l^i(\bar{x})$, $l = \overline{1, k_i}$, i.e.,

$$\bar{\bar{\psi}}_i(\bar{x}) = \sum_{l=1}^{k_i} a_l^i \psi_l^i(\bar{x})$$

where a_l^i are the unknown coefficients. The representation for a composite basis function may also be redundant.

Thus, the redundant representation that specifies an unknown relationship has the form:

$$\sum_{j=1}^{L_1} a_j \bar{\psi}_j(\bar{x}) + \sum_{i=1}^{L_2} \sum_{l=1}^{k_i} a_l^i \psi_l^i(\bar{x}) \tag{8.47}$$

where a_j, a_l^i, $j = \overline{1, L_1}$, $i = \overline{1, L_2}$, $l = \overline{1, k_i}$, are the unknown coefficients, some of which may be zeros (the representation is redundant).

Method 1a is an obvious modification of Method 1. Indeed, we can rewrite (8.47) in the form

$$\sum_{j=1}^{L_1} a_j \bar{\psi}_j(\bar{x}) + \sum_{i=1}^{L_2} \bar{\bar{\psi}}_i(\bar{x}), \quad I_1 = \{\bar{\psi}_i(\bar{x})\}, \quad I_2 = \left\{\bar{\bar{\psi}}_i(\bar{x})\right\}$$

Let I denote the set of basis functions specifying the desired relationship. First, we exclude from (8.47) the composite basis function $\bar{\bar{\psi}}_1(\bar{x})$. If the functional of the corresponding LPP is zero, then we finally exclude $\bar{\bar{\psi}}_1(\bar{x})$ from the representation (8.47). Otherwise, we implement Method 1 successively for the basis functions $\psi_l^1(\bar{x})$, $l = \overline{1, k_1}$. As a result, we obtain a true set of basis functions that define $\bar{\bar{\psi}}_1(\bar{x})$ in the set I. In a similar way we solve the problem of sequential exclusion or the construction of true sets of composite basis functions in the set I. After that, we use Method 1 to exclude the basis functions $\bar{\psi}_j(\bar{x})$ not included in the set I from the redundant representation (8.47). After the final set I determination, we estimate the coefficients a_j, a_l^i of the basis functions belonging to the set I with the help of an LPP. We have reconstructed the true relationship.

Method 2, Deterministic ($P \leq L$, $K \leq P - 1$) Let us consider preconditions for an efficient use of Method 2. The experiments number P should not be much smaller than the number of basis functions L in the redundant representation of a true relationship

$$\sum_{i=1}^{L} a_i \psi_i(\bar{x}),\ I = \{\psi_i(\bar{x})\},\ i = \overline{1, L},\ \ I_1 = \left\{\psi_{j_l}(\bar{x})\right\},\ \ l = \overline{1, k}$$

where I_1 is the set of basis functions that define the true relationship. Mandatory requirement: $K \leq P - 1$; $K \leq P - 2$ is desirable.

Stage 1 We randomly construct a sequence of basis functions:

$$\psi_{i_1}(\bar{x}),\ \psi_{i_2}(\bar{x}),\ \ldots,\ \psi_{i_L}(\bar{x}). \tag{8.48}$$

We exclude functions $\psi_{i_P}(\bar{x}),\ \ldots,\ \psi_{i_L}(\bar{x})$ from the set I. We have obtained the sequence of basis functions

$$\psi_{i_1}(\bar{x}),\ \ldots,\ \psi_{i_{P-1}}(\bar{x}). \tag{8.49}$$

We solve the following LPP for the set (8.49):

$$\min y \tag{8.50}$$

$$-y \leq f\left(\bar{x}^l\right) - \sum_{\forall j_k} a_{j_k} \psi_{j_k}\left(\bar{x}^l\right) \leq y,\quad y \geq 0,\ \ l = \overline{1, P} \tag{8.51}$$

If the value of (8.50) is greater than zero in the optimal solution of the LPP, then the set of basis functions (8.49) does not contain the set I_1. We randomly generate the next sequence of $P - 1$ basis functions. Then we formulate and solve the LPP of the form (8.50)–(8.51) for this sequence. Stage 1 of the algorithm that implements Method 2 terminates for the first LPP with zero optimal solution. The number of iterations at Stage 1 is finite. If we could not find such an LPP, then $K \geq P$. Otherwise, we go to Stage 2.

Stage 2 We apply Method 1 for the sequence of basis functions for which the LPP of the form (8.50)–(8.51) has zero optimal solution.

Method 3, Stochastic ($P > L$, $K \leq L$) Suppose that the value of $f\left(\bar{x}^l\right)$ is measured with an error (a random variable) in experiments. In [1], we have experimentally found limitations on the error of $f\left(\bar{x}^l\right)$ calculation. Method 3 statistically efficiently reconstructs the desired relationship if we meet the limitations. The redundant representation has the form

$$\sum_{i=1}^{L} a_i\psi_i(\bar{x}),\quad I=\{\psi_i(\bar{x})\},\quad i=\overline{1,L}$$

where $I_1=\left\{\psi_{i_l}(\bar{x})\right\}$, $l=\overline{1,k}$, is an unknown set of true basis functions.

The current iteration of the algorithm that implements Method 3 is the following modification of the current iteration of the algorithm used for Method 1. We alternately exclude each basis component from the set of basis functions considered at the current iteration. Then, we solve the corresponding LPP for the remaining components. After that, we return the excluded basis function to the basis functions set of the current iteration and exclude the next basis function, etc.

As a result of the current iteration, we finally exclude the basis function with the smallest functional value of the corresponding LPP from the current set of basis functions.

We terminate the method in the case of a sharp increase in the minimum value of the functional value in the LPP built on the current iteration, in comparison with the minimum values in LPPs built on the previous iterations. We presented in [1] the experimental studies of the method and recommendations for its application.

Remark 8.9 Modification of Method 3 for the case $P\le L$, $K\le P-2$ is obvious.

Remark 8.10 We can easily verify the reliability of the results obtained by Methods 1, 1a, 2 by additional experimental data which did not participate in the true relationship construction.

Remark 8.11 It is desirable to satisfy $P\ge L+T_1$ in Methods 1 and 3, $P>L_1+\sum_{i=1}^{L_2}k_i+T_1$, $T_1>2$ in Method 1a, and to start the construction in Method 2 with a subsequence of the form $\psi_{i_1}(\bar{x}),\ \ldots,\ \psi_{i_{P-T_2}}(\bar{x})$, $T_2>1$. The rationale is the guaranteed exclusion of a false zero possibility in an optimal solution of an LPP.

Remark 8.12 Method 3 is heuristic. We can justify a guarantee for its obtained result reliability only on the basis of a statistical experiment with a sufficient data quantity.

Remark 8.13 Suppose that we use Methods 1–3 to reconstruct an overall goal using its approximating values $\approx f(A_j)$, $j=\overline{1,P}$, obtained by the modified AHP. Here, A_j is an alternative of any nature (not necessarily a real number vector). Then, the redundant representation for an overall goal $f(\cdot)$ has either the form $\sum_{i=1}^{L} a_i\psi_i(A_j)$ or $\sum_{t=1}^{L_1} a_t\bar{\psi}_t(A_j)+\sum_{i=1}^{L_2}\sum_{l=1}^{K_i} a_l^i\psi_l^i(A_j)$, $\forall j=\overline{1,P}$, and we introduce the values of basis functions for alternatives A_j, $j=\overline{1,P}$, into all linear programming models.

Remark 8.14 All the above models of linear programming can be reformulated in an obvious way as models of the least squares method.

References

1. Zgurovsky, M.Z., Pavlov, A.A.: Prinyatie Resheniy v Setevyh Sistemah s Ogranichennymi Resursami (Принятие решений в сетевых системах с ограниченными ресурсами; Decision Making in Network Systems with Limited Resources). Naukova dumka, Kyiv (2010) (in Russian)
2. Kovács, A.: Novel models and algorithms for integrated production planning and scheduling. Dissertation. Computer and Automation Research Institute, Budapest (2005)
3. Emelianov, S.V. (ed.).: Upravlenie gibkimi proizvodstvennymi sistemami. Modeli i algoritmy (Управление гибкими производственными системами. Модели и алгоритмы; Flexible Production Systems Management. Models and Algorithms). Mashinostroenie, Moscow; Technik, Berlin (1987) (in Russian)
4. Sokolicyn, S.A. (ed.): Mnogourovnevaya Sistema Operativnogo Upravleniya GPS v Mashinostroenii (Многоуровневая система оперативного управления ГПС в машиностроении; Multi-level System of Operational Control Over FPS in Engineering). Politehnika, Saint Petersburg (1991) (in Russian)
5. Petrov, V.A., Maslennikov, A.N., Osipov, L.A.: Planirovanie gibkih proizvodstvennyh sistem (Планирование гибких производственных систем; Flexible production systems planning). Mashinostroenie, Leningrad (1985) (in Russian)
6. Pavlov, A.A., Telenik, S.F.: Informacionnye tehnologii i algoritmizaciya v upravlenii (Информационные технологии и алгоритмизация в управлении; Information technologies and algorithmization in management). Tehnika, Kyiv (2002) (in Russian)
7. Ohno, T.: Toyota Production System: Beyond Large-Scale Production. Productivity Press, Cambridge (1988)
8. Bitran, G.R., Tirupati, D.: Hierarchical production planning. In: Graves, S.C., Rinnooy Kan, A.H.G., Zipkin, P.H. (eds.) Logistics of Production and Inventory. Handbooks in Operations Research and Management Science, vol. 4, pp. 523–568. Elsevier Science Publishers B. V., Amsterdam (1993). https://doi.org/10.1016/s0927-0507(05)80190-2
9. Graves, S.C.: Manufacturing planning and control. In: Pardalos P., Resende, M. (eds.) Handbook of Applied Optimization, pp. 728–746. Oxford University Press, New York (2002)
10. Pervin, Y.A., Portugal, V.M., Semenov, A.I.: Planirovanie melkoseriynogo proizvodstva v ASUP (Планирование мелкосерийного производства в АСУП; Small-series production planning in automated control systems). Nauka, Moscow (1973) (in Russian)
11. Efetova, K.F., Podchasova, T.P., Portugal, V.M., Trinchuk, B.E.: Planirovanie proizvodstva v usloviyah ASU (Планирование производства в условиях АСУ; Production planning in an automated control system conditions). Tehnika, Kyiv (1984) (in Russian)
12. Pavlov, A.A., Pavlova, L.A.: Osnovy metodologii proektirovaniya PDS-algoritmov dlya trudnoreshaemyh kombinatornyh zadach (Основы методологии проектирования ПДС-алгоритмов для труднорешаемых комбинаторных задач; Fundamentals of PDC-algorithms design methodology for intractable combinatorial problems). Problemy informatiki i upravleniya **4**, 135–141 (1995) (in Russian)
13. Bitran, G.R., Haas, E.A., Hax, A.C.: Hierarchical production planning: a single stage system. Oper. Res. **29**,717–743 (1981). https://doi.org/10.1287/opre.29.4.717
14. Bitran, G.R., Hax, A.C.: Disaggregation and resource allocation using convex knapsack problems with bounded variables. Manag. Sci. **27**, 431–441 (1981). https://doi.org/10.1287/mnsc.27.4.431
15. Bitran, G.R., Hax, A.C.: On the design of hierarchical production planning systems. Decis. Sci. **8**, 28–55 (1977). https://doi.org/10.1111/j.1540-5915.1977.tb01066.x
16. Hax, A.C., Meal, H.C.: Hierarchical integration of production planning and scheduling. In: Geisler, M.A. (ed.) Logistics. Studies in Management Sciences, vol. 1, pp. 53–69. North Holland/American Elsevier, New York (1975)

17. Jain, A.S., Meeran, S.: Deterministic job-shop scheduling: past, present and future. Eur. J. Oper. Res. **113**, 390–434 (1999). https://doi.org/10.1016/s0377-2217(98)00113-1
18. Cook, S.A.: The complexity of theorem-proving procedures. In: Proceedings of the 3rd Annual ACM Symposium on Theory of Computing STOC '71, Shaker Heights, Ohio, 03–05 May 1971, pp. 151–158. ACM, New York (1971). https://doi.org/10.1145/800157.805047
19. Karp, R.M.: Reducibility among combinatorial problems. In: Miller, R.E., Thather, J.W. (eds.) Complexity of Computer Computations, pp. 85–103. Plenum press, New York (1972). https://doi.org/10.1007/978-1-4684-2001-2_9
20. Garey, M.R., Johnson, D.S.: Computers and Intractability: A Guide to the Theory of NP-Completeness. W.H. Freeman and Co, San Francisco (1979). https://doi.org/10.1137/1024022
21. Johnson S.M.: Optimal two- and three-stage production schedules with set-up times included. Naval Res. Logist. Quart **1**(1), 61–68 (1954). https://doi.org/10.1002/nav.3800010110
22. Akers Jr., S.B.: Letter to the editor—a graphical approach to production scheduling problems. Oper. Res. **4**(2), 244–245 (1956). https://doi.org/10.1287/opre.4.2.244
23. Hefetz, N., Adiri, I.: An efficient optimal algorithm for the two-machines unit-time job-shop schedule-length problem. Math. Oper. Res. **7**(3), 354–360 (1982). https://doi.org/10.1287/moor.7.3.354
24. Jackson, J.R.: An extension of Johnson's result on job lot scheduling. Naval Res. Logistics Q. **3**(3), 201–203 (1956). https://doi.org/10.1002/nav.3800030307
25. Pinedo, M.L.: Scheduling: Theory, Algorithms, and Systems. Springer, Cham (2016). https://doi.org/10.1007/978-3-319-26580-3
26. Lawler, E.L., Lenstra, J.K., Rinnooy Kan, A.H.G., Shmoys, D.B.: Sequencing and scheduling: Algorithms and complexity. In: Graves, S.C., Rinnooy Kan, A.H.G., Zipkin, P.H. (eds.) Logistics of Production and Inventory. Handbook in Operations Research and Management Science, vol. 4, pp. 445–522. North-Holland, Amsterdam (1993). https://doi.org/10.1016/s0927-0507(05)80189-6
27. Fisher, H., Thompson, G.L. Probabilistic learning combinations of local job-shop scheduling rules. In: Muth, J.F., Thompson, G.L. (eds.) Industrial Scheduling, pp. 225–251. Prentice-Hall, Englewood Cliffs (1963)
28. Lawrence, S.: Supplement to resource constrained project scheduling: An experimental investigation of heuristic scheduling techniques. GSIA, Carnegie-Mellon University, Pittsburgh (1984)
29. Panwalkar, S.S., Iskander, W.: A survey of scheduling rules. Oper. Res. **25**(1), 45–61 (1977). https://doi.org/10.1287/opre.25.1.45
30. Grabot, B., Geneste, L.: Dispatching rules in scheduling: A fuzzy approach. Int. J. Prod. Res. **32**(4), 903–915 (1994). https://doi.org/10.1080/00207549408956978
31. Werner, F., Winkler, A.: Insertion techniques for the heuristic solution of the job-shop problem. Discr. Appl. Math. **58**(2), 191–211 (1995). https://doi.org/10.1016/0166-218x(93)e0127-k
32. Morton, T.E., Pentico, D.W.: Heuristic Scheduling Systems: With Applications to Production Systems and Project Management. Wiley, New York (1993)
33. Sabuncuoglu, I., Bayiz, M.: A beam search based algorithm for the job shop scheduling problem. Research Report IEOR-9705. Bilkent University, Bilkent (1997)
34. Fisher, M.L., Rinnooy Kan, A.H.G.: The design, analysis and implementation of heuristics. Manag. Sci. **34**(3), 263–265 (1988). https://doi.org/10.1287/mnsc.34.3.263
35. Glover, F., Greenberg, H.J.: New approaches for heuristic search: A bilateral linkage with artificial intelligence. Eur. J. Oper. Res. **39**(2), 119–130 (1989). https://doi.org/10.1016/0377-2217(89)90185-9
36. Rodammer, F.A., White, K.P., Jr.: A recent survey of production scheduling. IEEE Trans. Syst. Man Cybern. **18**(6), 841–851 (1988). https://doi.org/10.1109/21.23085

37. Sergienko, I.V., Kapitonova, Y.V., Lebedeva, T.T.: Informatika v Ukraine: Stanovlenie, Razvitie, Problemy (Информатика в Украине: становление, развитие, проблемы. Informatics in Ukraine: Formation, Development, Problems). Naukova dumka, Kyiv (1999) (in Russian)
38. Johnson, D.S., Papadimitriou, C.H., Yannakakis, M.: How easy is local search? J. Comput. Syst. Sci. **37**(1), 79–100 (1988). https://doi.org/10.1016/0022-0000(88)90046-3
39. Yannakakis, M.: The analysis of local search problems and their heuristics. In: Choffrut, C., Lengauer, T. (eds.) STACS 90. Lecture Notes in Computer Science, vol. 415, pp. 298–311. Springer, Berlin (1990). https://doi.org/10.1007/3-540-52282-4_52
40. Sergienko, I.V.: O primenenii metoda vektora spada dlya resheniya zadach optimizacii kombinatornogo tipa (О применении метода вектора спада для решения задач оптимизации комбинаторного типа; On the application of the recession vector method to solve optimization problems of combinatorial type). Upravlyayuschie sistemy i mashiny **2**, 86–94 (1975) (in Russian)
41. Bykov, A.Y., Artamonova, A.Y.: Modifikaciya metoda vektora spada dlya optimizacionno-imitacionnogo podhoda k zadacham proektirovaniya sistem zaschity informacii (Модификация метода вектора спада для оптимизационно-имитационного подхода к задачам проектирования систем защиты информации; A modified recession vector method based on the optimization-simulation approach to design problems of information security systems). Nauka i Obrazovanie of Bauman MSTU **1**, 158–175 (2015). https://doi.org/10.7463/0115.0754845 (in Russian)
42. Shilo, V.P.: Metod globalnogo ravnovesnogo poiska (Метод глобального равновесного поиска; Global equilibrium search method). Cybern. Syst. Anal. **35**(1), 74–81 (1999) (in Russian)
43. Adams, J., Balas, E., Zawack, D.: The shifting bottleneck procedure for job-shop scheduling. Manag. Sci. **34**(3), 391–401 (1988). https://doi.org/10.1287/mnsc.34.3.391
44. Balas, E., Lancia, G., Serafini, P., et al.: Job-shop scheduling with deadlines. J. Comb. Optim. **1**(4), 329–353 (1998). https://doi.org/10.1023/a:1009750409895
45. Balas, E., Vazacopoulos, A.: Guided local search with shifting bottleneck for job-shop scheduling. Manag. Sci. **44**(2), 262–275 (1998). https://doi.org/10.1287/mnsc.44.2.262
46. Caseau, Y., Laburthe, F.: Disjunctive scheduling with task intervals. LIENS Technical Report 95-25, Laboratoire d'Informatique de l' Ecole Normale Superieure, Paris (1995)
47. Demirkol, E., Mehta, S., Uzsoy, R.: A computational study of shifting bottleneck procedures for shop scheduling problems. J. Heuristics **3**(2), 111–137 (1997). https://doi.org/10.1023/a:1009627429878
48. Yamada, T., Nakano, R.: Job-shop scheduling by simulated annealing combined with deterministic local search. In: Osman, I.H., Kelly, J.P. (eds.) Meta-Heuristics: Theory and Applications, pp. 237–248. Springer, Boston (1996). https://doi.org/10.1007/978-1-4613-1361-8_15
49. Falkenauer, E., Bouffouix, S.: A genetic algorithm for job-shop. In: Proceedings of the IEEE International Conference on Robotics and Automation, Sacramento, 9–11 Apr 1991. https://doi.org/10.1109/robot.1991.131689
50. Nakano, R., Yamada, T.: Conventional genetic algorithm for job-shop problems. In: Kenneth, M.K., Booker, L.B. (eds.) Proceedings of the 4th International Conference on Genetic Algorithms and their Applications, San Diego (1991)
51. Aarts, E.H.L., Van Laarhooven, P.J.M., Ulder, N.L.J.: Local search based algorithms for job-shop scheduling. Working Paper. University of Technology, Eindhoven (1991)
52. Dorndorf, U., Pesch, E.: Evolution based learning in a job-shop scheduling environment. Comput. Oper. Res. **22**(1), 25–40 (1995). https://doi.org/10.1016/0305-0548(93)e0016-m
53. Grefenstette, J.J.: Incorporating problem specific knowledge into genetic algorithms. In: Davis, L. (ed.) Genetic Algorithms and Simulated Annealing, pp. 42–60. Pitman, London (1987)

54. Moscato, P.: On evolution, search, optimization, genetic algorithms and martial arts: towards memetic algorithms. C3P Report 826: Caltech Concurrent Computation Program, Caltech (1989)
55. Ulder, N.L.J., Aarts, E.H.L., Bandelt, H.-J., et al.: Genetic local search algorithms for the travelling salesman problem. Lect. Notes Comput. Sci. **496**, 109–116 (1991). https://doi.org/10.1007/bfb0029740
56. Fox, M.S.: Constraint-directed search: a case study of job shop scheduling. Dissertation, Carnegy Mellon University, Pittsburgh (1983)
57. Nuijten, W.P.M., Le Pape, C.: Constraint-based job-shop scheduling with ILOG SCHEDULER. J. Heuristics **3**(4), 271–286 (1998). https://doi.org/10.1023/a:1009687210594
58. Pesch, E., Tetzlaff, U.A.W.: Constraint propagation based scheduling of job shops. INFORMS J. Comput. **8**(2), 144–157 (1996). https://doi.org/10.1287/ijoc.8.2.144
59. Sadeh, N.: Look-ahead techniques for micro-opportunistic job shop scheduling. Dissertation, Carnegie Mellon University, Pittsburgh (1991)
60. Foo, S.Y., Takefuji, Y.: Integer linear programming neural networks for job-shop scheduling. In: IEEE International Conference on Neural Networks, San Diego, 24–27 July 1988. https://doi.org/10.1109/icnn.1988.23946
61. Foo, S.Y., Takefuji, Y.: Stochastic neural networks for solving job-shop scheduling: part 1. Problem representation. In: IEEE International Conference on Neural Networks, San Diego, 24–27 July 1988. https://doi.org/10.1109/icnn.1988.23939
62. Foo, S.Y., Takefuji, Y.: Stochastic neural networks for solving job-shop scheduling: part 2. Architecture and simulations. In: IEEE International Conference on Neural Networks, San Diego, 24–27 July 1988. https://doi.org/10.1109/icnn.1988.23940
63. Sabuncuoglu, I., Gurgun, B.: A neural network model for scheduling problems. Eur. J. Oper. Res. **93**(2), 288–299 (1996). https://doi.org/10.1016/0377-2217(96)00041-0
64. Zhou, D.N., Cherkassky, V., Baldwin, T.R., et al.: Scaling neural networks for job-shop scheduling. In: Proceedings of International Joint Conference on Neural Networks (IJCNN'90), San Diego, 17–21 June 1990, pp. 889–894. https://doi.org/10.1109/ijcnn.1990.137947
65. Zhou, D.N., Cherkassky, V., Baldwin, T.R., et al.: A neural network approach to job-shop scheduling. IEEE Trans. Neural Netw. **2**(1), 175–179 (1991). https://doi.org/10.1109/72.80311
66. Dorigo, M.: Optimization, learning and natural algorithms. Dissertation, Politecnico di Milano (1992)
67. Donati, A.V., Darley, V., Ramachandran, B.: An ant-bidding algorithm for multistage flowshop scheduling problem: optimization and phase transitions. In: Siarry P., Michalewicz Z. (eds.) Advances in Metaheuristics for Hard Optimization, pp.111–136. Springer, Berlin (2008). https://doi.org/10.1007/978-3-540-72960-0_6
68. Prabhakar, B., Dektar, K.N., Gordon, D.M.: The regulation of ant colony foraging activity without spatial information. PLoS Comput. Biol. **8**(8), e1002670 (2012). https://doi.org/10.1371/journal.pcbi.1002670
69. Brucker, P., Hurink, J., Werner, F.: Improving local search heuristics for some scheduling problems—I. Discr. Appl. Math. **65**(1–3), 97–122 (1996). https://doi.org/10.1016/0166-218x(95)00030-u
70. Brucker, P., Hurink, J., Werner F.: Improving local search heuristics for some scheduling problems. Part II. Discr. Appl. Math. **72**(1–2), 47–69 (1997). https://doi.org/10.1016/s0166-218x(96)00036-4
71. Lourenco, H.R.D.: A computational study of the job-shop and the flow-shop scheduling problems. Dissertation, Cornell University (1993)
72. Lourenco, H.R.D.: Job-shop scheduling: computational study of local search and large-step optimization methods. Eur. J. Oper. Res. **83**(2), 347–364 (1995). https://doi.org/10.1016/0377-2217(95)00012-f

73. Lourenco, H.R.D., Zwijnenburg, M.: Combining the large-step optimization with tabu-search: application to the job-shop scheduling problem. In: Osman, I.H., Kelly, J.P. (eds.) Meta-Heuristics: Theory and Applications, pp. 219–236. Springer, Boston (1996). https://doi.org/10.1007/978-1-4613-1361-8_14
74. Martin, O., Otto, S.W., Felten, E.W.: Large-step Markov chains for traveling salesman problem. Complex Syst. **5**(3), 299–326 (1989)
75. Martin, O., Otto, S.W., Felten, E.W.: Large-step Markov chains for TSP incorporating local search heuristics. Oper. Res. Lett. **11**(4), 219–224 (1992). https://doi.org/10.1016/0167-6377(92)90028-2
76. Glover, F.: Future paths for integer programming and links to artificial intelligence. Comput. Oper. Res. **13**(5), 533–549 (1986). https://doi.org/10.1016/0305-0548(86)90048-1
77. Glover, F.: Heuristics for integer programming using surrogate constraints. Decis. Sci. **8**(1), 156–166 (1977). https://doi.org/10.1111/j.1540-5915.1977.tb01074.x
78. Glover, F.: Tabu search—part I. ORSA J. Comput. **1**(3), 190–206 (1989). https://doi.org/10.1287/ijoc.1.3.190
79. Glover, F.: Tabu search—part II. ORSA J. Comput. **2**(1), 4–32 (1990). https://doi.org/10.1287/ijoc.2.1.4
80. Glover, F., Laguna, M.: Tabu Search. Springer, Boston (1997). https://doi.org/10.1007/978-1-4615-6089-0
81. Nowicki, E., Smutnicki, C.: A fast taboo search algorithm for the job-shop problem. Manag. Sci. **42**(6), 797–813 (1996). https://doi.org/10.1287/mnsc.42.6.797
82. Taillard, E.: Parallel taboo search technique for the job-shop scheduling problem. Internal Research Report ORWP89/11. Ecole Polytechnique Federale de Lausanne, Lausanne (1989)
83. Edelkamp, S., Schrödl, S.: Heuristic Search: Theory and Applications. Morgan Kaufmann Publishers, Waltham, MA (2012). https://doi.org/10.1016/c2009-0-16511-x
84. Resende, M.G.C.: A GRASP for job shop scheduling. In: INFORMS National Meeting, pp. 23–31, San Diego, CA, 4–7 May 1997
85. Matsuo, H., Suh, C.J., Sullivan, R.S.: A controlled search simulated annealing method for the general job-shop scheduling problem. Working Paper. University of Texas, Austin (1988)
86. Van Laarhoven, P.J.M., Aarts, E.H.L., Lenstra, J.K.: Job-shop scheduling by simulated annealing. Report OS-R8809. Centrum voor Wiskunde en Informatica, Amsterdam (1988)
87. Dechter, R.: Constraint Processing. Morgan Kaufmann, San Francisco (2003)
88. Michel, L., Hentenryck, P.V.: Activity-based search for black-box constraint programming solvers. In: Beldiceanu, N., Jussien, N., Pinson, É. (eds.) Integration of AI and OR Techniques in Constraint Programming for Combinatorial Optimization Problems. CPAIOR 2012. Lecture Notes in Computer Science, vol. 7298, pp. 228–243. Springer, Berlin (2012). https://doi.org/10.1007/978-3-642-29828-8_15
89. Schulte, C., Tack, G., Lagerkvist, M.Z.: Modeling and programming with Gecode. http://www.gecode.org/doc-latest/MPG.pdf (2018). Accessed 02 Apr 2018
90. Flajolet, P., Sedgewick, R.: Analytic Combinatorics. Cambridge University Press, Cambridge (2009). https://doi.org/10.1017/cbo9780511801655
91. Zgurovsky, M.Z., Pavlov, O.A., Misura, E.B. PDS-algoritmy i trudnoreshaemye zadachi kombinatornoi optimizacii (ПДС-алгоритмы и трудноpешаемые задачи комбинаторной оптимизации; PDC-algorithms and intractable combinatorial optimization problems). Syst. Res. Inform. Technol. **2009**(4), 14–31 (2009) (in Russian)
92. Kolisch, R., Hartmann, S.: Heuristic algorithms for the resource-constrained project scheduling problem: Classification and computational analysis. In: Węglarz, J. (ed.) Project Scheduling. International Series in Operations Research & Management Science, vol. 14, pp. 147–178. Springer, New York (1999). https://doi.org/10.1007/978-1-4615-5533-9_7
93. Baptiste, P., Le Pape, C., Nuijten, W.: Constraint-based Scheduling: Applying Constraint Programming to Scheduling Problems Series. International Series in Operations Research & Management Science, vol. 39. Springer, Boston (2001). https://doi.org/10.1007/978-1-4615-1479-4

94. Fox, M.S., Smith, S.F, Allen, B.P., et al.: ISIS: A constraint-directed reasoning approach to job-shop scheduling. In: Proceedings of the IEEE Trends and Applications Conference, pp. 76–81. Washington, May 1983
95. Wallace, M.G.: Practical applications of constraint programming. Constraints **1**(1–2), 139–168 (1996). https://doi.org/10.1007/bf00143881
96. Vollmann, T.E., Berry, W.L., Whybark, D.C.: Manufacturing Planning and Control Systems. Irwin, McGraw-Hill (1997)
97. Vernikov, G.: Osnovy sistem klassa MRP-MRPII (Основы систем класса MRP-MRPII; Fundamentals of MRP-MRP II class systems). https://www.cfin.ru/vernikov/mrp/mrpmine.shtml (1999). Accessed 02 Apr 2018 (in Russian)
98. MRP i MRP II (MRP и MRP II; MRP & MRP II). Planeta KIS. http://bigc.ru/publications/other/logistics/mpr_and_mpr2.php (1999). Accessed 02 Apr 2018 (in Russian)
99. Schonberger, R.J.: Japanese Manufacturing Techniques: Nine Hidden Lessons in Simplicity. The Free Press, New York (1982)
100. Chaya, V.T., Chupahina, N.I.: Konceptualnye osnovy razvitiya upravlencheskogo ucheta (Концептуальные основы развития управленческого учета; Conceptual bases of management accounting development). Ekonomicheskii analiz: teoriya i praktika **2009**(3), 25–31 (2009) (in Russian)
101. Hans, E.W.: Resource loading by branch-and-price techniques. Dissertation, Twente University, Enschede (2001)
102. Kis, T.: A branch-and-cut algorithm for scheduling of projects with variable-intensity activities. Math. Progr. **103**(3), 515–539 (2005) https://doi.org/10.1007/s10107-004-0551-6
103. Leachman, R.C., Dincerler, A., Kim, S.: Resource constrained scheduling of projects with variable intensity activities. IIE Trans. **22**(1), 31–40 (1990). https://doi.org/10.1080/07408179008964155
104. Vasil'ev, K.A., Gorohov, M.M.: Sistemy klassov MRP i MRPII (Системы классов MRP и MRPII; Systems of MRP & MRPII class). In: Proceedings of the 10th International Scientific-Practical Conference "Problemy Effektivnogo Ispolzovaniya Nauchnogo Potenciala Obschestva", Chelyabinsk, 10 December 2017, vol. 4, pp. 14–18. Aeterna, Ufa (2017) (in Russian)
105. Turdyshov, D.H.: Osobennosti postroeniya informacionnyh sistem upravleniya (Особенности построения информационных систем управления; Specifics of information management systems building). Sovremennye problemy nauki i obrazovaniya **2013**(1). https://www.science-education.ru/ru/article/view?id=8187 (2013). Accessed 02 Apr 2018 (in Russian)
106. Gastek, D.M.: Managing MRP-II. Manag. Autom. **1**, 39–41 (1986)
107. Fedyaev, A.A., Fedyaeva, E.M.: K voprosu o razvitii sovremennyh ERP-sistem (К вопросу о развитии современных ERP-систем; On the development of modern ERP systems). Molodoi uchenyi **17**, 26–30. https://moluch.ru/archive/97/21837/ (2015). Accessed 02 Apr 2018 (in Russian)
108. Sheina, Y.V.: Informacionnaya komponenta reinjiniringa logisticheskih processov krupnoformatnyh torgovyh organizacii (Информационная компонента реинжиниринга логистических процессов крупноформатных торговых организаций; Information component of logistics processes reengineering for large-format trading organizations). Ekonominfo **8**, 66–69 (2007) (in Russian)
109. Krylovich, A.: ERP-sistemy pozvolyayut planirovat v rynochnyh usloviyah (ERP-системы позволяют планировать в рыночных условиях; ERP-systems allow you to plan in market conditions). BOSS **5**. http://bigc.ru/publications/other/logistics/erp_systems_planir_v_rinochn_uclov.php (2000). Accessed 02 Apr 2018 (in Russian)
110. Ultimate humanless enterprises CIS: Ultima Solutions. https://www.ultimatebusinessware.ru/solutions/ (2018). Accessed 02 Apr 2018 (in Russian)
111. SAP Website CIS. http://www.sap.com/cis (2018) Accessed 02 Apr 2018 (in Russian)

112. 2011 Guide to ERP Systems and Vendors. An Independent Research Report. Panorama Consulting. https://www.webcitation.org/65BhgKyzp?url=http://panorama-consulting.com/Documents/2011-Guide-to-ERP-Systems-and-Vendors.pdf (2011). Accessed 04.04.2018
113. ERP: Wikipedia Russian. https://ru.wikipedia.org/wiki/ERP (2017). Accessed 02 Apr 2018 (in Russian)
114. Smirnov, N.: ERP bez kompleksov (ERP без комплексов; ERP without complexes). Computerworld Russia **17**. http://www.osp.ru/cw/2013/17/13036445/ (2013). Accessed 02 Apr 2018 (in Russian)
115. Mescheryakov, V.: Rossiiskii rynok ERP: 1S rastet bystree vseh (Российский рынок ERP: 1С растет быстрее всех; Russian ERP market: 1C grows faster than all). CNews. RosBusinessConsulting. http://www.cnews.ru/news/top/index.shtml?2011/09/19/455890 (2011). Accessed 02 Apr 2018 (in Russian)
116. Information Technologies Website. IT-Enterprise. http://www.it.ua (2018) Accessed 01 Aug 2012 (in Russian)
117. ERP-rynok glazami veterana (ERP-рынок глазами ветерана; ERP-market through the eyes of a veteran). PC Week Ukrainian Edition. SK-Press. http://www.pcweek.ua/themes/detail.php?ID=135600 (2011). Accessed 02 Apr 2018 (in Russian)
118. Jelvicky, D.: Razrushenie stereotipov (Разрушение стереотипов; Stereotypes destruction). Computerworld Russia **17**. http://www.osp.ru/cw/2013/17/13036460/ (2013). Accessed 02 Apr 2018 (in Russian)
119. Lastovirya, V.N., Gladkov, E.A., Konovalov, A.V.: Optimizaciya v Avtomatizirovannom Proektirovanii Svarochnogo Proizvodstva (Оптимизация в Автоматизированном Проектировании Сварочного Производства; Optimization in the Automated Design of Welding Production). MSIU, Moscow (2008) (in Russian)
120. Zagidullin, R.R.: Planirovanie Mashinostroitel'nogo Proizvodstva (Планирование машиностроительного производства; Engineering Production Planning). TNT, Staryi Oskol (2015) (in Russian)
121. Zagidullin, R.R.: Upravlenie Mashinostroitel'nym Proizvodstvom s Pomosch'yu Sistem MES, APS, ERP (Управление машиностроительным производством с помощью систем MES, APS, ERP; Engineering Production Management with the Help of MES, APS, ERP Systems). TNT, Staryi Oskol (2011) (in Russian)
122. Advanced Planning & Scheduling: Wikipedia Russian. https://ru.wikipedia.org/wiki/Advanced_Planning_&_Scheduling (2018). Accessed 02 Apr 2018 (in Russian)
123. Aldzhanov, V.: IT-arhitektura. Prakticheskoe Rukovodstvo ot A do Ya (ИТ-архитектура. Практическое руководство от А до Я; IT Architecture. A Practical Guide from A to Z). Litres, Moscow (2018)
124. Werner, L., Yong, M.S.: A survey of commercial production scheduling software. In: Lean Aerospace Initiative. Plenary Workshop. Massachusetts Institute of Technology, Cambridge (1999)
125. Yong, M.S.: Simulation of real-time scheduling policies in multi-product, make-to-order semiconductor fabrication facilities. Master thesis, Massachusetts Institute of Technology, Cambridge (2001)
126. Saver, J.: Knowledge-based design of scheduling systems. Intell. Autom. Soft Comput. **7**(1), 55–62 (2001). https://doi.org/10.1080/10798587.2001.10642804
127. Dorn, J., Froeschl, K.A.: Scheduling of Production Processes. Ellis Horwood, Saddle River (1993)
128. Sauer, J., Bruns, R.: Knowledge-based scheduling systems in industry and medicine. IEEE Expert **12**(1), 24–31 (1997). https://doi.org/10.1109/64.577410
129. Smith, S.F.: Knowledge-based production management: approaches, results and prospects. Prod. Plan. Control **3**(4), 350–380 (1992). https://doi.org/10.1080/09537289208919407
130. Zweben, M., Fox, M.: Intelligent Scheduling. Morgan Kaufmann Publishers, San Francisco (1994)
131. Kempf, K.G., Le Pape, C, Smith, S.F., Fox, B.R.: Issues in the design of AI-based schedulers: a workshop report. AI Magazine **11**(5), 37–46 (1991)

132. Oleynik, P.P.: Korporativnye Informacionnye Sistemy (Корпоративные Информационные Системы; Corporate Information Systems). Piter Publishing House, Saint Petersburg (2012)
133. Pavlov, A.A. (ed.): Konstruktivnye Polinomialnye Algoritmy Resheniya Individualnyh Zadach iz Klassa NP (Конструктивные полиномиальные алгоритмы решения индивидуальных задач из класса NP; Constructive Polynomial Algorithms for Solving Individual Problems from the Class NP). Tehnika, Kyiv (1993) (in Russian)
134. Zgurovsky, M.Z., Pavlov, A.A.: Ierarhicheskoe planirovanie v sistemah, imeyuschih setevoe predstavlenie tehnologicheskih processov i ogranichennye resursy, kak zadacha prinyatiya resheniy (Иерархическое планирование в системах, имеющих сетевое представление технологических процессов и ограниченные ресурсы, как задача принятия решений; Hierarchical planning in systems that have a network representation of technological processes and limited resources, as a decision making problem). Syst. Res. Inform. Technol. **2009**(3), 70–75 (2009) (In Russian)
135. Saaty, T.L.: The Analytic Hierarchy Process. McGraw Hill, New York (1980)
136. Pavlov, A.A., Kalashnik, V.V.: Rekomendacii po vyboru zony provedeniya aktivnogo eksperimenta dlya odnomernogo polinomialnogo regressionnogo analiza (Рекомендации по выбору зоны проведения активного эксперимента для одномерного полиномиального регрессионного анализа; Recommendations for choosing an active experiment area for univariate polynomial regression analysis). Visnyk NTUU KPI Inform. Oper. Comput. Sci. **60**, 41–45 (2014) (in Russian)
137. Pavlov, A.A., Kalashnik, V.V., Kovalenko, D.A.: Postroenie mnogomernoi polinomialnoi regressii. Regressiya s povtoryayuschimisya argumentami vo vhodnyh dannyh (Построение многомерной полиномиальной регрессии. Регрессия с повторяющимися аргументами во входных данных; Multivariate polynomial regression construction. Regression with repeated arguments in the input data). Visnyk NTUU KPI Inform. Oper. Comput. Sci. **62**, 57–61 (2015) (in Russian)
138. Pavlov, A.A., Ivanova, A.A.: Models and methods of project selection in terms of non-formalized global target. In: Management of the Distribute Projects and Programmes Resources: Monograph, pp. 122–150. Torubara V.V., Nikolayev (2015)
139. Saaty, T.L. Kearns, K.: Analytical Planning: The Organization of Systems. Pergamon Press, Oxford (1985). https://doi.org/10.1016/c2013-0-03782-6
140. Andreychikov, A.V., Andreychikova, O.N.: Analiz, Sintez, Planirovanie Resheniy v Ekonomike (Анализ, Синтез, Планирование Решений в Экономике; Analysis, Synthesis, Planning of the Economy Decisions). Finansy i Statistika, Moscow (2000)
141. Saaty, T.L.: Decision Making with Dependence and Feedback: The Analytic Network Process. RWS Publications, Pittsburgh (1996)
142. Hudson, D.J.: Lectures on Elementary Statistics and Probability, vol. 1. CERN Reports 63 (29). CERN, Geneva (1963). https://doi.org/10.5170/cern-1963-029
143. Hudson, D.J.: Statistics Lectures, Vol. 2: Maximum Likelihood and Least Squares Theory. CERN Reports 64(18). CERN, Geneva (1964). https://doi.org/10.5170/cern-1964-018
144. Draper, N.R., Smith H.: Applied Regression Analysis, 3rd edn. Wiley, New York (1998). https://doi.org/10.1002/9781118625590
145. Flom, P.L., Cassell, D.L.: Stopping stepwise: why stepwise and similar selection methods are bad, and what you should use. In: NESUG 2007 Proceedings. NorthEast SAS Users Group (2007)
146. Harrell Jr., F.E.: Regression Modeling Strategies: With Applications to Linear Models, Logistic and Ordinal Regression, and Survival Analysis. Springer International Publishing Switzerland (2015). https://doi.org/10.1007/978-3-319-19425-7
147. Foster, D.P., George, E.I.: The risk inflation criterion for multiple regression. Ann. Stat. **22** (4), 1947–1975 (1994). https://doi.org/10.1214/aos/1176325766
148. Jonathan, M., Goldberg, M.: Multiple regression analysis and mass assessment: a review of the issues. Appraisal J. **56**(1), 89–109 (1988)

149. Ivahnenko, A.G.: Induktivnyi Metod Samoorganizacii Modelei Slojnyh Sistem (Индуктивный метод самоорганизации моделей сложных систем; Inductive Method of Self-Organization of Complex System Models). Naukova dumka, Kyiv (1982)
150. Ivahnenko, A.G., Stepashko, V.S.: Pomehoustoichivost Modelirovaniya (Помехоустойчивость моделирования; Simulation's Immunity to Interference). Naukova dumka, Kyiv (1985)
151. Meyer, C.D.: Matrix Analysis and Applied Linear Algebra Book and Solutions Manual. Society for Industrial and Applied Mathematics, Philadelphia (2000). https://doi.org/10.1137/1.9780898719512

Chapter 9
Algorithms and Software of the Four-Level Model of Planning and Decision Making

Abstract We give an interrelated description for all algorithms which implement a procedure of scheduling and operative planning on the basis of formal procedures and expert solutions. We have achieved this result sequentially solving the following problems: (1) we formalize the network representation of a technological process; develop a language which implements the dynamics of the process functioning, define a set of practical optimality criteria for the operational plan; (2) we obtain a formal representation of the technological process, adapt the general procedures of its two-level aggregation; (3) we substantiate the scheduling problem reduction to the approximating problem of the total weighted completion times minimization on a single machine with precedence relations between the tasks; (4) we formalize the procedure of the coordinated planning which defines the processing order of products (or product series) corresponding to a single basic criterion; (5) we formalize the process of an operational plan construction based on results of the coordinated planning; (6) we develop two algorithms for an operational plan adjustment; (7) we substantiate a procedure for a decision maker for the expert evaluation and adjustment both of a potential portfolio of orders and the due dates of the products (or product series) included in it, for the approval of the final operational plan on the basis of analysis of plans obtained by the formal methods. The last section describes an informational software system implementing the hierarchical model in a specific practical area: for solving the problems of scheduling and planning for small-series productions.

9.1 The Structure of the Planning System

This chapter provides a general methodology for implementing the algorithms of the four-level model of planning (including operative planning) and decision making (FLM). We briefly described the FLM in Sect. 8.4. Also we give in this chapter an extended example of solving the multi-stage network scheduling problem (MSNSP).

M. Z. Zgurovsky and A. A. Pavlov, *Combinatorial Optimization Problems in Planning and Decision Making*, Studies in Systems, Decision and Control 173,
https://doi.org/10.1007/978-3-319-98977-8_9

The planning system under consideration works with a virtual network of a special type. The network should have the properties described in Sect. 9.1. It consists of structural units called elements which are machines of different types. Also, the network contains completed and uncompleted jobs and relations between them. Experts build the model of the virtual network based on data from a real organizational or manufacturing process of a discrete type. More than 80% of the world's production systems are discrete type productions. In particular, experts may use the data of the production technology, design documentation, capacity properties, the order portfolio. We assume that a job moves from one machine to another without interrupting the process. We consider only such real technologies which we can represent by such virtual network with specified properties. We break the physical equipment into a number of virtual machines. The number of physical machines is actually smaller than the virtual machines number because the same physical machine may process a job (a job group) after processing other jobs (job groups), in accordance with the technology. To solve the problem, experts build a network with the minimum necessary amount of physical equipment.

In this chapter:

- we formalize the network representation of a technological process, develop a language which implements the dynamics of the process functioning, define a set of practical optimality criteria for the operational plan (five basic criteria and all possible linear combinations of them);
- we obtain a formal representation of the technological process, adapt the general procedures of its two-level aggregation [1]: combining jobs to the aggregated units (the first level of aggregation) and combining critical paths of products with intersection on their common vertices of the precedence graph with combining the set of resources into a single unit (the second level of aggregation);
- for the second level of aggregation, we substantiate the scheduling problem reduction to the approximating problem of the total weighted completion times minimization on a single machine with precedence relations between the tasks; we propose an algorithm for its solving based on the ideas of PSC-algorithm for this problem from the seventh chapter of the book;
- we formalize the procedure of the coordinated planning which is executed on the aggregated network of the first level of aggregation; in this procedure, the solution obtained at the second level of aggregation defines the processing order of products (or product groups) corresponding to a single basic criterion;
- based on results of the coordinated planning (the second level of the planning model), we formalize the process of an operational plan construction (the third level of the FLM):
    - for a basic optimality criterion, we formulate and solve the operational planning problem as a MSNSP (the completion times of products (product groups) become the due dates in the scheduling problem). An efficient method of its solution uses PSC–algorithms for single-stage scheduling problems (Chaps. 2 and 3 of the monograph);

- for a synthetic optimality criterion (a linear combination of basic criteria), we propose the method which is implemented by four alternative algorithms for operational plan construction;

- we develop two algorithms for an operational plan adjustment (the fourth level of the model—operative planning level):
  - the first algorithm is a modification of the algorithmic procedure of planning on the levels 1–3 of the model. We solve the problem for one of four basic criteria or their linear combinations. It should be noted that customer sets again the weight coefficients of basic optimality criteria for the operative planning problem. We need to distinguish the problem of an operational plan construction to execute the potential order portfolio and setting weight coefficients as the tardiness costs with respect to due dates of products (product groups) on the already approved operational plan. In the first case, due dates of products or product groups may be adjusted during coordination planning, with the profit correction corresponding to the optimality criterion;
  - the second algorithm of operative planning contains within single framework a formal algorithm and expert planning methods. If its implementation is possible, it efficiently (in terms of the optimization criterion) solves the problem of operative adjustment of the operational schedule obtained at the third level of planning. The formal algorithm includes PSC-algorithms for intractable combinatorial optimization problems (Chaps. 4–6 of this book);

- we substantiate a procedure for a decision maker for the expert evaluation and adjustment both of a potential order portfolio and the due dates of the products (or product groups) included in it, for the final operational plan approval on the basis of analysis of plans obtained by formal methods;
- in Sect. 9.9, we describe a software system which implements the FLM to solve scheduling and planning problems in small-series productions.

This chapter contains our new original results based on our earlier work [1–11].

9.1.1 Structural Elements of the Network Model

Let us formulate the basic elements for an initial network model synthesis. The model will specify the operational technology to process a given order portfolio.

We present graphical notations of elements of type 1–5 at Fig. 9.1.

Element of type 1 (Fig. 9.1a): one machine processes one job. l is the job's processing time, r is its readiness time (the time point when the job is ready to be processed on the machine), C is its completion time, d is its due date. We denote products' completion times by a letter $\mathbb{C}$. $C \leq d (\mathbb{C} \leq d)$ is true in a feasible schedule. A symbol —| in a network indicates the completion of a terminal job (the one without successors).

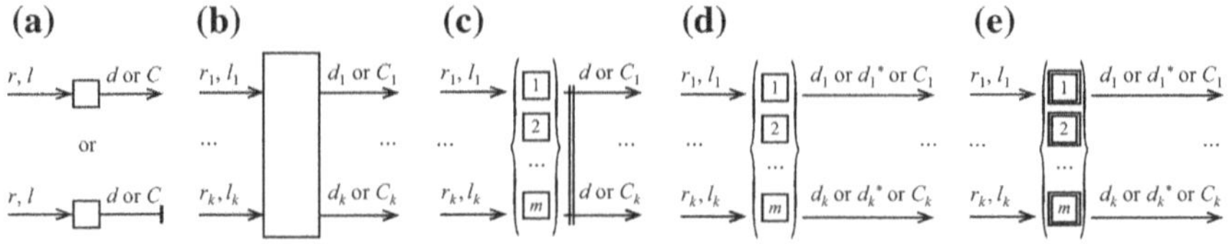

Fig. 9.1 Network elements of: **a** type 1, **b** type 2, **c** type 3, **d** type 4, **e** type 5

An arrow *entering* a circle ○ indicates a completed job. An arrow *coming out* of a circle is a new uncompleted job. A square □ designates a machine (resource) the job is processed on. Machine notations for elements of type 1–5 are different. We identify machines and jobs by a unique number. The first digit of a job number indicates its network layer. No other job precedes a job at the first layer. A dotted circle ◌ indicates a fictitious job corresponding to a product completion. Under a product, we understand either one physical product or a series of physically identical products.

Element of type 2 (Fig. 9.1b): one machine processes k jobs successively without interruptions. r_i is the readiness time of job i on this machine, l_i is its processing time, and d_i is its due date for this machine. C_i is the completion time of job i, $C_i \le d_i$ is fulfilled in a feasible schedule.

Element of type 3 (Fig. 9.1c): m independent (identical) parallel machines process k jobs without interruptions. d is their common due date. $C_i \le d$, $i = \overline{1,k}$,[1] is fulfilled in a feasible schedule. The start times of all machines must be the same. Jobs following an element of type 3 may start only after the completion of all jobs on this element.

Element of type 4 (Fig. 9.1d): m independent (identical) parallel machines of *equal* productivities process k jobs without interruptions. d_i are their due dates which may be different. Values of $d_i^* = [d_i - \Delta_i,\ d_i]$, $\Delta_i > 0$, $i = \overline{1,k}$, define an acceptable range of completion times in a feasible schedule (allowable earliness): either $C_i \le d_i$ or $C_i \in d_i^*$ must be satisfied. The start times of machines may be different.

Element of type 5 (Fig. 9.1e): m independent (unrelated) parallel machines of *various* productivities process k jobs without interruptions. d_i are their due dates which may be different. Values of $d_i^* = [d_i - \Delta_i,\ d_i]$, $\Delta_i > 0$, $i = \overline{1,k}$, define an acceptable range of completion times in a feasible schedule (allowable earliness): either $C_i \le d_i$ or $C_i \in d_i^*$ must be satisfied. The start times of machines may be different.

In the example in Sect. 9.6, for the sake of simplicity, we consider m machines in elements of type 5 as *proportional* (also called *uniform* in the literature) rather than arbitrary (*unrelated*). We specify productivities of the machines p_i, $i = \overline{1,m}$,

---

[1]Here and after, $\overline{a,b}$ denotes the interval of integer numbers from a to b, that is, $\overline{a,b} = \mathbb{Z} \cap [a,b] = a,\ a+1,\ \ldots,\ b$.

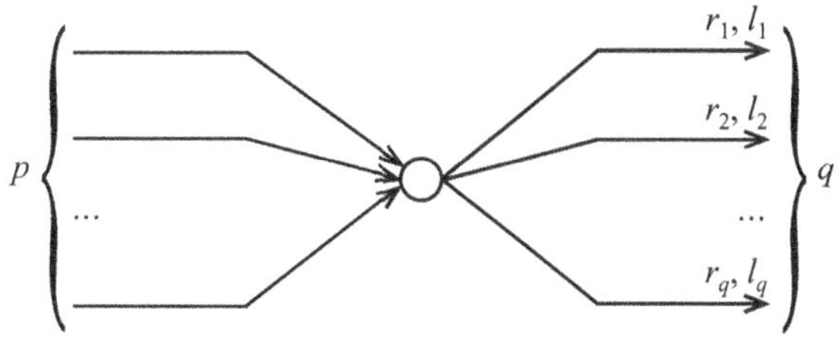

Fig. 9.2 An element of buffer type

not the matrix of each job processing time on each machine. This case covers most practical applications, however it is not a limitation for the general planning methodology.

We call a *buffer* an element of the form shown in Fig. 9.2. A buffer reflects the appearance of a new set of $q \geq 1$ uncompleted jobs as a result of the previous $p \geq 1$ jobs completion. A buffer describes the precedence relations between jobs but is not any combination of jobs. It may, in particular, describe a unit assembling from several parts or a unit disassembling to several parts. A buffer must always follow an element of type 3, according to the element's definition.

We use the two following general rules for buffer type elements. The due date of jobs to the left of a buffer is equal to the minimum start time of jobs to the right of it. And the earliest start time of jobs to the right of a buffer is equal to the maximum completion time of jobs to the left of it.

A network model should exactly reflect a real production process. Each job has its own number corresponding to index i of its parameters l_i, d_i, r_i, etc. Job numbers are different for each element and determined by the production technology.

9.1.2 *Properties of the Technological Process Specified by a Network*

The subject of our study is the following class of network models:

1. A network is a combination of elements of type 1–5. It is an oriented acyclic graph similar to flow network with several sources and several sinks.
   Elements of type 2–5 correspond in the network to machines (machine sets) which must process a set of jobs in random order without interruption. At any time point, a single machine may process only one job. Element of type 1 corresponds to a machine which processes only one job.
2. The network is an oriented graph with two types of vertices. A vertex of type 1 is represented by a circle ○, a vertex of type 2 is a square □ or a different machine notation. An arrow (arc) *entering* a circle indicates a completed job. An arrow *coming out* of a circle is a new uncompleted job ready for processing. An uncompleted job must be immediately followed by a vertex of type 2 (a square)

which is a machine the job is processed on. A vertex of type 1 may be immediately followed only by a vertex of type 2.

3. Direction of all arrows is from the left to the right.
4. We can divide the network into three parts.

4.1. The *final part* of the network. It may terminate with:

- elements of type 1: $\bigcirc\!\to\!\square\!\dashv C_i$ where C_i is the completion time of a terminal job (which does not have successors);
- a job completed on an element of type 2–5.

Let d_i denote the due date of a job i. A feasible schedule must satisfy $C_i \le d_i$ for jobs immediately preceded by elements of type 1–3 and $C_i \in d_i^*$ for jobs immediately preceded by elements of type 4, 5.

The final part of the network is an arbitrary combination of elements of type 1–5 which meets the following condition. Values of d_i must unambiguously specify the due dates of jobs for elements of type 1–3; values of d_i^* must unambiguously specify the ranges of completion times in a feasible schedule (allowable earliness) for elements of type 4, 5. Jobs are processed in random order on elements of type 2–5. Each job does not have elements of type 2–5 immediately following it in the final part of the network (without taking into account elements of type 1).

For some products, we may specify terminal jobs by elements of type 2–5 not necessarily of the last network layer. The last layer of the network may relate to the one before last or to any intermediate network layer.

Terminal jobs may converge to one terminal assembly job through an element of buffer type.

4.2. The *intermediate part* of the network. This is an arbitrary combination of elements of type 1–5 satisfying the following condition. The latest readiness times of jobs on elements of type 2–5 unambiguously specify the due dates for the immediately preceding elements of type 2–5. Immediate preceding does not takes into account elements of type 1 which may appear between elements of type 2–5. This means that between an element of type 2–5 and the immediately preceding element of type 2–5 may exist an oriented path containing only vertices corresponding to elements of type 1.

Intermediate part of a network may contain elements of type 2–5 immediately preceding more than one element of type 2–5.

Elements of type 2–5 of any network layer may relate to elements of type 2–5 of any previous network layer immediately or through elements of type 1.

4.3. The *initial part* of the network. This is an arbitrary combination of elements of type 1–5 satisfying the following conditions:

- there are all initial jobs which are the vertices in the form: $\bigcirc\!\to\!\boxed{1j}$ where $j = \overline{1,p}$ and the digit 1 means the first layer of the network (this job does not have any predecessors);
- an arbitrary element of type 1–5 is immediately preceded only by initial jobs;

- consider a complete group of elements of type 2–5 in the network. The start times of all elements in this group unambiguously specify the start times of all the initial jobs in this group.

We show a simplified example of a network model in Fig. 9.3.

Remark 9.1 A network model should reflect a real production process.

Remark 9.2 The jobs processing may be spaced in time. Therefore, different vertices of type 2 may present in the network the same physically existing machines.

Remark 9.3 The network should be built so that we are able to see the production technology of each product (the path from each initial job to a terminal job).

Remark 9.4 All output arrows of elements of type 3 must pass to the next network layer through a buffer. This is the working requirement for elements of type 3.

The requirement for elements of type 2–5 to work without interrupts is not always strict. Suppose we know in advance that the same machines should process different job groups which are spaced in time. Then we replace physical machines (machine groups) in the resulting MSNSP model with a virtual set of machines (machine groups) operating without interruption.

The assumption that job groups are spaced in time allows us to exclude the following configuration of the considered network. An output arrow(s) of any element of type 2–5 cannot immediately (or through elements of type 1) relate to a terminal job in the network when the remaining arrows immediately (or through elements of type 1) relate to the input jobs of elements of type 2–5. That is, we break physically existing equipment into virtual elements, some of them are in the initial and intermediate parts of the network, and some are in the final part.

Algorithmic ware of the FLM consists of five units. We determine the actions of experts in Decision Making Unit (Unit 1) and implement the formal models in Units 2–5:

- Unit 1: Decision Making Unit (DMU).
- Unit 2: the first level model construction, preliminary planning.
- Unit 3: the second level model construction, coordinated planning.
- Unit 4: the third level model construction, exact operational planning.
- Unit 5: the fourth level of the model, operative planning.

Functions of the DMU are:

1. The order portfolio composition.
2. The planning criterion specification: the choice of objective; defining the functional by coefficients; setting the minimum expected profit.
3. Specification of the technology to process the order portfolio. The topology associated with the technology is specified in the form of N separate subnets where N is the number of basic criteria in the overall objective function. Subnets can have common equipment. For a synthetic criterion, therefore, we consider a special case when the initial network is represented as independent subnets with common virtual machines.

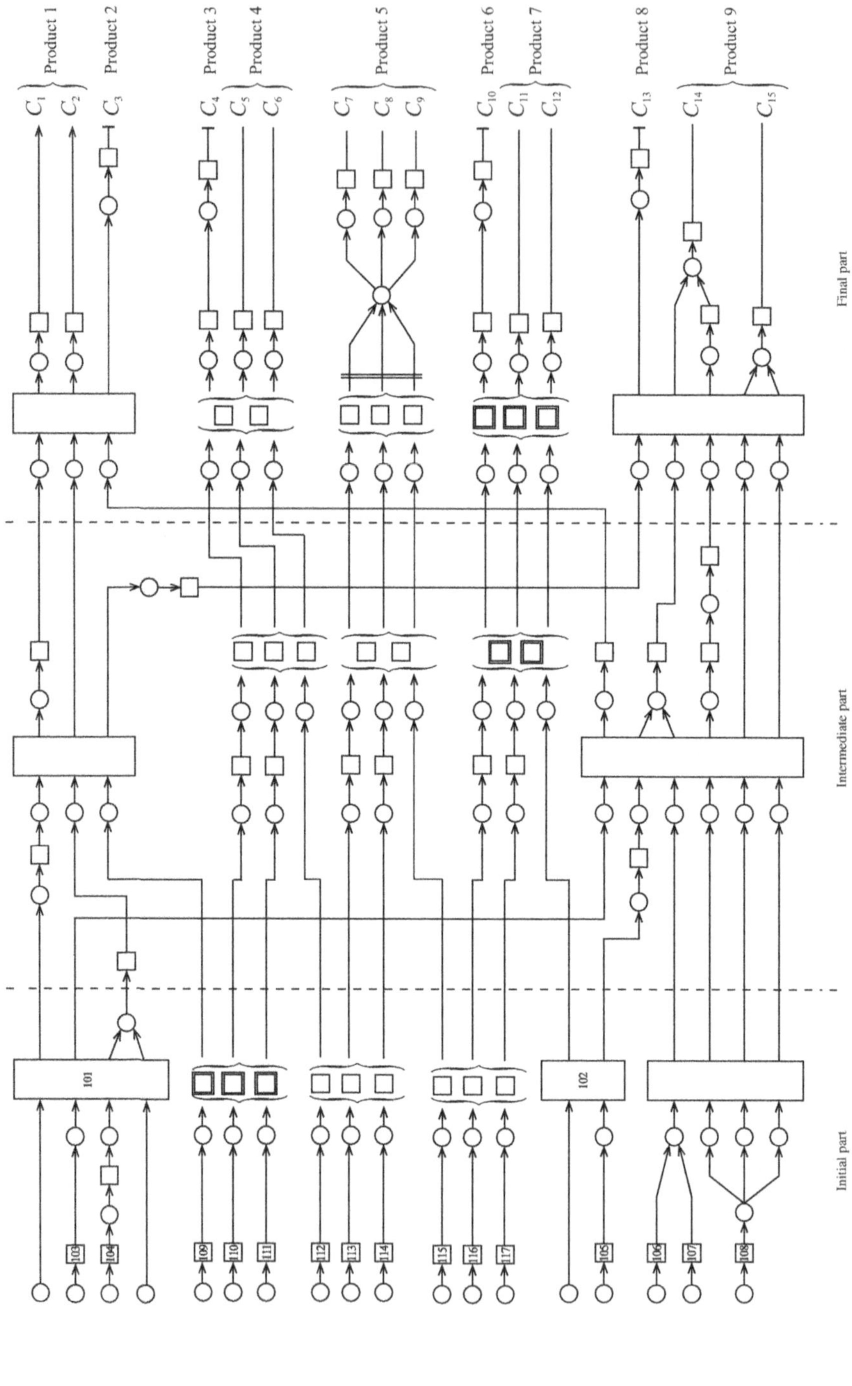

Fig. 9.3 An example of initial technological network model

4. Achieving an agreement with the customer on all changes to the plan (if a feasible schedule is not obtained).
5. A new order portfolio composition (if the obtained profit is less than the specified minimum expected value).
6. Choosing the best plan according to the criterion of minimizing the risk of not obtaining the expected profit. We choose the plan from the set of alternative plans obtained at the third level of planning.
7. Organization of interaction between Units 2–5.
8. Operative planning execution using models of Unit 5.
9. If necessary, experts set the desired completion times for products during planning by a basic optimality criterion. The times are passed to the model of the third level of planning as the due dates.
10. Determination of the operational plan adjustment time at the fourth level and setting new evaluation criteria in regard to the approved due dates.

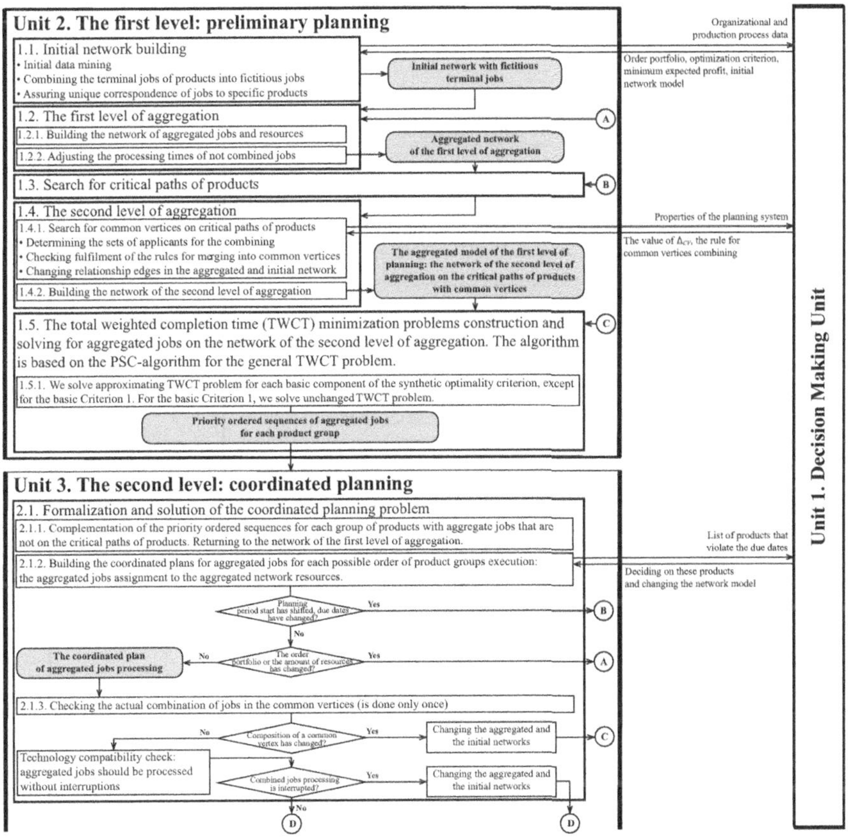

Fig. 9.4 Scheme of the algorithmic ware of the FLM, part 1

We considered the procedure for a well-grounded choice of a production plan from the set of feasible plans in [1]. The procedure was developed for uncertainty conditions and is based on a hierarchy tree construction. We propose the further development of this procedure in Sect. 8.5.

We show the general scheme of the algorithmic ware of the FLM in Figs. 9.4 and 9.5. The numbers in the scheme correspond to the step numbers of the methodology presented in Sects. 9.2–9.7.

We formalize the model of the second level taking into account the solution obtained for the model of the first level. Similarly, we formalize the model of the third level considering the solution obtained for the model of the second level. This is why we combine a model formulation and its solution into a single block.

We will now describe the algorithmic support of blocks 2–5 in detail by steps.

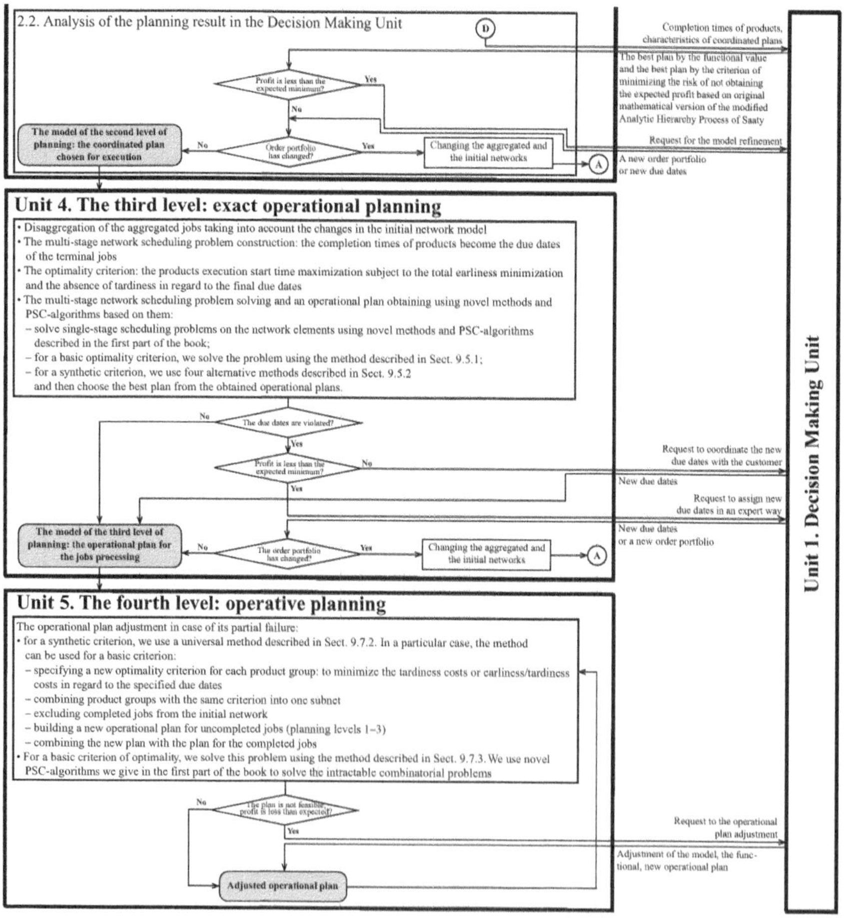

Fig. 9.5 Scheme of the algorithmic ware of the FLM, part 2

9.2 Unit 2. Model of the First Level of Planning

In Unit 2, we construct an aggregated planning model. We solve on it optimization problems by the criterion of minimizing the total weighted completion time of aggregated jobs (TWCT).

9.2.1 *The Problem Formulation. The Formalization of Optimality Criteria*

Step 1.1. The Problem Formulation Here we specify:

- a technological network that defines the technology for a set of interrelated jobs packages (products) J; under a product we may also mean the entire series of uniform products; specifying the technology guarantees that the technological network is relevant to a real technological process;
- one of the 31 criteria of optimality (five basic criteria plus 26 synthetic criteria which are all possible linear convolutions of the basic ones); we list all the criteria below;
- a minimum expected profit of the enterprise;
- for each job i: a unit processing time l_i for a single product in a series; start times s_i and completion times C_i are calculated during the problem solving;
- for each product i: the number of products in the order portfolio (in the series) N_i, weight ω_i, due date d_i, and range of completion times in a feasible schedule (allowable earliness) Δ_i for those products where the elements of type 4 and 5 are in their production technology; the completion time $\mathbb{C}_i$ for each product i will be calculated during the problem solving;
- for each resource: the type of the element and the number of parallel machines;
- for each element of type 5: productivities p_i of machines (we specify the processing time l_i for a machine with unit productivity).

Each functional we give below is unambiguously specified by the completion times of products. The completion time of a product should correspond to the completion time of a terminal job. But one product (product series) may correspond to many terminal jobs in the initial network. In this case, we combine all such terminal jobs into one fictitious job. We assign to it the due date corresponding to the due date of the product. The processing time of a fictitious job is zero. Figure 9.6 shows an example of initial network of the first level in which we have combined terminal jobs into fictitious jobs denoted by dotted circles ◌. We specify the problem of the first level of planning with a network like this.

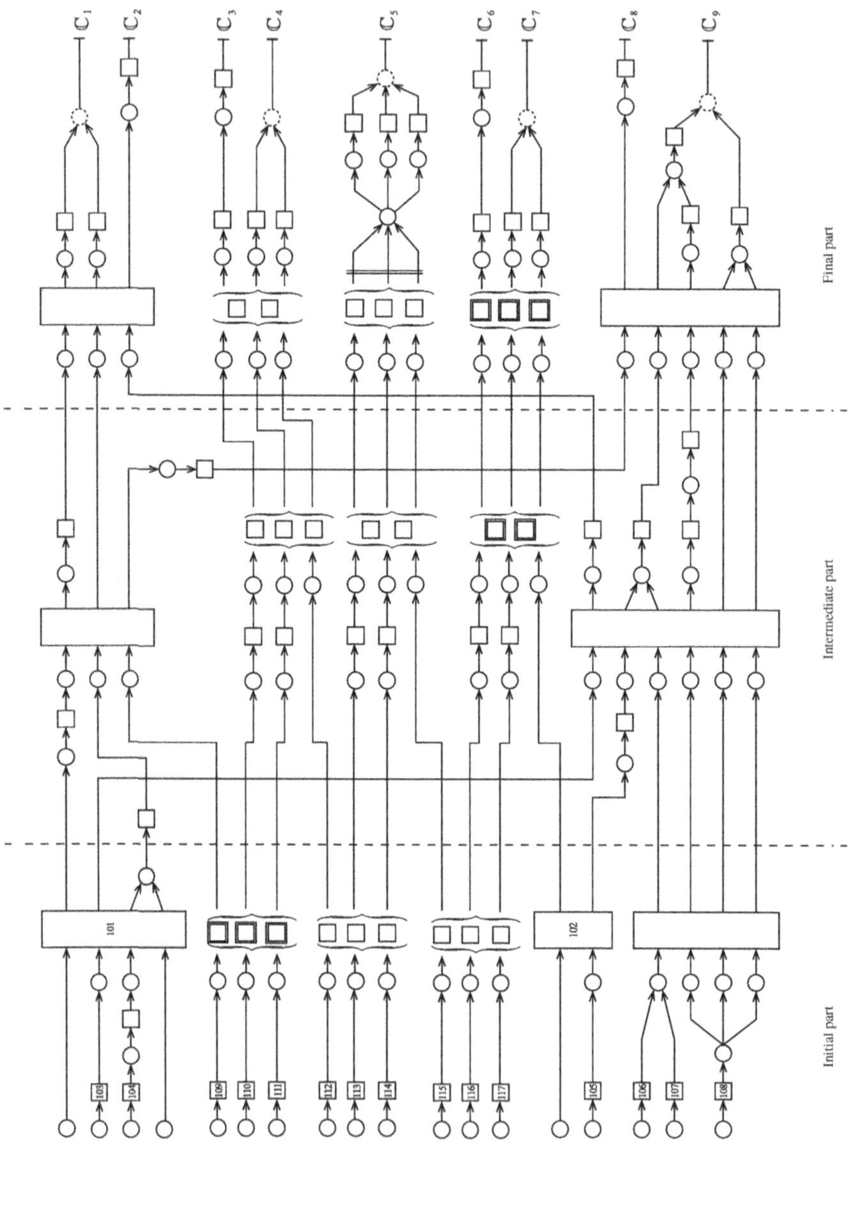

Fig. 9.6 An example of the model of the first level of planning

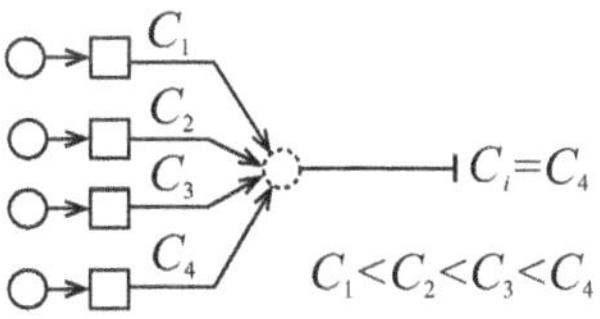

Fig. 9.7 The completion time determination for a fictitious job

Remark 9.5 We should be able to unambiguously determine from production technology the terminal job that completes later than the others. We exclude all terminal jobs included into a fictitious job, except for the latest one, from consideration during planning, as they are obviously complete before the due date. An example is at Fig. 9.7.

For planning by a basic optimality criterion, the network should have the properties described in Sect. 9.1 with the following restriction. Several virtual machines that correspond to the same physical machine (we denote them by the same element number) may appear in the network only if they are spaced in time by the technology. This makes impossible different jobs processing at the same time interval on the same physical machine. To satisfy this restriction, we can use, e.g., a buffer between the elements corresponding to the same physical machine to separate the sets of jobs they process.

If this restriction is satisfied, we can use efficient solving methods given in Sects. 9.5.1 and 9.7.3. The methods allow us to obtain a solution directly moving from the terminal jobs to the initial ones using mathematically exact solutions of single stage scheduling problems. For an arbitrary virtual network, we can solve the problem using heuristic methods (see Sects. 9.5.2 and 9.7.2), with a lesser guarantee for the solution optimality.

For planning by a synthetic optimality criterion, we should divide the products into $N \in [1, 5]$ separate groups. N corresponds to the number of basic components in the synthetic criterion. No product can be a part of two different groups. We specify the topology associated with the technology as N separate subnets corresponding to N separate product groups. Each subnet has its own basic criterion. The properties of each subnet should correspond to Sect. 9.1. The subnets cannot have precedence relations with each other but may be interrelated through a common equipment. This prevents us from scheduling jobs processing in subnets independently of each other. In addition, the common equipment (virtual machines corresponding to the same physical machine) may be inside each subnet. The planning methodology for a synthetic criterion presented below allows to space in time jobs processed on a common equipment. Therefore, the methods can be applied on an acyclic network of arbitrary type.

Basic Optimality Criteria are the criteria 1–5 below.2

Criterion 1 The planning system's total profit maximization in the absence of the products' due dates.

The time factor is important for an enterprise's profitability. Who provides the fastest fulfillment of orders and releases new products earlier will have an advantage. In the absence of due dates, the profit from the sale of product i (product series i) is a function of time and is equal to $\omega_i(T) \cdot (T - \mathbb{C}_i)$. Here, $\omega_i(T)$ is the weight coefficient of product i (product series i) determined in an expert way; T is the planning period; $\mathbb{C}_i \leq T$ is the completion time of product i (product series i) corresponding to the completion time of its (fictitious) terminal job. The planning system's total profit maximization criterion in this case is defined by the formula

$$\sum_{i=1}^{n} \omega_i(T) \cdot (T - \mathbb{C}_i) + P - C$$
$$= T \cdot \sum_{i=1}^{n} \omega_i(T) - \sum_{i=1}^{n} \omega_i(T) \cdot \mathbb{C}_i + P - C \rightarrow \max$$

where n is the number of products optimized; P is the guaranteed minimum income from the sale of all n products or from fulfillment of all n orders; C is the total cost. The maximized function has the form

$$\max\left\{\sum_{i=1}^{n} \omega_i(T) \cdot (T - \mathbb{C}_i)\right\} + P - C.$$

Thus, *Criterion 1* is

$$\min \sum_{i=1}^{n} \omega_i(T) \cdot \mathbb{C}_i. \tag{9.1}$$

The criterion (9.1) is equivalent to the total weighted completion time of products minimization criterion (TWCT) with a partial order given by an oriented acyclic graph constructed on the products' critical paths.

Criterion 2 The planning system's total profit maximization subject to the condition: each product $i \in I$ has a due date d_i that must not be violated (just in time planning):

$$\max\left\{\sum_{i=1}^{n} \omega_i U_i\right\} - C \quad \text{where} \quad U_i = \begin{cases} 1, \mathbb{C}_i = d_i \\ 0, \mathbb{C}_i \neq d_i \end{cases};$$

ω_i is the income from the sale of the product i if it completes just in time; C is the total cost. Thus, *Criterion 2* has the form:

$$\max\left\{\sum\nolimits_{i=1}^{n} \omega_i U_i\right\}. \tag{9.2}$$

Criterion 3 The planning system's total profit maximization subject to the conditions: each product $i \in I$ has a due date d_i; the total weighted tardiness of products in regard to their due dates must be minimized:

$$\max\left\{P - \sum_{i=1}^{n} \omega_i \max(0, \mathbb{C}_i - d_i)\right\} - C$$

where P is the guaranteed minimum income from the sale of all n products if they are not tardy; C is the total cost; ω_i is the cost of (penalty for) the tardiness of product i in relation to its due date per time unit. Thus, *Criterion 3* has the form:

$$\min\left\{\sum_{i=1}^{n} \omega_i \max(0,\ \mathbb{C}_i - d_i)\right\}. \tag{9.3}$$

The value of $\omega_i \max(0, \mathbb{C}_i - d_i)$ is the reduction of income P in the case if product i has the tardiness $\mathbb{C}_i - d_i$. Experts decide in the DMU on whether to process such products or refuse to process them.

Criterion 4 Each task $i \in I$ has a due date d_i and a given absolute value of profit ω_i for its processing. The profit does not depend on the completion time of the task if it is not tardy in regard to its due date. Otherwise, the planning system's profit for this task is zero. The problem is to maximize the total profit of the planning system:

$$\max\left\{\sum_{i=1}^{n} \omega_i U_i\right\} - C \quad \text{where} \quad U_i = \begin{cases} 1, \mathbb{C}_i \le d_i \\ 0, \mathbb{C}_i > d_i \end{cases};$$

ω_i is the income from the sale of the product i if it is not tardy in regard to its due date; C is the risk of the profit loss due to the due date violating. Thus, *Criterion 4* has the form:

$$\max\left\{\sum_{i=1}^{n} \omega_i U_i\right\}. \tag{9.4}$$

Criterion 5 All tasks have due dates d_i. We need to minimize the total cost (penalty for the planning system) both for earliness and tardiness in regard to the due dates:

$$\max\left\{P - \sum_{i=1}^{n} \omega_i |\mathbb{C}_i - d_i|\right\} - C$$

where P is the guaranteed minimum income from the sale of all n products if they are not tardy or early; C is the total cost; ω_i is the cost of (penalty for) the deviation of the completion time of product i from the due date per time unit. Thus, *Criterion 5* has the form:

$$\min \sum_{i=1}^{n} \omega_i |\mathbb{C}_i - d_i|. \tag{9.5}$$

The value of $\omega_i|\mathbb{C}_i - d_i|$ is the reduction of income P in the case if product i has the tardiness $\mathbb{C}_i - d_i$ or the earliness $d_i - \mathbb{C}_i$. Experts decide in the DMU on whether to process such products or refuse to process them.

Synthetic Criteria are the linear convolutions of all possible combinations of the basic criteria. The costs C for a synthetic criterion are determined by an expert, since a common equipment process the products under different basic criteria.

Criterion 6 (Criteria 1 and 2 Combination) The planning system's total profit maximization subject to the condition: some tasks $i \in \overline{1,k}$ have due dates that must not be violated, other tasks do not have due dates:

$$\max\left\{\sum_{i=1}^{k} \omega_i U_i - \sum_{i=k+1}^{n} \omega_i(T)\cdot \mathbb{C}_i\right\} + P - C \text{ where } U_i = \begin{cases} 1, \mathbb{C}_i = d_i \\ 0, \mathbb{C}_i \neq d_i \end{cases};$$

P is the guaranteed minimum income from the sale of products $i = \overline{k+1,n}$; C is the total cost; ω_i is the income from the sale of the product i if it completes just in time; $\omega_i(T)$ is the weight coefficient of product i (has the same meaning as in Criterion 1). Thus, *Criterion 6* has the form:

$$\max\left\{\sum_{i=1}^{k} \omega_i U_i - \sum_{i=k+1}^{n} \omega_i(T)\cdot \mathbb{C}_i\right\}. \tag{9.6}$$

Criterion 7 (Criteria 2 and 3 Combination) Corresponds to the Criterion 3 with the additional condition: due dates of some tasks $i \in \overline{1,k}$ must not be violated:

$$\max\left\{\sum_{i=1}^{k} \omega'_i U_i - \sum_{i=k+1}^{n} \omega''_i \max(0, \mathbb{C}_i - d_i)\right\} + P - C \quad \text{where } U_i = \begin{cases} 1, \mathbb{C}_i = d_i \\ 0, \mathbb{C}_i \neq d_i \end{cases};$$

ω'_i is the income from the sale of the product i if it completes just in time; ω''_i is the cost of (penalty for) the tardiness of product i in relation to its due date per time unit; P is the guaranteed minimum income from the sale of products $i = \overline{k+1,n}$ if they are not tardy; C is the total cost. Thus, *Criterion 7* has the form:

$$\max\left\{\sum_{i=1}^{k}\omega'_iU_i-\sum_{i=k+1}^{n}\omega''_i\max(0,\mathbb{C}_i-d_i)\right\}. \quad (9.7)$$

The value of $\omega''_i\max(0,\mathbb{C}_i-d_i)$ is the reduction of income P in the case if product i has the tardiness $\mathbb{C}_i-d_i$. Experts decide in the DMU on whether to process such products or refuse to process them.

Similarly to criteria 6 and 7, we define the remaining synthetic criteria 8–31 which are the combinations of basic criteria 1–5.

Criterion 8 (Criteria 1 and 3 Combination) has the form:

$$\max\left\{-\sum_{i=1}^{k}\omega_i(T)\cdot\mathbb{C}_i-\sum_{i=k+1}^{n}\omega_i\max(0,\ \mathbb{C}_i-d_i)\right\} \quad (9.8)$$

where $\omega_i(T)$ has the same meaning as in Criterion 1 and ω_i as in Criterion 3.

Criterion 9 (Criteria 1 and 4 Combination) has the form:

$$\max\left\{-\sum_{i=1}^{k}\omega_i(T)\cdot\mathbb{C}_i+\sum_{i=k+1}^{n}\omega_iU_i\right\},\ U_i=\begin{cases}1,\mathbb{C}_i\le d_i\\0,\mathbb{C}_i>d_i\end{cases} \quad (9.9)$$

where $\omega_i(T)$ has the same meaning as in Criterion 1 and ω_i as in Criterion 4.

Criterion 10 (Criteria 1 and 5 Combination) has the form:

$$\max\left\{-\sum_{i=1}^{k}\omega_i(T)\cdot\mathbb{C}_i-\sum_{i=k+1}^{n}\omega_i|\mathbb{C}_i-d_i|\right\} \quad (9.10)$$

where $\omega_i(T)$ has the same meaning as in Criterion 1 and ω_i as in Criterion 5.

Criterion 11 (Criteria 2 and 4 Combination) has the form:

$$\max\left\{\sum_{i=1}^{k}\omega'_iU_i+\sum_{i=k+1}^{n}\omega''_iV_i\right\},U_i=\begin{cases}1,\mathbb{C}_i=d_i\\0,\mathbb{C}_i\ne d_i\end{cases},V_i=\begin{cases}1,\mathbb{C}_i\le d_i\\0,\mathbb{C}_i>d_i\end{cases} \quad (9.11)$$

where ω'_i has the same meaning as ω_i in Criterion 2 and ω''_i as ω_i in Criterion 4.

Criterion 12 (Criteria 2 and 5 Combination) has the form:

$$\max\left\{\sum_{i=1}^{k}\omega'_iU_i-\sum_{i=k+1}^{n}\omega_i|\mathbb{C}_i-d_i|\right\},\quad U_i=\begin{cases}1,\mathbb{C}_i=d_i\\0,\mathbb{C}_i\ne d_i\end{cases} \quad (9.12)$$

where ω'_i has the same meaning as ω_i in Criterion 2 and ω''_i as ω_i in Criterion 5.

Criterion 13 (Criteria 3 and 4 Combination) has the form:

$$\max\left\{\sum_{i=1}^{k}\omega_i' U_i - \sum_{i=k+1}^{n}\omega_i''\max(0, \mathbb{C}_i - d_i)\right\},\quad U_i = \begin{cases}1, \mathbb{C}_i \le d_i\\ 0, \mathbb{C}_i > d_i\end{cases} \tag{9.13}$$

where ω_i' has the same meaning as ω_i in Criterion 4 and ω_i'' as ω_i in Criterion 3.

Criterion 14 (Criteria 3 and 5 Combination) has the form:

$$\min\left\{\sum_{i=1}^{k}\omega_i'\max(0,\ \mathbb{C}_i - d_i) + \sum_{i=k+1}^{n}\omega_i''|\mathbb{C}_i - d_i|\right\} \tag{9.14}$$

where ω_i' has the same meaning as ω_i in Criterion 3 and ω_i'' as ω_i in Criterion 5.

Criterion 15 (Criteria 4 and 5 Combination) has the form:

$$\max\left\{\sum_{i=1}^{k}\omega_i' U_i - \sum_{i=k+1}^{n}\omega_i''|\mathbb{C}_i - d_i|\right\},\quad U_i = \begin{cases}1, \mathbb{C}_i \le d_i\\ 0, \mathbb{C}_i > d_i\end{cases} \tag{9.15}$$

where ω_i' has the same meaning as ω_i in Criterion 4 and ω_i'' as ω_i in Criterion 5.

Criterion 16 (Criteria 1, 2, and 3 Combination) has the form:

$$\begin{gathered}\max\left\{-\sum_{i=1}^{k}\omega_i(T)\cdot\mathbb{C}_i + \sum_{i=k+1}^{r}\omega_i' U_i - \sum_{i=r+1}^{n}\omega_i''\max(0, \mathbb{C}_i - d_i)\right\},\\ U_i = \begin{cases}1, \mathbb{C}_i = d_i\\ 0, \mathbb{C}_i \ne d_i\end{cases}\end{gathered} \tag{9.16}$$

where $\omega_i(T)$ has the same meaning as in Criterion 1, ω_i' as ω_i in Criterion 2, and ω_i'' as ω_i in Criterion 3.

Criterion 17 (Criteria 1, 2, and 4 Combination) has the form:

$$\begin{gathered}\max\left\{-\sum_{i=1}^{k}\omega_i(T)\cdot\mathbb{C}_i + \sum_{i=k+1}^{r}\omega_i' U_i + \sum_{i=r+1}^{n}\omega_i'' V_i\right\},\\ U_i = \begin{cases}1,\ \mathbb{C}_i = d_i\\ 0,\ \mathbb{C}_i \ne d_i\end{cases},\ V_i = \begin{cases}1,\ \mathbb{C}_i \le d_i\\ 0,\ \mathbb{C}_i > d_i\end{cases}\end{gathered} \tag{9.17}$$

where $\omega_i(T)$ has the same meaning as in Criterion 1, ω_i' as ω_i in Criterion 2, and ω_i'' as ω_i in Criterion 4.

Criterion 18 (Criteria 1, 2, and 5 Combination) has the form:

$$\max\left\{-\sum_{i=1}^{k}\omega_i(T)\cdot\mathbb{C}_i+\sum_{i=k+1}^{r}\omega_i' U_i-\sum_{i=r+1}^{n}\omega_i''|\mathbb{C}_i-d_i|\right\},\quad U_i=\begin{cases}1, \mathbb{C}_i=d_i\\ 0, \mathbb{C}_i\neq d_i\end{cases} \tag{9.18}$$

where $\omega_i(T)$ has the same meaning as in Criterion 1, ω_i' as ω_i in Criterion 2, and ω_i'' as ω_i in Criterion 5.

Criterion 19 (Criteria 1, 3, and 4 Combination) has the form:

$$\max\left\{-\sum_{i=1}^{k}\omega_i(T)\cdot\mathbb{C}_i-\sum_{i=k+1}^{r}\omega_i'\max(0,\ \mathbb{C}_i-d_i)+\sum_{i=r+1}^{n}\omega_i'' U_i\right\},$$
$$U_i=\begin{cases}1, \mathbb{C}_i\leq d_i\\ 0, \mathbb{C}_i> d_i\end{cases} \tag{9.19}$$

where $\omega_i(T)$ has the same meaning as in Criterion 1, ω_i' as ω_i in Criterion 3, and ω_i'' as ω_i in Criterion 4.

Criterion 20 (Criteria 1, 3, and 5 Combination) has the form:

$$\max\left\{-\sum_{i=1}^{k}\omega_i(T)\cdot\mathbb{C}_i-\sum_{i=k+1}^{r}\omega_i'\max(0,\ \mathbb{C}_i-d_i)-\sum_{i=r+1}^{n}\omega_i''|\mathbb{C}_i-d_i|\right\} \tag{9.20}$$

where $\omega_i(T)$ has the same meaning as in Criterion 1, ω_i' as ω_i in Criterion 3, and ω_i'' as ω_i in Criterion 5.

Criterion 21 (Criteria 1, 4, and 5 Combination) has the form:

$$\max\left\{-\sum_{i=1}^{k}\omega_i(T)\cdot\mathbb{C}_i+\sum_{i=k+1}^{r}\omega_i' U_i-\sum_{i=r+1}^{n}\omega_i''|\mathbb{C}_i-d_i|\right\},\quad U_i=\begin{cases}1, \mathbb{C}_i\leq d_i\\ 0, \mathbb{C}_i> d_i\end{cases} \tag{9.21}$$

where $\omega_i(T)$ has the same meaning as in Criterion 1, ω_i' as ω_i in Criterion 4, and ω_i'' as ω_i in Criterion 5.

Criterion 22 (Criteria 2, 3, and 4 Combination) has the form:

$$\max\left\{\sum_{i=1}^{k}\omega_i U_i - \sum_{i=k+1}^{r}\omega_i'\max(0,\ \mathbb{C}_i - d_i) + \sum_{i=r+1}^{n}\omega'' V_i\right\},$$
$$U_i = \begin{cases}1, \mathbb{C}_i = d_i\\ 0, \mathbb{C}_i \neq d_i\end{cases}, V_i = \begin{cases}1, \mathbb{C}_i \leq d_i\\ 0, \mathbb{C}_i > d_i\end{cases} \tag{9.22}$$

where ω_i has the same meaning as in Criterion 2, ω_i' as ω_i in Criterion 3 and ω_i'' as ω_i in Criterion 4.

Criterion 23 (Criteria 2, 3, and 5 Combination) has the form:

$$\max\left\{\sum_{i=1}^{k}\omega_i U_i - \sum_{i=k+1}^{r}\omega_i'\max(0,\ \mathbb{C}_i - d_i) - \sum_{i=r+1}^{n}\omega_i''|\mathbb{C}_i - d_i|\right\},$$
$$U_i = \begin{cases}1, \mathbb{C}_i = d_i\\ 0, \mathbb{C}_i \neq d_i\end{cases} \tag{9.23}$$

where ω_i has the same meaning as in Criterion 2, ω_i' as ω_i in Criterion 3, and ω_i'' as ω_i in Criterion 5.

Criterion 24 (Criteria 2, 4, and 5 Combination) has the form:

$$\max\left\{\sum_{i=1}^{k}\omega_i U_i + \sum_{i=k+1}^{r}\omega' V_i - \sum_{i=r+1}^{n}\omega_i''|\mathbb{C}_i - d_i|\right\},$$
$$U_i = \begin{cases}1, \mathbb{C}_i = d_i\\ 0, \mathbb{C}_i \neq d_i\end{cases}, V_i = \begin{cases}1, \mathbb{C}_i \leq d_i\\ 0, \mathbb{C}_i > d_i\end{cases} \tag{9.24}$$

where ω_i has the same meaning as in Criterion 2, ω_i' as ω_i in Criterion 4, and ω_i'' as ω_i in Criterion 5.

Criterion 25 (Criteria 3, 4, and 5 Combination) has the form:

$$\max\left\{\sum_{i=1}^{k}\omega' U_i - \sum_{i=k+1}^{r}\omega_i'\max(0, \mathbb{C}_i - d_i) - \sum_{i=r+1}^{n}\omega_i''|\mathbb{C}_i - d_i|\right\},$$
$$U_i = \begin{cases}1, \mathbb{C}_i = d_i\\ 0, \mathbb{C}_i \neq d_i\end{cases} \tag{9.25}$$

where ω_i has the same meaning as in Criterion 4, ω_i' as ω_i in Criterion 3, and ω_i'' as ω_i in Criterion 5.

Criterion 26 (Criteria 1, 2, 3, and 4 Combination) has the form:

$$\max\left\{-\sum_{i=1}^{k}\omega_i(T)\cdot\mathbb{C}_i+\sum_{i=k+1}^{r}\omega_iU_i-\sum_{i=r+1}^{s}\omega_i'\max(0,\ \mathbb{C}_i-d_i)+\sum_{i=s+1}^{n}\omega''V_i\right\},$$
$$U_i=\begin{cases}1,\mathbb{C}_i=d_i\\0,\mathbb{C}_i\neq d_i\end{cases},V_i=\begin{cases}1,\mathbb{C}_i\leq d_i\\0,\mathbb{C}_i>d_i\end{cases} \tag{9.26}$$

where $\omega_i(T)$ has the same meaning as in Criterion 1, ω_i as in Criterion 2, ω_i' as ω_i in Criterion 3, and ω_i'' as ω_i in Criterion 4.

Criterion 27 (Criteria 1, 2, 3, and 5 Combination) has the form:

$$\max\left\{-\sum_{i=1}^{k}\omega_i(T)\cdot\mathbb{C}_i+\sum_{i=k+1}^{r}\omega_iU_i-\sum_{i=r+1}^{s}\omega_i'\max(0,\ \mathbb{C}_i-d_i)-\sum_{i=s+1}^{n}\omega_i''|\mathbb{C}_i-d_i|\right\} \tag{9.27}$$

where $U_i=\begin{cases}1,\mathbb{C}_i=d_i\\0,\mathbb{C}_i\neq d_i\end{cases}$, $\omega_i(T)$ has the same meaning as in Criterion 1, ω_i as in Criterion 2, ω_i' as ω_i in Criterion 3, and ω_i'' as ω_i in Criterion 5.

Criterion 28 (Criteria 1, 2, 4, and 5 Combination) has the form:

$$\max\left\{-\sum_{i=1}^{k}\omega_i(T)\cdot\mathbb{C}_i+\sum_{i=k+1}^{r}\omega_iU_i+\sum_{i=r+1}^{s}\omega'V_i-\sum_{i=s+1}^{n}\omega_i''|\mathbb{C}_i-d_i|\right\},$$
$$U_i=\begin{cases}1,\mathbb{C}_i=d_i\\0,\mathbb{C}_i\neq d_i\end{cases},V_i=\begin{cases}1,\mathbb{C}_i\leq d_i\\0,\mathbb{C}_i>d_i\end{cases} \tag{9.28}$$

where $\omega_i(T)$ has the same meaning as in Criterion 1, ω_i as in Criterion 2, ω_i' as ω_i in Criterion 4, and ω_i'' as ω_i in Criterion 5.

Criterion 29 (Criteria 1, 3, 4, and 5 Combination) has the form:

$$\max\left\{-\sum_{i=1}^{k}\omega_i(T)\cdot\mathbb{C}_i-\sum_{i=k+1}^{r}\omega_i\max(0,\ \mathbb{C}_i-d_i)+\sum_{i=r+1}^{s}\omega'U_i-\sum_{i=s+1}^{n}\omega_i''|\mathbb{C}_i-d_i|\right\} \tag{9.29}$$

where $U_i=\begin{cases}1,\mathbb{C}_i\leq d_i\\0,\mathbb{C}_i>d_i\end{cases}$, $\omega_i(T)$ has the same meaning as in Criterion 1, ω_i as in Criterion 3, ω_i' as ω_i in Criterion 4, and ω_i'' as ω_i in Criterion 5.

Criterion 30 (Criteria 2, 3, 4, and 5 Combination) has the form:

$$\max\left\{\sum_{i=k+1}^{r}\omega_i U_i - \sum_{i=r+1}^{s}\omega_i'\max(0,\ \mathbb{C}_i - d_i) + \sum_{i=s+1}^{m}\omega'' V_i - \sum_{i=m+1}^{n}\omega_i'''|\mathbb{C}_i - d_i|\right\},$$
$$U_i = \begin{cases}1, \mathbb{C}_i = d_i \\ 0, \mathbb{C}_i \neq d_i\end{cases}, V_i = \begin{cases}1, \mathbb{C}_i \leq d_i \\ 0, \mathbb{C}_i > d_i\end{cases} \tag{9.30}$$

where ω_i has the same meaning as in Criterion 2, ω_i' as ω_i in Criterion 3, ω_i'' as ω_i in Criterion 4, and ω_i''' as ω_i in Criterion 5.

Criterion 31 (Criteria 1, 2, 3, 4, and 5 Combination) has the form:

$$\max\left\{-\sum_{i=1}^{k}\omega_i(T)\cdot\mathbb{C}_i + \sum_{i=k+1}^{r}\omega_i U_i - \sum_{i=r+1}^{s}\omega_i'\max(0,\ \mathbb{C}_i - d_i) + \sum_{i=s+1}^{m}\omega'' V_i - \sum_{i=m+1}^{n}\omega_i'''|\mathbb{C}_i - d_i|\right\},$$
$$U_i = \begin{cases}1, \mathbb{C}_i - d_i \\ 0, \mathbb{C}_i \neq d_i\end{cases}, V_i = \begin{cases}1, \mathbb{C}_i \leq d_i \\ 0, \mathbb{C}_i > d_i\end{cases} \tag{9.31}$$

where $\omega_i(T)$ has the same meaning as in Criterion 1, ω_i as in Criterion 2, ω_i' as ω_i in Criterion 3, ω_i'' as ω_i in Criterion 4, and ω_i''' as ω_i in Criterion 5.

The due dates violation is allowed during the planning stage. The coefficients ω_i in the basic criteria 3 and 5 define the reduction in the estimated profit. And those in the basic criteria 2 and 4 define the penalties for the due dates violation.

The constructed topology network must uniquely define (for product's critical paths search purposes) which product this or that completed job on each buffer belongs to. Therefore, if the jobs to the left of a buffer belong to different products, then we must specify a one-to-one correspondence. For the rest of the jobs, we determine their belonging to a particular product in an obvious way by the original topology. A product's processing may begin from a buffer: initial job(s) of the product may appear in the buffer's set of uncompleted jobs. In this case, the maximum completion time of the completed jobs of the buffer must be considered as different (non-zero) beginning time of the planning period for this product.

Remark 9.6 We see that the planning period may begin at different time points for different products. Also, there may be jobs on the elements at the beginning of the planning period which belong to the previous planning period. In these cases, for the first, second and third planning levels use the modified method described in Sect. 9.7.2. The modification refers to the case when there are ongoing jobs at the time of the operational plan adjustment.

We will now consider the procedure for the problem solving by a basic optimality criterion (when the number of subnets $N = 1$). Next, we will separately describe the planning by a synthetic criterion ($N \in [2,\ 5]$).

9.2.2 *The Construction of the Model of the First Level of Planning*

9.2.2.1 The Technological Network Aggregation into a Single Machine Model with Precedence Relations Specified by an Oriented Acyclic Graph

Step 1.2. The First Level of Aggregation Here we construct an aggregated representation of initial network through an aggregation of jobs and resources.

Step 1.2.1. An aggregated job is the set of jobs that belong to one product and are processed in the same aggregated resource within single pass through the resource. We depict aggregated resources on schemes as elements of type 1 and 2.

We aggregate jobs and resources the following way. All elements of type 3–5 become aggregated elements of type 2. If element(s) with the same number sequentially process a job chain of one product, then we combine the jobs into one aggregated job. If, after that, only one job remains on the element of type 2, then we transform this element into that of type 1. We assign the processing time for the aggregated job this way:

- for jobs belonging to elements of type 2, it is the sum of processing times of jobs included into the aggregated job;
- for jobs belonging to elements of type 3, 4, it is the sum of processing times of jobs included into the aggregated job, divided by the number of machines in the element;
- for jobs belonging to elements of type 5, it is the sum of pseudo-processing times of jobs included into the aggregated job. We calculate the pseudo-processing time of each job by the formula [12, p. 76]:

$$l = \frac{1}{\sum_{i=1}^{m} \frac{1}{l_i}} \tag{9.32}$$

where l_i is the processing time of the job on a machine i; m is machine count. We can arbitrarily specify the processing time of a job j on a machine i in the element of type 5 or calculate it as l_j/p_i where p_i is the productivity of the machine i (we specify the processing time l_j of a job for a machine with unit productivity).

Remark 9.7 In cases where the processing times of combined jobs significantly differ from each other, the following restriction may be necessary: the processing time of the aggregated job must not be less than the maximum processing time of combined jobs.

Remark 9.8 As a result of aggregation, some terminal jobs of products may be merged into one terminal job. This is allowable. The rule is: we must not combine jobs coming out of a buffer with jobs that does not come out of it, as this can lead to the appearance of arcs that were originally absent from the network graph.

Step 1.2.2 We aggregate parallel machines (elements of type 3–5) into one machine (elements of type 1–2). Due to this fact, the processing time of the jobs remaining on the elements and not combined with other jobs changes. The processing time of these jobs is calculated according to the same rules: it does not change for elements of types 1 and 2; is divided by the number of machines for elements of types 3 and 4; is calculated by Eq. (9.32) for an element of type 5.

As a result, the aggregated network of the first level of aggregation may contain both aggregated jobs and non-aggregated jobs (with initial or changed processing time) if they are not included into the aggregated jobs' composition.

We show example of an aggregated network in Fig. 9.8.

Step 1.3. The Search for Products' Critical Paths Despite the smaller dimension of the problem in the aggregated representation, it still remains hard to solve. Therefore, we introduce the following restrictions in order to simplify the model:

- the flow time of each product should be determined by its critical path;
- common aggregated jobs of different products lie on their critical paths and are processed in one resource (element).

Thus, we search for the critical path of each product, i.e., the aggregated jobs route with maximum total processing time from the beginning of the product manufacturing to its completion.

To implement an algorithm for critical paths search, we must be ensured that the network is acyclic. Also, we create two empty temporary arrays. The first one is the list containing two fields: N_Z denotes a successor job number at the critical path and $L_{\max}$ is the maximum total processing time from the successor to the (fictitious) terminal job of the product. The second array is a stack, i.e., a list of the "last input —first output" type. It also has two fields: N_P is a predecessor job number and N_Z is a successor job number, that is each record of the second array lists an arc on the critical path.

Unlike the shortest path search algorithms, we cannot avoid traversing the entire network in this case. But the search should be directed. For tree type networks, a simple recursive traversal procedure of a path search in a subtree (or a supertree, for in-trees) allows you to traverse all paths in the network and at the same time visit each job only once. For general form networks, we must take into account possible revisiting a job already considered and, therefore, have to try and reduce the search at a second hit. We can store the longest path from a given job to the terminal job of the product in the second array using the job number. We can create a fictitious job with zero number as the predecessor of all the aggregated jobs of the first network layer for this product. Such zero job would be similar to a fictitious terminal job, but in the initial part of the network. Then, we could store the critical path of the product in the second array in its zero number record.

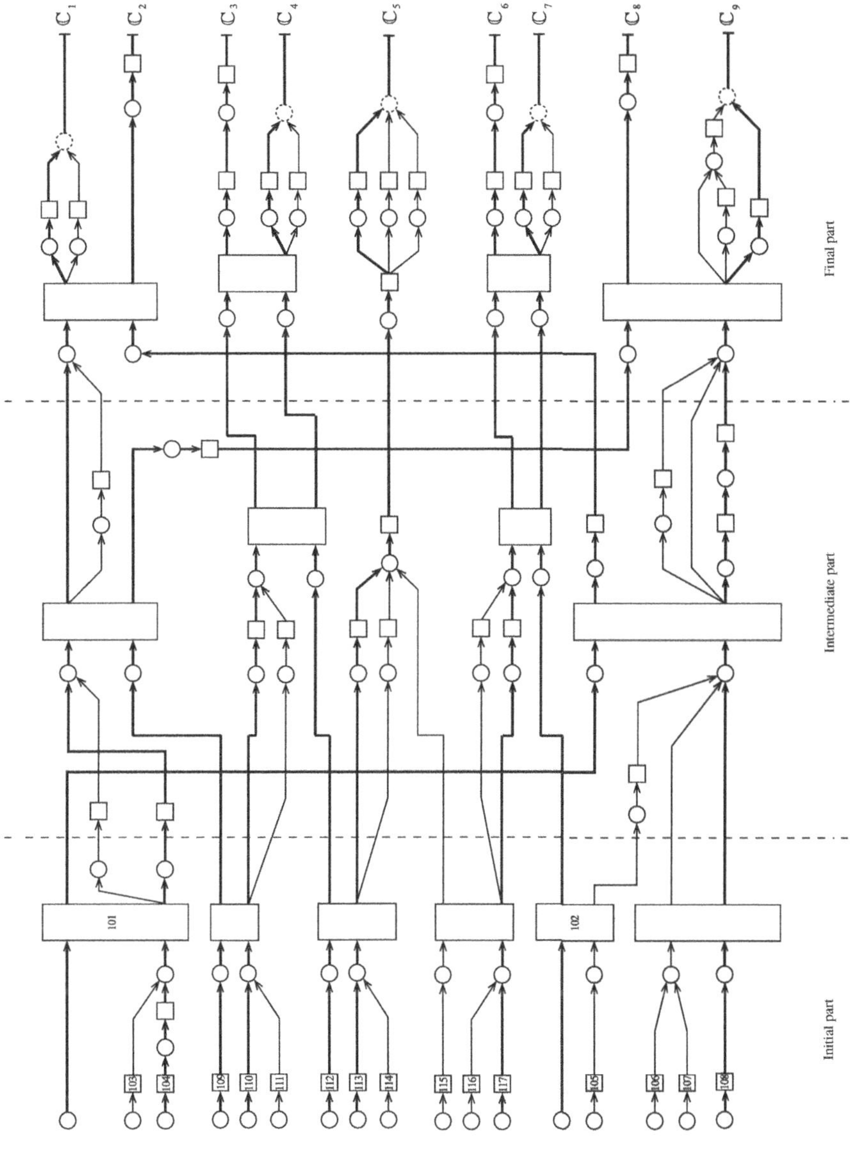

Fig. 9.8 Example of an aggregated network of the first level of aggregation

If, after a job revisiting, the length of the longest path from it to the terminal job changes, then obviously the length of the longest paths of all jobs in its supergraph changes by the same amount. The search problem would be solved if we could recalculate the lengths in the entire supergraph of any job without revisiting all the jobs. But a supergraph traversal requires recursive consideration of each job predecessors, which is similar to an exhaustive search. However, there is a way to reduce the search. If the job we revisit is on the longest path found so far (stored in the zero record of the array), then it is sufficient to increase all lengths only along this path, and not in the entire supergraph. In this case, revisiting jobs that are not on this path will simply result in a re-traversal, and recalculating the length will give the correct characteristic, regardless of its previous value.

Theoretically, there are two type of algorithms for graph traversal: forward and backward methods. In the first case, we traverse through successors. In the second case, through predecessors.

Critical Path Search Algorithm for one product (backward method):

1. Put the (fictitious) terminal job of the product into the list with $N_Z = 0$ and $L_{\max} = 0$. Put its predecessors (terminal jobs of the product) into the stack as pairs of N_P and $N_Z = 0$.
2. Process the stack top:
   - 2.1. If N_P is not in the list, then put it into the list with N_Z from the stack top and $L_{\max} = L_{\max}(N_Z) + l_{N_P}$ where l_j is the processing time of an aggregated job j.
   - 2.2. Otherwise, if $\Delta = L_{\max}(N_Z) + l_{N_P} - L_{\max}(N_P) > 0$, then replace N_Z in the list with the number of considered job and $L_{\max}$ with $L_{\max}(N_Z) + l_{N_P}$. Review the list (the path from zero fictitious job to the terminal fictitious job of the product) to search for job N_P among predecessors. If the job N_P is on this path (we revisit it), then recalculate the path again increasing the length $L_{\max}$ on it by Δ. Else, put predecessors of the job N_P into the stack for revisiting.
3. Remove the stack top. If we have put job N_P there for the first time, then put all its predecessors into the stack.
4. If the stack is empty, then go to step 5, else go to step 2.
5. The critical path is found. It is the sequence of aggregated jobs in the list starting from zero record.

In the example of an aggregated network of the first level of aggregation (see Fig. 9.8), we have marked critical paths with bold lines.

Step 1.4. The Second Level of Aggregation is the construction of a network on products' critical paths with common vertices which are the common aggregated jobs belonging to different products of the same group. This network is smaller in size because it only contains aggregated jobs that belong to the critical paths and also because some aggregated jobs are combined into the common vertices.

Step 1.4.1. Merging Jobs Into Common Vertices on Products' Critical Paths We combine into common vertices the aggregated jobs belonging to different critical paths and processed within a single pass through a resource. The requirement to such integration is determined by the production technology. Such merging should shorten the flow time of products or improve the production process in another way. For example, this should reduce the setup times, increase the overall energy savings, or reduce the total work costs. The processing time of a common vertex is equal to the sum of processing times of the aggregated jobs it combines.

We have developed heuristic rules for constructing common vertices. The rules take into account the optimality criteria, priorities of the products[2] which the combining aggregated jobs belong to, the due dates and the expected readiness time of the aggregated job for processing in this resource. We determine the readiness time on the basis of the product's due date and its critical path. The rules are based on two thesis:

- one of the main planning criteria is the maximum closeness of each product's flow time to the length of its critical path;
- a common property of the functionals we research is to execute in the first place a highest priority product, in order to implement a heuristic strategy of achieving an optimal functional value (see Step 1.5).

To prioritize the products, we construct an approximating TWCT problem for each of the optimality criteria. Its solution is a priority-ordered sequence of aggregated jobs which determines the processing order of the products.

Therefore, it is expedient to combine into common vertices the aggregated jobs that belong to products with similar priorities. However, in the case of significant setup time values, a common vertex may unite aggregated jobs with different priorities. The setups are initially included into the processing time of jobs. Jobs merging reduces the number of setups and, hence, shortens the critical paths of the combined products.

A common requirement for all the functionals under consideration is also the comparability of the combined aggregated jobs' start times. Otherwise, the waiting time of the aggregated jobs following the common vertices grows, which, as a result, increases the flow time of the merged products.

The aggregated jobs being combined into common vertices must meet the following requirements.

1. If the setup time of the resource does not exceed the maximum processing time of the aggregated jobs being combined, then the setup time does not significantly affect the priority of the common vertex.[3] In this case, the priorities of

---

[2]A product's priority is the ratio of the product's weight to its critical path length.

[3]We define an aggregated job priority as the priority of the product it belongs to. A common vertex priority is the total weight of the products the combined aggregated jobs belong to, divided by the total length of their critical paths.

products containing the aggregated jobs being combined must coincide or differ by an insignificant value Δ_{CV} ("delta of common vertices"). Δ_{CV} specifies the priority range for the products with common vertices. Numerical studies have estimated the most efficient Δ_{CV} within 8–10% of the lower priority. Such value of Δ_{CV} would correspond to a minimal correction in common vertices composition during the coordinated planning. The greater Δ_{CV}, the more jobs are included into common vertices and the less number of setups. However, the greater Δ_{CV}, the more likely the inclusion of an aggregated job belonging to a low priority product into a common vertex. This would cause a decrease in the priority of the common vertex and, as a consequence, postpone its processing.

2. We determine the start times of the aggregated jobs being combined by the product's critical path. The start time of an aggregated job is the length of the path from the start time of the product to the job, taking into account the beginning time of the planning period for this product. The start times of aggregated jobs being combined must coincide or differ by not more than the value of Δ_{CV} determined on the basis of experiments. Δ_{CV} is expressed in time units and determined by the setup times for production problems or by the condition that the jobs merging reduces products' flow time or their total production costs. The smaller Δ_{CV}, the greater number of setups in resources. The greater Δ_{CV}, the greater number of aggregated jobs will be part of common vertices and the less the number of setups, on the one hand. This reduces the products' flow time. On the other hand, the greater Δ_{CV}, the more possible the increase in the products' flow time due to the increase in the waiting time of aggregated jobs.
3. We define the due dates of aggregated jobs being combined as the product's due date minus the processing times of jobs which follow the job being combined through the critical path till the terminal job. The due dates must be equal or differ by $\Delta_{CV} \approx$ 5–7% of the smaller due date. The value of Δ_{CV} is determined on the basis of experiments. The smaller Δ_{CV}, the greater the number of setups and, as a consequence, the products' flow time increases. The greater Δ_{CV}, the less the number of setups, but the more possible the increase in the readiness time of the aggregated jobs being combined. As a consequence, the products' flow time increases, which is unacceptable if our criterion is just in time production.
4. If the setup time significantly exceeds the maximum processing time of the aggregated jobs being combined, then the common vertex may combine different priority jobs. The reason is: the setups are initially included into the processing times of jobs. Combining aggregated jobs decreases the setups and hence the length of the critical paths of the corresponding products. However, the common vertex priority should differ from the maximum priority of the aggregated jobs it combines by no more than $\Delta_{CV} \approx$ 8–10% of the maximum priority. The value of Δ_{CV} is determined on the basis of experiments.

Experts in the DMU specify the value of Δ_{CV} (priority range or time interval) and the rule for merging aggregated jobs into common vertices. For functionals without due dates, they should apply the requirement 2 and the requirement 1 or 4, depending on setup times. If the setup time in the resource exceeds the maximum processing time of the combined jobs, then the requirement 4 is reasonable, otherwise the requirement 1 should be used. For problems with due dates, experts should apply the requirements 2 and 3 and, depending on setup times, the requirement 1 or 4. In the case of synthetic functionals that include several optimality criteria, the corresponding requirements 1–4 are used for each criterion.

We use the following algorithm to determine the set of aggregated jobs combined into a common vertex. The algorithm takes into account the decrease in the critical paths lengths of products the combined aggregated jobs belongs to (and, therefore, the increase in the priority of the "combined" products).

The base of the algorithm is the following heuristic. The products "combined" with a common vertex must immediately follow one another in the priority ordered sequence obtained as a TWCT problem solution. Otherwise, the products' flow times in the coordinated plan built according to the sequence significantly increase.

Candidates Selection to Include Into a Common Vertex contains seven steps.

1. Arrange the critical paths of the products of the set J in non-increasing order of their priorities (sequence σ^{ord}).
2. Choose the critical path of a product J_p with the highest priority. Select on the path the aggregated jobs processed in the resources that require significant setup time. Arrange these jobs in increasing order of their start times. The start time of an aggregated job is the length of the path from the start time of the product to the job, taking into account the beginning time of the planning period for this product. We have got a sequence σ^1.
3. Choose from the sequence σ^1 the earliest (by the start time) aggregated job $j_r \in J_p$ that requires processing in a resource i_s.
4. Sequentially review the critical paths of the products following the product J_p in the sequence σ^{ord}. Search for the aggregated jobs that require processing in the resource i_s.
5. Check the following two conditions for each found critical path. The conditions allow us to determine the possibility of combining with aggregated job $j_r \in J_p$ into a common aggregated job. Suppose that an aggregated job j_u processed by the resource i_s is on the critical path of a product J_t. The first condition is:

$$\omega_{J_t} \sum_{q=p+1}^{t-1} L_{J_q} + l_{i_s}^{sp} \sum_{q=t+1}^{n} \omega_{J_q} > \left(L_{J_t} - l_{i_s}^{sp}\right) \sum_{q=p+1}^{t-1} \omega_{J_q} \tag{9.33}$$

where ω_{J_q} is the weight of the product J_q; L_{J_q} its critical path length; $l_{i_s}^{sp}$ the setup time in the resource i_s.

The left-hand side of inequality (9.33) reflects the decrease in TWCT functional value if we place the product J_t (i.e., the aggregated jobs on its critical path) after the product J_p in the sequence σ^{ord}. Also, it shows a decrease in the number of setups. The right-hand size is the increase in the functional value as a result of shifting the products $\overline{p+1, t-1}$ to later positions. The second condition:

$$\frac{\omega_{J_p}}{L_{J_p}} - \frac{\omega_{J_p} + \omega_{J_t}}{L_{J_p} + L_{J_t} - l_{i_s}^{sp}} \leq \Delta_{CV} \tag{9.34}$$

where $\Delta_{CV} \approx 8$–10% of the maximum priority (the requirements 1 and 4).

According to (9.34), the priority of the common vertex must not be less than the priority of the product J_p by more than Δ_{CV}. If the merging is not performed, then the processing time of the aggregated job includes the setup time $l_{i_s}^{sp}$.

We check the conditions (9.33) and (9.34) sequentially for all critical paths of the products in the sequence σ^{ord} where we have aggregated jobs that require processing in the resource i_s. We search for the possibility of combining the jobs with the aggregated job $j_r \in J_p$.

6. Choose from the sequence σ^1 the aggregated job j_v with the next start time. Repeat steps 2–5 for each such job j_v searching for the possibility of merging it with the aggregated jobs of the products which follow the product J_p.
7. Move on to the next product J_p in the sequence σ^{ord} which has on its critical path aggregated jobs processed in resources that require significant setup time. If we have considered the entire sequence, then terminate. Otherwise, go to step 2 to determine the possibility to combine the aggregated jobs with the jobs of products that follow the product J_p in the sequence σ^{ord}.

If the conditions (9.33) and (9.34) are met, we determine the sets of aggregated job-candidates to merge into common aggregated jobs. We also require some additional conditions to construct common vertices.

Based on the above requirements, we have developed rules for common vertices construction. We can divide the rules into two groups depending on the setup time. We understand a setup in non-production systems as a necessary preparatory work. The two groups of the rules are:

1. The setup time is greater than the maximum processing time of the aggregated jobs being combined (rules 1 and 2).
2. The opposite case (rules 3 and 4).

In each group of the rules, we in turn distinguish the rules for combining the aggregated jobs that belong to products without due dates and with due dates.

Rule 1 (products do not have due dates). We combine aggregated jobs if the conditions (9.33) and (9.34) are met and the difference in path lengths from the start time of the corresponding products to the jobs being combined does not exceed the setup time in the resource (the requirement 4). The start times should take into account the beginning time of the planning period for each product. Critical paths of some products may have several common vertices corresponding to the requirement 2. Then, we combine aggregated jobs if conditions (9.33) and (9.34) are met and the difference in the beginning path lengths for jobs with the earliest start time does not exceed the total setup time for the combined jobs in this interval.

Rule 2 (products have due dates). We combine aggregated jobs if the conditions (9.33) and (9.34) are met and the due dates of the aggregated jobs are equal or differ no more than by the value of Δ_{CV} (the requirement 3). The due date of an aggregated job is defined as the product's due date minus the processing times of jobs following the combined job through the critical path till the terminal job.

Rule 3 (products do not have due dates). We combine aggregated jobs if the condition (9.34) is met and the difference of start times of the jobs being combined does not exceed the setup time in the resource (the requirement 2). For non-manufacturing problems, such merging should shorten the products' flow time or reduce their cost. Critical paths of some products may have several common vertices corresponding to the requirement 2. Then, we combine the aggregated jobs if the difference in the beginning path lengths for jobs with the earliest start time does not exceed the total setup time of the jobs being combined in this interval.

Rule 4 (products have due dates). We combine aggregated jobs if the condition (9.34) is met and due dates of the jobs being combined are also equal or differ no more than by the value of Δ_{CV} (the requirement 3). The due date of an aggregated job is defined as the product's due date minus the processing times of jobs following the job being combined through the critical path till the terminal job.

We use rules 1 and 3 to construct common vertices in the problems with the functionals based on the completion times of products (basic Criterion 1). Rules 2 and 4 are for the problems with the functionals based on products' due dates (basic criteria 2–5).

Remark 9.9 We must not combine the terminal (weighted) jobs of products into common vertices. Also, we must not combine non-completed jobs coming out of a buffer (if $q > 1$) except for the case if this does not lead to an appearance of arcs that were originally absent from the technology network. E.g., there must not appear the precedence relation to a new common vertex of a job that did not precede to any of the combined jobs.

Example 9.1 We show a part of an aggregated subnet in Fig. 9.9. We must not combine jobs 310 and 311 there, since then job 312 would precede this common vertex and so precede job 309. There was no such arc in the technology. But we can combine jobs 211 and 212, since new relationships do not appear.

We have got the following formal rule in a heuristic way: we must not combine jobs coming out of a buffer (if $q > 1$) with jobs that does not come out of it.

A subnet may contain a common equipment (two or more virtual machines with the same number, that is, corresponding to the same physical machine). In this case, we can consider combining aggregated jobs on all virtual machines with the same number if they satisfy the combining conditions, i.e., if their merging is beneficial.

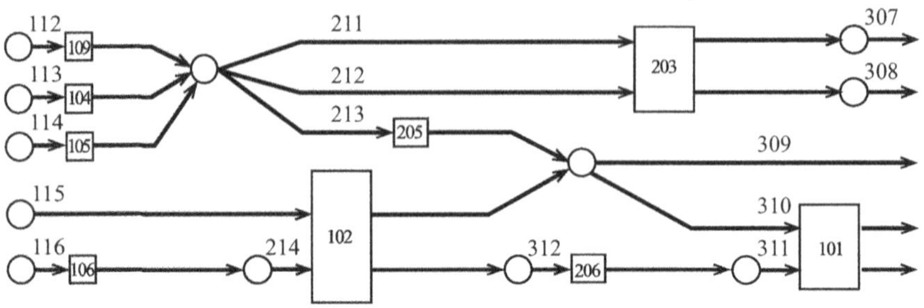

Fig. 9.9 Checkup for new arcs appearance

If we combine them, then we should change the aggregated and initial network according to the new relationships between the jobs.

Example 9.2 (Fig. 9.10). Suppose that aggregated job 107 (jobs 107 and 108 in the initial network) has merged into a common vertex with aggregated jobs 104 and 106 (jobs 104, 105, and 106 in the initial network). But aggregated job 109 (jobs 109 and 110 in the initial network) did not merge into one common vertex with them. Then, we redistribute jobs between elements with number 205. The combined jobs 104, 106, 107 (and, accordingly, jobs 104–108 in the initial network) will be processed on one element. And the job 109 not combined with them (jobs 109 and 110 in the initial network) will be on the other element. Other relationships will hold. Processing times of the aggregated jobs will be recalculated.

To simplify the example we show in Sect. 9.6, we do not consider in it a possibility of such recombination.

Step 1.4.2. Building a Network on the Products' Critical Paths with common vertices. We combine all critical paths into one graph based on common vertices. At the same time we remove jobs that do not lie on products' critical paths. But we

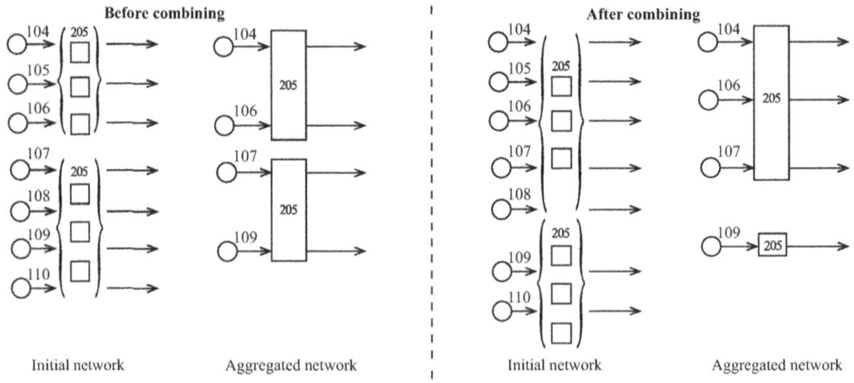

Fig. 9.10 Changing the subnet when combining jobs on a common equipment

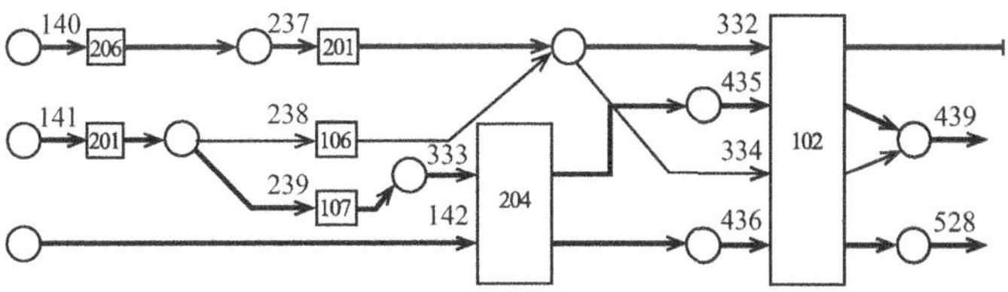

Fig. 9.11 Checking the relationships of jobs being removed

hold their relationships to the remaining jobs, so that the precedence relations that originally existed in the network are not violated.

Example 9.3 We show a part of an aggregated network in Fig. 9.11. We want to remove job 238. But the job 238 and, consequently, job 141 (product 2) precede job 332 (product 1). Also, we need to remove job 334. But job 237 (product 1) precedes job 334 and, consequently, job 439 (product 2). Therefore, we should hold arcs 141–332 and 237–439 in the critical paths network.

Not all terminal jobs of products lie on their critical paths. So, we may remove some of them from the list of products' terminal jobs when building the critical paths network.

We show an example of a network on the products' critical paths with common vertices in Fig. 9.12. We denote common vertices by the double circle symbol ◎.

9.2.2.2 TWCT Problems Construction for the Corresponding Basic Components of a Synthetic Optimality Criterion

Step 1.5 Building and solving the optimization problems on the network of the second level of aggregation (see Step 1.4). The optimization criterion is the total weighted completion time (TWCT) of aggregated jobs minimization.

Step 1.5.1 We solve an approximating TWCT problem for each basic component of the synthetic optimality criterion included in the specified functional, except for the basic Criterion 1. The Criterion 1 is exactly the TWCT problem, so for the basic Criterion 1 we solve the TWCT problem without changing its parameters.

We describe the algorithm for the problem solving in Sect. 9.2.3. Now we consider approximating TWCT problems construction for basic criteria 2–5.

We construct the approximating TWCT problem for *Criterion 2* replacing its functional $\max \sum_{i=1}^{n} \omega_i U_i$ with $\min \sum_{i=1}^{n} \omega_i \mathbb{C}_i$ where $\mathbb{C}_i$ is the completion time of product i, ω_i is the income from the sale of the product i if it completes just in time. Let us give a justification for the possibility of such problem reduction.

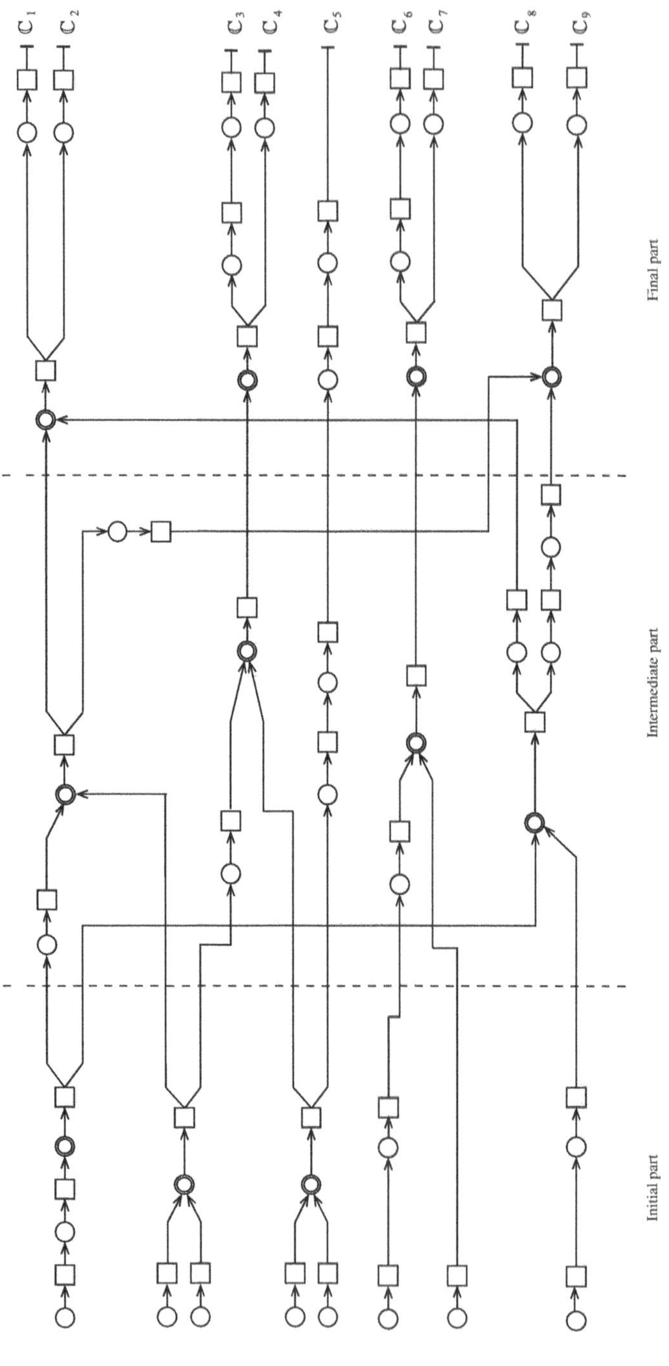

Fig. 9.12 An example of critical paths graph with common vertices

1. Consider the special case when the products are independent. According to Theorem 3.10 from [12] (taken from [13]), we should at first process the product with the maximum ratio of the profit (in accordance with Criterion 2) to the processing time. In other words, it should have the maximum unit profit per time unit of the production cycle. Therefore, it is logical for Criterion 2 at first to process the product with the maximum unit profit per time unit of production cycle ω_1/l_1. This would maximize the efficient use of the production cycle time. Next, we should process the product with the next maximum unit profit $\omega_2/l_2 \leq \omega_1/l_1$, etc. An alternative to this heuristic is only a full enumeration of various orders of products processing. Thus, the goal is to obtain a priority ordered schedule. We achieve this by the TWCT problem solving.
2. The general case. A partial order is specified by the constructed oriented graph on the products' critical paths. In this case, an optimal solution to the TWCT problem is a priority ordered schedule in which the products sequence is divided into the maximum priority sets G_i, $i = \overline{1,k}$, $p(G_1) \geq p(G_2) \geq \cdots \geq p(G_k)$. Thus, if we at first process jobs of the optimal schedule corresponding to the set G_1, then the initial segment of the production cycle with the length $\sum l_i$, $i \in G_1$ contains the products with the maximum averaged unit profit per time unit of the production cycle:

$$p(G_1) = \sum_{i \in G_1} \omega_i \Big/ \sum_{i \in G_1} l_i.$$

The justification for using an optimal schedule of the TWCT problem in the general case for Criterion 2 is the same as in the special case considered above. We should only replace the term "maximum unit profit per time unit of production cycle" with the term "maximum averaged unit profit per time unit of the production cycle".

The justification for the possibility of approximating TWCT problems construction for the remaining basic criteria (3–5) is similar to the above.

For *Criterion 3*, we replace the functional $\min \sum_{i=1}^{n} \omega_i \max(0, \mathbb{C}_i - d_i)$ with $\min \sum_{i=1}^{n} \omega_i \mathbb{C}_i$ where ω_i is the cost of (penalty for) the tardiness of product i in relation to its due date per time unit.

For *Criterion 4*, we replace the functional $\max \sum_{i=1}^{n} \omega_i U_i$ with $\min \sum_{i=1}^{n} \omega_i \mathbb{C}_i$ where ω_i is the absolute value of profit from the product i. The profit does not depend on the completion time of the product if it is not tardy in regard to its due date, otherwise the enterprise's profit for this product is zero.

For *Criterion 5*, we construct the approximating TWCT problem replacing the functional $\min \sum_{i=1}^{n} \omega_i |\mathbb{C}_i - d_i|$ with $\min \sum_{i=1}^{n} \omega_i \mathbb{C}_i$ where ω_i is the penalty for the deviation of the completion time of product i from the due date per time unit.

As an example, consider the approximating TWCT problems construction to determine the aggregated jobs sequence for Criterion 7. The basic criteria for this synthetic criterion are the products completion just in time (Criterion 2) and the total weighted tardiness minimization (Criterion 3):

$$\max\left\{\sum_{i=1}^{k}\omega_i' U_i - \sum_{i=k+1}^{n}\omega_i''\max(0, \mathbb{C}_i - d_i)\right\}. \quad (9.35)$$

Therefore, we should solve two approximating TWCT problems for synthetic criterion 7 with a partial order given by the graph on the critical paths:

$$\min\sum_{i=1}^{k}\omega_i'\mathbb{C}_i \text{ and } \min\sum_{i=k+1}^{n}\omega_i''\mathbb{C}_i$$

where ω_i' has the same meaning as in Criterion 2 and ω_i'' as in Criterion 4.

9.2.3 Algorithm for TWCT Problem Solving When Nonzero Weights Are Specified Only for the Terminal Vertices of the Oriented Graph (The TWCTZ Problem)

The TWCTZ problem is a special case of the total weighted completion time minimization problem for work with different weights on a single machine with a partial order given by an oriented acyclic graph (TWCT). The problem is NP-hard. We give the research results and the efficient PSC-algorithm for TWCT problem solving in Chap. 7. We formulate the TWCTZ problem using the terminology adopted in Chap. 7, for ease of understanding.

The Problem Statement A partial order on the set of tasks $J = (j_1, j_2, \ldots, j_n)$ is given by an oriented acyclic graph G. We know a processing time l_j for each task j of the graph G. Each terminal vertex (without successors) of the graph has a weight ω_j. Other vertices have zero weight. We need to find a sequence of tasks that minimizes the functional: $\sum\omega_j C_j \to \min$. In this section, notation ω_i means $\omega_i(T)$ (see Sect. 9.2.1). The tasks mean aggregated jobs. G is the graph constructed on the products' critical paths. The common vertices in the graph indicate common aggregated jobs for different products (see Step 1.4). Elements of buffer type should be taken into account when considering the precedence relations.

For the problem in such formulation, we have developed an efficient fast polynomial heuristic algorithm based on the study of the TWCT problem properties. The algorithm allows to solve real practical large dimension problems.

Main Definitions and Notation Let i or j_i be the number of task corresponding to the indexation specified by the functional; $j_{[g]}$ be the number of a task in position g in some considered schedule; $j^p_{[k]}$ be a task in position k in the sequence which has the number p in the set J. We introduce the following definitions.

Definition 9.1 Suppose that a partial order is given by an oriented acyclic graph on the set of tasks J. If j_k precedes j_l, we denote this fact as $j_k \prec j_l$. If $j_k \prec j_l$ and there does not exist a task j_z such that $j_k \prec j_z$, $j_z \prec j_l$, then we will call j_k an *immediate predecessor* of j_l and denote this fact as $j_k \prec\!\prec j_l$.

Definition 9.2 A sequence of tasks $\sigma = \left(j_{[1]}, j_{[2]}, \ldots, j_{[g]}\right)$ is called *feasible* if for any $j_{[k]} \prec j_{[l]}$ processing of the task $j_{[k]}$ completes before processing of the task $j_{[l]}$, according to this sequence.

Definition 9.3 A sequence of tasks $\sigma^k \in \sigma$ is called *ready for processing* in a given time point d in relation to initial sequence in the time interval $[0, d]$ if it is feasible, all predecessors of the first task in this sequence have completed till the time point d, and all predecessors of each task $j \in \sigma^k$ have completed before the start time of the task j.

Definition 9.4 A subsequence of tasks $\alpha\left(j_{[g]}\right) = \left(j_{[r]}, j_{[r+1]}, \ldots, j_{[g-1]}, j_{[g]}\right)$ in a sequence σ is called an *elementary construction of tasks* if each task $j_{[l]} \in \alpha\left(j_{[g]}\right)$ satisfies: $j_{[l]} \prec j_{[g]}, j_{[r-1]} \overline{\prec} j_{[g]}$ for $g > l$, $l = \overline{r, g-1}$. Here, $j_{[g]}$ is a task that occupies the latest position g in the elementary construction of tasks $\alpha\left(j_{[g]}\right)$; l and r are the numbers of position occupied by tasks $j_{[l]}$, $j_{[r]}$ in the sequence σ.

Thus, an elementary construction combines dependent tasks in a certain order of precedence.

Definition 9.5 *Priorities* of a task j and of an elementary construction α are the values p_j, p_α defined the following way:

$$\forall j_{[g]} \in J_t,\ \exists p_{j_{[g]}} = \frac{\omega_{j_{[g]}}}{l_{j_{[g]}}};\quad p_\alpha = \frac{\Omega_{\alpha\left(j_{[g]}\right)}}{L_{\alpha\left(j_{[g]}\right)}}$$

where J_t is the set of graph's terminal vertices; $\omega_{j_{[g]}}$, $l_{j_{[g]}}$ are the weight and the processing time of a task $j_{[g]}$, respectively; $\Omega_{\alpha\left(j_{[g]}\right)} = \sum_{i=r}^{g} \omega_{j_{[i]}}$ is the weight of an elementary construction of tasks $\alpha\left(j_{[g]}\right)$; $L_{\alpha\left(j_{[g]}\right)} = \sum_{i=r}^{g} l_{j_{[i]}}$ is its processing time.

Definition 9.6 A sequence of tasks σ on the set of tasks J with the precedence order given by an oriented acyclic graph is called *priority ordered* if it satisfies the following conditions. First positions in σ occupies a subsequence σ_1 ready for processing at the time point zero and having the highest priority of all possible feasible subsequences on the set of tasks J ready for processing at the time point zero. If the subsequence σ_1 contains independent groups of interrelated tasks, they should be arranged in non-increasing order of their priorities. The subsequence σ_1 is followed by a subsequence σ_2 constructed according to similar rules, ready for processing after σ_1 completion and having the highest priority of all feasible subsequences on the set of tasks $J \backslash \sigma_1$. Next positions in σ takes a subsequence σ_3

defined by the same rules. The process continues until we consider the entire set of tasks J.

Definition 9.7 Two elementary constructions of tasks $\alpha(j_{[g]})$, $\beta(j_{[f]})$ are called *independent* (from each other) if any pair of tasks $j_{[k]} \in \alpha(j_{[g]})$ and $j_{[l]} \in \beta(j_{[f]})$ does not exist in the precedence relations.

Definition 9.8 An *interval of move* in a sequence σ for an individual task $j_{[k]}$ is the maximum time interval satisfying the condition: if we move the task $j_{[k]}$ by this interval into an earlier position, then it does not violate the feasibility of the sequence σ.

For example, the interval of move of a task $j^*_{[g]}$ into a position $p+1$ in a sequence $\sigma = \left(j^1_{[1]}, j^2_{[2]}, \ldots, j^p_{[p]}, j^{p+1}_{[p+1]}, j^{p+2}_{[p+2]}, \ldots, j^{g-1}_{[g-1]}, j^*_{[g]}\right)$ is defined by the value of $I_{j^*_{[g]}} = \sum_{k=p+1}^{g-1} l_{j_{[k]}}$ where $\overline{p+1, g-1}$ are the positions of tasks that do not precede $j^*_{[g]}$: $j_{[k]} \overline{\prec} j^*_{[g]}$ $\forall k = \overline{p+1, g-1}$. After the move, the sequence of tasks σ is transformed into a sequence $\sigma_1 = \left(j^1_{[1]}, j^2_{[2]}, \ldots, j^p_{[p]}, j^*_{[p+1]}, j^{p+1}_{[p+2]}, \ldots, j^{g-2}_{[g-1]}, j^{g-1}_{[g]}\right)$.

The interval of move for an elementary construction of tasks $\alpha(j_{[g]}) = (j_{[r]}, j_{[r+1]}, \ldots, j_{[g-1]}, j_{[g]})$ is $I_\alpha = \sum_{k=p+1}^{r-1} l_{j_{[k]}}$.

Definition 9.9 An *interval of efficient move* for an individual task $j_{[g]}$ or an elementary construction of tasks $\alpha(j_{[g]})$ is the interval of move for which the functional value (the total weighted completion time) maximally decreases.

The interval of efficient move for an individual task $j_{[g]}$ is defined as

$$I^{eff}_{j_{[g]}} = \sum_{k=\rho}^{g-1} l_{j_{[k]}} : \left(\Omega_{j_{[g]}} \sum_{k=\rho}^{g-1} l_{j_{[k]}} - l_{j_{[g]}} \sum_{k=\rho}^{g-1} \omega_{j_{[k]}} > 0 \right) \to \max$$

where ρ is the position we move the task $j_{[g]}$ to.

The interval of efficient move for an elementary constructions of tasks $\alpha(j_{[g]})$:

$$I^{eff}_{\alpha(j_{[g]})} = \sum_{k=q}^{r-1} l_{j_{[k]}} : \left(\Omega_{\alpha(j_{[g]})} \sum_{k=q}^{r-1} l_{j_{[k]}} - L_{\alpha(j_{[g]})} \sum_{k=q}^{r-1} \omega_{j_{[k]}} > 0 \right) \to \max$$

where q is the position we move the first task $j_{[r]} \in \alpha(j_{[g]})$ of the construction to; $\alpha(j_{[g]}) = (j_{[r]}, j_{[r+1]}, \ldots, j_{[g-1]}, j_{[g]})$.

We solve the TWCTZ problem and construct the priority ordered schedule on the basis of PSC-algorithm for TWCT problem (see Chap. 7). At each iteration k, we construct a structure of interrelated tasks (elementary, simple or nested construction) on the basis of a current task $j_{[g]}$. We then insert the construction in accordance with its priority into earlier positions in a feasible sequence without violating the entire sequence feasibility. The algorithm we propose is not exact. If an optimal schedule building requires a combinatorial search of various types of constructions, then we do not perform this search. Despite this, the algorithm remains efficient because, according to the results of statistical studies, such a search of constructions occurs rarely. The proposed algorithm takes into account the specific properties of the problem. This significantly affects its speed and allows solving real practical problems with tens of thousands of tasks.

Algorithm A for the Problem Solving contains three units: ordering unit (Algorithm A1), optimization unit (Algorithm A2), and the unit of reordering on intervals of efficient move (Algorithm A1.1).

Algorithm A1 (Ordering Unit) has the following eight steps.

1. Arrange terminal vertices j of the graph G for which $\omega_j > 0$ in non-increasing order of ω_j. Denote the obtained sequence as σ_{ord}.
2. Consider the sequence σ_{ord} from the beginning.
3. Select the first task $j_{[1]}$ of the sequence σ_{ord} and move it into the first position in the new sequence σ_{ord}^{new}.
4. Select from the set of tasks $J \backslash j_{[1]}$ the tasks $j_{[r]}$, $r = \overline{2, n}$ that immediately precede $j_{[1]}$.
5. Move the tasks $j_{[r]}$ and put them before the task $j_{[1]}$ according to their processing times (in non-decreasing order of processing times, without violating the feasibility of the sequence σ_{ord}^{new}).
6. For each moved task in turn find immediate predecessors and in a similar way (see steps 4, 5) insert them into the new sequence.
7. Repeat steps 4–6 for the remaining tasks of the sequence σ_{ord} until all the tasks necessary for completing the task $j_{[1]}$ are included.
8. Proceed to the next task $j_{[2]}$ of the sequence σ_{ord}. If this task is not in the sequence σ_{ord}^{new}, include it in this sequence after the task $j_{[1]}$. Go to step 4 and in a similar way insert before the task $j_{[2]}$ its predecessors excluding the tasks that exist at the earlier positions in the sequence σ_{ord}^{new}.
9. If all the tasks of the sequence σ_{ord} are considered, terminate. We have constructed the sequence σ_{ord}^{new}.

The sequence σ_{ord}^{new} is feasible by construction. We perform the optimization unit to build a priority ordered sequence.

Investigation of Properties of the Sequence σ_{ord}^{new} Consider the sequence σ_{ord}^{new} obtained as a result of the ordering unit. The following properties are satisfied for this sequence.

Property 9.1 Elementary constructions of tasks $\alpha(j_{[k]})$, $k = \overline{1, n}$, in the sequence σ_{ord}^{new}, include all tasks $j_{[l]}$, $l = \overline{r, k-1}$ that satisfy: $\omega_{j_{[l]}} = 0$, $j_{[l]} \prec j_{[k]}$, $j_{[r-1]} \not\prec j_{[k]}$.

According to Definition 9.4, a subsequence of tasks $\alpha(j_{[k]}) = (j_{[r]}, j_{[r+1]}, \ldots, j_{[k]})$ in a sequence σ is called an elementary construction of tasks if each task $j_{[l]} \in \alpha(j_{[k]})$ satisfies: $j_{[l]} \prec j_{[k]}$, $k > l$, $l = \overline{r, k-1}$. Here, $j_{[k]}$ is a task that occupies the latest position k in the elementary construction of tasks $\alpha(j_{[k]})$; l and r are the numbers of position occupied by tasks $j_{[l]}, j_{[r]}$ in the sequence σ.

Consequently, we construct the elementary constructions of tasks $\alpha(j_{[k]})$ in the sequence σ_{ord}^{new} as follows. We consider the sequence σ_{ord}^{new} from position k back to the first position. We include in the elementary construction $\alpha(j_{[k]})$ the tasks $j_{[l]}$ that satisfy: $\omega_{j_{[l]}} = 0,\ l = \overline{r, k-1};\ j_{[l]} \prec j_{[k]},\ j_{[r-1]} \overline{\prec} j_{[k]}$.

The rule for building the elementary construction remains valid for all modifications of the sequence σ_{ord}^{new} obtained as a result of the above permutations.

Property 9.2 The interval of move into earlier positions for an elementary construction of tasks $\alpha(j_{[g]})$ is the interval defined by the positions $\overline{p+1, g-1}$ that satisfies: $j_{[l]} \overline{\prec} \alpha(j_{[g]}),\ l = \overline{p+1, g-1}$. Here, p is the position of a common vertex in the critical paths graph, $j_{[p]} \prec \alpha(j_{[g]})$. According to Definition 9.8, the interval of move is the maximum time interval satisfying the condition: if the elementary construction $\alpha(j_{[g]})$ is moved by this interval, then it does not violate the feasibility of the sequence. From the problem statement, the weights of all tasks except for the terminal vertices of the graph are zero. We have: $j_{[s]} \overline{\prec} j_{[k]}$ if $\omega_{j_{[s]}} > 0,\ \omega_{j_{[k]}} > 0$. In accordance with permutation rules, a zero-weighted task can take an earlier position, in relation to an elementary construction of tasks that has a precedence relation with it, only if the task is a common vertex.

Algorithm A2 (*Optimization Unit*) has the following 23 steps.

1. Consider the sequence σ_{ord}^{new} from the beginning. Suppose that the current weighted task occupies a position g, $\omega_{j_{[g]}} > 0$. Consider the tasks occupying the previous positions within interval $\overline{1,\ g-1}$ and build an elementary construction of dependent tasks $\alpha(j_{[g]}) = (j_{[r]}, j_{[r+1]}, \ldots, j_{[g-1]}, j_{[g]})$. We have $\omega_{j_{[k]}} = 0$ $\forall j_{[k]},\ k = \overline{r, g},\ j_{[r-1]} \overline{\prec} j_{[g]}$.
2. Determine the parameters of the elementary construction of tasks $\alpha(j_{[g]})$: $\Omega_{\alpha(j_{[g]})} = \sum_{j \in \alpha(j_{[g]})} \omega_j;\ L_{\alpha(j_{[g]})} = \sum_{j \in \alpha(j_{[g]})} l_j$.
3. Determine the interval of move for the elementary construction of tasks $\alpha(j_{[g]})$. For this purpose, consider the tasks within interval $\overline{1,\ r-1}$ that are common vertices. Search for the task preceding the elementary construction $\alpha(j_{[g]})$ and occupying the latest position.
4. Determine the interval of efficient move: $I_{\alpha(j_{[g]})}^{eff} = \sum_{k=q}^{r-1} l_{j_{[k]}}$ where q is the position that maximizes the difference

$$\Omega_{\alpha(j_{[g]})} \sum_{k=q}^{r-1} l_{j_{[k]}} - L_{\alpha(j_{[g]})} \sum_{k=q}^{r-1} \omega_{j_{[k]}} > 0, \quad \omega_{j_{[q]}} = 0.$$

Suppose that this difference is maximal in a position $f + 1$. Go to step 5.

5. Move the elementary construction into the position $f + 1$ determined as the interval of efficient move so that the first task $j_{[r]}$ of the construction $\alpha(j_{[g]})$ occupies the position $f + 1$. The following cases are possible:
   - an elementary construction $\beta(j_{[f]}) = (j_{[s]}, j_{[s+1]}, \ldots, j_{[f]})$, $\beta(j_{[f]}) \prec \alpha(j_{[g]})$, occupies the positions $\overline{s, f}$;
   - a task $j_{[f]} \overline{\prec} \alpha(j_{[g]})$ occupies the position f.

   In the first case, go to step 6. Otherwise, go to step 16 considering there an elementary construction $\alpha(j_{[g]})$ as the construction K_1.
6. Check the condition:
$$\frac{\Omega_{\beta(j_{[f]})}}{L_{\beta(j_{[f]})}} < \frac{\Omega_{\alpha(j_{[g]})}}{L_{\alpha(j_{[g]})}}.$$
   If it is true, go to step 7. Otherwise, go to step 16 considering there an elementary construction $\alpha(j_{[g]})$ as the construction K_1.
7. Build an elementary construction of tasks $\rho(j_{[f+n_1]})$ including elementary constructions $\beta(j_{[f]})$ and $\alpha(j_{[f+n_1]})$ where n_1 is the number of tasks in the elementary construction $\alpha(j_{[g]})$. Determine the priority of the elementary construction $\rho(j_{[f+n_1]})$:
$$\frac{\Omega_{\rho(j_{[f+n_1]})}}{L_{\rho(j_{[f+n_1]})}} = \frac{\omega_{j_{[f+n_1]}} + \omega_{j_{[f]}}}{L_{\alpha(j_{[f+n_1]})} + L_{\beta(j_{[f]})}}.$$
8. Determine the interval of efficient move for the elementary construction $\rho(j_{[f+n_1]})$ by the rules described in steps 3 and 4.
9. Move the elementary construction $\rho(j_{[f+n_1]})$ into positions determined by the interval of efficient move. As a result of this move, the elementary construction $\rho(j_{[f+n_1]})$ transforms into an elementary construction $\rho(j_{[f+n_1-n_2]})$ where n_2 is the number of tasks $j \in I^{eff}_{\rho(j_{[f+n_1]})}$.

Property 9.3 Only elementary constructions of tasks built on terminal vertices of the graph G may increase the priority of the elementary construction $\rho(j_{[f+n_1-n_2]})$.

10. Reorder all the tasks that follow the elementary construction $\rho(j_{[f+n_1-n_2]})$ according to Algorithm A1.1 given below.
11. Build the construction of tasks K_1 by the following rules. For each position k, $\omega_{j_{[k]}} > 0$, within the interval $\overline{f + n_1 - n_2 + 1,\ g}$, find the value of $M(k)$ from the expression

$$M(k) = \frac{\sum_{l=f+n_1-n_2+1}^{k} \omega_{j_{[l]}} + \Omega_{\rho\left(j_{[f+n_1-n_2]}\right)}}{\sum_{l=f+n_1-n_2+1}^{k} l_{j_{[l]}} + L_{\rho\left(j_{[f+n_1-n_2]}\right)}} - \frac{\Omega_{\rho\left(j_{[f+n_1-n_2]}\right)}}{L_{\rho\left(j_{[f+n_1-n_2]}\right)}}$$

and determine $M = \max_k M(k)$. Suppose that $M(k)$ has the maximum in a position t. Go to step 12. If you could not increase the priority of the elementary construction ρ including additional tasks in it, go to step 16.

12. Build the construction of tasks K_1. It includes the elementary construction $\rho\left(j_{[f+n_1-n_2]}\right)$ and elementary constructions of tasks within the interval $\overline{f+n_1-n_2+1,\, t}$ built on the terminal vertices of the graph G. For the latter elementary constructions, the tasks belonging to the elementary construction $\rho\left(j_{[f+n_1-n_2]}\right)$ should be common vertices in the graph or relate by the precedence relations to other common vertices which are part of the above elementary constructions. Determine the priority of the construction K_1.
13. Determine the interval of efficient move for the construction K_1 by the rules described in steps 3 and 4.
14. Move the construction K_1 into a position determined by the interval of efficient move.
15. Reorder all the tasks that follow the construction K_1 up to the task $j_{[g]}$ in accordance with Algorithm A1.1. Proceed to step 5 considering there the construction K_1 as an elementary construction $\alpha\left(j_{[g]}\right)$.
16. Organize a new construction of tasks K_2 as follows.

- sequentially consider the tasks that occupy the previous positions in regard to the construction K_1 and are common vertices with precedence relations to the construction K_1;
- search for a task $j_{[z]} \prec K_1$.

Property 9.4 The task $j_{[z]}$ is a common vertex in the critical paths graph G.

The following cases are possible:

- a task $j_{[z]} \prec K_1$ is not found. Go to step 23;
- the task $j_{[z]}$ is found and occupies the position z. Go to step 17.

17. Organize an elementary construction of tasks $\beta\left(j_{[z]}\right)$.
18. Determine the parameters of an elementary construction of tasks $\beta\left(j_{[z]}\right)$:

$$\Omega_{\beta\left(j_{[z]}\right)} = \omega_{j_{[z]}}; \quad L_{\beta\left(j_{[z]}\right)} = \sum_{j \in \beta\left(j_{[z]}\right)} l_j.$$

19. Check the condition:

$$\frac{\Omega_{\beta(j_{[z]})}}{L_{\beta(j_{[z]})}} < \frac{\Omega_{K_1}}{L_{K_1}}. \tag{9.36}$$

If the condition is false, go to step 16 and search for the next task $j_{[z]} \prec K_1$ satisfying Eq. (9.36). If Eq. (9.36) is true, go to step 20. If such a task is not found, go to step 23.

20. Build a construction K_2 including the construction K_1, the elementary construction $\beta(j_{[z]})$, and the tasks in the intermediate positions between K_1 and $\beta(j_{[z]})$.
21. Determine the interval of efficient move for the construction K_2 by the rules described in steps 3 and 4. Move the construction K_2 into positions determined by the interval of efficient move.
22. Reorder all the tasks that follow the construction K_2 up to the task $j_{[g]}$ in accordance with Algorithm A1.1. Proceed to step 5 considering there the construction K_2 as an elementary construction $\alpha(j_{[g]})$.
23. Move on to the next task that occupies the position $g + 1$. If $\omega_{j_{[g+1]}} > 0$, go to step 2, new iteration begins. If $\omega_{j_{[g+1]}} = 0$, move on to the position $g + 2$ and so on. If $g+1 > n$, execute Algorithm AMPSS (Sect. 7.3.4) to partition the sequence into the maximum priority subsequences, terminate.

Property 9.5 We perform all the permutation procedures of Algorithm A only for the weighted structures (weighted task, elementary construction of tasks, constructions K_1 and K_2).

The use of Properties 9.1–9.5 in the above algorithm allowed us to significantly reduce its complexity and solve real practical problems with tens of thousands of tasks.

Algorithm A1.1 (*Unit of Reordering on Intervals of Efficient Move*) The purpose is to organize tasks within the interval of efficient move after performing an efficient move.

Sequentially consider tasks within the interval of efficient move I^{eff} starting with the task that occupies the earliest position. Suppose that the current task we consider is $j_{[k]} \in I^{eff}$.

1. Organize an elementary construction of dependent tasks $\alpha(j_{[k]})$.
2. Determine the parameters of the elementary construction of tasks $\alpha(j_{[k]})$:

$$L_{\alpha(j_{[k]})} = \sum_{j \in \alpha(j_{[k]})} l_j; \quad \Omega_{\alpha(j_{[k]})} = \sum_{j \in \alpha(j_{[k]})} \omega_j.$$

3. Determine the interval of move for the elementary construction of tasks $\alpha(j_{[k]})$.
4. Determine the interval of efficient move for the elementary construction of tasks $\alpha(j_{[k]})$.
5. Move the elementary construction $\alpha(j_{[k]})$ into a position determined by the interval of efficient move.
6. Proceed to the next task $j_{[k+1]} \in I^{eff}$, go to step 1. If all the tasks $j \in I^{eff}$ are considered, the algorithm terminates.

Step 1.5.2 The TWCT problem solution is an optimal (suboptimal, if the solution is approximate) sequence of the products processing. We use the sequence to determine the order of the products processing during the coordinated planning.

For planning by a *synthetic optimality criterion*, the initial network is a set of N separate subnets. The subnets cannot have precedence relations with each other, but may be interrelated through a common equipment (virtual machines in different subnets). This prevents us from scheduling jobs processing in subnets independently of each other. Each subnet corresponds to its own basic criterion. In this case, we perform the procedures described in Steps 1.2–1.5.2 separately for each subnet. The result would be N optimal (suboptimal) sequences of products processing, one for each subnet.

9.3 Unit 3. The Model of the Second Level of Planning

9.3.1 The Formalization of the Model of the Second Level of Planning Using the Model of the First Level

In Unit 3, we formalize and solve the problem of coordinated planning of aggregated jobs on the network of the first level of aggregation taking into account the specified optimality criterion.

Step 2.1 The coordinated planning problem statement *for a basic criterion*: to build a schedule (coordinated plan) of aggregated jobs processing by aggregated elements on the network of the first level of aggregation (Step 1.2). The criterion is the total flow time of jobs minimization (equivalent to minimizing the equipment idle time). The constraint is: the priority ordered sequence obtained as a result of the TWCT problem solving defines the order of products processing (we take the information from Unit 2).

The coordinated planning problem statement *for a synthetic criterion*: there are N separate subnets of the first level of aggregation (see Step 1.2). The subnets may be interrelated through a common equipment (virtual machines corresponding to the same physical equipment). This prevents us from building schedules in subnets independently of each other. Each subnet corresponds to its own basic optimality criterion. We need to build a schedule (coordinated plan) of aggregated jobs processing by aggregated elements on the set of N subnets. The criterion is the total flow time of jobs minimization (equivalent to minimizing the equipment idle time) subject to the constraints:

- we assign the aggregated jobs group by group; the job groups correspond to the subnets of the aggregated network (we take the information from Unit 2);
- the order of products processing within each product group is defined by a priority ordered sequence; the sequence is the solution of the TWCT problem corresponding to the product group.

We construct an approximating TWCT problem for each subnet in accordance with Step 1.5.1. Customer in advance specifies the products' weights in accordance with the optimization criterion for each product group.

Idle times may remain at elements after assigning all jobs of a product group (in accordance with the sequence of their assignment for processing). We can use the idle times to assign jobs of the next product group (in accordance with the sequence of their assignment in this group). But it must not lead to significant storage costs for unfinished products and to a significant deviation in the flow time of products from the critical path lengths.

9.3.2 Coordinated Planning Algorithms. Properties, Heuristics

Let us first consider the technique of coordinated planning for a basic optimality criterion, and then for a synthetic criterion.

Step 2.1.1 We need to restore the full production technology of products to determine the order of aggregated jobs assignment within each product group. So, we complement the priority ordered sequence obtained as a solution of the TWCT problem by aggregated jobs that lie outside of the products' critical paths, in accordance with technological relationships. We insert into the sequence, before each aggregated job on a critical path, the supergraph of its predecessors not yet included in the sequence. We break common vertices into their components (including them in the sequence in the form of a chain). We can start the following aggregated job processing immediately after each component completion. The result is the supplemented sequence σ^*. We also restore the list of products' terminal jobs to the form obtained at the first level of aggregation.

Step 2.1.2. The Coordinated Plan Construction for Aggregated Jobs The coordinated planning implementation is based on the following *principles*.

1. Maximum approximation of each product's flow time to its critical path, which allows:
    - to reduce the amount of work in progress;

- to reduce product flow times;
- to reduce idle times and thereby increase the production capacity;
- to increase the profitability of the planning system.

2. We assign products without due dates to be processed in order defined by the priority ordered sequence, from its beginning to the end. The rule is: the higher the product's priority, the earlier it must be assigned for processing.
3. We assign products with due dates to be processed in order defined by the priority ordered sequence as close to the due dates as possible. So, the order is from the end of the sequence to the beginning, from the terminal vertices to the initial jobs of a product. The rule is: the higher the product's priority, the smaller the deviation of the product from its due date.

We have developed the following *distribution algorithms* to implement the aggregated jobs distribution (assignment) to aggregated resources separately for each group of products:

- compact schedules construction (Distribution Algorithm 1);
- nondelay schedules construction (Distribution Algorithm 2);
- construction of schedules that ensure highest priority products processing in the specified due dates (Distribution Algorithm 3).

Definition 9.10 A schedule is *compact* (also known as active or compressed schedule) if none of jobs in it allows a left shift, i.e., any decrease in the time at which the job starts that does not require an increase in the starting time of any job [12, 14]. A compact schedule is obtained by a feasible compression with job shift or move operations.

Definition 9.11 A schedule is *nondelay* if there is no instance in which a job is delayed when the machine that is to process the next job is available and idle [12].

We apply one of the developed algorithms for each product group depending on the specified basic criterion. We use Distribution Algorithms 1 and 2 for the products optimized by the basic Criterion 1 and apply Distribution Algorithm 3 for aggregated jobs distribution for products with due dates (basic criteria 2–5). If we plan according to a basic criterion, we distribute all products by a single algorithm.

We have built Distribution Algorithms 1 and 2 using algorithms of Giffler and Thompson [12, 14]. We can show that any optimal schedule for a regular optimality criterion (non-decreasing with respect to the completion times) is compact. Not in every problem there is an optimal schedule among the nondelay schedules [12]. Yet the nondelay schedules are a very important class because they are easy to generate, and there is strong empirical evidence that while they may lack an optimal schedule, on the whole the nondelay schedules are better than the remainder of the compact schedules.

Compact and nondelay schedules minimize the idle time between aggregated jobs. Consequently, the products' flow time is minimized. We choose Distribution Algorithm 1 or 2 depending on the initial data. If the processing time of the aggregated jobs included in the products varies insignificantly (the processing time is of the same order of magnitude), we select Distribution Algorithm 2. Otherwise, we use Distribution Algorithm 1.

During the coordinated planning, we distribute aggregated jobs of sequence σ^* to aggregated elements. The structure of σ^* is as follows. First positions in σ^* occupies a maximum priority subsequence (MPSS) σ_1 (MPSS number 1). σ_1 is ready for processing at the time point zero and has the highest priority of all possible feasible subsequences on the set of aggregated jobs J which are ready for processing at the time point zero. The subsequence σ_1 is followed by a subsequence σ_2 (MPSS number 2) constructed according to similar rules, ready for processing after σ_1 completion, and having the highest priority of all feasible subsequences on the set of aggregated jobs $J \backslash \sigma_1$. Next positions in σ^* takes a subsequence σ_3 (MPSS number 3) defined by the same rules. The process continues until we consider the entire set of aggregated jobs J.

Each of the distribution algorithms described below is based on sequential assignment of aggregated jobs of subsequences σ_i to elements' reserves (the idle time intervals between previously distributed aggregated jobs). We use the elements of types 1 and 2 of the aggregated network of the first level of aggregation obtained in Step 1.2. Before the distribution, we create an empty list of elements without any assigned job and reserve the time for jobs assignment from zero point (or from the beginning time of the planning period) to the end of the planning period T. Aggregated jobs assignment begins either with jobs of the first network layer, i.e., having no predecessors on the aggregated network (in Distribution Algorithms 1 and 2), or with the terminal jobs of products, i.e., having no successors (in Distribution Algorithm 3). We put the jobs into the list of unassigned jobs in the order determined by the sequence σ^*. Then, while the list of unassigned jobs is not empty, we select the next aggregated job. According to certain rules, we search such a reserve of the element for its placement, so that the assignment of the job does not violate technological constraints. If there are two or more reserves with a sufficient length for the job assignment (e.g., if the system provides uniform parallel elements), we choose a shorter reserve. If a required reserve is found, we move the job from the list of unassigned jobs to the list of assigned aggregated jobs, i.e. we assign the job to the found reserve of the element. If we could not find a required reserve, it means that this product violates its due date, so we should make a decision on this product in the DMU. After the aggregated job assignment, we add all its successors (in Distribution Algorithms 1 and 2) or predecessors on the aggregated network (in Distribution Algorithm 3) to the list of unassigned jobs and perform their assignment under similar rules. We search the predecessors (successors) through the network of the first level of aggregation. The coordinated plan of aggregated jobs processing is a list of assigned aggregated jobs with information about their start and completion times on the elements.

We use the following general rule for aggregated job assignments in each distribution algorithm. The start time of an aggregated job cannot be less than the maximum completion time of its predecessors in the network; and the completion time of an aggregated job cannot be greater than the minimum stat time of its successors in the network.

Below we give an algorithm for compact schedules construction. Then, we present its modification for building nondelay schedules.

Distribution Algorithm 1 Compact schedules construction is based on the following heuristic. At each stage, we select from the sequence σ^* the next aggregated job with the minimum MPSS number. We assign it to an element which will have the minimum release time after this job processing, also we must not violate the technological constraints. The earliest possible release time of an element is the earliest start time of the aggregated job plus its processing time. The earliest start time of the aggregated job is equal either to the maximum completion time of all its predecessors determined during the predecessors' assignment, or to the beginning time of the planning period. To assign an aggregated job to an element, we seek the earliest free reserve within the element's time fund. The time fund consists of idle time gaps available after all previous jobs' assignment. We must place the aggregated job into the reserve not earlier than its earliest start time. This makes impossible processing any aggregated job at a later time without postponing the start time of another aggregated job.

Notation Used in Distribution Algorithms Let us denote by

P_t	a partial schedule of $t - 1$ assigned jobs;
S_t^k	a set of unassigned jobs ready to be assigned to element k at the stage t;
S_{j_k}	the earliest possible start time of a job $j_k \in S_t^k$ (equal to the maximum completion time of predecessors or the beginning time of the planning period for a job without predecessors);
R_{ik}	the i-th interval of the time reserve array on element k not occupied by a job assigned at the previous stages;
$R_k = \{R_{ik}\ \forall i\}$	the reserve array for element k;
R_{ik}^{st}	the beginning of i-th reserve interval;
R_{ik}^{end}	the end of i-th reserve interval;
$s_{j_k} = \max\left(R_{ik}^{st}, S_{j_k}\right)$	the start time of a job j_k assigned to the interval R_{ik} of element k;
C_{j_k}	the completion time of the assigned job j_k: $C_{j_k} = s_{j_k} + l_{j_k}$;
$s_{j_k}^0$	the earliest possible start time of a job $j_k \in S_t^k$: $s_{j_k}^0 = \min\limits_k \min\limits_{j \in S_t^k} s_{j_k}$ $\forall k, i : s_{j_k}^0 + l_{j_k} \le R_{ik}^{end}$;
$C_{j_k}^0$	the earliest possible completion time of a job $j_k \in S_t^k$: $C_{j_k}^0 = \min\limits_k \min\limits_{j \in S_t^k} C_{j_k}$ $\forall k, i : R_{ik}^{end} - l_{j_k} \le R_{ik}^{st}$.

Distribution Algorithm 1 has the following four steps.

1. Set $t = 1$, $i = 1$, $P_1 = \varnothing$, $\sigma_i = \sigma_1$ where σ_i is the i-th MPSS including all predecessors on the graph G. Create sets S_1^k of jobs that do not have predecessors and require processing on each element k, $k = \overline{1, m}$. Initialize the time reserve arrays R_k of all elements.

2. Select for assignment from the set S_t^k a job j_k with the minimum completion time.
3. Move on to the next stage the following way:

   3.1. Put the job j_k to P_t and, hence, create P_{t+1}.
   3.2. Create S_{t+1}^k, $k = \overline{1,m}$ by adding to one of the sets S_t^k a job which immediately follows the job j_k, if only j_k is not the last job in σ_i. Leave all other sets S_t^k without changes.
   3.3. Change the reserve array R_k of element k.
   3.4. Set $t = t + 1$.

4. If there are unassigned jobs in a subsequence σ_i, then go to step 2. Otherwise, proceed to the next MPSS (set $i = i + 1$) and to step 2. If all jobs were assigned, terminate.

Let us analyze the work of Distribution Algorithm 1. Step 1 initializes t, P_1, S_1^k, R_k, $k = \overline{1,m}$. In step 2, we analyze the reserves of each element and assign a job that has the minimum completion time from the set of jobs ready to be assigned. In step 3, we move from stage to stage, and in step 4 we check whether the schedule is complete.

We now show that Distribution Algorithm 1 constructs compact schedules, that is, those in which no job can be started earlier without causing the delay in other jobs processing or violating technological constraints.

At each stage of the algorithm, we select a job with completed predecessors. We assign it in such a way that does not violates the technological constraints. The procedure of a job assignment searches for a free time interval on element k at the earliest possible time. This guarantees that we cannot process any job at an earlier time without postponing the start time of any other job.

Distribution Algorithm 2 The following modification of Distribution Algorithm 1 yields nondelay schedules. Distribution Algorithm 2 is based on the following heuristic. At each stage, we assign the next aggregated job from the sequence σ^* that has the minimum start time and the minimum MPSS number. The minimum start time of a job is defined as the maximum completion time of its predecessors in the network or the beginning time of the planning period.

Distribution Algorithm 2 has the following four steps.

1. Set $t = 1$, $i = 1$, $P_1 = \varnothing$, $\sigma_i = \sigma_1$. Create sets S_1^k of jobs that do not have predecessors and require processing on an element k, $k = \overline{1,m}$. Initialize the reserve arrays R_k of all elements.
2. Select for assignment from the set S_t^k a job j_k with the earliest possible start time $s_{j_k} = \max\left(R_{ik}^{st}, S_{j_k}\right)$; $s_{j_k}^0 = \min\limits_k \min\limits_{j \in S_t^k} s_{j_k}$ $\forall k, i : s_{j_k}^0 + l_{j_k} \le R_{ik}^{end}$.
3. Move on to the next stage the following way:

   3.1. Put the job j_k to P_t and, hence, create P_{t+1}.

3.2. Create S_{t+1}^k, $k = \overline{1,m}$ by adding to one of the sets S_t^k a job which immediately follows the job j_k, if only j_k is not the last job in σ_i. Leave all other sets S_t^k without changes.

3.3. Change the reserve array R_k of element k.

3.4. Set $t = t + 1$.

4. If there are unassigned jobs in a subsequence σ_i, then go to step 2. Otherwise, proceed to the next MPSS (set $i = i + 1$) and to step 2. If all jobs were assigned, terminate.

The constructed schedule is nondelay by definition.

An advantage of compact and nondelay schedules is the elements' idle times minimization. Consequently, they minimize the products' flow time, reduce the amount of work in progress and improve the system efficiency.

Distribution Algorithm 3 is based on the following heuristic. We consider the sequence σ^* from the beginning, i.e., from products of a higher priority. At the first stage, we choose the aggregated jobs from the list of terminal jobs (in the order they appear in σ^*). We assign them so that their completion times correspond to the product's due date. If a range Δ_i of completion times in a feasible schedule is specified, then we assign the job into the interval $[d_i - \Delta_i,\ d_i]$. If such assignment is impossible, that is, the required element is occupied by other jobs, then we should assign the terminal job as close as possible to the due date (or the interval $[d_i - \Delta_i,\ d_i]$). For products optimized according to the basic Criterion 2 or 3, the completion time should not be later than the due date (or the interval $[d_i - \Delta_i,\ d_i]$). At each next stage, we assign to the element the next aggregated job with the maximum start time and the minimum MPSS number. The maximum start time of an aggregated job is the latest possible completion time of the aggregated job minus its processing time in the element. The latest possible completion time of an aggregated job is the minimum start time of its successors in the network determined during the successors' assignment. To assign an aggregated job, we search for the most recent free time reserve within the element's time fund. The time fund consists of idle time gaps available after all previous jobs' assignment. We must place the aggregated job into the reserve not later than its latest possible completion time. If we could not find a reserve with required length, it means that we cannot distribute jobs of this product to existing resources. We should continue the distribution procedure for this product, assigning aggregated jobs before the beginning time of the planning period (with negative start times). Then, we need to determine in the DMU what to do with the products for which we could not find time reserves. For example, experts may reduce the number of products in the order portfolio or remove them from the order portfolio completely, change the due date of a product, increase the resources amount etc. Almost all the solutions require solving the entire problem again, except for a shift in the beginning time and a change of the due date. The two latter options require common vertices revision in the critical paths network, re-solving the TWCT problem, and re-distribution.

By such distribution, we first of all assign the highest priority products as closely as possible to their due dates without their violation. For the tardy products, we minimize the amount of the total penalty for their tardiness. This allows the enterprise to fulfill the maximum number of orders without due dates violation and maximize the total profit.

Distribution Algorithm 3 has the following seven steps.

1. Initialization: create empty lists of assigned and unassigned jobs and of elements without assigned jobs.
2. Choose all the terminal jobs of products. Put them into the list of unassigned jobs in the order determined by the sequence σ^*. The value of the latest possible completion time for a terminal job of a product is equal to its due date.
3. If the list of unassigned jobs is empty, then go to step 7, else go to step 4.
4. Find a job with the minimum MPSS number and the maximum start time in the list of unassigned jobs. The maximum start time of a job is its latest possible completion time (i.e., the minimum start time of its successors) minus its processing time with a possible setup time. Remember its number for later use.
5. Assign the found job, i.e., move it from the list of unassigned jobs to the list of assigned jobs, as follows. Look through the list of the assigned jobs of the element in which the job is to be processed. Search for a free time reserve in the "real time" of the element (taking into account work hours) in the order from the end to the beginning. The reserve must end not later than the latest possible completion time of the unassigned job. The reserve's length must not be less than the job's processing time with the setup time (if the equipment's setup is required). If the job appears immediately before the assigned job that requires a similar setup, then we eliminate the setup for the already assigned job (the time interval at the element is released for subsequent filling). If a reserve is found, go to step 6. If we could not find the required reserve within the time fund of this element, then we popup a user message about due date violation by this product; mark the product as tardy; go to the DMU to determine the action for this product; return to step 1.
6. Select all the predecessors of the remembered job. Add them to the list of unassigned jobs by general rules, calculating their latest possible completion times as the difference of the remembered job's completion time and its processing time. Go to step 3.
7. Create the coordinated plan of aggregated jobs processing. It is a list of assigned jobs with information about their start and completion times.

Coordinated Planning Specifics Coordinated planning for each basic criterion of optimality has its own specifics. Now we consider the planning methodology for each of the basic criteria 1–5.

Criterion 1 We distribute the priority ordered sequence σ^* obtained as a solution of the TWCT problem and complemented by aggregated jobs which lie outside of products' critical paths. We use Distribution Algorithm 1 or 2. This minimizes the total flow time of each MPSS and thus provide the minimum completion time for the most profitable products. Such implementation of the problem solving maximizes the total profit of the enterprise for all products.

Criterion 2 We distribute the aggregated jobs of the sequence σ^* using Distribution Algorithm 3. In such implementation of the problem solving, we assign the products according to the priority of the subsequence in which they are included. The priority of the subsequence is understood as the ratio of the total weight of aggregated jobs included in it to their total processing time. The higher the priority of the subsequence, the earlier we assign the jobs that belong to it. These aggregated jobs occupy earlier positions in the sequence σ^*. Such assignment ensures that the highest priority products will be processed at the first place and in the specified due date. This, consequently, maximizes the total profit of the enterprise.

If there are products that violate the due date, then we pass the schedule to the DMU for adjustment. We either exclude these products from execution, or make a decision to purchase additional resources.

Criterion 3 The distribution procedure is hybrid. First of all, Distribution Algorithm 3 distributes the highest priority products. We assign them to be processed at the planning period intervals as close to the due dates as possible, without their violation. We thus free up the reserves at earlier time intervals for the remaining products. If a product violates its due date, we re-distribute it by Distribution Algorithm 1 or 2 in the order determined by the sequence σ^*.

Such distribution assigns the highest priority products, i.e., the most profitable, to be processed without tardiness. For tardy products, we minimize the total tardiness cost. Thus, we maximize the total profit of the enterprise over all products.

Criterion 4 We distribute the sequence σ^* using Distribution Algorithm 3. If we cannot distribute some product without its due date violation, then we pass the schedule to the DMU for adjustment. Such distribution first assigns the highest priority products as close to the due dates as possible, without their violation. This allows to fulfill the maximum number of orders without a tardiness and, consequently, maximize the total profit according to the optimality criterion.

Criterion 5 We distribute the sequence σ^* using Distribution Algorithm 3. We assign each product so that its completion time corresponds to its due date d_i or is less than d_i no more than by a specified value Δ_i. We place the products which we could not distribute without due dates violation into a separate list. We sort the list by the due dates and re-distribute it using Distribution Algorithm 1 or 2.

Such distribution will ensure processing the most profitable products without due dates violation or as close to them as possible. For tardy products, we minimize the deviation of the completion time from the due date. So, we minimize the total earliness and tardiness costs and maximize the total profit of the enterprise.

Ergo, we try to tie up the products completion times and the due dates for the basic criteria 2–5. Consequently, such products, as a rule, start their processing not from the beginning time of the planning period. This is not an error but corresponds to the planning methodology.

Remark 9.10 Some variations are possible in the distribution direction determination depending on the basic optimization criterion. The functional value of the basic criteria 1, 3, and 4 does not get worse when a product is manufactured before the due date. The resources quantity may be sufficient to process all products planned by criteria 1, 3, and 4 from the beginning time of the planning period. Someone can prove in an expert way (or by a particular distribution of aggregated jobs) that such distribution may improve the functional value or another schedule quality measure (capacity load, the total idle time through the critical paths, etc.). Then, we can plan product groups corresponding to basic criteria 1, 3, and 4 starting from the beginning of the planning period, and those corresponding to criteria 2 and 5 from the due dates. If this is the case, we should use rules 1 and 3 for combining the aggregated jobs into common vertices (Step 1.4) for product groups corresponding to criteria 1, 3, and 4 and the rules 2 and 4 for criteria 2 and 5. During the coordinated planning, we should use Distribution Algorithm 1 or 2 for criteria 1, 3, and 4 and Distribution Algorithm 3 for criteria 2 and 5.

Step 2.1.3 We should also *check the actual combination of jobs in the common vertices* in any distribution algorithm implementation. We assign components of a common vertex successively to the same element if the time gap between them is less than the value of Δ_{CV} corresponding to the rule used for the common vertex combination. The gap is the difference between the completion time of one component and the start time of another component. If it turns out that we cannot form the common vertex because the resource is busy (the gap between its components is greater than Δ_{CV}), then we need to change the technology. We will have to break such common vertices in the network of the second level of aggregation into components. We should restore the initial parameters of aggregated jobs and the arcs of the network obtained at the first level of aggregation (see Step 1.2). There may be more than two components in a common vertex, and a gap greater than Δ_{CV} may exist between only some of them (and we may not be able to eliminate the gap by permuting the components among themselves). Then we should leave in the common vertex only the components that have a gap smaller than Δ_{CV}. The processing time of such common vertex decreases up to the sum of the processing times of the components. Also, we should restore the relationships with excluded components of the common vertex which were removed during aggregation and add those arcs into the aggregated network of the second level of aggregation.

All changes made to the aggregated network of the second level of aggregation at the coordinated planning stage should also be reflected in the aggregated network of the first level of aggregation and the initial network. We should do it after the coordinated plan approval, before passing it to the third level of planning.

When we break down the common vertices, we distribute their components, together with their precedence relations, to the elements with the same number for processing at different times. The components correspond to aggregated jobs (in the aggregated network) and initial jobs (in the initial network). The condition of processing without interruptions now holds for separate job groups, not for a single group which the jobs represented earlier being processed by a single element. An example of such an element breakdown for a basic criterion is in Fig. 9.13a (element 201). Examples for a synthetic criterion are in Fig. 9.13b (element 201) and Fig. 9.13c (element 102). We will give more explanations later.

A change in the composition of at least one common vertex (that is, a modification of the aggregated network of the second level of aggregation) requires to re-solve the TWCT problem (Step 1.5) and re-distribute the aggregated jobs (Step 2.1). This should be done after we have distributed the entire set of aggregated jobs and checked all common vertices. We need to disaggregate all common vertices that were broken into components during the distribution, and only then re-solve the TWCT and coordinated planning problems. Doing this after each common vertex disaggregation may lead to a loop with a constant change in the set of common vertices at each iteration.

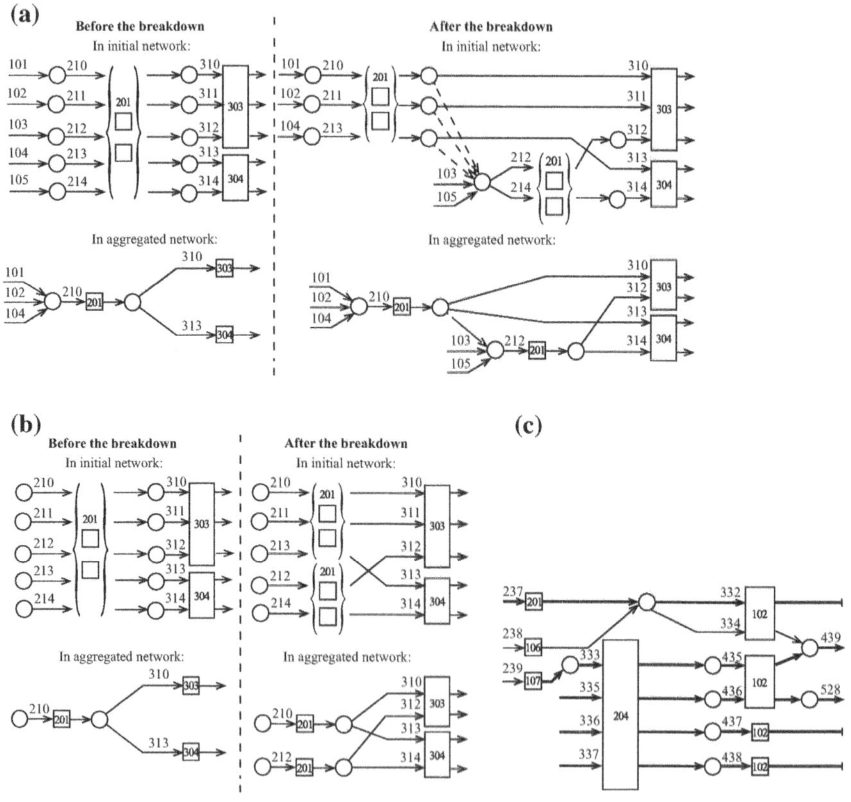

Fig. 9.13 Changing the initial and the aggregated model after a common vertex breakdown: **a** a basic or a synthetic criterion, **b** a synthetic criterion, **c** a synthetic criterion (the aggregated network after a common vertex breakdown on element 102)

Because of the relatively large complexity of the distribution procedure, we can check the actual combination of jobs in the common vertices by distributing only the aggregated jobs belonging to the products' critical paths, without supplementing them with other aggregated jobs.

We also perform a *technology compatibility check* after the aggregated jobs distribution. Some aggregated jobs (job groups) should be processed without interruptions on their element, according to the aggregated network of the first level of aggregation. We should if a time gap appears between some jobs (job groups) after their distribution. If we cannot eliminate the gap without other aggregated jobs redistribution, then we should change the aggregated network of the first level of aggregation (and, accordingly, the initial network). We have to indicate that these jobs are processed on the same element but with a time gap. This does not require solving the problem again, since the critical paths of products do not change, and the composition of common vertices is already adjusted during the distribution. However, we must use the new model at the following planning levels.

During planning *by a basic criterion*, we need to observe the restriction that we cannot overlap jobs processing intervals in the network. Therefore, we should forcibly space in time the sets of jobs processed on the broken elements by inserting a buffer between them. We show an example of such a partition in Fig. 9.13a. We can also use a buffer for a synthetic criterion planning (in a network of an arbitrary form). In this case, the buffer guarantees that job sets are spaced in time, so that the set of common vertices and the corresponding network model could not change later.

Example 9.4 for a Basic or a Synthetic Criterion Consider Fig. 9.13a. Jobs 210–214 (aggregated job 210) are on the same element 201 in both the initial and the aggregated networks. But after the coordinated planning there are two separate job groups processed without interruptions: first, jobs 210, 211, and 213, and later, jobs 212 and 214. Then, we change this part of the aggregated network and, accordingly, the initial network, as shown in Fig. 9.13a.

Jobs 101–105 in Fig. 9.13a are the immediate predecessors of jobs 210–214. We use general rules when merging new common vertices, including the restrictions established for output jobs of a buffer. The dashed lines in Fig. 9.13a indicate fictitious jobs with zero processing time. Such jobs reflect only the precedence relations between the buffers. Here, jobs 212 and 214 on the second element #201 can begin only after the completion of jobs 210, 211, and 213 on the first element #201. Thus, we space in time the job sets.

Now we give two examples for a synthetic criterion.

Example 9.5 The network model for the example 9.4 described above (see Fig. 9.13a) can be simplified as shown in Fig. 9.13b if it is not necessary to force the job sets separation.

Example 9.6 Consider Fig. 9.13c. Jobs 332, 334, and 435–438 are on the same element 102 in both the initial and the aggregated network. But after the coordinated planning there are two separate job groups processed without interruptions.

The first group contains jobs 332 and 334, the second group is jobs 435 and 436. Jobs 437 and 438 were not combined into the common vertex. Then, we change this part of the aggregated network and, accordingly, the initial network, as shown in Fig. 9.13c. If, in such a case, we want to use buffers to space jobs in time, then we should thrice use the technique shown in Fig. 9.13a.

We perform the *coordinated planning by a synthetic criterion* on $N \in [2, 5]$ subnets of the first level of aggregation for N product groups (see Step 1.2). We combine the subnets into a single common network. The subnets do not have precedence relationships and common vertices between them. They are interrelated only by common equipment. This allows us to solve the problem sequentially alternating product groups assignment order to choose the best solution. We examine all possible ordering combinations of the sequential assignment of product groups. The number of combinations is $N!$, that is, from 1 to 120. In this case, we perform the procedures described in Steps 2.1.1–2.1.3 $N!$ times, according to various orders of product groups assignment. The next product group distribution does not begin until we finish the distribution of the previous one. As a result, we get $N!$ coordinated plans of the aggregated jobs processing. We will choose from them the best plan at the next stage.

For a synthetic criteria, the distribution procedure is hybrid. We apply one of the developed algorithms for each product group, depending on the given basic component of the synthetic criterion. We use Distribution Algorithms 1 and 2 to distribute product groups optimized by the basic Criterion 1. We choose from Distribution Algorithm 1 or 2 according to the initial data characteristics. Distribution Algorithm 3 is for product groups with due dates (basic criteria 2–5).

The reserves within the time fund of the elements are the idle times of the elements available for jobs assignment. They remain after the assignment of all jobs of the previous product groups, according to their assignment sequence, and the previous assigned jobs from the currently distributed product group, according to their assignment sequence in this group.

Remark 9.11 During the aggregation and coordinated planning stages, we don't take into account the assumption that jobs from different product groups may be processed on the same element in the same time interval. We have to check this situation only during the exact planning. For example, if an equipment setup is required, we can combine jobs of different product groups distributed into the same time interval to the same element, to improve the plan.

As an example, consider the coordinated planning of aggregated jobs of some products with numbers 1 and 2. Suppose that they should be planned according to the basic Criterion 2 ("just in time"). We present the product's critical paths graph in Fig. 9.14. Suppose that the priority of product 1 is significantly higher than that of product 2. Then, the sequence σ^* begins from the aggregated jobs of product 1 and their predecessors belonging to other products: in this case, to product 2.

We show the distribution diagram for products 1 and 2 in Fig. 9.15. As we can see, product 1 cannot be executed in time, since its start time is less than the beginning time of the planned period. Therefore, we should adjust the plan or the due date for this product in the DMU.

Coordinated Planning Heuristic Justification for a Basic Criterion We perform the coordinated planning in accordance with the aggregated jobs sequence obtained as a solution of the approximating TWCT problem. Such sequence of the jobs assignment maximizes the corresponding component of the synthetic functional, up to its value relevance to the approximating TWCT problem functional.

Coordinated Planning Heuristic Justification for a Synthetic Criterion We implement the coordinated planning as sequential assignment of the product groups. Each group is defined by the corresponding component of the synthetic criterion. We examine all possible ordering combinations of the product groups. The order of products assignment in each group maximizes the corresponding component of the synthetic functional, up to its value relevance to the approximating TWCT problem functional. This heuristic allowed us to reduce the initial planning problem by 31 criteria to a single multi-stage network scheduling problem (MSNSP). Each of the solutions by 31 criteria corresponds to its own due dates for the terminal jobs. The heuristic is efficient if it is relevant to a real technological process.

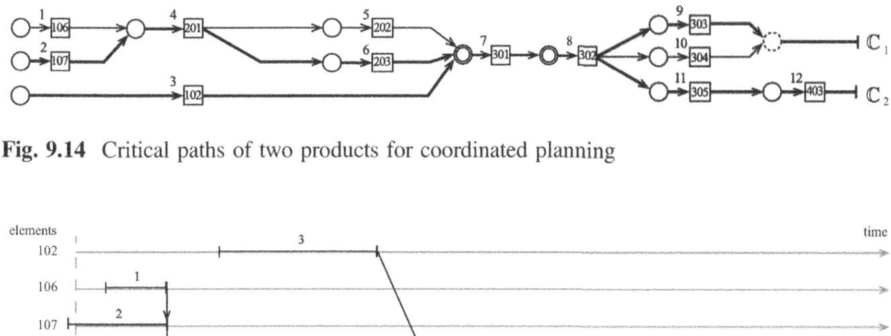

Fig. 9.14 Critical paths of two products for coordinated planning

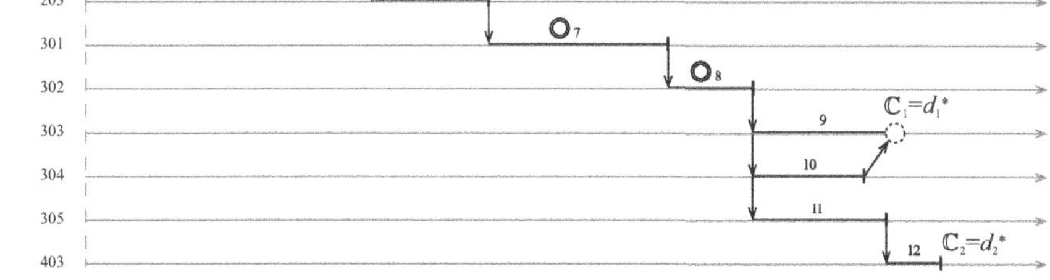

Fig. 9.15 Distribution of aggregated jobs of the two products

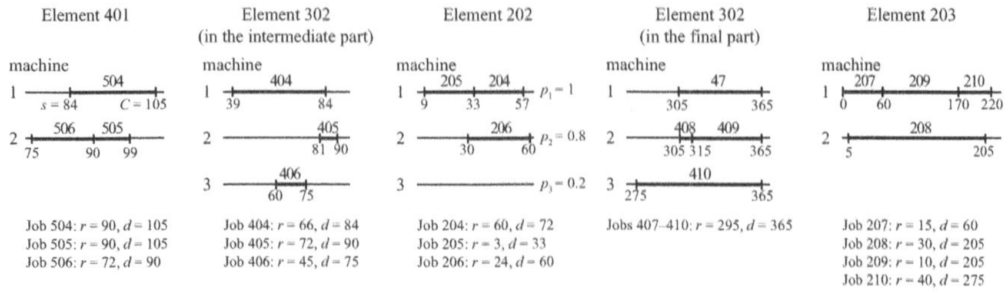

Fig. 9.16 Single stage problems solution on elements of type 3–5 (preliminary stage)

9.3.3 *The Coordinated Plan Analysis in Decision Making Unit*

Step 2.2 After we have constructed all coordinated plans, we calculate the value of the initial functional for each possible sequence of product groups assignment. The functional values are based on the products' completion times $\mathbb{C}_i$ obtained during the coordinated planning. The completion time for each product is the maximum completion time of its terminal jobs. We choose the best of these plans by the functional values. If the profit, according to the chosen plan, is less than the

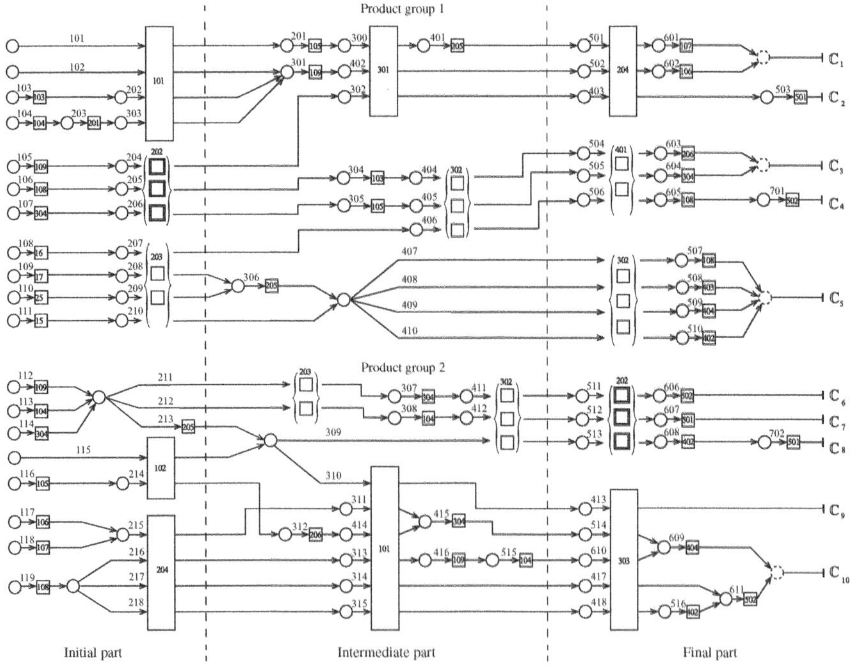

Fig. 9.17 Initial network (products 1–10)

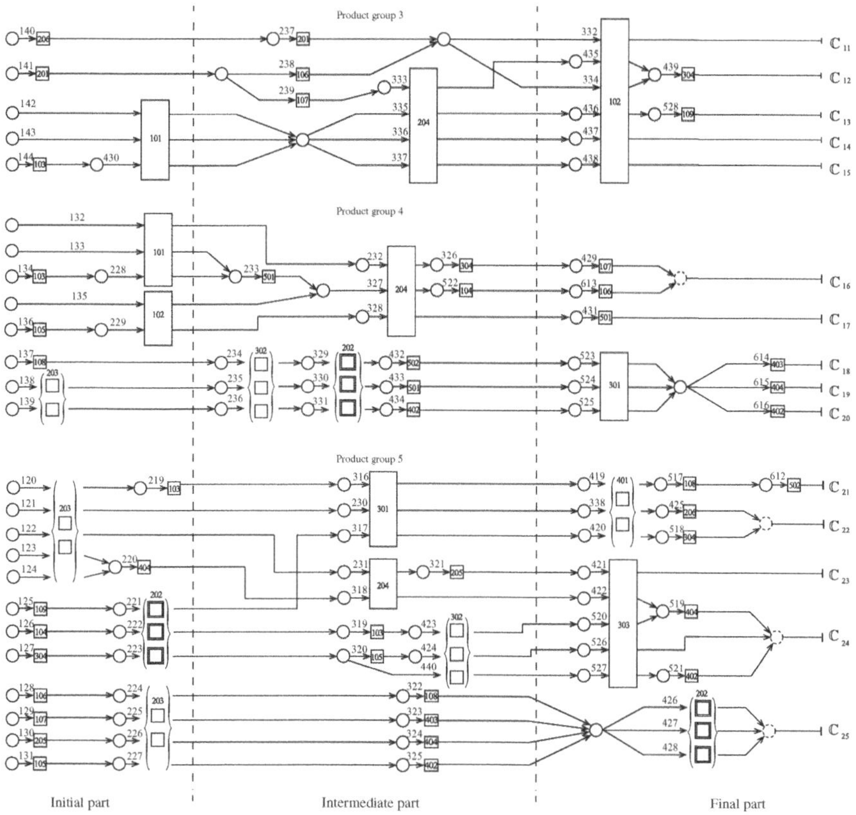

Fig. 9.18 Initial network (products 11–25)

expected minimum, we send the plan to the DMU to form a new order portfolio. Otherwise, we pass the plan to the next planning level.

We should also evaluate the deviation of the products' flow time from critical paths in the chosen plan. If the deviation exceeds the value specified by experts, we should redistribute the product groups according to the principle of spacing the resources in time and choose a new best plan.

During planning according to the basic criteria 2 or 4 (or the synthetic criterion including the basic criterion 2 and/or 4), if the profit is not less than the expected minimum, we may exclude not profitable products from the order portfolio. In this case, we can pass to the third level a truncated plan with a corresponding change in the technological network. New due dates agreement with the customer in the DMU without products exclusion from the order portfolio may serve as another solution.

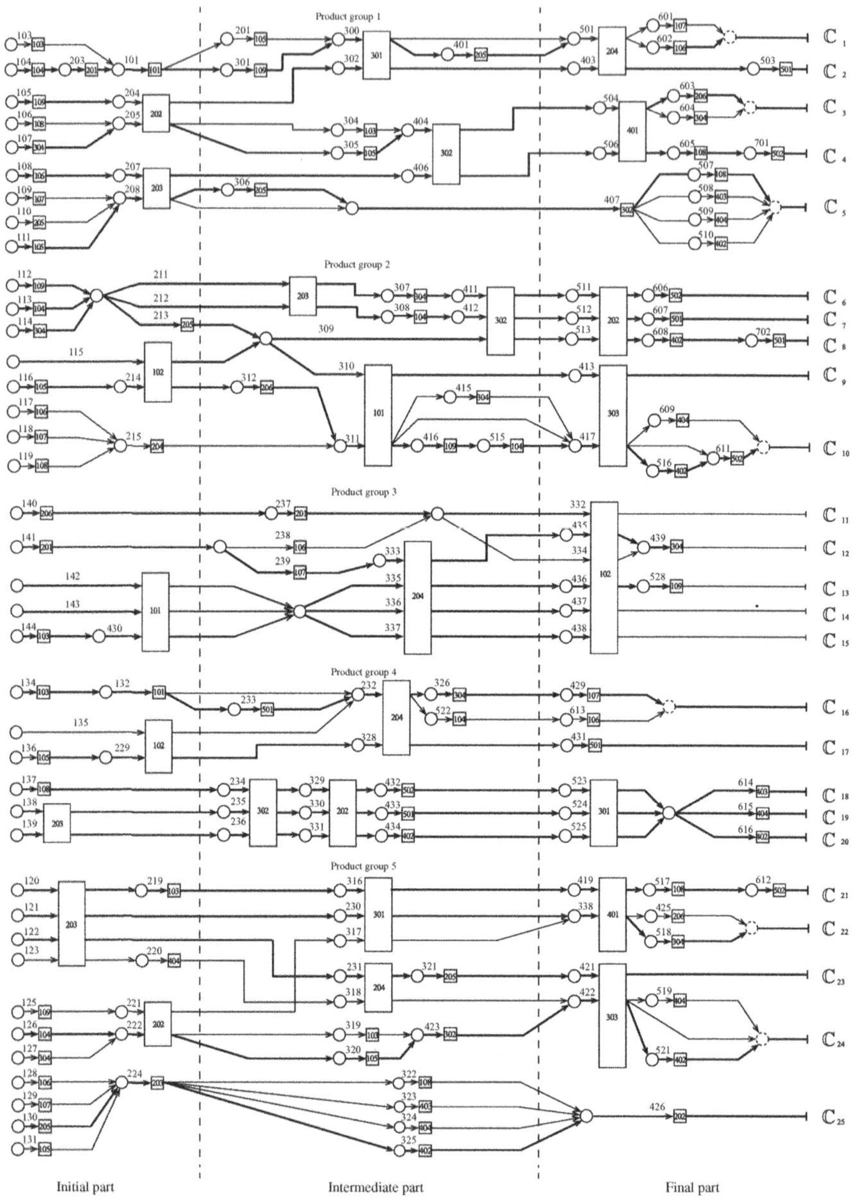

Fig. 9.19 Aggregated network of the first level of aggregation with critical paths (products 1–25)

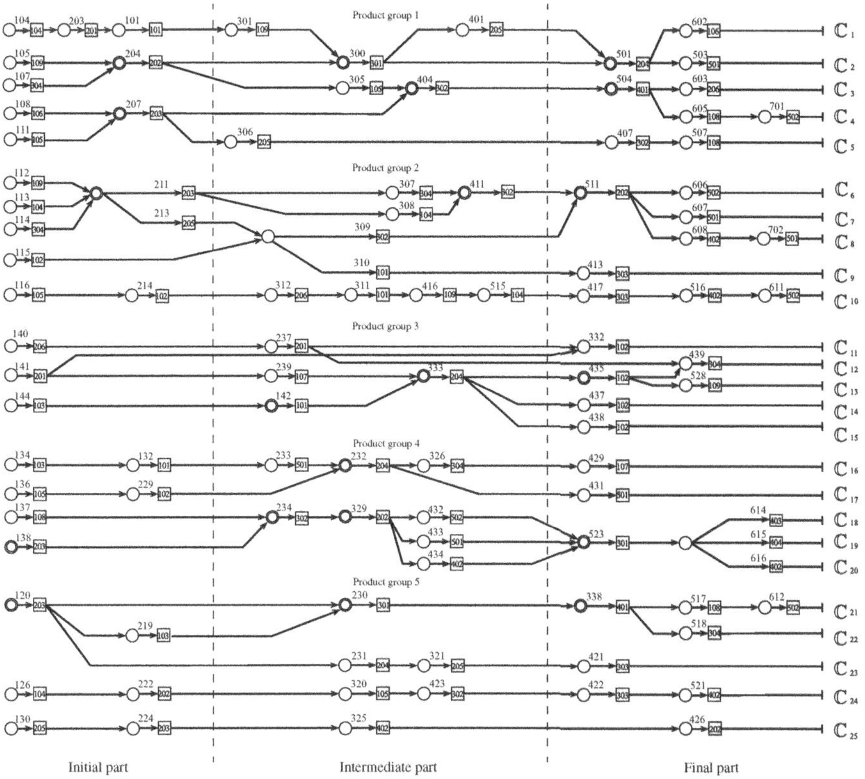

Fig. 9.20 Network on the products' critical paths with common vertices

We can implement alternative processing plans for different order portfolios at the third planning level. Each plan has its own risk of not obtaining the expected profit after its implementation. Therefore, we choose the best plan in the DMU from the set of alternative plans according to the criterion of minimizing the risk of not obtaining the expected profit (see Sect. 8.5).

9.4 Unit 4. The Model of the Third Level of Planning (Multi-stage Network Scheduling Problem Construction)

In Unit 4, based on the data we obtained in Unit 3, we construct the problem of the third level of the FLM, a multi-stage network scheduling problem (MSNSP).

We have approved some coordinated plan of aggregated jobs in Unit 3. After that, we disaggregate the aggregated jobs, i.e., return to the initial network model,

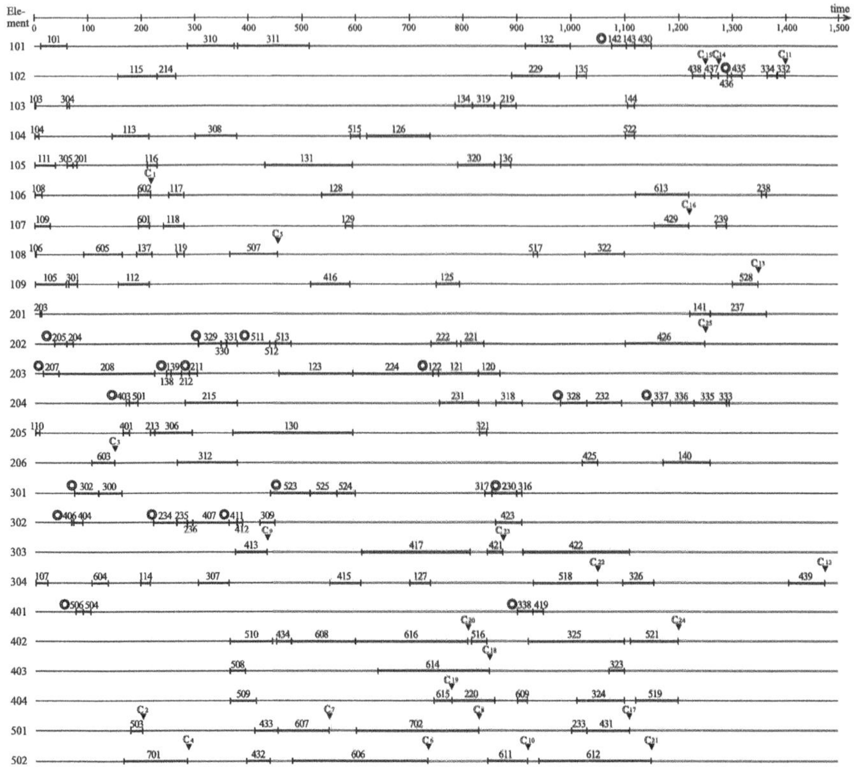

Fig. 9.21 Gantt chart: aggregated jobs distribution to elements (Ordering 1)

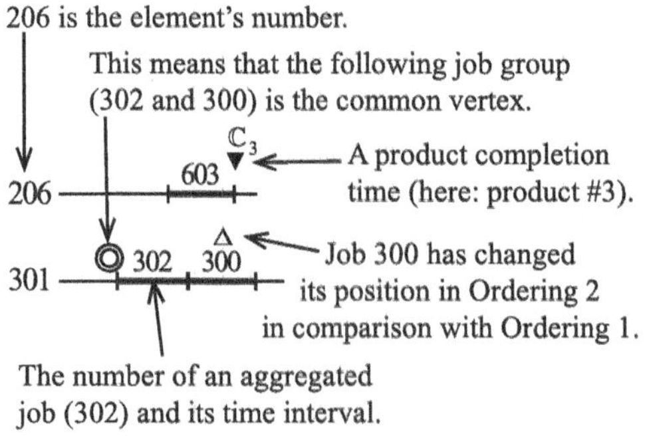

Fig. 9.22 Notation used on the Gantt charts

taking into account the changes made during the coordinated planning. Some elements could be split into parts in the approved coordinated plan. In this case, we replace each such element on the initial technological network with similar independent elements of the same type. We appropriately redistribute the input and output arcs between the new elements. Each output arc should enter the same vertex as in the original technological model.

For example, we break element 201 into two similar ones in the fragment of the source network shown in Fig. 9.13b. We give another examples (taking into account the specifics of the methodology for basic and synthetic optimality criteria) earlier in Fig. 9.13a, c.

We construct the MSNSP at the third level of planning by setting the due dates of the terminal jobs on the modified initial network. The completion times $\mathbb{C}_i$ of the products determined during the coordinated planning specify the actual due dates d_i of the terminal jobs for the next planning level. If a product has several terminal jobs, then we choose the maximum completion time, and it becomes a single due date for all terminal jobs of the product. For products planned by Criterion 5 (minimizing the costs of earliness/tardiness in relation to the due dates), we select one terminal job. The job should have the latest completion time determined during the coordinated planning, compared to the completion time of the other terminal jobs of this product. We set its completion time as the due date for determining the earliness/tardiness costs in the functional of the model of the third level. For the rest of the terminal jobs of this product, this time point is the due date, but we can complete these jobs earlier, without a penalty for earliness. This way, we formulate the optimization problems for each element of type 2–5 of the initial network. Each problem is to find a feasible (not violating the due dates) schedule of jobs processing on each machine of a given element that processes jobs without interruptions. We give the optimality criterion below.

9.5 The Methodology of Solving the Multi-stage Network Scheduling Problem (The Model of the Third Level of Planning)

9.5.1 *A Method to Solve the Multi-stage Network Scheduling Problem Corresponding a Basic Optimality Criterion*

For planning by a basic optimality criterion, the network should satisfy the properties described in Sect. 9.1. We propose the method of the problem solving for the case when due dates are given for all the terminal jobs of the network, and completion of a terminal job before its due date is allowable.

Definition 9.12 A schedule is *feasible* in the MSNSP problem if the completion times of all terminal jobs in it do not exceed their due dates.

Let us formulate the optimality conditions for a feasible solution of a MSNSP constructed for a basic optimality criterion.

Definition 9.13 A schedule is optimal if it meets the following conditions:

1. The schedule is feasible.
2. Of all the feasible schedules, only those may be optimal which minimize the total earliness in relation to the due dates of terminal jobs.
3. A feasible schedule satisfying items 1 and 2 is optimal if it maximizes one of the following functionals on the set of initial jobs:

(a) $\max\left(\min_j s_{1j}\right), \quad j = \overline{1,p};$

(b) $\max\left(\min_j s_{1j}\right) = s_{1j_1}, \quad j = \overline{1,p},$

$$\max\left(\min_j s_{1j}\right) = s_{1j_l},\ j = \overline{1,p},\ l = \overline{2,p},\ j \neq j_k,\ k = \overline{1,l-1}.$$

Here, $s_{1j} \geq 0$ are the start times of the initial jobs of the network (the first jobs on the elements $1j$ of the first network layer); p is the number of such elements.

Remark 9.12 Suppose that a terminal job must be completed just in time. The method presented here is based on the PSC-algorithms described in Chaps. 2–3 of this book and in [15]. It suggests to achieve the minimum possible deviation from the due dates for a feasible schedule. Thus, we reasonably believe that in this case we can neglect the earliness costs.

We now describe the method step by step.

1. Where necessary, calculate the due dates of output jobs on elements of type 2–5 of the last network layer by the due dates of the terminal jobs. In the general case, the output jobs may be at the layer before last or at the one that immediately relates with the terminal jobs. This follows the properties of the technological network corresponding to the planning model for a basic criterion.

1.1. Solve one of the following single stage scheduling problems for each element of type 2 of the last network layer (in this section, we call a job of the initial network a task):

1.1.1. The total earliness of tasks minimization for a feasible schedule on a single machine at the latest start time of tasks and arbitrary due dates (see Chap. 2).

1.1.2. The total earliness of tasks minimization for a feasible schedule on a single machine with arbitrary due dates (see Chap. 2).

Remark 9.13 It is better to use the problem 1.1.2 for the case when the total earliness costs of tasks are significant.

1.2. Solve the following scheduling problem for each element of type 3 of the last network layer: the total earliness of tasks minimization for a feasible schedule on identical parallel machines with a common due date (see [15]).

Remark 9.14 Work [15] shows that the obtained schedule corresponds to the latest simultaneous start time of the machines.

1.3. Solve one of the following two scheduling problems for each element of type 4 of the last network layer (see Chap. 3):

1.3.1. Given m independent parallel machines with equal productivities to process n tasks without interruptions. l_i is the processing time of task i, $i = \overline{1,n}$, its due date is d_i. The start times of machines are arbitrary. We need to build a feasible schedule that maximizes the criterion:

$$\begin{aligned} s_{j_1} &= \max\left\{ \min_{j=\overline{1,m}} s_j \right\} \\ s_{j_l} &= \max\left\{ \min_{j} s_j, j = \overline{1,m}, j \neq j_k, k = \overline{1,l-1} \right\}, l = \overline{2,m} \end{aligned} \quad (9.37)$$

where s_j is the start time of machine j; j_1 is the number of machine with the earliest start time in an optimal schedule (it is the latest for all feasible schedules); j_l, $l = \overline{2,m}$, is the number of machine with the next earliest start time after machines j_k, $k = \overline{1,l-1}$ (it is the latest for all feasible schedules with fixed s_{j_k}, $k = \overline{1,l-1}$).

1.3.2. The problem 1.3.1 with different criterion:

$$s_{j_1} = \max\left\{ \min_{j=\overline{1,m}} s_j \right\} \quad (9.38)$$

where s_j is the start time of machine j; j_1 is the number of machine with the earliest start time in an optimal schedule (it is the latest for all feasible schedules).

1.4. Solve one of the following two scheduling problems for each element of type 5 of the last network layer (see Chap. 3):

1.4.1. Given m independent parallel machines with various productivities to process n tasks without interruptions. l_i^j is the processing time of a task i on a machine j. The due date of a task i is d_i. Processing times of tasks are independent (it is possible that one task is processed faster than another on some machine but slower than another on a different machine). The start times of machines are arbitrary. We need to build a feasible schedule that maximizes the criterion (9.37).

1.4.2. The problem 1.4.1 with the criterion (9.38).

Remark 9.15 The choice of the problem (1.3.1 or 1.3.2; 1.4.1 or 1.4.2) is determined by the optimality criterion chosen for the MSNSP.

2. The schedules obtained for elements of type 2–5 of the last network layer unambiguously specify the latest readiness times of input jobs of the elements of type 2–5. These times, due to the technological network properties, unambiguously specify the due dates for the immediately preceding elements of type 2–5 (in the layer before last or any related network layer). Immediate preceding does not takes into account elements of type 1. The same situation applies to elements of type 2–5 of any network layer. For elements of type 2–5 of the layer before last, implement step 1 of this general method.

For buffer type elements, use the two following general rules when determining jobs' due dates. The due date of jobs to the left of a buffer is equal to the minimum start time of jobs to the right of it. And the earliest start time of jobs to the right of a buffer is equal to the maximum completion time of jobs to the left of it.

We perform steps 1 and 2 until we obtain the schedules for elements of type 2–5 of the first network layer.

3. Schedules for elements of type 2–5 of the first network layer, by virtue of its properties, unambiguously specify the schedule of all jobs in the initial part of the network. We have built a suboptimal schedule.

Remark 9.16 The presented method has some properties of a PSC-algorithm for elements of types 2 and 3 in the final part of the network (the total earliness minimization for the terminal jobs in relation to the due dates specified by these elements).

Remark 9.17 For other components of the optimality criterion of the MSNSP, this method is heuristic. We implement in it the optimization by lexicographic or scalar criterion (maximizing the beginning time of the technological process) sequentially from the final part of the network to the beginning part. We use PSC-algorithms to obtain locally optimal schedules for each element of type 2–5.

Example 9.7 illustrates the order of single stage scheduling problems solving in the case of an element breakdown. Consider the initial network in Fig. 9.13a. Jobs 210–214 are on the same element 201 in the initial networks. But after the coordinated planning there are two separate job groups processed without interruptions: first, jobs 210, 211, and 213, and later, jobs 212 and 214. According to this, we break the element 201 in two elements. Let us number the first element as 201A and the second one as 201B. The dashed lines in Fig. 9.13a indicate fictitious jobs with zero processing time. Such jobs reflect only the precedence relations between the buffers. Here, jobs 212 and 214 on the element 201B can begin only after the completion of jobs 210, 211, and 213 on the element 201A. Thus, we space these job sets in time.

When solving the problem of the third level of planning, having the completion times of jobs on elements 303 and 304 ($C_{310}\ldots C_{314}$), we find optimal (suboptimal) schedules for these jobs processing on the elements 303 and 304. As a result, we obtain the start times of jobs 310–314 ($s_{310}\ldots s_{314}$).

Move on to the previous layer. First we solve a single stage problem for element 201B, then for element 201A because the due dates of jobs on element 201A will be no greater than the start times of jobs on element 201B.

Start times of jobs 312 and 314 automatically become the due dates for jobs 212 and 214. We solve a single stage problem on element 201B and obtain the start times of jobs 212 and 214 (s_{212} and s_{214}). Let $s_{201\text{B}} = \min(s_{212}, s_{214})$ be the start time of the element 201B (the element starts when its first jobs are ready for processing). It is clear that $s_{201\text{B}}$ is the due date for jobs 103 and 105 that precede jobs 212 and 214.

Next, we compare the time point $s_{201\text{B}}$ with the start times of jobs 310, 311, and 313. We determine the due dates for jobs 210, 211, 213 on element 201A, according to the specified precedence relations and buffer work rules, by the following expressions:

$$d_{210} = \min(s_{310}, s_{201\text{B}}); d_{211} = \min(s_{311}, s_{201\text{B}}); d_{213} = \min(s_{313}, s_{201\text{B}}).$$

Having due dates of jobs 210, 211, and 213, we solve a single stage problem on element 201A and obtain the start times of jobs 210, 211, and 213. The start times, in turn, become the due dates for the preceding jobs 101, 102, and 104.

9.5.2 *Methods to Solve the Multi-stage Network Scheduling Problem Corresponding a Synthetic Optimality Criterion*

During planning by a synthetic optimality criterion, the initial network is a set of independent subnets, each satisfying the properties described in Sect. 9.1. One or several subnets may have a common equipment (several virtual machines corresponding to the same physical machine). So, we cannot use the method described in Sect. 9.5.1 without changes. A schedule obtained for one subnet may not allow us to obtain a schedule for another subnet if the equipment is already busy processing the schedule of a previous product group. Therefore, we propose four alternative methods to solve the MSNSP and build an operational plan.

The First Method is the fastest. We can use it if single stage problems solving on elements does not lead to a significant reduction in the start times of jobs on these elements obtained during the coordinated planning. Then, the optimality criterion for planning by a synthetic criterion is to obtain an operational plan which is as close as possible to the plan obtained at the coordinated planning stage. We can also use this method for planning by a basic optimality criterion, as well as for a network of an arbitrary form.

In this first alternative method, we set the completion times of aggregated jobs from the approved coordinated plan as the due dates of corresponding disaggregated jobs on elements of the initial network. Start times of aggregated jobs in the coordinated plan determine the minimum start times of jobs on the elements. Due to the coordinated planning, jobs processed on the elements with the same number are spaced in time. We solve the MSNSP from the end of the planning period to its beginning, starting with the elements of the last network layer, in an arbitrary order of the network elements consideration.

To build an operational plan, we perform disaggregation of aggregated jobs. We add to the coordinated plan of aggregated jobs the jobs of the initial network, which were previously part of the aggregated job, marking them as unassigned. The processing time of all the jobs is changed according to the initial network. For each element of type 2–5, we solve single stage scheduling problems with the aim to find a feasible (not violating the due dates) schedule of jobs optimal by the following criterion. For elements of type 2 and 3, we maximize the start time of machine(s) (see Chap. 2 and [15]). For elements of type 4 and 5, we minimize the lexicographic criterion of optimality (see Chap. 3). Such a solution for elements of type 3–5 indirectly minimizes shifts of predecessors if they are necessary.

We will show an example of single stage scheduling problems solving on elements of type 3–5 in Fig. 9.16 (Sect. 9.6) when presenting an example of the planning methodology implementation. Jobs processing on an element of type 2 looks like a chain of jobs processed without interrupts on a single physical machine.

We obtain the start times of jobs on the network element as a solution of its single stage problem. We set a new due date for each preceding job: it is equal to the minimum of the preceding job's completion time and the start time of its successors. Then we proceed to the previous layer and continue to solve similar problems on elements. If the due date of a job on the element-predecessor has decreased, we shift the work interval on it toward a decrease in the maximum completion time. If the shift is necessary, we can try to reduce it setting the due dates of jobs without taking into account their completion times in the coordinated plan, as the minimum start time of their successors in the network.

The Second Method has a greater complexity than the first one. But it is also universal, we can use it also for planning by a basic optimality criterion and in a network of an arbitrary kind. According to the second method, we perform the entire coordinated planning procedure on the initial network (see Step 2.1 of the methodology). We use Distribution Algorithm 3, from the end of the planning period to its beginning. We start with the approved due dates of products and traverse the initial network in accordance with the precedence relations. We consider only one order of product groups processing, the one corresponding to the approved coordinated plan. The distribution has the following differences in implementation:

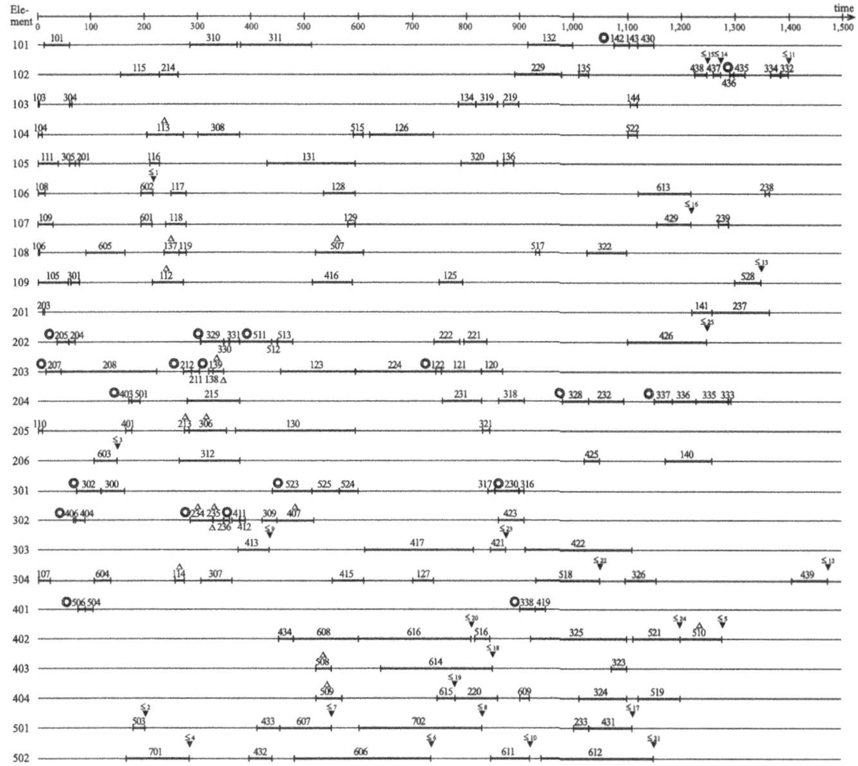

Fig. 9.23 Gantt chart: aggregated jobs distribution to elements (Ordering 2)

1. We assign initial jobs, not aggregated, to the resources of elements (physical machines), not to aggregated resources.
2. We perform the assignment of jobs on elements of type 2–5 by solving single stage optimization problems (see above) when due dates of all jobs on the element become known after their successors' assignment.
3. After the single stage problem solving, if the resource is busy, we move the assigned jobs to an appropriate time interval.
4. We do not perform common vertices and technology compatibility checks.

Distribution Algorithm for the Second Method We apply the algorithm for each product group in the order of groups corresponding to the coordinated plan chosen at the second level of planning.

1. Initialization: create empty lists of assigned and unassigned jobs and of machines on all elements without assigned jobs.
2. Choose all the terminal jobs of products. Put them into the list of unassigned jobs in the order determined by the sequence σ^*. The due date (the value of the

latest possible completion time) for a terminal job of a product is equal to its completion time in the chosen coordinated plan.

3. If the list of unassigned jobs is empty, go to step 7, else go to step 4.
4. Choose the next job from the list of unassigned jobs. Mark it as ready for assignment (i.e., we have already determined a due date for it). Go to step 5.
5. Suppose that all unassigned jobs of the network element are ready for assignment. We denote this set of jobs by I. Solve a single stage optimization problem on this element. The problem solution will specify the start and completion times of the jobs of the set I. If there are jobs conflicting with job(s) of the set I after the single stage problem solving, then shift the job(s) of the set I to the left so that to avoid a conflict in time with other jobs on this element. Shift each job separately, except for the element of type 2 where the entire set of jobs I should be shifted. To perform a shift, look through the list of the assigned jobs of the element in which the unassigned job is to be processed. Search for a free reserve in the "real time" of the element (taking into account work hours and weekends) in the order from the end to the beginning. The reserve must end not later than the due date of the unassigned job. The reserve's length must not be less than the job's processing time. If you could not find the required reserve within the time fund of this element, then continue the search before the beginning time of the planning period. If the start time of a job will be earlier than the beginning time of the planning period, we will shift the job after this algorithm to the right, up to the beginning time, using existing reserves.
6. Move all the assigned jobs from the list of unassigned jobs to the list of assigned jobs. Select all the predecessors of the assigned jobs. Add them to the list of unassigned jobs by general rules, setting their due dates as the start times of their assigned successors according to the precedence relations. Go to step 3.
7. Create the operational plan of jobs processing. It is a list of assigned jobs with information about their start and completion times.

The Third Alternative Method is a modification of the method described in Sect. 9.5.1, with the following changes:

- We perform the method for all product groups sequentially in the order corresponding to the chosen coordinated plan.
- We apply steps 1 and 2 of the method described in Sect. 9.5.1 unchanged if the resource is available for jobs assignment. In this case, we assign the jobs in accordance with the schedule obtained as a solution of the single stage optimization problem. For the case when a resource may be partially occupied by previously assigned conflicting jobs, we rewrite the steps 1 and 2 as follows:

1. Where necessary, calculate the due dates of output jobs on elements of type 1–5 of the last network layer from the due dates of the terminal jobs. In the general case, the output jobs may be at the layer before last or the one that immediately relates with the terminal jobs. This follows the properties of the technological network corresponding to the planning model for a basic criterion. If the jobs

assignment is prevented by the previously assigned conflicting jobs, then perform one of the following adjustments, the one which is more advantageous:

- either reduce the due date calculated for the jobs being assigned making it equal to the start time of the conflicting jobs;
- or reduce the completion times of the conflicting jobs making them equal to the anticipated start time of the jobs being assigned, with an appropriate correction of the due dates of the conflicting jobs' predecessors.

2. The schedules obtained for elements of type 2–5 of the last network layer unambiguously specify the latest readiness times of input jobs of the elements of type 1–5. These times, due to the properties of the technological network, unambiguously specify the due dates for the immediately preceding elements of type 1–5 (in the layer before last or any related network layer). Immediate preceding does not takes into account elements of type 1. If the jobs assignment is prevented by the previously assigned conflicting jobs, then perform one of the following adjustments, the one which is more advantageous:

   - either reduce the due date calculated for the jobs being assigned making it equal to the start time of the conflicting jobs;
   - or reduce the completion times of the conflicting jobs making them equal to the anticipated start time of the jobs being assigned, with an appropriate correction of the due dates of the conflicting jobs' predecessors.
   The same situation applies to elements of type 1–5 of any network layer. For elements of type 1–5 of the layer before last, implement step 1 of this general method.

- If you change the due dates of a conflicting job's predecessors, you should also re-solve single stage problems on the elements of type 2–5 of the previous network layer, if such problems have already been solved. Therefore, adjusting the positions of conflicting jobs may be more time consuming than adjusting the due dates of jobs being assigned, although the first change may be more advantageous.

Which of the two adjustments is more advantageous is determined by the value of decrease in the due date. Let us consider an example for two jobs. Suppose that an job 1 being assigned has a processing time l_1 and the latest possible completion time d_1. But a conflicting job 2 has been assigned to the same resource with the completion time d_2 and the start time $s_2 = d_2 - l_2$. Adjusting the due date of the job being assigned changes it by $\Delta_{as} = d_1 - s_2$. Adjusting the conflicting job's due date reduces it by $\Delta_{cf} = d_2 - (d_1 - l_1)$. We choose the adjustment corresponding to the least change in the due date:

- if $\Delta_{as} \le \Delta_{cf}$, i.e., $d_1 - s_2 \le d_2 - (d_1 - l_1)$ or $2(d_1 - d_2) \le l_1 - l_2$, which is the same, then reduce the due date of the job being assigned by $d_1 - s_2$;
- otherwise, reduce the due date of the conflicting job and of all its assigned predecessors by $d_2 - (d_1 - l_1)$.

In the general case, suppose that we should space in time a set of jobs I of a network element and a set K of conflicting jobs. This third method states that:

- either the schedule of *all* jobs of the network element must be shifted to the left in order to drift apart from the set K of conflicting jobs;
- or the conflicting jobs and their predecessors must be shifted to the left in order to drift apart from the set of jobs I.

Under the set of conflicting jobs we understand the set of all jobs on the element that has the conflicting jobs.

Therefore, we solve the single stage problem on the element without taking into account conflicting jobs. Then we check if

$$\text{either } d_i \geq d_k \text{ and } d_i - d_k < l_i, \text{ or } i \in I, k \in K, \tag{9.39}$$

for at least one job from the set of jobs I of the network element. In these cases, we reduce:

- either the start and completion times of *all* the jobs on the element being assigned by $\max_i d_i - \min_k(d_k - l_k)$, $i \in I$, $k \in K$;
- or the start and completion times of *all* the jobs on the conflicting element and of all their predecessors by $\min_i(d_i - l_i) - \max_k d_k$, $i \in I$, $k \in K$.

We choose the smaller shift. We perform such shift until we find a reserve for the entire schedule of jobs of the element being assigned.

The Fourth Method is a modification of the third one. It uses three jobs spacing options:

- if we cannot assign *a majority* of the jobs being assigned because of previously assigned conflicting jobs, then we shift the entire schedule of jobs on the element;
- otherwise (if *a minority* of the jobs being assigned intersects with the conflicting jobs), we use a shift of:
  - either all jobs on the element being assigned (jobs of the set I),
  - or all jobs on the conflicting element (jobs of the set K),

  depending on which shift is smaller. Thus, we minimize an undesirable shift of already assigned jobs schedule.

So, if a majority of jobs from the set I on a network element satisfies (9.39), then we reduce the start and completion times of all jobs of the element *being assigned* by the value of $\max_i d_i - \min_k(d_k - l_k)$, $i \in I$, $k \in K$. We perform such a shift until a free reserve is found for all jobs of the element. If, on the contrary, (9.39) is true for a minority of jobs from the set I, then we reduce:

- either the start and completion times of all jobs on the element *being assigned* by the value of $\max_i d_i - \min_k(d_k - l_k)$, $i \in I$, $k \in K$;

- or the start and completion times of all jobs on the *conflicting* element and of their predecessors by the value of $\min_i(d_i - l_i) - \max_k d_k$, $i \in I$, $k \in K$ (with the single stage problems re-solving on the elements-predecessors).

We choose the smaller shift. We perform such shift until we find a reserve for the entire schedule of jobs of the element being assigned.

We use the following general rule for buffer type elements in all the four methods when determining the due dates of jobs. The due date of all jobs to the left of a buffer is equal to the minimum start time of jobs to the right of it.

We can use a breakdown of jobs in batches in all the methods to more efficiently solve single stage problems on elements of type 3–5. This may help to increase the start time of jobs on the element and to reduce the shift of the predecessors. E.g., if we need to manufacture three pieces of a product, then we can break all or some of the product's jobs in three batches of the same size. The size is one third of the initial job's processing time for the whole product series. Then, we can place the batches on different machines of the element of type 3–5 if this improves the functional value of the single stage scheduling problem on this element.

If some jobs start before the beginning time of the planning period after start times calculation, then we should shift them to the right, up to the beginning time, towards increasing the processing interval on the elements-successors. In this case, we should keep the order of all jobs' processing and minimally shift the start times of network elements of type 2–5. We have to meet the condition of processing without interruptions coupling the problems on the elements according to the principle of spacing in time. The maximum completion time of jobs on a preceding element must be less than or equal to the minimum start time of jobs on an element-successor. We should begin such a shift with jobs that have the earliest start time. Free time reserves available on the elements will sometimes allow us to shift jobs processing to the right without violating the due dates of products. After we perform the appropriate shifts, we obtain the final operational plan.

We choose the best plan from the operational plans obtained by all alternative methods. The choice is done by the highest functional value or by other parameters (in the case of equal functional values). Then we analyze the completion times of products' terminal jobs on the elements of the last network layer according to the chosen operational plan.

If the solution of the MSNSP is feasible (does not violate the due dates specified by the customer), then we have an operational plan for this order portfolio.

If the solution is not feasible (some terminal jobs complete after the due dates specified by the customers), then we recalculate the profit for this schedule by the original functional. If the new calculated profit is not less than the expected minimum, then we need to go to the DMU to coordinate the new due dates with the customer. If the profit is less than the expected minimum, then in case of planning according to a basic optimality criterion, we implement the above listed function 9 of the DMU. If the criterion is synthetic, then we change the order portfolio directly in the DMU. If the new profit becomes less than the expected minimum after the implementation of the DMU's function 9, then we also change the order portfolio in the DMU.

The FLM implementation is statistically significant and almost always efficient in the case of a basic planning criterion. For a synthetic criterion, it is efficient assuming that the heuristic of the sequential assignment of product groups corresponds to a real production process. Then, due to a TWCT problem solving and the coordinated planning, obtaining the products' completion times becomes formalized and close to the optimal, unlike the informal implementation of the DMU's function 9. And only in the case when the product groups assignment heuristic is not relevant to a real technological process, the implementation of the function 9 becomes necessary.

9.6 Example of the Methodology Implementation for the Model of the Third Level of Planning Construction from the Initial Data

We now show the implementation of levels 1–3 of the algorithmic ware in more detail by example.

Unit 2 The first level model construction.
Step 1.1. The Problem Statement We build an example of a network model on the basis of a real small-series type production technology. We will have in our example a synthetic optimality criterion consisting of $N = 5$ basic criteria. Therefore, we have 5 independent product groups, i.e., the source network has 5 subnets that do not have precedence relations with each other. Each subnet has 5 products.

Figures 9.17 and 9.18 show the initial network with fictitious jobs which specifies the technology of products manufacturing. To simplify the scheme, we write the numbers of uncompleted jobs on the arrows, instead of the parameters of the jobs. Element numbers will be shown inside their symbols, as well. We present the parameters of elements and jobs, respectively, in Tables 9.1 and 9.2. We measure unit processing times l_i of jobs for one product in minutes.

For each element of type 5 (in our case, there is the only element #202), we specify the machine productivities p_i $(i = \overline{1, m})$: 1; 0.8; 0.2.

Table 9.3 specifies product specifications: the number N_i of products i in the order portfolio, the weight ω_i, the due date d_i, the range Δ_i of completion times in a feasible schedule (specified for those products that have elements of types 4 and 5 in manufacturing technologies).

We determine the completion time of a product as the maximum completion time of its terminal jobs. For example, $\mathbb{C}_1 = \max(C_{601}, C_{602})$ for the first product.

In order to uniquely determine the critical paths of products, we need to know which product this or that completed job on each buffer belongs to. Therefore, for

Table 9.1 Parameters of elements

#	101	102	103	104	105	106	107	108	109	201	202	203	204
type	2	2	1	1	1	1	1	1	1	1	5	4	2
m	1	1	1	1	1	1	1	1	1	1	3	2	1
#	205	206	301	302	303	304	401	402	403	404	501	502	
type	1	1	2	4	2	1	4	1	1	1	1	1	
m	1	1	1	3	1	1	2	1	1	1	1	1	

Table 9.2 Parameters of jobs

i	101	102	103	104	105	106	107	108	109	110	111	112	113	114	115	116
l_i	1	4	2	5	15	1	8	3	3	1	4	4	14	2	5	4
i	117	118	119	120	121	122	123	124	125	126	127	128	129	130	131	132
l_i	6	8	3	8	10	4	5	23	3	12	4	4	1	15	11	6
i	133	134	135	136	137	138	139	140	141	142	143	144	201	202	203	204
l_i	10	7	4	2	2	4	4	6	8	3	1	3	4	2	1	6
i	205	206	207	208	209	210	211	212	213	214	215	216	217	218	219	220
l_i	8	8	12	20	11	5	2	6	1	7	3	8	3	6	3	8
i	221	222	223	224	225	226	227	228	229	230	231	232	233	234	235	236
l_i	6	8	2	4	7	7	2	1	9	3	15	2	6	9	12	3
i	237	238	239	300	301	302	303	304	305	306	307	308	309	310	311	312
l_i	7	2	4	7	9	12	18	2	4	7	4	16	9	6	8	23
i	313	314	315	316	317	318	319	320	321	322	323	324	325	326	327	328
l_i	6	4	3	1	1	5	4	7	3	5	2	6	12	12	11	5
i	329	330	331	332	333	334	335	336	337	338	401	402	403	404	405	406
l_i	6	4	4	1	1	4	6	3	7	2	7	15	2	15	3	3
i	407	408	409	410	411	412	413	414	415	416	417	418	419	420	421	422
l_i	6	1	5	9	3	6	4	6	12	15	4	22	4	2	6	8
i	423	424	425	426	427	428	429	430	431	432	433	434	435	436	437	438
l_i	4	9	2	4	7	9	13	6	8	3	9	3	4	1	1	5
i	439	440	501	502	503	504	505	506	507	508	509	510	511	512	513	514
l_i	14	2	6	2	6	7	3	6	9	3	5	8	8	4	6	3
i	515	516	517	518	519	520	521	522	523	524	525	526	527	528	601	602
l_i	4	6	1	8	8	5	9	4	5	7	5	3	4	5	11	12
i	603	604	605	606	607	608	609	610	611	612	613	614	615	616	701	702
l_i	15	11	15	17	19	12	4	12	15	21	20	14	7	21	24	23

elements where such a relation is ambiguous, we specify one-to-one correspondence in Table 9.4. For the rest of the jobs, we determine their belonging to a particular product in an obvious way by the initial network.

We take the most complex synthetic Criterion 31 as the optimality criterion. It is a combination of five basic criteria, in each group of five products we use one basic criterion, namely:

Table 9.3 Parameters of products

i	Terminal jobs	N_i	ω_i	d_i	Δ_i	i	Terminal jobs	N_i	ω_i	d_i	Δ_i
1	601, 602	2	1			14	437	15	80	1275	
2	503	4	2		10	15	438	5	100	1250	
3	603, 604	3	3		5	16	429, 613	5	200	1220	
4	701	5	4		10	17	431	10	400	1110	
5	507–509, 510	10	5		15	18	614	15	600	850	5
6	606	15	200	735	5	19	615	5	800	780	10
7	607	5	400	550	10	20	616	10	1000	810	15
8	702	10	600	830	15	21	612	10	20	1150	5
9	413	15	800	435		22	425, 518	15	40	1050	10
10	609, 611	5	1000	920		23	421	5	60	875	15
11	332	15	20	1400		24	519, 521, 526	10	80	1200	5
12	439	5	40	1475		25	426–428	15	100	1250	10
13	528	10	60	1350							

Table 9.4 Job-to-product correspondence for buffer type elements

j	112	113	114	115	136	137	142	143	213	237	238	430	523	524	525
P_j	6	7	8	9	17	18	13	14	8	11	12	15	18	19	20

j job's number, P_j product's number

- for the product group 1 (products 1–5): the basic Criterion 1;
- for the product group 2 (products 6–10): the basic Criterion 2;
- for the product group 3 (products 11–15): the basic Criterion 5;
- for the product group 4 (products 16–20): the basic Criterion 4;
- for the product group 5 (products 21–25): the basic Criterion 3:

$$\max\left\{-\sum_{i=1}^{5}\omega_i(T)\cdot\mathbb{C}_i+\sum_{i=6}^{10}\omega_i U_i-\sum_{i=11}^{15}\omega_i'\,|\mathbb{C}_i-d_i|+\sum_{i=16}^{20}\omega''V_i-\sum_{i=21}^{25}\omega_i'''\max(0,\,\mathbb{C}_i-d_i)\right\} \tag{9.40}$$

where $U_i=\begin{cases}1, & \mathbb{C}_i=d_i\\ 0, & \mathbb{C}_i\neq d_i\end{cases}$; $V_i=\begin{cases}1, & \mathbb{C}_i\le d_i\\ 0, & \mathbb{C}_i>d_i\end{cases}$; $\omega_i(T)$ is the weight coefficient of product series i determined in an expert way; T is the planning period; ω_i is the income from the sale of the product i if it completes just in time; ω_i' is the cost of (penalty for) the deviation of the completion time of product i from the due date per time unit; ω_i'' is the income from the sale of the product i if it is not tardy in regard to its due date; ω_i''' is the cost of (penalty for) the tardiness of product i in relation to its due date per time unit.

Let us specify the minimum expected profit of the enterprise at the level of 20,000 nominal units. It is the minimum value of the criterion (9.40) that the enterprise should get for the chosen and approved plan implementation after the planning problem solving. The nominal unit, or the profit measurement unit, is the same as for the weight coefficients $\omega_i(T)$, ω_i, ω_i', ω_i'', and ω_i''', taking into account their multiplication by the time unit.

Step 1.2. The First Level of Aggregation is an aggregated representation of the initial network.

Step 1.2.1 We show the aggregated network of the first level of aggregation in Fig. 9.19. For convenience, we keep the numbers of jobs and elements: the number of an aggregated job is the number of the first of combined jobs).

We calculated unit processing times of aggregated (combined) jobs according to the above rules and summarized them in Table 9.5. For example, we sum the unit processing times of jobs 101, 102, 202, and 303 (product 1) processing on the same element of type 2. This results in job 101 with the unit processing time 25.

As a result of aggregation, some terminal jobs of products were combined into a single terminal job. For product 24, terminal job 526 became part of aggregated job 422. For product 25, terminal jobs 426–428 were combined into a single aggregated job 426.

Step 1.2.2 Due to the fact that we aggregate elements of type 3–5 into elements of type 1–2, the processing time of the jobs remaining on the elements and not combined with other jobs changes (Table 9.6).

Table 9.5 Aggregated jobs

i	Combined job numbers	l_i	i	Combined job numbers	l_i
101	101, 102, 202, 303	25	311	311, 313–315, 414	27
123	123, 124	14	338	338, 420	2
132	132, 133, 228	17	404	404, 405	6
205	205, 206	8	407	407–410	7
208	208–210	18	417	417, 418, 514, 610	41
215	215–218	20	422	422, 520, 526, 527	20
222	222, 223	5	423	423, 424, 440	5
224	224–227	10	426	426–428	10
232	232, 327	13	501	501, 502	8
300	300, 402	22	504	504, 505	5

Table 9.6 Changed processing times

i	120	121	122	138	139	204	207	211	212	221	234	235	236
l_i	4	5	2	2	2	3	6	1	3	3	3	4	1
i	309	329	330	331	406	411	412	419	506	511	512	513	
l_i	3	3	2	2	1	1	2	2	3	4	2	3	

For example, job 406 has the processing time 3 and is processed on element 302 (type 4, number of machines $m = 3$). So, its new processing time is 3/3 = 1.

Step 1.3. The Search for Products' Critical Paths We show the critical paths of products as bold lines in Fig. 9.19.

Step 1.4. The Second Level of Aggregation is the construction of a network on products' critical paths with common vertices.

Step 1.4.1. Merging Jobs Into Common Vertices on Products' Critical Paths In our example, we specify $\Delta_{CV} = 70$. This value corresponds to about 10% of the maximum critical path length ($L_{CP} = 710$ for product 10). We will use the second rule for combining into common vertices for product group 1 and the third rule for the remaining product groups. As a result, all common vertices specified by the technology were combined except those mentioned in Remark 9.9 (see after the rules for combining jobs into common vertices).

Step 1.4.2. Building a Network on the Products' Critical Paths We combine all critical paths into a single network.

Not all terminal jobs of products lie on their critical paths. So, we have removed some of them from the list of products' terminal jobs when building the critical paths network:

- only job 602 has remained for product 1 (job 601 has been removed);
- job 603 has remained for product 3 (job 604 has been removed);
- job 507 has remained for product 5 (jobs 508–510 have been removed);
- job 611 has remained for product 10 (job 609 has been removed);
- job 429 has remained for product 16 (job 613 has been removed);
- job 518 has remained for product 22 (job 425 has been removed);
- job 521 has remained for product 24 (jobs 422 and 519 have been removed).

We show the constructed network of the second level of aggregation in Fig. 9.20. Double circle symbols ◎ denote the common vertices. We present their processing times in Table 9.7, we already take into account the number of products

Table 9.7 Common vertices

i	Combined job numbers	l_i	i	Combined job numbers	l_i
120	120–122	125	329	329– 331	75
138	138, 139	30	333	333, 335–337	145
142	142, 143, 430	75	338	338, 419	50
204	204, 205	36	404	404, 406	23
207	207, 208	210	411	411, 412	25
211	211, 212	30	435	435, 436	30
230	230, 316	55	501	501, 403	24
232	232, 328	115	504	504, 506	30
234	234–236	75	511	511–513	100
300	300, 302	92	523	523–525	160

Table 9.8 Initial sequence for TWCT problem solving: product group 1

i	l_i	C_i	ω_i	f_i	i	l_i	C_i	ω_i	f_i	i	l_i	C_i	ω_i	f_i
105	60	60			504	30	450			300	92	862		
107	24	84			603	45	495	3	1485	401	14	876		
204	36	120			605	75	570			501	24	900		
305	12	132			701	120	690	4	2760	503	24	924	2	1848
108	15	147			104	10	700			306	70	994		
111	40	187			203	2	702			407	70	1064		
207	210	397			101	50	752			507	90	1154	5	5770
404	23	420			301	18	770			602	24	1178	1	1178

Table 9.9 Initial sequence for TWCT problem solving: product group 2

i	l_i	C_i	ω_i	f_i	i	l_i	C_i	ω_i	f_i	i	l_i	C_i	ω_i	f_i
112	60	60			307	60	555			416	75	1185		
113	70	130			411	25	580			515	20	1205		
114	20	150			309	30	610			417	205	1410		
213	10	160			511	100	710			516	30	1440		
115	75	235			607	95	805	4	3220	611	75	1515	10	15,150
310	90	325			116	20	825			608	120	1635		
413	60	385	8	3080	214	35	860			702	230	1865	6	11,190
211	30	415			312	115	975			606	255	2120	2	4240
308	80	495			311	135	1110							

Table 9.10 Initial sequence for TWCT problem solving: product group 3

i	l_i	C_i	ω_i	f_i	i	l_i	C_i	ω_i	f_i	i	l_i	C_i	ω_i	f_i
141	40	40			437	15	310	8	2480	237	105	610		
239	20	60			438	25	335	10	3350	439	70	680	4	2720
144	15	75			435	30	365			332	15	695	2	1390
142	75	150			528	50	415	6	2490					
333	145	295			140	90	505							

in the order portfolio N_i. A common vertex number is, for simplicity, the number of the first of the combined jobs.

For example, we have combined aggregated jobs 204 (product 2) and 205 (product 3) into common vertex 204. The number of products 2 in the order portfolio is $N_2 = 4$ and $N_3 = 3$. The unit processing times of aggregated jobs 204 and 205 are, respectively, $l_{204} = 3$ and $l_{205} = 8$. Therefore, the processing time of the common vertex 204 is $l_{204}N_2 + l_{205}N_3 = 3 \cdot 4 + 8 \cdot 3 = 36$.

Table 9.11 Initial sequence for TWCT problem solving: product group 4

i	l_i	C_i	ω_i	f_i	i	l_i	C_i	ω_i	f_i	i	l_i	C_i	ω_i	f_i
138	30	30			523	160	490			233	30	995		
137	30	60			615	35	525	8	4200	232	115	1110		
234	75	135			616	210	735	10	7350	431	80	1190	4	4760
329	75	210			136	20	755			614	210	1400	6	8400
433	45	255			229	90	845			326	60	1460		
434	30	285			134	35	880			429	65	1525	2	3050
432	45	330			132	85	965							

Table 9.12 Initial sequence for TWCT problem solving: product group 5

i	l_i	C_i	ω_i	f_i	i	l_i	C_i	ω_i	f_i	i	l_i	C_i	ω_i	f_i
120	125	125			518	120	500	4	2000	320	70	1445		
231	75	200			130	225	725			423	50	1495		
321	15	215			224	150	875			422	200	1695		
421	30	245	6	1470	325	180	1055			521	90	1785	8	14,280
219	30	275			426	150	1205	10	12,050	517	10	1795		
230	55	330			126	120	1325			612	210	2005	2	4010
338	50	380			222	50	1375							

Table 9.13 Priority ordered sequence (TWCT problem solution): product group 1

i	MPSS	l_i	C_i	ω_i	f_i	i	MPSS	l_i	C_i	ω_i	f_i
108	1	15	15			603	1	45	725	3	2175
111	1	40	55			605	1	75	800		
207	1	210	265			701	1	120	920	4	3680
306	1	70	335			104	2	10	930		
407	1	70	405			203	2	2	932		
507	1	90	495	5	2475	101	2	50	982		
105	1	60	555			301	2	18	1000		
107	1	24	579			300	2	92	1092		
204	1	36	615			401	2	14	1106		
305	1	12	627			501	2	24	1130		
404	1	23	650			503	2	24	1154	2	2308
504	1	30	680			602	2	24	1178	1	1178

Step 1.5 Building and solving the optimization problems by TWCT criterion on the network of the second level of aggregation (see Fig. 9.20).

Step 1.5.1 We build and solve an approximating TWCT problem for each basic component of the synthetic optimality criterion included in the specified functional, except for the basic Criterion 1. For the basic Criterion 1, we solve the TWCT

Table 9.14 Priority ordered sequence (TWCT problem solution): product group 2

i	MPSS	l_i	C_i	ω_i	f_i	i	MPSS	l_i	C_i	ω_i	f_i
112	1	60	60			417	2	205	990		
113	1	70	130			516	2	30	1020		
114	1	20	150			611	2	75	1095	10	10,950
213	1	10	160			211	3	30	1125		
115	1	75	235			308	3	80	1205		
310	1	90	325			307	3	60	1265		
413	1	60	385	8	3080	411	3	25	1290		
116	2	20	405			309	3	30	1320		
214	2	35	440			511	3	100	1420		
312	2	115	555			607	3	95	1515	4	6060
311	2	135	690			608	3	120	1635		
416	2	75	765			702	3	230	1865	6	11,190
515	2	20	785			606	4	255	2120	2	4240

Table 9.15 Priority ordered sequence (TWCT problem solution): product group 3

i	MPSS	l_i	C_i	ω_i	f_i	i	MPSS	l_i	C_i	ω_i	f_i
141	1	40	40			435	1	30	365		
239	1	20	60			528	1	50	415	6	2490
144	1	15	75			140	2	90	505		
142	1	75	150			237	2	105	610		
333	1	145	295			439	2	70	680	4	2720
437	1	15	310	8	2480	332	2	15	695	2	1390
438	1	25	335	10	3350						

Table 9.16 Priority ordered sequence (TWCT problem solution): product group 4

i	MPSS	l_i	C_i	ω_i	f_i	i	MPSS	l_i	C_i	ω_i	f_i
138	1	30	30			614	1	210	945	6	5670
137	1	30	60			136	2	20	965		
234	1	75	135			229	2	90	1055		
329	1	75	210			134	2	35	1090		
433	1	45	255			132	2	85	1175		
434	1	30	285			233	2	30	1205		
432	1	45	330			232	2	115	1320		
523	1	160	490			431	2	80	1400	4	5600
615	1	35	525	8	4200	326	2	60	1460		
616	1	210	735	10	7350	429	2	65	1525	2	3050

Table 9.17 Priority ordered sequence (TWCT problem solution): product group 5

i	MPSS	l_i	C_i	ω_i	f_i	i	MPSS	l_i	C_i	ω_i	f_i
120	1	125	125			325	3	180	1055		
231	1	75	200			426	3	150	1205	10	12,050
321	1	15	215			126	4	120	1325		
421	1	30	245	6	1470	222	4	50	1375		
219	2	30	275			320	4	70	1445		
230	2	55	330			423	4	50	1495		
338	2	50	380			422	4	200	1695		
518	2	120	500	4	2000	521	4	90	1785	8	14,280
130	3	225	725			517	5	10	1795		
224	3	150	875			612	5	210	2005	2	4010

problem without changing its parameters. We describe the algorithm for the problem solving in Sect. 9.2.3.

To solve the TWCT problem, we build an initial feasible sequence for each product group using Algorithm A1. It includes the jobs of the critical paths network taking into account the specified precedence relations. We arrange the products in the initial sequences in non-increasing order of their priorities (ratios of the weight to the critical path' length). The initial sequences are in Tables 9.8, 9.9, 9.10, 9.11, and 9.12. To simplify the calculations for the TWCT problems solving, we have proportionally reduced product weights in Tables 9.9, 9.10, 9.11, 9.12, 9.14, 9.15, 9.16, and 9.17. For product groups 2 and 4, we have made the weights 100 times less, and for product groups 3 and 5, 10 times less. This only affects the functional value, but not the order of jobs in a sequence.

The functional values for the initial sequences in Tables 9.8, 9.9, 9.10, 9.11, 9.12 for product groups 1–5, respectively, are 13,041, 36,880, 12,430, 27,760, and 33,810.

Step 1.5.2 The solutions of the TWCT problems for each product group are the sequences of aggregated jobs of the critical paths network (Tables 9.13, 9.14, 9.15, 9.16, 9.17).

The functional values for the priority ordered sequences for product groups 1–5, respectively, are 11,816, 35,520, 12,430, 25,870, and 33,810.

Unit 3 The second level model construction.
Step 2.1. Coordinated Plan Construction for the aggregated jobs processing taking into account the optimality criterion of each product group.
Step 2.1.1 Table 9.18 shows the supplemented sequences σ^* for all product groups, i.e., the information on the MPSS numbers and the processing times of the aggregated jobs.

Table 9.18 Aggregated job sequences for coordinated planning

i	MPSS	l_i	i	MPSS	l_i	i	MPSS	l_i	i	MPSS	l_i	i	MPSS	l_i
Product group 1			Product group 2			Product group 3			Product group 4			Product group 5		
108	1	15	112	1	60	141	1	40	138	1	10	120	1	40
111	1	40	113	1	70	239	1	20	139	1	20	121	1	75
109	1	30	114	1	20	238	1	10	137	1	30	122	1	10
110	1	10	213	1	10	144	1	15	234	1	45	231	1	75
207	1	30	115	1	75	142	1	30	235	1	20	321	1	15
208	1	180	310	1	90	143	1	15	236	1	10	421	1	30
306	1	70	413	1	60	430	1	30	329	1	45	219	2	30
407	1	70	116	2	20	333	1	5	330	1	10	230	2	45
508	1	30	214	2	35	335	1	60	331	1	20	125	2	45
509	1	50	312	2	115	336	1	45	433	1	45	221	2	45
510	1	80	117	2	30	337	1	35	434	1	30	317	2	15
507	1	90	118	2	40	437	1	15	432	1	45	338	2	30
105	1	60	119	2	15	438	1	25	523	1	75	419	2	20
107	1	24	215	2	100	435	1	20	524	1	35	425	2	30
106	1	3	311	2	135	436	1	10	525	1	50	518	2	120
204	1	12	416	2	75	528	1	50	615	1	35	130	3	225
205	1	24	515	2	20	140	2	90	616	1	210	128	3	60
305	1	12	415	2	60	237	2	105	614	1	210	129	3	15
304	1	6	417	2	205	334	2	20	136	2	20	131	3	165
404	1	18	516	2	30	439	2	70	229	2	90	224	3	150
406	1	5	609	2	20	332	2	15	134	2	35	325	3	180
504	1	15	611	2	75				132	2	85	322	3	75
506	1	15	211	3	15				233	2	30	323	3	30
604	1	33	212	3	15				135	2	20	324	3	90
603	1	45	308	3	80				232	2	65	426	3	150
605	1	75	307	3	60				328	2	50	126	4	120
701	1	120	411	3	15				431	2	80	127	4	40
104	2	10	412	3	10				326	2	60	222	4	50
203	2	2	309	3	30				522	2	20	320	4	70
103	2	4	511	3	60				613	2	100	319	4	40
101	2	50	512	3	10				429	2	65	423	4	50
301	2	18	513	3	30							123	4	140
201	2	8	607	3	95							220	4	80
300	2	44	608	3	120							318	4	50
302	2	48	702	3	230							422	4	200
401	2	14	606	4	255							519	4	80
501	2	12										521	4	90
403	2	8										517	5	10
503	2	24										612	5	210
601	2	22												
602	2	24												

Table 9.19 Aggregated jobs distribution to aggregated elements (Ordering 1), part 1

i	j	P_j	s_j	C_j	i	j	P_j	s_j	C_j	i	j	P_j	s_j	C_j
101	101	1	12	62	105	320	14	790	860	202	331	20	360	380
101	310	9	285	375	105	136	17	870	890	202	511	6	380	440
101	311	10	380	515	106	108	4	0	15	202	512	7	440	450
101	132	16	915	1000	106	602	1	194	218	202	513	8	450	480
101	142	23	1075	1105	106	117	10	250	280	202	222	14	740	790
101	143	24	1105	1120	106	128	15	535	595	202	221	12	795	840
101	430	25	1120	1150	106	613	16	1120	1220	202	426	15	1100	1250
102	115	9	155	230	106	238	22	1355	1365	203	207	4	15	45
102	214	10	230	265	107	109	5	0	30	203	208	5	45	225
102	229	17	890	980	107	601	1	194	216	203	138	19	245	255
102	135	16	1010	1030	107	118	10	240	280	203	139	20	255	275
102	438	25	1225	1250	107	129	15	580	595	203	212	7	275	290
102	437	24	1260	1275	107	429	16	1155	1220	203	211	6	290	305
102	436	23	1290	1300	107	239	22	1270	1290	203	123	14	455	595
102	435	22	1300	1320	108	106	3	0	3	203	224	15	595	745
102	334	22	1365	1385	108	605	4	90	165	203	122	13	745	755
102	332	21	1385	1400	108	137	18	190	220	203	121	12	755	830
103	103	1	0	4	108	119	10	265	280	203	120	11	830	870
103	304	3	60	66	108	507	5	365	455	204	403	2	170	178
103	134	16	785	820	108	517	11	930	940	204	501	1	178	194
103	319	14	820	860	108	322	15	1025	1100	204	215	10	280	380
103	219	11	870	900	109	105	2	0	60	204	231	13	755	830
103	144	25	1105	1120	109	301	1	62	80	204	318	14	860	910
104	104	1	0	10	109	112	6	155	215	204	328	17	980	1030
104	113	7	145	215	109	416	10	515	590	204	232	16	1030	1095
104	308	7	300	380	109	125	12	750	795	204	337	25	1150	1185
104	515	10	590	610	109	528	23	1300	1350	204	336	24	1185	1230
104	126	14	620	740	201	203	1	10	12	204	335	23	1230	1290
104	522	16	1100	1120	201	141	22	1220	1260	204	333	22	1290	1295
105	111	5	0	40	201	237	21	1260	1365	205	110	5	0	10
105	305	3	60	72	202	205	3	36	60	205	401	1	164	178
105	201	1	72	80	202	204	2	60	72	205	213	8	215	225
105	116	10	210	230	202	329	18	305	350	205	306	5	225	295
105	131	15	430	595	202	330	19	350	360	205	130	15	370	595

Step 2.1.2 We consider all possible ordering combinations of product groups at the coordinated planning stage. Let us show examples of the aggregated jobs distribution to the elements for the following orderings of product groups:

- Ordering 1: product groups 1, 2, 5, 4, and 3, in this order;
- Ordering 2: product groups 2, 5, 4, 3, and 1, in this order.

Table 9.20 Aggregated jobs distribution to aggregated elements (Ordering 1), part 2

i	j	P_j	s_j	C_j	i	j	P_j	s_j	C_j
205	321	13	830	845	304	326	16	1095	1155
206	603	3	105	150	304	439	22	1405	1475
206	312	10	265	380	401	506	4	75	90
206	425	12	1020	1050	401	504	3	90	105
206	140	21	1170	1260	401	338	12	900	930
301	302	2	72	120	401	419	11	930	950
301	300	1	120	164	402	510	5	365	445
301	523	18	440	515	402	434	20	450	480
301	525	20	515	565	402	608	8	480	600
301	524	19	565	600	402	616	20	600	810
301	317	12	840	855	402	516	10	815	845
301	230	12	855	900	402	325	15	920	1100
301	316	11	900	910	402	521	14	1110	1200
302	406	4	67	72	403	508	5	365	395
302	404	3	72	90	403	614	18	640	850
302	234	18	220	265	403	323	15	1070	1100
302	235	19	265	285	404	509	5	365	415
302	236	20	285	295	404	615	19	745	780
302	407	5	295	365	404	220	14	780	860
302	411	6	365	380	404	609	10	900	920
302	412	7	380	390	404	324	15	1010	1100
302	309	8	420	450	404	519	14	1120	1200
302	423	14	860	910	501	503	2	178	202
303	413	9	375	435	501	433	19	410	455
303	417	10	610	815	501	607	7	455	550
303	421	13	845	875	501	702	8	600	830
303	422	14	910	1110	501	233	16	1000	1030
304	107	3	0	24	501	431	17	1030	1110
304	604	3	105	138	502	701	4	165	285
304	114	8	195	215	502	432	18	395	440
304	307	6	305	365	502	606	6	480	735
304	415	10	550	610	502	611	10	845	920
304	127	14	700	740	502	612	11	940	1150
304	518	12	930	1050					

Ordering 1 corresponds to the distribution at first the products without due date restrictions. Ordering 2 applies if we consider products that should not violate their due dates as more important.

In accordance with the principles presented in the above general methodology, we apply different distribution algorithms (Sect. 9.3.2) during the coordinated

Table 9.21 Aggregated jobs distribution (Ordering 2): jobs with changed positions

i	j	P_j	s_j	C_j	i	j	P_j	s_j	C_j	i	j	P_j	s_j	C_j
104	113	7	205	275	205	213	8	275	285	304	114	8	255	275
108	137	18	235	265	205	306	5	285	355	402	510	5	1200	1280
108	507	5	520	610	302	234	18	285	330	403	508	5	520	550
109	112	6	215	275	302	235	19	330	350	404	509	5	520	570
203	138	19	320	330	302	236	20	350	360					
203	139	20	330	350	302	407	5	450	520					

Table 9.22 Products' completion times $\mathbb{C}_i$

i	j_{term}	$\mathbb{C}_i$	d_i	Δ_i	i	j_{term}	$\mathbb{C}_i$	d_i	Δ_i	i	j_{term}	$\mathbb{C}_i$	d_i	Δ_i
1	602	218			10	611	920	920		19	615	780	780	10
2	503	202			11	332	1400	1400		20	616	810	810	15
3	603	150			12	439	1475	1475		21	612	1150	1150	5
4	701	285			13	528	1350	1350		22	518	1050	1050	10
5	507[a]	455[a]			14	437	1275	1275		23	421	875	875	15
6	606	735	735	5	15	438	1250	1250		24	521	1200	1200	5
7	607	550	550	10	16	429	1220	1220		25	426	1250	1250	10
8	702	830	830	15	17	431	1110	1110						
9	413	435	435		18	614	850	850	5					

[a]In the Ordering 2, the terminal job 510 was the last completed from the list of terminal jobs of product 5. Therefore, the completion time of product 5 is equal to $\mathbb{C}_5 = C_{510} = 1280$

Table 9.23 Coordinated plans indicators

Indicator	Ordering 1	Ordering 2
The value of the functional (9.40) in nominal units	21,013	16,888
Total flow time of products	9780	10,230
Total delay in jobs on the critical paths processing including:	2257	2707
Total delay for the first product group	132	957
Total delay for the rest of product groups	2125	1750
Total idle time of resources after jobs processing start	16,831	16,516

planning in our example. We use Distribution Algorithm 2 for product group 1 and Distribution Algorithm 3 for the rest of product groups.

We show the distribution of product groups to aggregated elements in accordance with Ordering 1 in Tables 9.19 and 9.20. Graphical representation of the distribution (Gantt chart) is in Fig. 9.21. The notation we use on the chart is in Fig. 9.22. The distribution in accordance with Ordering 2 is in Table 9.21. We only show the jobs with changed positions in comparison with Ordering 1. The graphical representation is in Fig. 9.23. In the tables, i is element's number, j job's number, P_j product's number, s_j start time of job j, C_j its completion time.

Table 9.24 Distribution of aggregated jobs of product group 1

i	j	P_j	s_j	C_j	i	j	P_j	s_j	C_j	i	j	P_j	s_j	C_j
101	101	1	12	62	109	105	2	0	60	301	300	1	120	164
103	103	1	0	4	109	301	1	62	80	302	406	4	67	72
103	304	3	60	66	201	203	1	10	12	302	404	3	72	90
104	104	1	0	10	202	205	3	36	60	302	407	5	295	365
105	111	5	0	40	202	204	2	60	72	304	107	3	0	24
105	305	3	60	72	203	207	4	15	45	304	604	3	105	138
105	201	1	72	80	203	208	5	45	225	401	506	4	75	90
106	108	4	0	15	204	403	2	170	178	401	504	3	90	105
106	602	1	194	218	204	501	1	178	194	402	510	5	365	445
107	109	5	0	30	205	110	5	0	10	403	508	5	365	395
107	601	1	194	216	205	401	1	164	178	404	509	5	365	415
108	106	3	0	3	205	306	5	225	295	501	503	2	178	202
108	605	4	90	165	206	603	3	105	150	502	701	4	165	285
108	507	5	365	455	301	302	2	72	120					

i element's number, j job's number, P_j product's number, s_j start time of job j, C_j its completion time

Step 2.1.3 We check the *actual combination of jobs in the common vertices* during the distribution, according to the critical paths network. In our case, all the common vertices determined by the technology were combined during the jobs assignment, we did not need any change in the aggregated and initial networks.

The *technology compatibility check:* in our case, the jobs of only one group (316, 317, and 230) have actually been assigned to the element sequentially. In the case of a job pair 135 and 229, there is 30 min gap. It can be eliminated by shifting job 135 by 30 min to the beginning: $s_{135} = 1010 - 30 = 980$. For other job groups (namely: 221 and 222; 231 and 318; 120–123; 332, 334, and 435–438), it is difficult to eliminate the gaps between non-consecutive jobs. This requires redistribution of some aggregated jobs which may lead to new incompatibilities of the distribution with the original technology. Therefore, we make changes to the aggregated network of the first level of aggregation (see Fig. 9.19) and the initial network. For example, part of the aggregated network of the first level of aggregation and the initial network for the product group 3 on element 102 changes as shown in Fig. 9.13c. This does not require the problem to be re-solved, since the products' critical paths do not change.

Step 2.2 The result of the distribution are the completion times of the products' terminal jobs. The product's completion time $\mathbb{C}_i$ is the completion time of its terminal job with the number j_{term} (see Table 9.22). The tardiness of each product has appeared to be zero. We use the completion times to calculate the functional value in Eq. (9.40).

Table 9.25 Preliminary operational plan for product group 1

i	j	P_j	s_j	C_j	i	j	P_j	s_j	C_j	i	j	P_j	s_j	C_j
101	101	1	12	14	109	301	1	62	80	301	402	1	134	164
101	102	1	14	22	201	203	1	10	12	302	406	4	60	75
101	202	1	22	26	202	205	3	9	33	302	404	3	39	84
101	303	1	26	62	202	206	3	30	60	302	405	3	81	90
103	103	1	0	4	202	204	2	33	57	302	407	5	305	365
103	304	3	33	39	203	207	4	0	60	302	408	5	305	315
104	104	1	0	10	203	208	5	5	205	302	409	5	315	365
105	111	5	0	40	203	209	5	60	170	302	410	5	275	365
105	305	3	60	72	203	210	5	205	255	304	107	3	0	24
105	201	1	72	80	204	403	2	170	178	304	604	3	105	138
106	108	4	−15	0	204	501	1	178	190	401	504	3	84	105
106	602	1	194	218	204	502	1	190	194	401	506	4	75	90
107	109	5	−25	5	205	110	5	0	10	401	505	3	90	99
107	601	1	194	216	205	401	1	164	178	402	510	5	365	445
108	106	3	0	3	205	306	5	205	275	403	508	5	365	395
108	605	4	90	165	206	603	3	105	150	404	509	5	365	415
108	507	5	365	455	301	302	2	72	120	501	503	2	178	202
109	105	2	−27	33	301	300	1	120	134	502	701	4	165	285

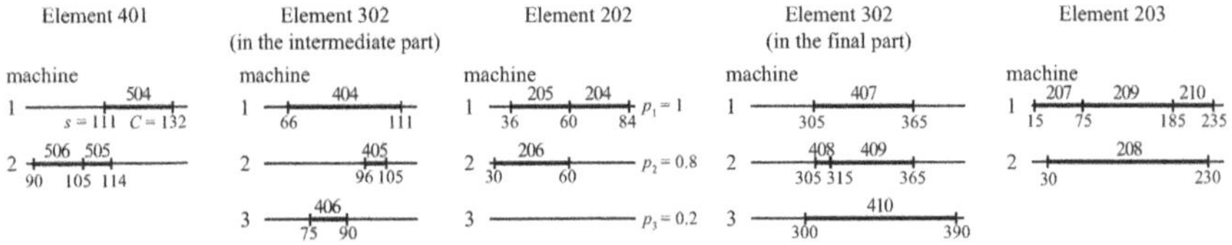

Fig. 9.24 Single stage problems solution on elements of type 3–5 (final stage)

We have summarized some indicators of our coordinated plans in Table 9.23. The number of products completed just in time, for both distribution orderings, is equal to 20 (we did not specify due dates for the first product group). The total critical paths' length for all products is 7523. In the functional value calculations for the first product group, we took the planning period length T = 1300 min.

Indicators of the plan for the second ordering have worsened due to busyness of resources. Because of this, some jobs had to be processed much later than it was specified in the technology.

In our case, we have achieved the minimum expected profit of the enterprise (20,000 nominal units) only with the first distribution ordering of product groups. The value of the functional (9.40) is 21,013 > 20,000 there. Therefore, we choose the coordinated plan for the first ordering of product groups assignment to pass it to the next planning level.

Unit 4. The third level model construction In Unit 4, we build a MSNSP. We determine the due dates d_i (or ranges of completion times in a feasible schedule) of the products' terminal jobs as the maximum of their completion times obtained during the coordinated planning (see Table 9.22). The MSNSP's solution determines the operational plan of jobs processing on the initial network.

As an example, using the planning method for a synthetic criterion, we construct an operational plan for the case where the network contains only the first product group. We planned this group according to the basic Criterion 1. We give the coordinated plan for the aggregated jobs of this product group in Table 9.24.

Since all the common vertices of this product group have been combined under coordinated planning, the initial network does not change (see Fig. 9.17).

We perform the aggregated jobs disaggregation and complement the coordinated plan of the aggregated jobs processing with the initial network jobs which were previously parts of an aggregated job. We mark added jobs as unassigned and bring the processing times of all jobs into the correspondence with the initial network. Then we use the first alternative method (see Sect. 9.5.2) to build the operational plan.

We show the single stage scheduling problems solution on elements of type 3–5 in Fig. 9.16 (for clarity, we choose a different time scale for different elements). In the figure, r denotes job's readiness time (the maximum completion time of its predecessors), d is its due date (the minimum start time of its successors).

We obtain a *non-feasible* operational plan as a result of the unassigned jobs assignments to the machines and the corresponding shifts of the predecessors. You can see the plan in Table 9.25. Jobs 105, 108, and 109 start their processing before the beginning time of the planning period.

We should shift the jobs which start before the beginning time of the planning period to the right, up to zero time point. The shifting begins with jobs that have the earliest start time (in this case, with job 105, then we move jobs 109 and 108). Free time reserves available on the elements sometimes allow us to shift jobs processing to the right without violating the products' due dates. After the appropriate shifts we obtain the final operational plan (Table 9.26; Fig. 9.24).

Products 3, 4, and 5 violate the due dates in this plan: $\mathbb{C}_3 = 177$ (while $d_3 = 150$), $\mathbb{C}_4 = 300$ ($d_4 = 285$), $\mathbb{C}_5 = 470$ ($d_5 = 455$). Therefore, if the profit from this plan implementation is not less than the expected minimum, we need to go to the DMU to coordinate new due dates (177, 300, and 470) with the customer. Otherwise, we should use function 9 of the DMU.

Table 9.26 Final operational plan for product group 1

i	j	P_j	s_j	C_j	i	j	P_j	s_j	C_j	i	j	P_j	s_j	C_j
101	101	1	12	14	109	301	1	62	80	301	402	1	146	176
101	102	1	14	22	201	203	1	10	12	302	406	4	75	90
101	202	1	22	26	202	205	3	36	60	302	404	3	66	111
101	303	1	26	62	202	206	3	30	60	302	405	3	96	105
103	103	1	0	4	202	204	2	60	84	302	407	5	305	365
103	304	3	60	66	203	207	4	15	75	302	408	5	305	315
104	104	1	0	10	203	208	5	30	230	302	409	5	315	365
105	111	5	0	40	203	209	5	75	185	302	410	5	300	390
105	305	3	60	72	203	210	5	185	235	304	107	3	0	24
105	201	1	72	80	204	403	2	170	178	304	604	3	114	147
106	108	4	0	15	204	501	1	178	190	401	504	3	111	132
106	602	1	194	218	204	502	1	190	194	401	506	4	90	105
107	109	5	0	30	205	110	5	0	10	401	505	3	105	114
107	601	1	194	216	205	401	1	164	178	402	510	5	390	470
108	106	3	0	3	205	306	5	230	300	403	508	5	315	345
108	605	4	105	180	206	603	3	132	177	404	509	5	365	415
108	507	5	365	455	301	302	2	84	132	501	503	2	178	202
109	105	2	0	60	301	300	1	132	146	502	701	4	180	300

9.7 Unit 5. The Operative Planning Model (The Fourth Level of the Planning Model)

At the fourth level of the model, we perform operative planning for the event of a partial non-fulfillment of the operational plan obtained in Unit 4.

The operative adjustment (operative planning) problem arises if certain jobs of the operational plan complete later than they were planned to complete, as a result of the production process failure on some elements.

9.7.1 *The Operative Planning Model Formalization*

The Problem Statement Given an operational plan built at the third planning level for jobs processing on elements of the initial network. We need to perform an adjustment at an arbitrary time point specified in the DMU (we will call it the *adjustment time*). By this time, the plan is partially completed: a part of the network contains completed jobs, some jobs are not yet started, some are still being processed. Some of the completed jobs have been completed after the planned time: their actual completion time $C_j^{act} > C_j$ of the operational plan.

If there are jobs at the adjustment time that should have been completed but are still being processed, then we estimate their expected completion time. The operative adjustment problem is to build a new operational plan for all jobs processing which is optimal according to some criteria. We associate the criteria with the products' tardiness in relation to the previously approved due dates. In the new plan, we also should take into account the actual completion times of already completed jobs and the expected completion times of jobs in progress.

We minimized additive functionals in all models of the fourth level. The components of the functionals are the new weight coefficients: costs of (penalty for) earliness or tardiness of products or product groups in relation to the due dates (basic criteria 3 and 5). The customer specifies the weights of the basic optimality criteria again for the operative planning problem. It may be done at the third level along with the new due dates approval after choosing the operational plan. Then, we use the weights at the fourth planning level.

We specify the correspondence of the initial optimality criterion and the criterion on the fourth level as follows. If the initial basic criterion is Criterion 1 or Criterion 3, then we use Criterion 3 for this subnet at the fourth level. If the initial basic criterion for a subnet is Criterion 4 or Criterion 5, then it does not change at the fourth level of the model.

The optimization criterion that requires the products completion just in time (Criterion 2) is not considered at the fourth level. Obtaining a new adjusted operational plan with just in time completion of products is an unsolvable mathematical problem, in the general case. We must implement the operational plan obtained at

Table 9.27 The operational plan for product group 1 that needs an adjustment

i	j	P_j	s_j	C_j	C_j^{act}	i	j	P_j	s_j	C_j	C_j^{act}	i	j	P_j	s_j	C_j	C_j^{act}
108	106	3	0	3	3	203	207	4	15	75	75	204	501	1	178	190[a]	–
103	103	1	0	4	4	109	301	1	62	80	80	301	402	1	164	194[a]	–
104	104	1	0	10	10	105	201	1	72	80	80	204	502	1	190	194	–
205	110	5	0	10	10	202	204	2	60	84	84	501	503	2	178	202	–
201	203	1	10	12	12	302	406	4	75	90	90	107	601	1	194	216	–
101	101	1	12	14	14	401	506	4	90	105	105	106	602	1	194	218	–
106	108	4	0	15	15	302	405	3	96	105	105	203	208	5	30	250[a]	–
101	102	1	14	22	22	302	404	3	66	111	111	203	210	5	185	235	–
304	107	3	0	24	24	401	505	3	105	114	114	502	701	4	180	300	–
101	202	1	22	26	26	301	302	2	84	132	132	205	306	5	230	300	–
107	109	5	0	30	30	401	504	3	111	132	132	302	408	5	305	315	–
105	111	5	0	40	40	301	300	1	132	146	164	403	508	5	315	345	–
109	105	2	0	60	60	304	604	3	114	147	147	302	407	5	305	365	–
202	206	3	30	60	60	206	603	3	132	177	177	302	409	5	315	365	–
202	205	3	36	60	60	205	401	1	164	178	178	302	410	5	300	390	–
101	303	1	26	62	62	204	403	2	170	178	178	404	509	5	365	415	–
103	304	3	60	66	66	108	605	4	105	180	180	108	507	5	365	455	–
105	305	3	60	72	72	203	209	5	75	185	–	402	510	5	390	470	–

[a]The expected completion time (actual start time plus processing time)

the third level for product groups planned by Criterion 2. If this plan has already been violated, then experts in the DMU adjust the new operational plan so that it can be executed in the accepted due dates. They can change the technology (add an equipment), increase the capacity (productivity, number of shifts), etc. This interactive procedure must be performed before considering product groups according to criteria 1, 3, 4, and 5 which are subject to adjustment.

Example 9.8 Suppose that the operational plan given in Sect. 9.6 is partially completed. At the time point 180, job 300 was completed after its assigned time; jobs 402, 501, and 208 are delayed (Table 9.27).

The problem, in this case, is to adjust the start times s_j and completion times C_j of jobs starting from the adjustment time (180), so that to maximize the functional (Criterion 3).

9.7.2 *Universal Method of the Operative Planning Problem Solving Using a Modified Method of Planning*

It is clear that we cannot implement the problem solving by simply shifting the uncompleted jobs to the right. Such a shift does not take into account the specified optimization criteria. Therefore, we propose a modified method of preliminary, coordinated and exact planning as a solution. Its idea is that we build a new partial network of the uncompleted jobs. Then we repeat the general planning procedure at levels 1–3 of the model on this network taking into account the following methodology modifications:

1. We exclude the completed jobs and the arcs between them from the initial network. But we take into account equipment busyness by the excluded jobs when solving single stage scheduling problems at the third level. Start and completion times of the excluded jobs do not change during the solving.
2. If the optimization criterion is synthetic, then we do not include in the planning the product groups without violations. We should perform an interactive procedure to analyze which part of the network does not change until the products completion. It is the part of the jobs whose predecessors are completed without a delay and which are not related to an equipment with delayed jobs. However, when planning the remaining product groups during the coordinated and exact planning, we take into account the equipment busyness with the jobs of the product groups not included.
3. In the remaining part of the network, we add new conditions for uncompleted jobs. They cannot start before the beginning time of a new planning period, which is the adjustment time. The adjustment time is considered to be the zero time point for critical paths calculations, in the TWCT problem solving, coordinated and exact planning.

4. For a synthetic criterion of optimality, we combine the product groups planned by Criterion 3 into one subnet. Product groups planned by criteria 4 and 5 remain in their subnets. We solve own TWCT problem for each product group. Then we build the coordinated plan for one or several (if there are product groups according to different criteria) distribution orderings of product groups.
5. We break each ongoing (which is in process at the adjustment time) job in two parts. We consider the part of a job until the adjustment time as a completed job and exclude it from the planning. We formalize the part of the job after the adjustment time as a new fictitious uncompleted job. Its processing time (for the entire product series) equals to the planned or expected completion time minus the adjustment time. To determine the unit processing times of these jobs for the aggregated processing times calculation, we divide them by the number of products in the series.

Figure 9.25 shows an example of fictitious jobs inclusion into the initial network.

Jobs 501 and 402 in the initial network in Fig. 9.25 are in process at the adjustment time. So, we break them in two parts. We consider the first part to be a completed job (jobs 501 and 402). The second part is the fictitious uncompleted job (jobs 501^f and 402^f). Jobs 502, 601, 602, and 503 have not yet started. Job 403 is already completed, it does not participate in the planning.

We apply the following rules for fictitious uncompleted jobs:

- we do not combine them with uncompleted jobs during aggregation, but add them to the aggregated network separately. However, their processing time changes in accordance with the principles of an aggregated job's processing time calculation (see Step 1.2.1 of the general methodology);
- we take into account the fictitious jobs by general rules, together with uncompleted jobs, when:
  - calculating the products' critical paths, since the model assumes that the products processing begins at the beginning time of the planning period;
  - combining common vertices (see Step 1.4.1 of the methodology);
  - solving an approximating TWCT problem (see Step 1.5).

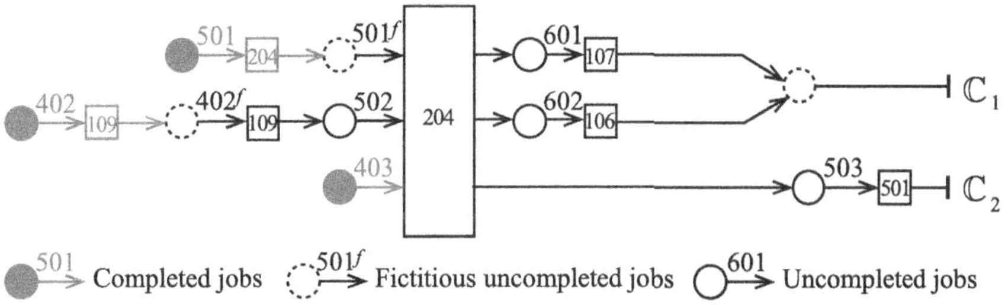

Fig. 9.25 Breakdown of ongoing jobs and formation of fictitious jobs

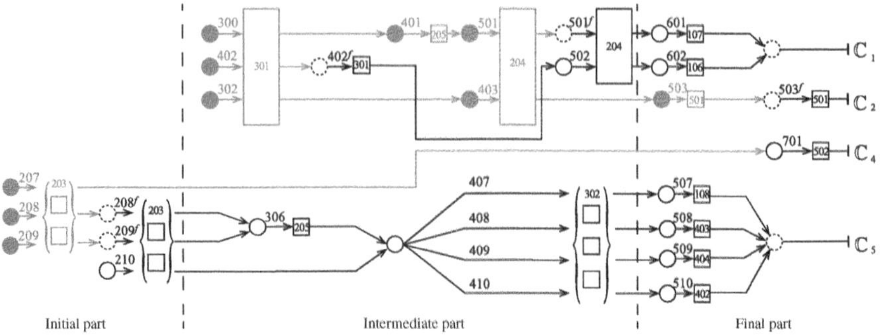

Fig. 9.26 The network we use during operational planning at the third level

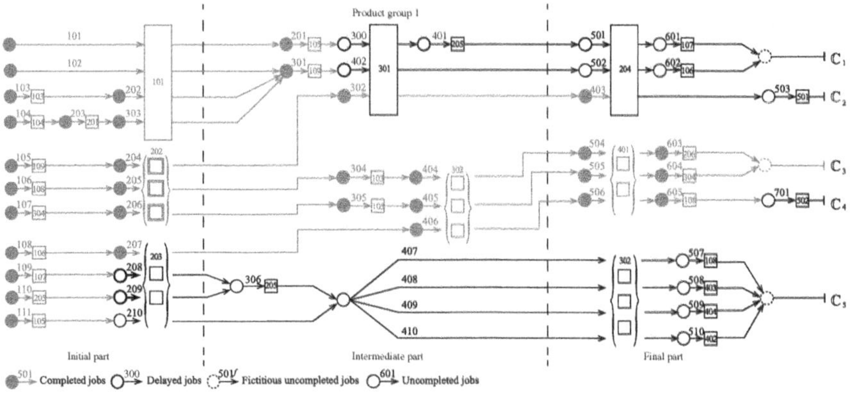

Fig. 9.27 The initial network for operative planning at the time point 180

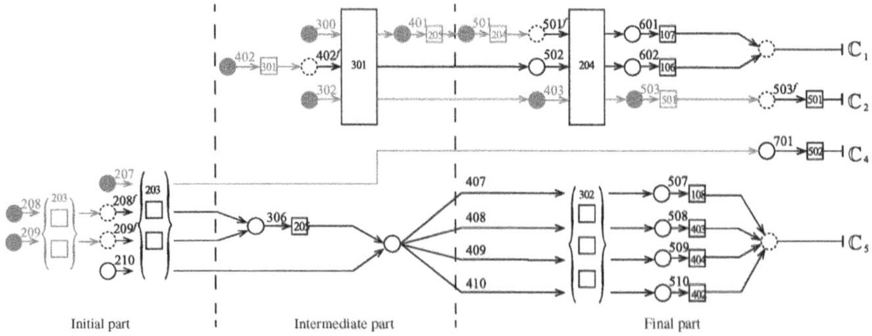

Fig. 9.28 Partial initial network of jobs that are subject to operative adjustment

The composition and processing time of some aggregated jobs may change (since we exclude the completed and ongoing jobs). The composition of some common vertices may alter because some jobs delay, etc.;

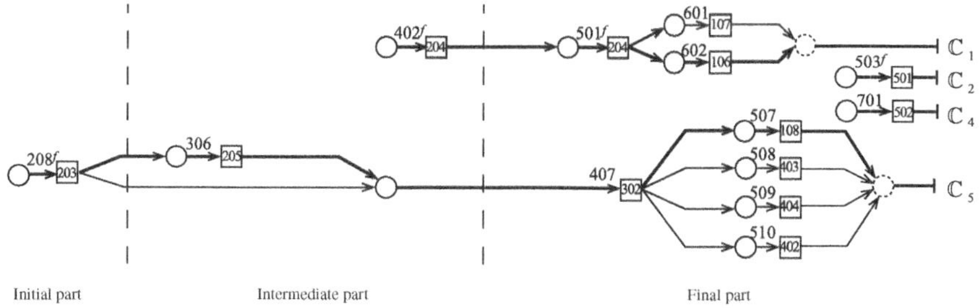

Fig. 9.29 The aggregated network for operative adjustment

- in the distribution algorithms at the coordinated planning level (see Step 2.1.2 of the methodology), we assign the fictitious jobs first, before the uncompleted aggregated jobs distribution. This should guarantee the equipment busyness with the ongoing previously processed jobs at the beginning of the planning period (immediately after the adjustment time);
- in the methods of the third level of planning (Unit 4), we leave fictitious jobs in their positions on the machines of the element where they processed in the operational plan we adjust. That is, the planned completion time of an ongoing job must coincide with its expected completion time in the plan we adjust.

6. According to Step 2.1.2 of the general methodology, we perform the coordinated planning (determining the start and completion times of aggregated jobs) according to Distribution Algorithm 3. If some aggregated jobs obtain negative start times (less than the adjustment time), then we redistribute these jobs and their successors in the network from the adjustment time by Distribution Algorithms 1 or 2. We use the two following general rules for aggregated job assignments in all the distribution algorithms. The start time of an aggregated job cannot be less than the maximum completion time of its predecessors in the network. And the completion time of an aggregated job cannot be greater than the minimum stat time of its successors in the network.
7. We may obtain several alternative plans as a result of the coordinated planning. We choose a plan with the minimum total weighted tardiness in regard to the due dates approved at the third level. The total profit obtained for product groups planned by Criterion 1 should be subtracted from the tardiness costs. The chosen plan will be considered at the third level of planning.
8. When we construct an operational plan according to a basic criterion, situations not provided in Sect. 9.5.1 are possible. The required resource of an element may be partially or fully occupied with fictitious jobs. In this case, we reduce the due dates of the jobs being assigned so that they become equal to the start times of fictitious uncompleted jobs. We can also use other universal methods described in Sect. 9.5.2 for the jobs assignment, since they are also suitable for a basic criterion.
9. At the third level of planning (Unit 4), both completed and uncompleted (or fictitious ongoing) jobs may be processed on an element at the adjustment time. In this case, we break such element in two successive elements of the same type

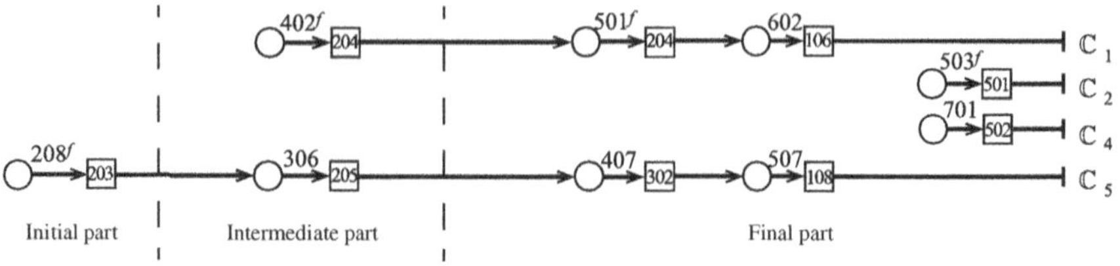

Fig. 9.30 Graph on the products' critical paths for operative adjustment

(e.g., see Fig. 9.26). We place completed jobs in the element-predecessor and uncompleted and fictitious ongoing jobs in the element-successor following it. The single stage scheduling problem for the element-predecessor is a feasible (satisfying the due dates) schedule construction for uncompleted jobs on the single or parallel machines on this element. The constraint: the start time of jobs on the machine(s) cannot be less than the expected completion times of fictitious ongoing jobs. The obtained schedule should not have any interruption in the uncompleted jobs' processing on each machine. We solve the problem by one of the PSC-algorithms (see Chaps. 2–3 and [15]). If we could not obtain a feasible schedule, then we shift some or all of the jobs to the right or to another machines of the element. The new start times of the assigned jobs must be not less than the start times of the fictitious jobs on the element and not less than the adjustment time. We aim to process jobs without interruption and the due dates violation. We can use a breakdown of jobs in batches. If we could not obtain a feasible schedule on the element, then experts correct it in the DMU. They can, for example, change the technology (add an equipment).

After we obtain a new operational plan for uncompleted jobs, we combine the fictitious (ongoing) jobs with their completed parts. Then, we obtain the complete operational plan combining the new plan with the initial plan for the completed and ongoing jobs.

If the profit calculated for the new operational plan is less than the expected minimum or the due dates have been violated, then experts in the DMU adjust the new operational plan so that it can be executed in the accepted due dates. They can change the technological network (add an equipment), increase the capacity (productivity, number of shifts), etc.

This method can be applied unchanged even when planning in an arbitrary type network both for a basic and for a synthetic optimality criterion.

Example 9.9 Consider the problem instance given in Example 9.8 (Sect. 9.7.1). In accordance with Table 9.27, we mark completed jobs ($C_j^{act} \leq 180$) and delayed jobs in different way on the initial network in Fig. 9.27.

Jobs 208, 209, 501, and 503 are still being processed at the adjustment time. Job 402 is still being processed because of a delay, and its expected completion time is 194. The expected completion time of job 208 is 250. Suppose that new weights (products' tardiness costs) were approved at the third level: $\omega_1 = 2$, $\omega_2 = 4$, $\omega_3 = 6$, $\omega_4 = 8$, $\omega_5 = 10$.

To solve the problem, we need to build a plan for the uncompleted jobs taking into account the completion times of already completed jobs and the expected completion times of the ongoing jobs. We break jobs 208, 209, 402, 501, and 503 in two parts: the completed job and the fictitious uncompleted one. Therefore, we use a partial initial network (see Fig. 9.28) to solve the problem.

We determine the processing times of fictitious jobs for a product series using the initial operational plan (Table 9.27): $l_{208^f} = 70$, $l_{209^f} = 5$, $l_{402^f} = 14$, $l_{501^f} = 10$, $l_{503^f} = 22$. To determine the unit processing times of these jobs for the aggregated processing times calculation, we divide them by the number of products in the series: $l_{208^f} = 70/10 = 7$, $l_{209^f} = 5/10 = 0.5$, $l_{402^f} = 14/2 = 7$, $l_{501^f} = 10/2 = 5$, $l_{503^f} = 22/4 = 5.5$.

The completed jobs 207–209, 300, 302, 401–403, 501, and 503 do not participate in the planning. They are left in the example to illustrate single stage scheduling problem solutions at the third level.

Figure 9.29 shows the aggregated network obtained at the first level of planning.

Unit processing times of aggregated jobs coincide with those obtained at the first level of the general model, except for the following: $l_{208^f} = 6.25$ (or 62.5 for the entire product series), $l_{402^f} = 7$ (14), $l_{501^f} = 7$ (14), $l_{503^f} = 22/4 = 5.5$ (22). We show the graph on the critical paths in Fig. 9.30.

As a result of the first level of planning, we obtain the TWCT problem solution presented in Table 9.28.

The TWCT functional value for this sequence is 6477. At the second level we obtain a partial coordinated plan for aggregated jobs (Table 9.29) using Distribution Algorithm 3. We used the due dates approved at the third level, namely: $d_1 = 218$, $d_2 = 202$, $d_4 = 300$, $d_5 = 470$. We assigned jobs starting from the products' due dates.

Table 9.28 Priority ordered sequence of aggregated jobs

i	MPSS	l_i	C_i	ω_i	f_i
503^f	1	22	22	4	88
701	2	120	142	8	1136
402^f	3	14	156		
501^f	3	14	170		
602	3	24	194	2	388
208^f	4	62.5	256.5		
306	4	70	326.5		
407	4	70	396.5		
507	4	90	486.5	10	4865

Table 9.29 Distribution of uncompleted aggregated jobs

i	j	P_j	s_j	C_j	i	j	P_j	s_j	C_j	i	j	P_j	s_j	C_j
204	503^f	2	180	202	106	62	1	208	232	403	508	5	382.5	412.5
502	71	4	180	300	203	208^f	5	180	242.5	404	509	5	382.5	432.5
301	402^f	1	180	194	205	306	5	242.5	312.5	402	510	5	382.5	462.5
204	501^f	1	194	208	302	407	5	312.5	382.5					
107	61	1	208	230	108	507	5	382.5	472.5					

Due dates of products 1 and 5 have been violated in the coordinated plan for aggregated jobs: $\mathbb{C}_1 = 232 > 218$, $\mathbb{C}_5 = 472.5 > 470$. But, given that the plan in the aggregated units is approximate, and the due dates violation may be leveled by exact planning methods, we pass it to the third level.

At the third level, according to the modified procedure described above, we break the elements containing both completed or fictitious ongoing jobs and uncompleted jobs in two parts. Thus, we perform the third level planning on the network shown in Fig. 9.26.

We exclude the completed jobs from consideration. The fictitious ongoing jobs do not participate in the optimization, but their completion times specify restrictions on the start times of uncompleted jobs when solving single stage scheduling problems on the elements.

At the third level, when solving a single stage problem on element 204 (of type 2), we have the following situation: ongoing job 501^f should complete at the time point 190. However, job 502 cannot begin earlier than time point 194, since its predecessor is delayed. We obtain a gap in the schedule, which should not be on an element of type 2. Therefore, we break the element 204 in the initial network in two similar elements (Fig. 9.31).

Thus, we obtain a partial operational plan shown in Table 9.30 and in Fig. 9.32. In this plan, $\mathbb{C}_1 = 222$, $\mathbb{C}_2 = 202$, $\mathbb{C}_4 = 300$, $\mathbb{C}_5 = 490$. Thus, the completion time of product 5 has increased by 20 and is equal to 490. Product 5 violates its due date. Experts' decision is to increase the number of machines on element 302 to four. This allows to use a breakdown in batches for job 410. Product 5 has ten pieces in its series, so we can break job 410 in batches which number is a divisor of ten. Let us break it into two batches, both with the processing time 45. We can place one batch on machine 3 and one on machine 4 of element 302. We obtain the

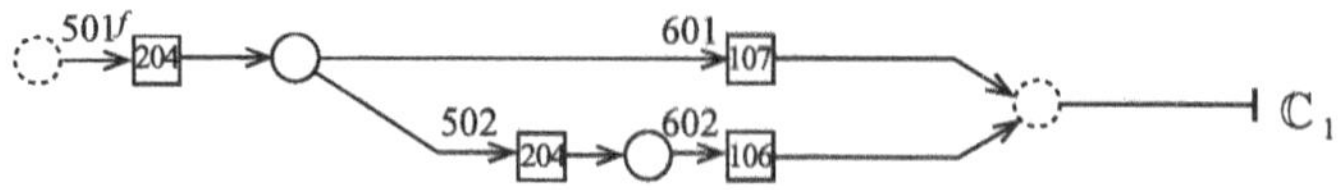

Fig. 9.31 The breakdown of element 204 as a result of a single stage problem solving

Table 9.30 The operational plan for uncompleted jobs

i	j	l_i	P_j	s_j	C_j	i	j	l_i	P_j	s_j	C_j	i	j	l_i	P_j	s_j	C_j
203	209^f	5	5	180	185	203	208^f	70	5	180	250	302	409	50	5	330	380
204	501^f	10	1	180	190	203	210	50	5	185	235	302	410	90	5	320	410
301	402^f	14	1	180	194	502	701	120	4	180	300	404	509	50	5	380	430
204	502	4	1	194	198	205	306	70	5	250	320	108	507	90	5	380	470
501	503^f	22	2	180	202	302	408	10	5	320	330	402	510	80	5	410	490
107	601	22	1	190	212	403	508	30	5	330	360						
106	602	24	1	198	222	302	407	60	5	320	380						

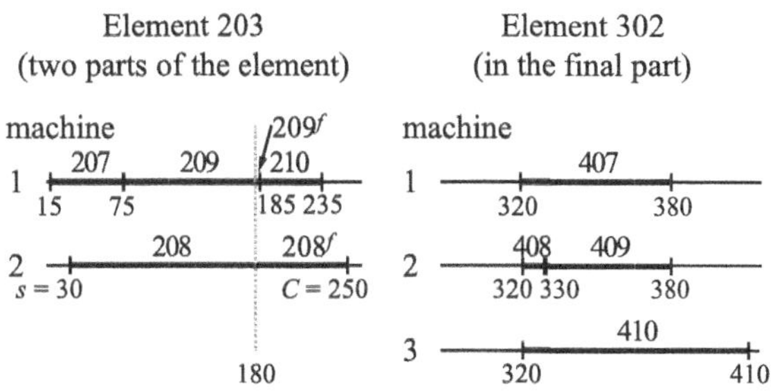

Fig. 9.32 Single stage problems solutions on elements of type 3–5 (the third level)

schedule of jobs on the element 302 shown in Fig. 9.33. In this case, job 410 competes at time point 365, and its successor, job 510, at time point 445, which is less than $d_5 = 470$.

Similarly, the due date for product 1 has been violated in the obtained operational plan: $d_1 = 218$, $\mathbb{C}_1 = C_{602} = 222$. The decrease in the functional value (total profit) for such a plan will be only 8 nominal units out of 21,013. However, since

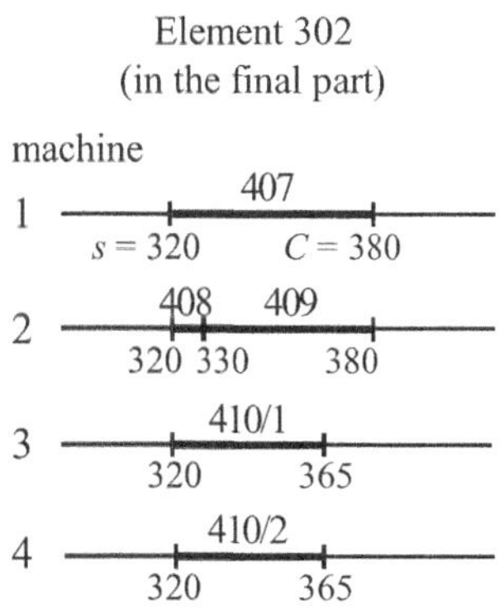

Fig. 9.33 A single stage problem solution with the breakdown of job 410 in two batches

Table 9.31 Adjusted operational plan for product group 1

i	j	P_j	s_j	C_j	C_j^{act}	i	j	P_j	s_j	C_j	C_j^{act}	i	j	P_j	s_j	C_j	C_j^{act}
108	106	3	0	3	3	203	207	4	15	75	75	301	302	2	84	132	132
103	103	1	0	4	4	105	201	1	72	80	80	204	502	1	194	198	–
104	104	1	0	10	10	109	301	1	62	80	80	501	503	2	178	202	–
205	110	5	0	10	10	202	204	2	60	84	84	106	602	1	198	210	–
201	203	1	10	12	12	302	406	4	75	90	90	107	601	1	190	212	–
101	101	1	12	14	14	302	405	3	96	105	105	203	210	5	185	235	–
106	108	4	0	15	15	401	506	4	90	105	105	203	208	5	30	250	–
101	102	1	14	22	22	302	404	3	66	111	111	502	701	4	180	300	–
304	107	3	0	24	24	401	505	3	105	114	114	205	306	5	250	320	–
101	202	1	22	26	26	401	504	3	111	132	132	302	408	5	320	330	–
107	109	5	0	30	30	301	300	1	132	164	164	403	508	5	330	360	–
105	111	5	0	40	40	206	603	3	132	177	177	302	410	5	320	365	–
109	105	2	0	60	60	205	401	1	164	178	178	302	407	5	320	380	–
202	205	3	36	60	60	204	403	2	170	178	178	302	409	5	330	380	–
202	206	3	30	60	60	108	605	4	105	180	180	404	509	5	380	430	–
101	303	1	26	62	62	203	209	5	75	185	–	402	510	5	365	445	–
103	304	3	60	66	66	204	501	1	178	190	–	108	507	5	380	470	–
105	305	3	60	72	72	301	402	1	164	194	–						

we must execute the plan within its due dates, we can break job 602 in two batches and convert element 106 to an element of type 2 with two machines. This would change processing time of job 602 to 12 and its completion time to 210.

We present the complete adjusted operational plan in Table 9.31.

9.7.3 *The Operative Planning Problem Solving Methodology if the Third Level Planning Problem Solved by a Basic Criterion*

We have obtained an operational plan as a result of the MSNSP solving by a basic Criterion 1, 3, 4 or 5. We will solve the problem of the plan's adjustment in the following formulation. Partially completed operational plan does not coincide with the operational plan obtained at the third level. Some or all earliest possible start times of elements of type 2–5 at the last network layer may appear to be later than the start times of the elements in the approved operational plan. Also, there may be jobs which processing order appears to be known and does not coincide with the operational plan of the third planning level. In this case, we perform the plan's adjustment as follows.

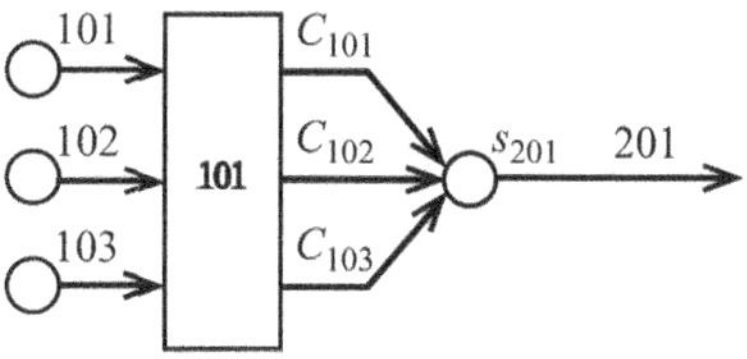

Fig. 9.34 An example of due dates adjustment

I. We clarify the job sets for processing in an arbitrary order on elements of type 2–5 of the last network layer. Readiness times or the processing order of the rest of jobs may turn out to be known and not coincide with the schedule obtained at the third level. Then we specify new due dates (if necessary) for this set of jobs, in accordance with the topology of the final part of the network.

Example 9.10 Suppose that the completion time C_{101} of job 101 (see Fig. 9.34) has increased and, as a result, the start time s_{201} of job 201 has increased. Then, we can increase the due dates of jobs 102 and 103 (no more than to s_{201}). This will increase the reserves for these jobs' processing.

II. The schedule of jobs which are not included in item I, for elements of type 2–5 at the last network layer, corresponds to the third planning level schedule.

III. The jobs on elements of type 2–5 of the last network layer included in item I may be processed in an arbitrary order. We solve the following problems for these elements:

III.1. For an element of type 2 (one machine, assigned jobs have different due dates), we build a schedule that minimizes the total or total weighted tardiness (see Chaps. 4 and 5) or the total earliness/tardiness (see Chap. 5) of jobs in relation to their due dates.

III.2. For an element of type 3, we construct a schedule for identical parallel machines minimizing the total tardiness of jobs in relation to their common due date (see Chap. 6) or minimizing the maximum deviation of jobs from the common due date (see [16]).

III.3. For elements of types 4 and 5 (parallel machines of equal or various productivity, different due dates of jobs), there are no efficient exact scheduling algorithms that minimize the total or total weighted tardiness. We propose the following interactive algorithm to solve such problem:

1. On each of the independent elements, leave a set of jobs corresponding to the operational plan obtained at the third planning level.
2. For each the element, build a schedule for identical parallel machines minimizing the total tardiness of jobs in relation to a common due date (see Chap. 6) or the maximum deviation of jobs from a common due date (see [16]).

3. Let experts analyze the obtained schedule for independent elements of types 4–5 in the DMU. They may accept the schedule or change in an expert way the set of jobs assigned for processing on each machine. If the schedule is accepted, go to step 4. Otherwise, go to step 2 (if necessary, steps 2 and 3 may be repeated several times).
4. Let experts in the DMU adjust the current operational plan in such a way that the schedule obtained for the elements of type 2–5 of the last network layer could be executed, in accordance with the technological process. They can change the technology (add an equipment), increase the capacity (productivity, number of shifts), etc.

Remark 9.18 The first part of the monograph presents efficient PSC-algorithms for these problems solving.

The method described in Sect. 9.7.3 is efficient if it is possible to implement the fourth step of the algorithm (the schedule of the last network layer elements adjustment by experts). If such an adjustment is not possible, the universal method described in Sect. 9.7.2 should be used to solve the problem by the basic criterion.

9.8 Single Stage Scheduling Problems Used in Hierarchical Planning and Decision Making Models in Network Systems with Limited Resources

We use the following models and methods described in the first part of the book to solve single stage scheduling problems in the FLM.

At the first level of the model, we solve the problem "The total weighted completion time of tasks minimization with precedence relations on a single machine" (see Chap. 7). The tasks in this problem statement are the aggregated jobs of the network of the second level of aggregation.

At the third level of the model, we solve the following problems (the tasks are the jobs of the initial network):

- the total earliness of tasks minimization in a feasible schedule on a single machine with arbitrary due dates and maximum start time of the machine (see Chap. 2);
- the total earliness of tasks minimization in a feasible schedule on identical parallel machines with a common due date (see [15]);
- the lexicographical or scalar criterion maximization in a feasible schedule on identical parallel machines of equal productivities with arbitrary due dates (see Chap. 3);
- the lexicographical or scalar criterion maximization in a feasible schedule on independent parallel machines of various productivities with arbitrary due dates (see Chap. 3).

At the fourth level of the model, we solve the following problems (the tasks are the jobs of the initial network):

- the total weighted tardiness of tasks minimization on a single machine (see Chap. 4);
- the total earliness/tardiness minimization on a single machine with arbitrary due dates (see Chap. 5);
- the total tardiness of tasks minimization on identical parallel machines with a common due date (see Chap. 6);
- the maximum earliness/tardiness of tasks minimization on identical parallel machines with a common due date (see [16]).

Remark 9.19 The methods for solving the problems considered in this book can also be used in other applied models, different from the FLM. For example [17], PSC-algorithms considered in Chaps. 2 and 4 of the book are used in two mathematical models of systems with discrete nature of technological process and wide practical applications:

- scheduling process automation for a two-level model of teams working on software development and maintenance;
- determining an efficient portfolio of orders, in terms of the expected profit, and automation of its execution planning process, based on an aggregated model of discrete systems of a wide class (for example, small-scale productions).

9.9 The Universal Informational System of Planning and Decision Making for Small-Series Type Productions

To implement the above presented algorithms, we have created [18] a universal hierarchical system of scheduling and operational planning for small-series type productions. The system satisfies the requirements described in Sect. 8.2. We have based it on the FLM and classify it as APS class system [19–21].

We used Microsoft .NET Framework, C# programming language, and Microsoft Visual Studio integrated development environment to create the system. MS SQL Server database management system was used for data storage. The database management system itself can be located both on an application server and on a separate database server. We can state the following basic system requirements for an application server: high-performance processor of Intel Xeon E5 class (this requirement is highly dependent on the size of the problems solved), 8 Gbytes of RAM. The client part of the system runs on MS Windows XP or higher and requires 2 Gbytes of RAM.

We have designed the system using the Unified Modeling Language [22]. It is a tool for defining, presenting, designing and documenting software, organizational,

Products

Product number	Basic criterion	Product count in portfolio	Product's weight	Product's due date	Allowable earliness	Terminal jobs
1	Criterion 1	2	1			602, 601
2	Criterion 1	4	2		10	503
3	Criterion 1	3	3		5	603, 604
4	Criterion 1	5	4		10	701
5	Criterion 1	10	5		15	507, 510, 508, 509
6	Criterion 2	15	200	735	5	606
7	Criterion 2	5	400	550	10	607
8	Criterion 2	10	600	830	15	702
9	Criterion 2	15	800	435		413
10	Criterion 2	5	1000	920		609, 611
11	Criterion 5	15	20	1400		332
12	Criterion 5	5	40	1475		439
13	Criterion 5	10	60	1350		528
14	Criterion 5	15	80	1275		437
15	Criterion 5	5	100	1250		438
16	Criterion 4	5	200	1220		613, 429
17	Criterion 4	10	400	1110		431
18	Criterion 4	15	600	850	5	614
19	Criterion 4	5	800	780	10	615
20	Criterion 4	10	1000	810	15	616
21	Criterion 3	10	20	1150	5	612
22	Criterion 3	15	40	1050	10	425, 518
23	Criterion 3	5	60	875	15	421
24	Criterion 3	10	80	1200	5	526, 521, 519
25	Criterion 3	15	100	1250	10	426, 427, 428

Fig. 9.35 Products list window

Fig. 9.36 Jobs list window

Jobs list

Job number	Resouce number	Processing time
101	101	1
102	101	4
202	101	2
303	101	18
132	101	6
133	101	10
142	101	3
143	101	1
228	101	1
310	101	6
311	101	8
313	101	6
314	101	4
315	101	3
414	101	6
430	101	6
115	102	5

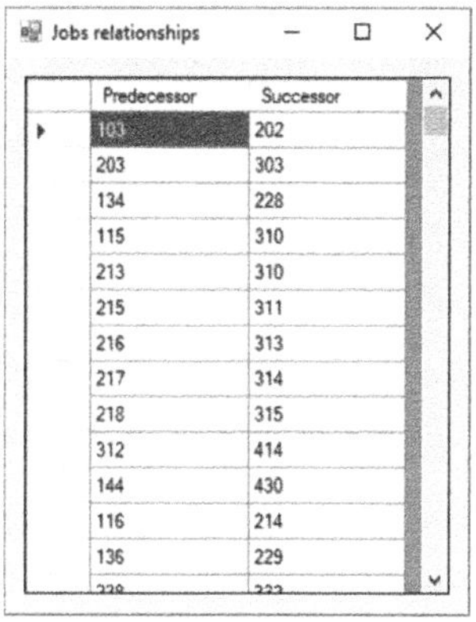

	Predecessor	Successor
▸	103	202
	203	303
	134	228
	115	310
	213	310
	215	311
	216	313
	217	314
	218	315
	312	414
	144	430
	116	214
	136	229

Fig. 9.37 Jobs relationships window

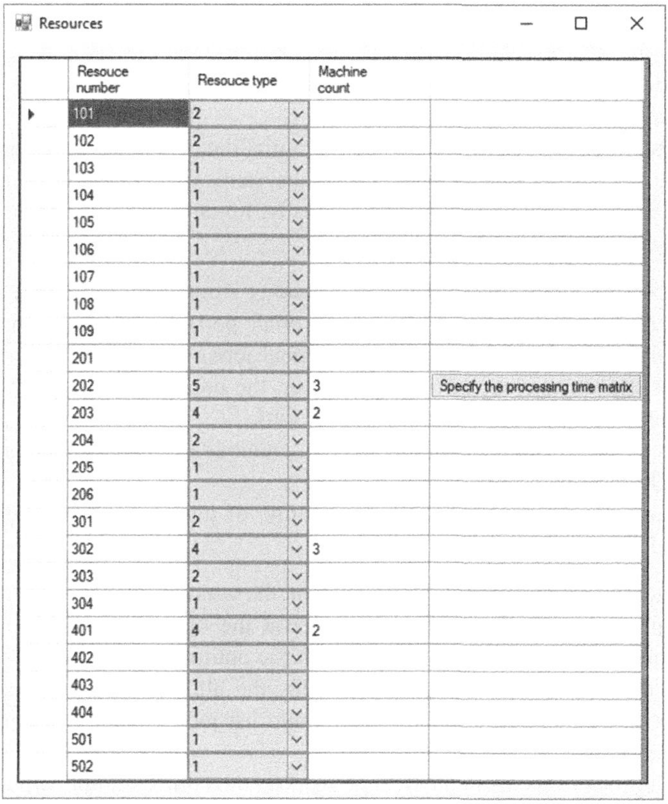

	Resouce number	Resouce type	Machine count	
▸	101	2		
	102	2		
	103	1		
	104	1		
	105	1		
	106	1		
	107	1		
	108	1		
	109	1		
	201	1		
	202	5	3	Specify the processing time matrix
	203	4	2	
	204	2		
	205	1		
	206	1		
	301	2		
	302	4	3	
	303	2		
	304	1		
	401	4	2	
	402	1		
	403	1		
	404	1		
	501	1		
	502	1		

Fig. 9.38 Resources list window

economic, technical and other systems of various nature. The consortium for technologies standardization, Object Management Group [23], supports and develops this language for systems description.

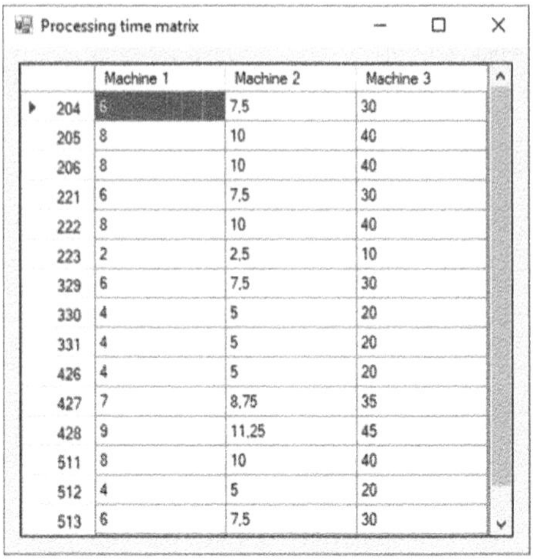

Processing time matrix

	Machine 1	Machine 2	Machine 3
204	6	7.5	30
205	8	10	40
206	8	10	40
221	6	7.5	30
222	8	10	40
223	2	2.5	10
329	6	7.5	30
330	4	5	20
331	4	5	20
426	4	5	20
427	7	8.75	35
428	9	11.25	45
511	8	10	40
512	4	5	20
513	6	7.5	30

Fig. 9.39 Processing time matrix window for an element of type 5

Working in the System Let us show how the developed system works with the data of example from Sect. 9.6.

The system allows user both to load the initial data from the stored arrays or to import them from the files of other ERP-subsystems of the enterprise. Also, user can enter or change the necessary data manually. Initially, the system must have the following data:

- products list (with the parameters: product's number, optimality criterion, number of products of this type in the order portfolio, product's weight, due date, allowable earliness, list of the terminal jobs);
- jobs list (job's number, resource's number, the processing time);
- jobs relationships (successor(s) for each job);
- resources list (resource's number, type, number of independent or identical parallel machines);
- for elements of type 5: the processing time of jobs on each of the unrelated parallel machines.

We enter information about the products in the window shown in Fig. 9.35. The product numbers must be unique. We choose the optimality criterion from the drop-down list: user can choose one of five basic optimality criteria. The number of products, the product's weight, the due date, the allowable earliness are integer fields. The last column "Terminal jobs" should list job numbers that complete the product, separated by commas.

We enter jobs of the technological network in the corresponding window shown in Fig. 9.36. We need to specify each job's number, the number of resource (element) on which it is processed, and its processing time.

Processing of all jobs corresponds to a network model defined by experts. The network should exactly reflect a real technological process in an oriented graph representation. To store the graph in the computer memory, we use a list of relations

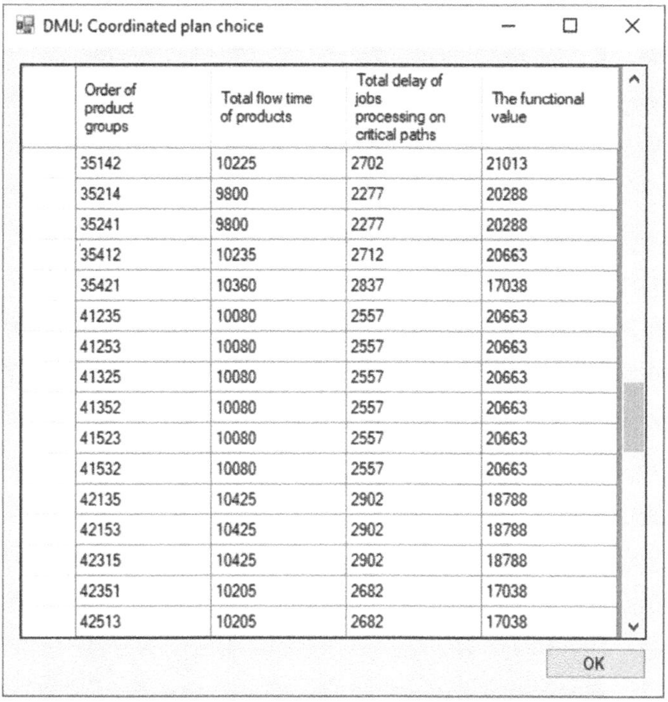

Order of product groups	Total flow time of products	Total delay of jobs processing on critical paths	The functional value
35142	10225	2702	21013
35214	9800	2277	20288
35241	9800	2277	20288
35412	10235	2712	20663
35421	10360	2837	17038
41235	10080	2557	20663
41253	10080	2557	20663
41325	10080	2557	20663
41352	10080	2557	20663
41523	10080	2557	20663
41532	10080	2557	20663
42135	10425	2902	18788
42153	10425	2902	18788
42315	10425	2902	18788
42351	10205	2682	17038
42513	10205	2682	17038

Fig. 9.40 Coordinated plan choice window

(arcs of the graph), that is, indicate the successors of each job. Thus, we need to enter the number of the predecessor and the number of successor for each arc of the graph (Fig. 9.37).

User enters the resources in the window shown in Fig. 9.38. For each resource, we need to enter its number and choose the type of the element. In accordance with the problem statement, elements of type 3–5 contain a certain number of parallel machines of equal or various productivities. For them, we enter the number of parallel machines in a separate column.

Elements of type 5 include m unrelated parallel machines of various productivities for processing k jobs without interruptions in different due dates. When we choose this type of resource, a button appears in a separate column for specifying the processing time matrix. We fill the cells of the matrix in a separate window (see Fig. 9.39).

The columns in this table are the numbers of machines with various productivities which constitute a single element. The rows are the jobs processed by this machine. A user must enter the processing time of each job on each machine in the cells. Typically, machines process jobs in a time proportional to the productivity of the machine. But for the general case, it is possible to specify the processing time individually for each job.

Coordinated plan table. Ordering: 3 - 5 - 2 - 4 - 1

Aggregated job	Aggregated resource	Start time	Completion time
103	103	0	4
304	103	60	66
219	103	790	820
319	103	820	860
134	103	880	915
144	103	1090	1105
104	104	0	10
113	104	205	275
308	104	350	430
515	104	590	610
126	104	620	740
522	104	1100	1120
111	105	0	40
305	105	60	72
201	105	72	80
116	105	210	230
131	105	430	595
320	105	790	860
136	105	870	890
108	106	0	15
602	106	194	218
117	106	250	280
128	106	535	595
613	106	1120	1220
238	106	1375	1385

Fig. 9.41 Coordinated plan table window

After entering or importing all the data, the user starts the process of solving the problem of an operational plan construction. First, the system performs preliminary planning based on the aggregated planning model construction. Next, it searches for critical paths on the obtained network and builds a network on products' critical paths with common vertices. After that, it constructs and solves an optimization TWCT problem for the aggregated jobs for each optimality criterion. Then, the system builds coordinated plans for the aggregated jobs processing for each possible product groups ordering. User chooses the best from the obtained plans, by the functional value, and then experts in the DMU choose the best plan minimizing the risk of not obtaining the expected profit (Fig. 9.40).

The system has a special window to display the schedule of aggregated jobs (Fig. 9.41). It shows information on start and completion times of all jobs and the resources they are processed on.

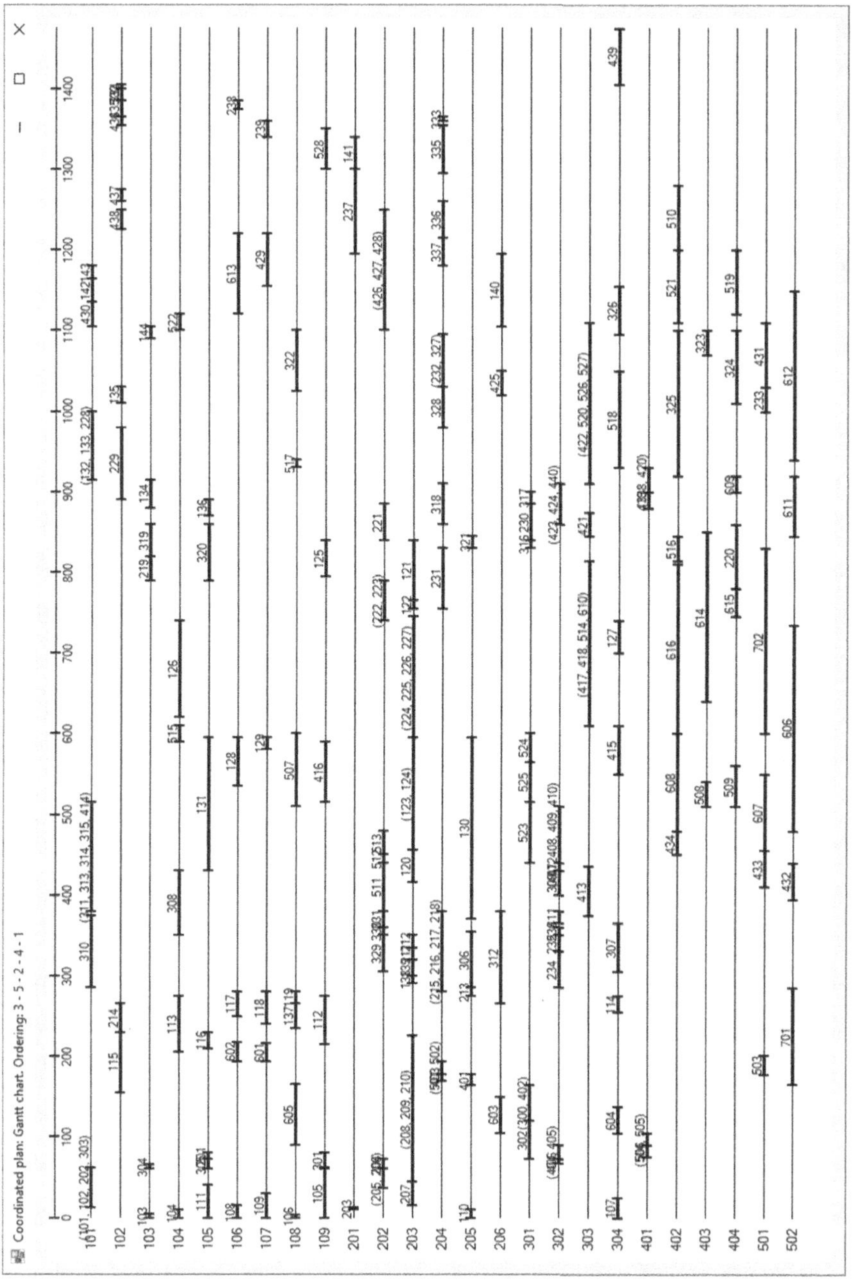

Fig. 9.42 Gantt chart window for a coordinated plan

To visualize the schedule of jobs, we use Gantt chart, a popular type of diagrams to illustrate plans and job schedules (Fig. 9.42). The diagram consists of horizontal axes that depict resources (elements) with their numbers. Each axis is the timeline that contains time segments showing the start and completion times of jobs. Also, we show the time axis at the top of the window for the reference to a "real" production time.

After choosing the best coordinated plan, the system continues the algorithm execution and proceeds to exact and operative planning.

References

1. Zgurovsky, M.Z., Pavlov, A.A.: Prinyatie Resheniy v Setevyh Sistemah s Ogranichennymi Resursami (Принятие решений в сетевых системах с ограниченными ресурсами; Decision Making in Network Systems with Limited Resources). Naukova Dumka, Kyiv (2010) (in Russian)
2. Pavlov, A.A., Misura, E.B., Melnikov, O.V., et al.: Informaciyna tehnologiya iyerarhichnogo planuvannya ta priyniattia rishen' v organizaciyno-virobnychih sistemah (Інформаційна технологія ієрархічного планування та прийняття рішень в організаційно-виробничих системах; The system of hierarchical planning and decision making in organizational and production systems). Visnyk NTUU KPI Inform. Oper. Comput. Sci. **52**, 3–14 (2010) (in Ukrainian)
3. Zgurovsky, M.Z., Pavlov, A.A., Misura, E.B., et al.: Metodologiya postroeniya chetyrehurovnevoi modeli planirovaniya, prinyatiya resheniy i operativnogo planirovaniya v setevyh sistemah s ogranichennymi resursami (Методология построения четырехуровневой модели планирования, принятия решений и оперативного планирования в сетевых системах с ограниченными ресурсами; Implementation of the methodology of the four-level model of planning, decision-making and operational management in networked systems with limited resources). Visnyk NTUU KPI Inform. Oper. Comput. Sci. **61**, 60–84 (2014) (in Russian)
4. Pavlov, A.A., Misura, E.B., Melnikov, O.V., et al.: Algoritmizaciya tret'ego i chetvertogo urovnei chetyrehurovnevoi modeli kalendarnogo i operativnogo planirovaniya i prinyatiya resheniy v setevyh sistemah s ogranichennymi resursami (Алгоритмизация третьего и четвертого уровней четырехуровневой модели календарного и оперативного планирования и принятия решений в сетевых системах с ограниченными ресурсами; Algorithmization of the third and fourth levels of the four-level model of scheduling, operational planning, and decision making in network systems with limited resources). Visnyk NTUU KPI Inform. Oper. Comput. Sci. **63**, 72–92 (2015) (in Russian)
5. Pavlov, A.A., Lisetsky, T.N.: Ispol'zovanie modificirovannogo metoda analiza ierarhiy pri priniatii resheniy v ierarhicheskom planirovanii (Использование модифицированного метода анализа иерархий при принятии решений в иерархическом планировании; Using modified analytic hierarchy process to make decisions in hierarchical planning). In: Paper Presented at the Ukrainian Scientific and Practical Conference on Information Control Systems and Technologies IUST-ODESA 2012, Papyrus Publishing House, Odesa, Sumy, 17–18 Oct 2012 (in Russian)
6. Pavlov, A.A., Lisetsky, T.N.: Nestacionarnyi metod analiza ierarhiy v zadachah ierarhicheskogo planirovaniya i prinyatiya resheniy (Нестационарный метод анализа иерархий в задачах иерархического планирования и принятии решений; Nonstationary method of analysis of hierarchies in hierarchical planning and decision-making). Visnyk NTUU KPI Inform. Oper. Comput. Sci. **54**, 82–86 (2011) (in Russian)

7. Pavlov, A.A., Sperkach, M.O., Halus, E.A.: Suboptimal'nyi polinomial'nyi algoritm resheniya odnogo klassa mnogoetapnyh setevyh zadach kalendarnogo planirovaniya (Субоптимальный полиномиальный алгоритм решения одного класса многоэтапных сетевых задач календарного планирования; Suboptimal polynomial algorithm for solving a multistage network scheduling problem of a single class). Visnyk NTUU KPI Inform. Oper. Comput. Sci. **57**, 51–55 (2012) (in Russian)
8. Pavlov, A.A., Shtankevich, A.S.: Matematicheskie modeli i metody v ierarhicheskih i setevyh sistemah podderzhki priniatiya resheniy (Математические модели и методы в иерархических и сетевых системах поддержки принятия решений; Mathematical models and methods in hierarchical and network decision support systems). Paper Presented at the Ukrainian Scientific and Practical Conference on Information Control Systems and Technologies IUST-ODESA 2012, Papirus publishing house, Odesa, Sumy, 17–18 Oct 2012 (in Russian)
9. Pavlov, A.A., Misura, E.B., Halus, E.A., et al.: Rezul'tiruyushchaya formalizaciya pervogo urovnya trehurovnevoi modeli operativnogo planirovaniya i priniatiya resheniy po kriteriyu minimizacii summarnogo operezheniya direktivnyh srokov (Результирующая формализация первого уровня трехуровневой модели оперативного планирования и принятия решений по критерию минимизации суммарного опережения директивных сроков; The resultant formalization of the first level of the three-level model of operational planning and decision making with the criterion of minimizing the total earliness in relation to the due dates). Visnyk NTUU KPI Inform. Oper. Comput. Sci. **56**, 56–57 (2012) (in Russian)
10. Pavlov A.A., Misura E.B., Melnikov O.V., et al.: Formal'noe opisanie trehurovnevoi modeli operativnogo planirovaniya sistem s setevym predstavleniem tehnologicheskih processov. Postanovka novyh zadach issledovaniya (Формальное описание трехуровневой модели оперативного планирования систем с сетевым представлением технологических процессов. Постановка новых задач исследования; The formal description of the three-level operational planning model for the systems with a network-based technological process. The new research tasks formulation). Visnyk NTUU KPI Inform. Oper. Comput. Sci. **55**, 5–10 (2012) (in Russian)
11. Pavlov, A.A., Misura, E.B., Lisetsky, T.N., et al.: Chetyrehurovnevaya model' plani-rovaniya, priniatiya resheniy i operativnogo upravleniya v setevyh sistemah s ogranichennymi resursami (Четырехуровневая модель планирования, принятия решений и оперативного управления в сетевых системах с ограниченными ресурсами; The four-level model of planning, decision making and operational control in the network systems with limited resources). Visnyk NTUU KPI Inform. Oper. Comput. Sci. **58**, 11–23 (2013) (in Russian)
12. Conway, R.W., Maxwell, W.L., Miller, L.W.: Theory of Scheduling. Addison-Wesley Pub. Co., Boston (1967)
13. Smith, W.E.: Various optimizers for single-stage production. Naval Res. Logistics Q. **3**(1–2), 59–66 (1956). https://doi.org/10.1002/nav.3800030106
14. Giffler, B., Thompson, G.L.: Algorithms for solving production-scheduling problems. Oper. Res. **8**(4), 487–503 (1960). https://doi.org/10.1287/opre.8.4.487
15. Pavlov, A.A., Sperkach, M.O.: Vypolnenie zadaniy s obshchim direktivnym srokom parallel'nymi priborami po kriteriyam optimal'nosti: minimizaciya summarnogo operezheniya otnositel'no direktivnogo stroka i maksimizaciya momenta zapuska zadaniy na vypolnenie (Выполнение заданий с общим директивным сроком параллельными приборами по критериям оптимальности: минимизация суммарного опережения относительно директивного строка и максимизация момента запуска заданий на выполнение; Tasks execution with a common due date on parallel machines with optimality criteria: total earliness minimization regarding the due date and the tasks' start time execution maximization). Visnyk NTUU KPI Inform. Oper. Comput. Sci. **62**, 89–92 (2015) (in Russian)
16. Pavlov, A.A., Zhdanova, E.G., Sperkach, M.O.: Zadacha skladannya rozkladu vikonannya zavdan' paralel'nymi priladami z metoyu minimizacii maksimumu vidhilennia vid direktivnogo terminu momentiv zavershennya priladami usih zavdan' (Задача складання

розкладу виконання завдань паралельними приладами з метою мінімізації максимуму відхилення від директивного терміну моментів завершення приладами усіх завдань; A scheduling problem for parallel machines to minimize the maximum deviation of the completion times of all tasks from the due date). Math. Comput. Model Ser. Tech. Sci. **10**, 148–158 (2014) (in Ukrainian)
17. Pavlov, A.A., Khalus, E.A., Borysenko, I.V.: Planning automation in discrete systems with a given structure of technological processes. In: Hu, Z., Petoukhov, S., Dychka, I., He M. (eds) Advances in Computer Science for Engineering and Education. ICCSEEA 2018. Advances in Intelligent Systems and Computing, vol. 754, pp. 177–185. Springer, Cham (2019). https://doi.org/10.1007/978-3-319-91008-6_18
18. Pavlov, O.A., Misura, O.B., Melnikov, O.V., Lisetsky, T.M.: Sistema kalendarnogo ta operativnogo planuvannya i priyniattia rishen' dlia pidpriemstv diskretnogo tipu (Система календарного та оперативного планування і прийняття рішень для підприємств дискретного типу; A system of scheduling, operational planning, and decision making for enterprises of discrete type). Paper Presented at the 19-th International Conference on System Analysis and Information Technology SAIT 2017, Institute for Applied System Analysis of National Technical University of Ukraine "KPI", Kyiv, 22–25 May 2017
19. Zagidullin, R.R.: Planirovanie Mashinostroitel'nogo Proizvodstva (Планирование машиностроительного производства; Engineering Production Planning). TNT, Staryi Oskol (2015) (in Russian)
20. Zagidullin, R.R.: Upravlenie Mashinostroitel'nym Proizvodstvom s Pomosch'yu Sistem MES, APS, ERP (Управление машиностроительным производством с помощью систем MES, APS, ERP; Engineering Production Management with the Help of MES, APS, ERP Systems). TNT, Staryi Oskol (2011) (in Russian)
21. Advanced Planning & Scheduling: Wikipedia Russian. https://ru.wikipedia.org/wiki/Advanced_Planning_&_Scheduling (2018). Accessed 02 Apr 2018 (in Russian)
22. Unified Modeling Language. http://uml.org (2018). Accessed 01 June 2018
23. Object Management Group. http://omg.org (2018). Accessed 01 June 2018

The manufacturer's authorised representative in the EU is Springer Nature Customer Service Centre GmbH, Europaplatz 3, 69115 Heidelberg, Germany. If you have any concerns regarding our products, please contact ProductSafety@springernature.com

Printed and bound by CPI Group (UK) Ltd, Croydon, CR0 4YY
30/09/2026
02225234-0017